MINOR CONSTITUENTS IN THE MIDDLE ATMOSPHERE

DEVELOPMENTS IN EARTH AND PLANETARY SCIENCES

MINOR CONSTITUENTS IN THE MIDDLE ATMOSPHERE

Tatsuo Shimazaki

Developments in Earth and Planetary Sciences

06

Terra Scientific Publishing Company
Tokyo, Japan

D. Reidel Publishing Company

A MEMBER OF THE
KLUWER ACADEMIC PUBLISHERS GROUP

Dordrecht / Boston / Lancaster

Library of Congress Cataloging in Publication Data

Main entry under title:

Shimazaki, Tatsuo.
Minor constituents in the middle atmosphere.
(Developments in earth and planetary sciences ; 06)
Bibliography: p.
Includes index.
1. Middle atmosphere. I. Title. II. Series.
QC881.2.M53S48 1985 551.5'1 85-14441
ISBN 90-277-2107-6

Published by Terra Scientific Publishing Company (TERRAPUB),
307 Shibuyadai-haim, 4-17 Sakuragaoka-cho, Shibuya-ku, Tokyo 150, Japan,
in co-publication with D. Reidel Publishing Company, Dordrecht, Holland

Sold and distributed in the U.S.A. and Canada
by Kluwer Boston Inc.,
190 Old Derby Street, Hingham, MA 02043, U.S.A.,
in Japan by Terra Scientific Publishing Company (TERRAPUB),
307 Shibuyadai-haim, 4-17 Sakuragaoka-cho, Shibuya-ku, Tokyo 150, Japan

In all other countries, sold and distributed
by Kluwer Academic Publishers Group,
P. O. Box 322, 3300 AH Dordrecht, Holland

D. Reidel Publishing Company is a member of the Kluwer Academic Publishers Group

Printed in Japan

PREFACE

The importance of the middle atmosphere was firmly established during an intensive study by the Climate Impact Assessment Program of the U.S. Department of Transportation in 1971–1974. The main objective of CIAP was to evaluate the impacts of supersonic transports (SST) on the stratospheric ozone layer. Although it was originally anticipated that the task could be achieved relatively easily and in a short time period by the energetic efforts of CIAP, it became evident by the end of CIAP that the problems of stratospheric ozone and the phenomena of the middle atmosphere were far more complicated than originally thought and that further long-term multidisplinary studies were necessary. Thus, the study by CIAP opened ''Pandora's box'', and touched off a new era of thorough studies of the middle atmosphere that is still going on.

The findings of CIAP and the later discovery of the potential threat to the stratospheric ozone layer of chlorofluoromethanes (CFM) released from aerosol spray cans and refrigerators led to the establishment of an interdisplinary and international ''Middle Atmosphere Program'' (MAP) starting in 1982. MAP plans to determine the structure and composition of the middle atmosphere by both observational and theoretical studies. It also intends to determine the interaction of radiation from the sun, the earth, and the atmosphere with the middle atmosphere, and to investigate the motions of the middle atmosphere on all scales.

The scientific disciplines involved in the study of the middle atmosphere are extremely diverse, and the progress of each field is very rapid; therefore, it is a rather difficult task to understand the overall picture fully and to keep it updated. An enormous number of scientitic papers are being published every month in various journals. Reviews of these results have been published occasionally in the form of workshop or committee reports or monographs by CIAP, NAS, NASA, WMO, JPL, FAA etc. Each of the review articles was written by specialists in that field, and is useful to scientists in the same field in updating their own work and in pointing it in the right direction.

The description and discussion in these reviews, however, are usually too specialized and detailed, and are not appropriate for students entering this field and non-specialists working in related areas. The purpose of this book is fill this gap and to act as a suitable reference or text book for such people. It will discuss the fundamentals of each field and clarify the relationships among the various fields. It is my hope that this book will be a worthwhile summary of the state of the art and serve as a basis for further studies in the coming years.

Since the fields are so diverse, it is impossible to cover all subjects in the same degree of detail. Naturally, the description and discussion of my own field, photochemical modeling, is the most comprehensive and accurate. Although I tried to cover the other related field as much as possible, some important subjects, such as the *detailed* discussion on stratospheric aerosols, have had to be omitted. The main concern of the book is atomic and molecular minor constituents in the middle atmosphere, their photochemical behavior, their transport, their impact on radiation (heating and cooling) and the impact of radiation in turn on the minor constituents and their transport.

The model assessments of the impact of various anthropogenetic perturbations on the stratospheric ozone are also discussed based mainly on the results from my own one-dimensional photochemical model. These are simply my own views and certainly do not represent by any means the views of the organization to which I belong. Precise answers or solutions to these questions can not be given, since so many unknowns and ambiguities are involved in the assessments. My purpose here is to provide helpful description and information so that the reader may understand the nature of the problems.

Part of the content of this book has been published in Japanese in two books; *Stratospheric Ozone* (1979) and *Lecture in Atmospheric Science, vol. 3*; *The Stratosphere and Mesosphere*, Chapter 2 (1981). Both books were published by the University of Tokyo Press. The present book has extended and updated much of these previous books. It covers the literature through 1982 and a part of early 1983. Because of the enormous number of publications in diverse journals, it is possible that some important references may have been overlooked. In addition, some lengthy references have also been omitted in order to keep the length of the book manageable. These include the reference list to most of laboratory measurements of chemical reaction rate constants, although the rate constant values determined by various panels as the best recommended values are given and discussed; the reader can find the complete references, however, in various workshop reports and monographs cited in the present book.

I am grateful to Prof. Tsuneji Rikitake, the editor of DEPS, for suggesting that I write this book and for his constant encouragement. Chapter

8 (Observations), chapter 9 (Dynamics), and chapter 12 (Radiation) have been kindly reviewed by Drs. Toshihiro Ogawa, Hideji Kida, and Takashi Sasamori, respectively. The author gratefully acknowledges their co-operative efforts. I am indebted to Mr. Keiji Oshida for his special effort in publishing this book, and Mrs. Yuko Yamasaki for her kind assistance in checking the paper and preparing the figures. I also wish to thank Mrs. Harue Onishi for typing the manuscript.

December 1983
Tatsuo Shimazaki

CONTENTS

Chapter 1

GENERAL INTRODUCTION

The term "middle atmosphere" was introduced rather recently in the scientific community. Meteorologists used to divide the atmosphere into the lower and upper atmosphere, and considered the entire region above the troposphere, i.e. above ~16 km over the equator and ~8 km over the poles, as the upper atmosphere. Radio scientists and aeronomers also divided the atmosphere into the lower and upper atmosphere, but they set the boundary at the mesopause (~80 km). In both cases the transition regions between the lower and upper atmosphere attracted little attention and interest. Being sandwiched between the two rapidly progressing domains in science, i.e. meteorology and space science, particularly aeronomy (dealing with physics and chemistry of the upper atmosphere), the study of the transition region has been relatively neglected for many years.

It was only about a decade and a half ago that the transition region became accepted as an interesting subject of science. It was triggered by the so-called *stratospheric ozone problems*. Scientists now distinguish this region from the lower and upper atmosphere and refer to it as the *middle atmosphere*; it is composed mainly of the stratosphere and the mesosphere, i.e. the region bounded below by the tropopause and above by the mesopause. In this book, however, we include the upper troposphere and the lower thermosphere in our discussion, since these two regions strongly interact with the stratosphere and the mesosphere, respectively. Thus, we will discuss in this book the region above the boundary layer near the earth's surface but below ~100 km (see Chapter 2 for more detailed definitions of each part of the atmosphere).

The stratosphere is a statically stable layer above the tropopause. The temperature generally increases with height in the stratosphere, first slowly to ~20 km and then much more rapidly arriving at a maximum temperature at ~50 km. This temperature distribution is associated with absorption of solar UV radiation by the ozone which is present in the stratosphere with much larger densities than in the troposphere. In May 1931, a Swiss physicist, A. Piccard, ascended to an altitude of 16 km above the surface of the earth on board an experimental balloon he had built; he became the first human being to enter the stratosphere. Having seen the "permanently cloudless and quiet" stratosphere, he dreamed of using this area of the atmosphere for the promotion of international airborne transportation. His dream is now realized, as supersonic transports (SST) fly through the stratosphere.

The main scientific objective of the balloon flights of Piccard was the observation of cosmic rays in the middle atmosphere; no attention was paid to the stratospheric ozone. At that time, however, the existence of ozone in the stratosphere was already known. In 1881, W. N. Hartley had predicted the existence of an ozone layer in the middle atmosphere, which he thought was responsible for the absence of UV radiation of wavelengths below ~ 300 nm (nanometer, 10^{-7} cm) in the solar spectrum observed on the surface. A theory of the production of the ozone layer in the middle atmosphere had first been presented by S. Chapman in 1930 at an international conference in Paris. The exact height of the ozone layer was not known then, but was estimated to be ~45 km. Later, it became clear that the ozone layer was at a much lower height (~25 km), but it is still higher than the altitude reached by Piccard's balloons.

Ozone, if its mixing ratio exceeds ~0.1 ppmv (parts per million by volume), has some toxic effects on the human body such as irritation to the eyes and damages to the respiratory organs. Inhaling air containing 5 ppm of ozone may endanger life. The mixing ratio of stratospheric ozone is as large as 15 ppm at the altitude of 25 km. The actual ozone density is not very large since the air itself is thin at these heights (~3% of the surface density). If the balloons of Piccard had risen to higher altitudes and stayed longer, however, there might have been some harmful impact on the human body. There is always a possibility that ozone invades the cabin, even though it is normally airtight. This is a problem even for modern SST vehicles. In fact, the main problems concerning SST's before the middle of the 1960's were thought to be the invasion of stratospheric ozone into cabins and corrosion of materials such as rubbers used in wheels, window frames and electrical insulation. It was only after the middle of the 1960's that the destruction of stratospheric ozone by exhaust gases from SST's became the major concern.

Ozone near the ground causes air pollution called "photochemical smog", which is very toxic to humans. If the entire ozone content of the stratosphere was to fall into the troposphere and be mixed uniformly, the ozone mixing ratio would be ~0.5 ppmv; if mixed within 1 km of the surface, the mixing ratio would reach ~4 ppmv. It is thus very fortunate that the stratospheric ozone stays up in the middle atmosphere. In addition to that, however, the ozone in the middle atmosphere has an invaluable impact on the protection of life on the earth's surface.

Ozone absorbs almost all solar ultraviolet (UV) radiation below ~300 nm and prevents this harmful radiation from arriving at the earth's surface. Without the protection of the ozone shield, even a lower form of animal life could not survive on the surface, since the harmful solar UV radiation would destroy the chromosomes of the cell nucleus, thus prohibiting cellular multiplication. Exposure to the proper amount of UV radiation has benefits such as the production of vitamin D in the human body, but excess exposure results in harmful effects such as sunburn and skin cancer. Nature protects surface life by maintaining the proper amount of ozone in the middle atmosphere, or turning the argument around, life has evolved by natural selection so that it matches the condition provided by the stratospheric ozone layer amongst others. Thus, if the amount of stratospheric ozone should change, life as we know it on the surface will greatly change. There are several manmade activities which may cause significant changes in the stratospheric ozone; these include nitrogen oxide (NO_x) emission from SST vehicles, the release of chlorofluoromethanes (CFM) from aerosol spray cans and refrigerators, and the increase of N_2O in the atmosphere due to fertilizing agricultural fields.

Stratospheric ozone is produced ultimately by the absorption of solar UV radiation below ~240 nm by atmospheric molecular oxygen (O_2). Photodissociation of O_2 produces atomic oxygen (O), whose three body recombination with O_2 then produces O_3. Many other minor constituents can also be produced by photochemical reactions in the middle atmosphere, and the ozone concentration is greatly affected through chemical reactions with these minor constituents, particularly with HO_x, NO_x and ClO_x. Although concentrations of these minor constituents are usually much smaller than the concentration of O_3, catalytic chain reactions can greatly reduce the O_3 concentration. Since concentrations of HO_x, NO_x and ClO_x themselves are affected by many other constituents, the number of reactions and constituents important to the stratospheric ozone chemistry are enormous. It is very important to measure in the laboratory the accurate reaction rate constants for these chemical reactions, including their temperature dependences.

Although problems of stratospheric ozone and the related photochemistry have provided a dramatic instance of the importance of the study of minor constituents in the middle atmosphere, there are many other aspects that stimulate research into the middle atmosphere today. The global distribution of O_3 in the lower stratosphere is directly and strongly affected by the stratospheric wind system, which is driven mainly by the geographical pattern of heating due to the absorption of solar UV radiation by O_3 and of cooling due to the emission of IR radiation from terrestrial minor constituents such as CO_2, O_3 and H_2O. Dynamics is also important for transporting many other minor constituents including man-made species such as CFM from their source regions to the regions where they undergo active photochemical reactions. Changes in the distributions of minor constituents then affect the radiation field which further influences the circulation. The above are only examples of coupling among photochemistry (composition, constituents), dynamics (transport, circulation) and radiation (heating and cooling, temperature). We will discuss some more details of the coupling in the following chapters.

It is becoming more and more evident that many meteorological events in the troposphere are related to stratospheric phenomena; therefore, changes in stratospheric conditions should also have a great influence on the global climate. Details of these effects, however, are not well known, and are certainly major issues for future investigation.

There are also many interesting scientific subjects in the mesosphere and lower thermosphere. They include production of ionized species and their impact on odd-nitrogen densities particularly for N and NO, ozone behavior affected by the HO_x chemistry, very cold temperatures and production of noctilucent clouds near the mesopause particularly at high latitudes in summer, and propagation of planetary and internal gravity waves.

This book is organized into 13 chapters including the present chapter (Chapter 1, general introduction). The remaining chapters discuss fundamentals of physical and chemical processes of minor constituents in the middle atmosphere. These discussions should be useful not only for a better understanding of each of the basic processes, but also for obtaining knowledge of mutual interaction among processes and an overall picture of the phenomena in the middle atmosphere.

The minor constituents of the middle atmosphere are produced mainly by interactions of atmospheric molecules with solar UV radiation. Chapter 2 discusses the structure and composition of the earth's atmosphere, whereas in Chapter 3 we discuss the nature of incoming solar UV radiation and its attenuation by molecular absorption and scattering in the middle atmosphere. Chapter 4 gives detailed information of experimental data on absorption cross-sections of various stratospheric molecules along with calculations of photodissociation coefficients.

Chapter 5 discusses the neutral chemistry in the middle atmosphere. After the theory of formation of an ozone layer in a pure oxygen atmosphere is described, various chemical reactions are discussed with special attention to the catalytic chain reactions due to HO_x, NO_x and ClO_x which reduce ozone density. Data of reaction rate constants determined by various panels at different times are reviewed and the latest values are given. Chapter 6 presents and discusses the photochemical time constants of individual constituents and their families.

Photochemical models are discussed in Chapter 7. A one-dimensional diurnally averaged steady state model and a time-dependent diurnal model are detailed. Multi-dimensional models are then briefly discussed. Chapter 8 is devoted to a review of the various techniques used to make field observations of stratospheric constituents and the results are summarized for each of the key constituents.

Chapter 9 deals with the dynamics of the middle atmosphere, particularly observations and theories of large scale transport of minor constituents. Eulerian and Lagrangian points of view for describing the atmospheric motion are outlined. Special discussion is then given to the eddy transports due to propagation of planetary waves and internal gravity waves.

The observed temporal variations in stratospheric ozone are the topic of Chapter 10. These include the long-term variation, i.e. the "ozone trend", and transient variations associated with special events such as nuclear explosions, solar proton events and solar eclipses. Finally, problems associated with the relationship between solar activity and stratospheric ozone are discussed. In Chapter 11 we present model predictions of potential changes in stratospheric ozone due to anthropogenetic perturbations. The impacts of these changes on UV radiation and life on the surface are evaluated. Evolution of the ozone layer is then discussed based on some model calculations. Finally, ozone in the Martian atmosphere is briefly discussed.

Chapter 12 involves radiation in the middle atmosphere. The main discussion here is the calculation of atmospheric heating due to the absorption of solar UV radiation by O_3, O_2, CO_2, H_2O and NO_2 and of atmospheric cooling due to the IR emissions from CO_2, O_3 and H_2O. Some important coupling mechanisms among radiation, chemistry and dynamics are discussed in the latter half of Chapter 12. The ionic constituents are treated in Chapter 13. The ion chemistry in the mesosphere and lower thermosphere is discussed in connection with the production of odd-nitrogen in these regions. Recent observations of various types of positive and negative cluster ions in the middle atmosphere are reviewed and their possible production mechanisms are discussed.

Chapter 2

ATMOSPHERIC STRUCTURE

A system of nomenclature of the structure of the atmosphere is essential for discussions of minor constituents in the middle atmosphere. Definitions of various regions of the atmosphere are first given based on the vertical profile of the atmospheric temperature. Discussions are then given of height variations of atmospheric pressure, density and composition. Various minor constituents observed in the middle atmosphere are classified into two groups, i.e. the primary (parent) and secondary constituents, according to their origins or production mechanisms. Chemical and dynamical effects on distributions of minor constituents are briefly discussed.

2.1 Temperature Structure and Nomenclature of the Atmosphere

It is convenient for general understanding as well as for scientific discussion to divide the atmosphere into subdivisions. Various nomenclatures for the atmosphere have been proposed, but the most commonly used one at present is the IUGG (International Union of Geodesy and Geophysics) nomenclature (NICOLET, 1960). We will follow this nomenclature, but it is sometimes interesting and instructive to discuss the differences among different nomenclatures (CHAPMAN, 1950; GERSON and KAPLAN, 1951; GOODY, 1954) in learning how scientists have interpreted the atmospheric structure in different ways at different times in the past.

The atmosphere can be divided into subdivisions based on the temperature structure (subdivisions based on other physical and chemical properties will be discussed in later sections). The mean vertical temperature variation observed at middle latitudes and the names of subdivisions according to IUGG nomenclature are shown in Fig. 2.1. The temperature in the lowest region decreases with height from the surface

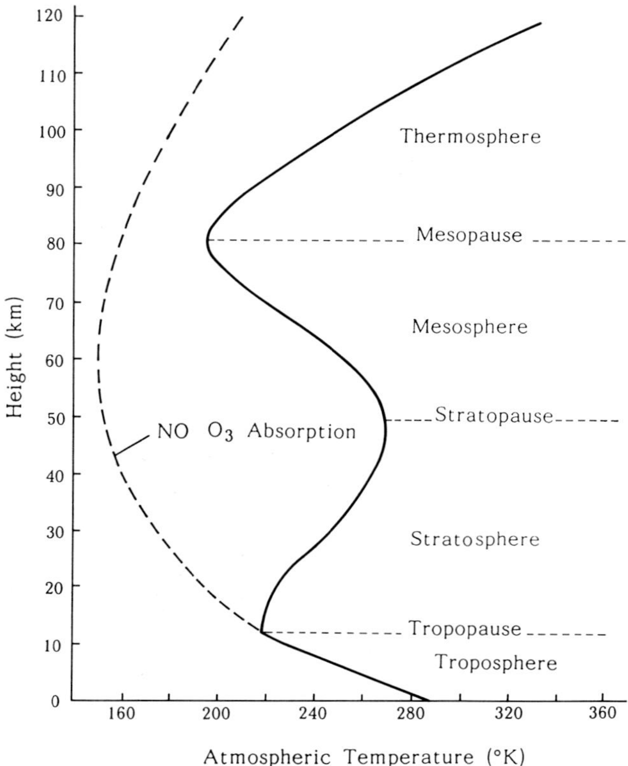

FIG. 2.1. Schematic illustration of a vertical profile of atmospheric temperature and names of subdivisions of the atmosphere. Dashed curve indicates a profile when there were no ozone absorption.

temperature (~290 K) at an almost constant rate. This region is called the *troposphere* (region of change). Strong air motions by convection and weather phenomena such as cloud formation and precipitation occur here.

When dry air parcels move upwards from the lower high-pressure region to the upper low-pressure region, the atmosphere performs some work by *adiabatic expansion*. Then, according to the first law of thermodynamics the internal energy, and therefore the temperature, must decrease. Conversely, the atmospheric temperature increases by *adiabatic compression* when the air moves downwards. As a result, the upper portion of the convection region is cooled, whereas the lower portion is heated. The rate of the temperature decreases with height by this process is called the *dry adiabatic lapse rate* and is expressed by (see, for instance, HESS, 1959)

$$-\frac{\partial T}{\partial z}=\Gamma_d=\frac{g}{C_p} \tag{2.1}$$

where g is the acceleration of gravity and $C_p \simeq 0.24$ cal gm^{-1}K^{-1} is the specific heat at constant pressure. The value of Γ_d in the troposphere is ~ 9.8 K/km.

In a wet atmosphere, the water vapor is condensed out as temperature decreases following the upward air motion, and the latent heat of condensation is released. This heating should cause the temperature lapse rate to decrease from Γ_d to

$$\Gamma_s=\Gamma_d\frac{1+\dfrac{JLw}{R^*T}}{1+\dfrac{Lw}{C_pE}\dfrac{\mathrm{d}E}{\mathrm{d}T}} \tag{2.2}$$

where L is the latent heat of condensation (~597.3 cal gm^{-1} at 0°C), w the amount of water vapor in gm, J the mechanical equivalent of heat (~4.184×10^7 erg cal^{-1}), R^* the absolute (or universal) gas constant (~8.314×10^7 erg mol^{-1} K^{-1}) and E the saturation water vapor pressure in mb given by

$$\log\frac{E}{6.106}=9.632(1-0.00035t)\cdot\frac{t}{T} \tag{2.3}$$

where $t=T-273$.

Γ_s is called the *wet adiabatic lapse rate*. It is always less than Γ_d but approaches Γ_d as the pressure increases or the temperature decreases. The value of Γ_s under normal temperature and pressure conditions in the troposphere is ~6.5 K/km and is close to the observed rate of the temperature decrease. The atmosphere is stable if the actual lapse rate of temperature

$$\Gamma=-\frac{\mathrm{d}T}{\mathrm{d}z} \tag{2.4}$$

is smaller than Γ_s, and is unstable if Γ is larger than Γ_d.

It is evident that the temperature cannot decrease with height indefinitely at the lapse rate of the troposphere. If the temperature of the upper atmosphere becomes too low, the cooling rate by long wave radiation to the space should decrease, and the atmosphere would be warmed up by radiation from below. Thus, the atmospheric temperature eventually ceases to decrease above a certain height even without ozone heating. This critical height is generally called the *tropopause*. In case of the earth's atmosphere, the temperature actually increases with height above the tropopause because of absorption by ozone (O_3) of solar UV radiation energy. This region of temperature increase is called the *stratosphere*. In the stratosphere the atmosphere is stable, since the warmer air lies over the cooler air (Γ is negative). Before the temperature tends to increase, there is a region in which the atmosphere is almost isothermal. In Chapman's nomenclature this isothermal region was called the stratosphere. Because of lack of data before 1950, it had been speculated that the isothermal region was much larger than it actually was, but subsequent observations revealed that it was a rather limited region except at high latitudes.

In the stratosphere the temperature is determined mainly by radiative equilibrium between heating by O_3 absorption of solar UV radiation and cooling by emission of IR radiation, mostly by CO_2. Since solar UV radiation decreases owing to O_3 absorption as it penetrates through the atmosphere, the height of maximum heating occurs slightly above the height of maximum ozone density (~25 km). The rate of temperature change due to atmospheric heating is calculated by dividing the absorbed energy by the specific heat of the atmosphere (ρC_p). Since the latter is proportional to the air density and decreases rapidly with height, the temperature maximum caused by O_3 absorption appears at a much higher altitude (~50 km) than the ozone density maximum. The height of the temperature maximum determines the upper limit of the stratosphere and is called the *stratopause*. The temperature at the stratopause is ~270 K.

It is sometimes convenient to divide the stratosphere further into the *lower*, *middle*, and *upper stratosphere*. They usually represent the regions below ~25 km, 25 - 40 km, and above ~40 km, respectively, but boundaries have not been defined precisely.

The temperature decreases above the stratopause, reaching a value less than 200 K near a height of 85 km, above which the temperature again rises through absorption by O_2 of the solar EUV radiation. The height of the temperature minimum is called the *mesopause*, and the region between the stratopause and the mesopause is the *mesosphere*. The region above the mesopause is called the *thermosphere*, which continues to the *exosphere* above ~600 km. In Chapman's nomenclature all regions of O_3 heating, i.e. the region from the upper end of the isothermal region (stratosphere in his nomenclature) to the mesopause, is defined as the mesosphere. The temperature maximum around 50 km, i.e. the stratopause in the IUGG nomenclature, was called the *mesopeak* in Chapman's nomenclature.

The middle atmosphere is usually defined as composed of the stratosphere and mesosphere, but we include the upper troposphere and the lower thermosphere in our discussion. The temperature in the middle at-

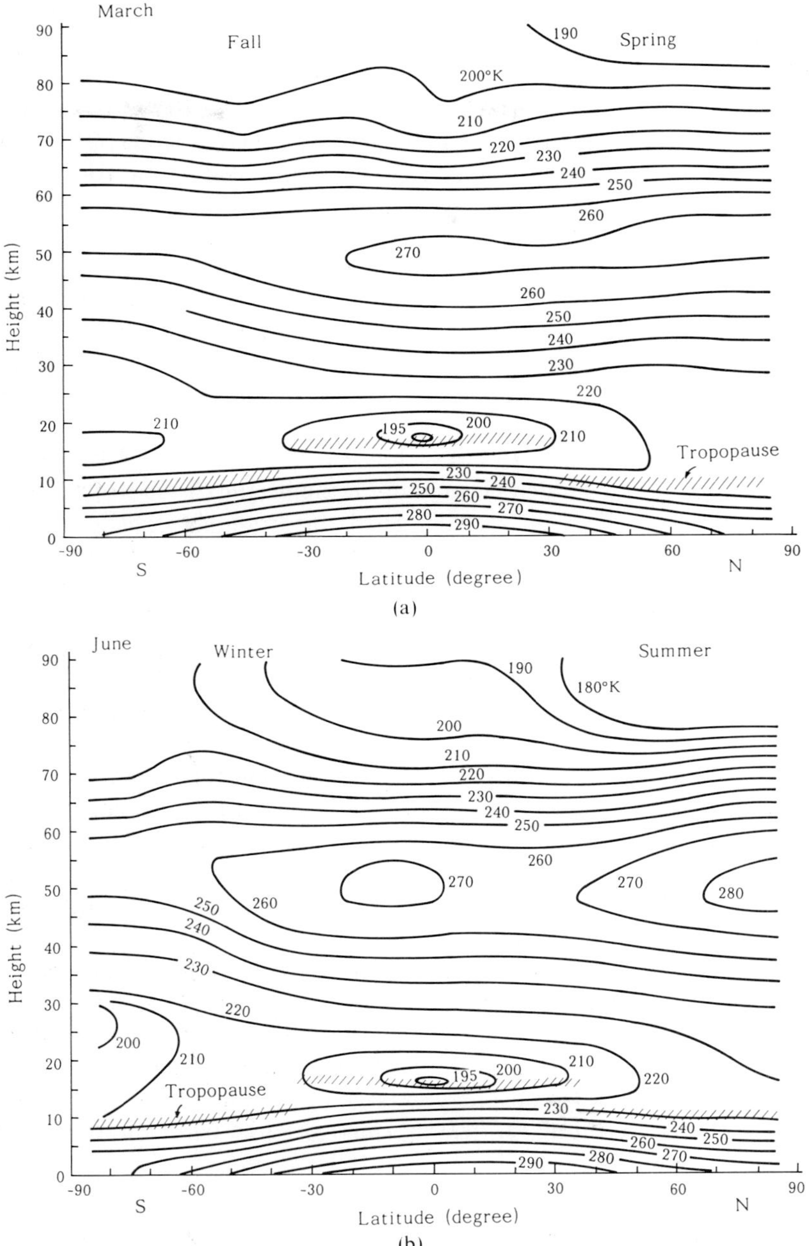

FIG. 2.2. (a) Contour plots of the atmospheric temperature in the meridional plane for March. Shaded area indicates the tropopause. (b) The same as Fig. 2.2 (a) but for June.

mosphere changes with seasons and latitude. The longitudinal and diurnal variations are relatively small. Figures 2.2 (a) and (b) illustrate the distribution of the mean temperature in the meridional plane for the March equinox and June solstice, respectively. The temperature data was compiled by C. Riegel (private communication, 1980), from OORT and RASMUSSEN (1971) for altitudes below 20 km and from rocket data (NASTROM and BELMONT, 1975) for higher regions.

The approximate height of the tropopause is shaded in Figs. 2.2 (a) and (b). It is sometimes difficult to determine precisely the height where the temperature ceases to decrease and starts to increase. This is particularly the case at high latitudes, where the temperature variation is very small. The WMO (World Meteorological Organization) defines the tropopause as the lowest height where the temperature lapse rate becomes less than 2 K/km.

The tropopause is at ~16 km at low latitudes and at ~8 km at high latitudes. There is a sudden large jump in the height of the tropopause in middle latitudes; this transition is called the *tropopause gap*. Since the temperature decreases with height at an almost constant rate in the troposphere, it continues to decrease up to the height of ~16 km at low latitudes, reaching ~195 K. On the other hand, at high latitudes the temperature decreases stops at ~8 km, and for some distance above this height the temperature remains almost constant at ~220 K. Thus, in regions of altitude 8–16 km, the temperature is lower at lower latitudes than at higher latitudes.

Above ~16 km, the atmosphere is heated by O_3 absorption more strongly at lower latitudes than at higher latitudes because the solar zenith angle is smaller; the temperature now becomes higher at lower latitudes than at higher latitudes. By the same reasoning the summer stratospheric temperature above ~16 km is higher than the corresponding winter temperature.

In the mesosphere, the seasonal variation in temperature is reversed, i.e. the summer mesosphere is colder than the winter mesosphere. The summer mesosphere is the coldest region of the atmosphere, having a temperature as low as 180 K. The reason for this is the large circulation of air in the upper part of the middle atmosphere; a large upward motion in the summer hemisphere due to strong O_3 heating causes a large temperature drop by adiabatic cooling.

2.2 *Atmospheric Pressure and Density*

In addition to temperature T, the physical state of the atmosphere is also determined by pressure p, and density as represented either by mass density

ρ or by number density (or concentration) n. The relationships among those variables are given by the ideal gas law

$$p=\rho RT \tag{2.5}$$

and

$$\rho=nm \tag{2.6}$$

where R is the specific gas constant per gram of the molecule being considered and m is the mean mass of the molecule in the atmosphere. R is equal to the universal gas constant R^* divided by the gram molecular weight (mole):

$$R=R^*/m_0. \tag{2.7}$$

(2.5) can also be written, by virtue of (2.6) and (2.7), as

$$p=n\frac{mR^*}{m_0}T=nkT \tag{2.8}$$

where k is the gas constant per molecule known as the *Boltzman constant* ($\sim 1.38\times 10^{-16}$ erg molecule^{-1} K^{-1}).

The atmospheric pressure decreases rapidly with height, and in a static atmosphere, in which molecules are in static equilibrium under the influence of gravitational force, the decrease in p per a vertical distance Δz may be expressed by

$$\Delta p=-\rho g\Delta z. \tag{2.9}$$

Dividing (2.9) by (2.8) or (2.5) we obtain

$$\frac{\Delta p}{p}=-\frac{\Delta z}{H} \tag{2.10}$$

where H is called the *scale height* of the atmosphere and is given by

$$H=\frac{nkT}{\rho g}=\frac{kT}{mg}=\frac{RT}{g}=\frac{R^*T}{m_0 g}. \tag{2.11}$$

Integration of (2.10) leads to the equation giving the height variation of p

$$p=p_0\exp\left(-\int_0^z\frac{\mathrm{d}z}{H}\right) \tag{2.12}$$

where p_0 indicates the surface pressure. From (2.8) and (2.10) we can derive a differential equation for the number density as

$$\frac{\partial n}{\partial z} = -\left(\frac{1}{T}\frac{\partial T}{\partial z} + \frac{1}{H}\right) \cdot n \tag{2.13}$$

whose integral form is

$$n = n_0 \frac{T_0}{T} \exp\left(-\int_0^z \frac{dz}{H}\right) \tag{2.14}$$

where the suffix o indicates the value at the ground.

The height variation in m and g are small in the lower and middle atmosphere; therefore, H changes mainly with the change in T. For an isothermal atmosphere in which the scale height is assumed to be constant at H_c, integration of the right-hand side of (2.14) can be done readily and we have

$$n = n_0 e^{-\frac{z}{H_c}}. \tag{2.15}$$

Thus, the number density of atmospheric molecules decreases by a factor of e^{-1} as we go up by the distance H_c.

If the temperature decreases at a constant rate as in the troposphere H would decrease linearly with height so that

$$H = H_0 - \gamma \cdot z \tag{2.16}$$

and (2.14) can be calculated as

$$n = n_0 \left(\frac{H_0}{H}\right)^{1-\frac{1}{\gamma}}. \tag{2.17}$$

The value of γ in the troposphere is ~0.19.

The values of physical parameters such as T, p, ρ, g, n, H and the mean molecular weight (the relative mass of a compound calculated on the basis of an atomic weight of 12 for carbon) at various heights are given in Appendix A as taken from the U. S. Standard Atmosphere, 1976.

2.3 *Atmospheric Composition*

The atmosphere is a mixed gas of various kinds of molecules and atoms as well as non-gaseous materials such as dust, aerosols and smoke. The chemical composition of the contemporary atmosphere near the surface is shown in Table 2.1. Major constituents are the nitrogen molecule N_2 and the oxygen molecule O_2. They are in a ratio of approximately 4 to 1, and make up 99% of the atmosphere. The atmospheric motion in the lower and

TABLE 2.1 Atmospheric composition near the surface.

Constituent	Molecular weight	Percent per volume	Percent per weight
Nitrogen (N_2)	28.01	78.088	75.527
Oxigen (O_2)	32.00	20.949	23.143
Argon (Ar)	39.94	0.93	1.28
Carbon Dioxide (CO_2)	44.01	0.03	0.046
Neon (Ne)	20.18	1.8×10^{-3}	1.25×10^{-3}
Helium (He)	4.00	5.24×10^{-4}	7.24×10^{-5}
Methane (CH_4)	16.05	1.4×10^{-4}	7.25×10^{-5}
Krypton (Kr)	83.7	1.14×10^{-4}	3.30×10^{-4}
Nitrous oxide (N_2O)	44.02	5×10^{-5}	7.6×10^{-5}
Hydrogen (H_2)	2.02	5×10^{-5}	3.48×10^{-6}
Ozone (O_3)	48.00	2×10^{-6}	3×10^{-6}
Water vapor (H_2O)	18.02	var. (0.1–1.0)	var.

middle atmospheres principally involves motions of these major constituents. However, minor constituents such as O_3, H_2O and CO_2 play important roles in radiative processes such as atmospheric heating and cooling, the driving forces of atmospheric motion. The minor constituents also play important roles in atmospheric chemistry.

2.3.1 Mean molecular weight

The atmospheric composition of chemically non-active species remains almost unchanged in the lower and middle atmospheres. The mean molecular weight is almost constant at ~28.964 up to the height near the mesopause; this value is determined by the constancy of the relative abundances of the major constituents N_2 and O_2. As is seen from the table in Appendix A, the mean molecular weight starts to decrease above ~90 km; this indicates that the effect of converting O_2 into O by photodissociation becomes appreciable above this height. This critical height is called the *homopause.* The region below the homopause, where there is no gross change in the mean molecular weight, is called the *homosphere*; the region above is called the *heterosphere.*

The mean molecular weight decreases even more rapidly above ~110 km, and it approaches ~16 in the ionospheric F region (250 - 400 km) where the oxygen atom (O) is the major constituent of the atmosphere. In even higher regions, lighter gases such as H_e and H tend to dominate and at 1,000 km the mean molecular weight becomes as low as ~4.

The separation of heavier and lighter gases takes place by the effect of molecular diffusion in the gravitational field. The molecular diffusion coefficient for the ith constituent is expressed by (CHAPMAN and COWLING, 1953)

$$D_i = \frac{3}{8n\sigma_i^2}\left\{\frac{kT(m+m_i)}{2\pi m m_i}\right\}^{1/2} \tag{2.18}$$

where σ_i is the collisional cross section between the ith constituent and the major constituent. Since D_i is inversely proportional to the atmospheric density, it is small in the lower regions. The effect of gravitational separation becomes appreciable only above a certain height, where D_i is greater than the eddy diffusion coefficient. This critical height is called the *turbopause* and is usually at ~110 km.

Above the turbopause, the height variation in n_i tends to follow the diffusive equilibrium distribution and is written in the form

$$\frac{\partial n_i}{\partial z} = -\left(\frac{1}{T}\frac{\partial T}{\partial z} + \frac{1}{H_i}\right)n_i \tag{2.19}$$

where H_i represents the scale height of the ith constituent as given by

$$H_i = \frac{kT}{m_i g} \tag{2.20}$$

where m_i is the mass of the ith constituent. Since H_i is in inverse proportion to m_i, H_i is larger for the lighter gas which has a smaller m_i. Because n_i of the lighter gas decreases more slowly with height than the heavier gas, the lighter constituent tends to dominate over the heavier constituent at higher altitudes. This phenomenon is called *diffusive separation*.

Below the turbopause the effect of atmospheric mixing by eddy diffusion is dominant, and the height variation in n_i for the chemically non-active species follows the equation for complete mixing (2.13) and is given by

$$\frac{\partial n_i}{\partial z} = -\left(\frac{1}{T}\frac{\partial T}{\partial z} + \frac{1}{H}\right)n_i. \tag{2.21}$$

Since H is common between (2.13) and (2.21), the ratio of n_i to n

$$f_i = n_i/n \tag{2.22}$$

is constant with height. f_i is called the *mixing ratio* or more precisely the *volume* or *mole mixing ratio*, since it represents the ratio of volumes at normal condition of moles or of number of molecules of gas to air.

The mixing ratio is sometimes defined by a *mass mixing ratio* which is the volume mixing ratio times the ratio of molecular weight of gas to air. The mixing ratio is a non-dimensional quantity, but the following notations are usually used:

		mixing ratio volume	mass
parts per million	(10^{-6})	ppmv	μg/g
parts per billion	(10^{-9})	ppbv	μg/kg
parts per trillion	(10^{-12})	pptv	$\mu\mu$g/g

In order to distinguish the volume mixing ratio from the mass mixing ratio a letter v is sometimes attached to the notations, e.g. ppmv (parts per million by volume).

2.3.2 *Chemical and dynamical effects*

Figure 2.3 illustrates the height variations in n_i for various chemical compounds observed in the middle atmosphere (ACKERMAN, 1979). Although these were compiled from observational data obtained at different times and places by different methods, it is adequate to show the general trend in the height variation for each constituent (however, the NO curve above ~ 50 km may not properly represent the details of the actual observed variation—see Chapter 13). The curves 10^{-1}, 10^{-2} etc. represent the variations in n_i for the case of a constant mixing ratio; if the molecule is completely

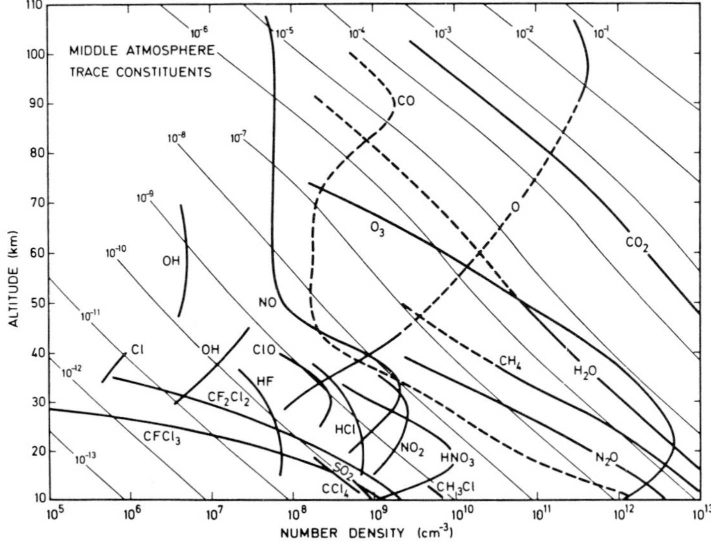

FIG. 2.3. Vertical profiles of observed number densities of the typical trace constituents in the middle atmosphere (ACKERMAN, 1979).

mixed, the *slope* of the variation in n_i should follow these curves. The actual position of the curve is determined by the mixing ratio of that constituent.

Two different kinds of molecule may be distinguished in Fig. 2.3; in one group n_i decreases monotonously with increasing height if not completely follows a curve of a constant mixing ratio, and the other group has a maximum of n_i somewhere in the middle atmosphere. The former group includes H_2O, CO_2, CH_4, SO_2, and various chlorofluoromethanes CFM (e.g. $CFCl_3$, CF_2Cl_2, CCl_4, CH_3Cl); these molecules have sources near the ground or in the soil and are transported into the middle atmosphere by atmospheric motions. We will call these molecules the *primary* or *parent constituents*. The latter group includes O_3, NO, NO_2, HNO_3, OH, Cl and ClO; these molecules or atoms are produced in the middle atmosphere by photochemical reactions. We will call them the *secondary constituents*. Many stratospheric molecules are classified in Table 2.2 either as parent or secondary constituents according to different sources for production. The secondary constituents may be produced either directly from a parent constituent by photodissociation, or by photochemical reactions between a secondary and a parent constituent, or between secondary constituents.

If a molecule is photochemically very active, the effect of motion on the n_i distribution may be neglected, and in the steady state n_i is determined by the equilibrium between the production and loss rates, i.e.

$$n_i = Q_i / L_i \tag{2.23}$$

where Q_i is the production rate and L_i is the loss coefficient for the ith constituent.

Some minor constituents such as excited atomic oxygen $O(^1D)$ are in

TABLE 2.2 Classification of middle atmosphere constituents according to their sources.

Classification	Constituents	Characteristics
Primary or parent consituents	N_2, O_2, N_2O, H_2O, CO_2, CH_4, SO_2, H_2, CFM	Sources are near the surface or in the soil; transported into the middle atmosphere
Secondary constituents	$O(^3P)$, $O(^1D)$, $O_2(^1\Delta_g)$ N, NO, H, OH, Cl, S, CO	Produced in the middle atmosphere by photo-dissociation of primary molecules
Secondary constituents	O_3, OH, NO, NO_2, NO_3, N_2O_5, HO_2, H_2O_2, HNO_2, HNO_3, HNO_4, ClO, HCl, OHCl, $ClONO_2$, COS	Produced in the middle atmosphere by photo-chemical reactions of a secondary molecule with the primary or other secondary molecules

photochemical equilibrium for the entire region of the middle atmosphere, but for some constituents the photochemical equilibrium condition can be applied only in part of the middle atmosphere. For instance, O_3 is in photochemical equilibrium in and above the middle stratosphere, but in the lower stratosphere the chemical reactions become too slow and the effect of atmospheric motion can no longer be neglected. In such a case, the variation in n_i may be governed by the *continuity equation*

$$\frac{\partial n_i}{\partial t} = Q_i - L_i n_i - \frac{\partial \phi_i}{\partial z} \tag{2.24}$$

where ϕ_i is the vertical *flux* of n_i due to the atmospheric mixing, and is usually expressed in terms of the *eddy diffusion coefficient K* as follows: (COLGROVE *et al.*, 1965; SHIMAZAKI, 1967)

$$\phi_i = -K\left\{\frac{\partial n_i}{\partial z} + \left(\frac{1}{T}\frac{\partial T}{\partial z} + \frac{1}{H}\right) n_i\right\}. \tag{2.25}$$

Assuming a steady state ($\partial/\partial t = 0$) and if the photochemical reactions are slow ($Q_i = L_i = 0$), (2.24) reduces to ϕ_i=constant. When there is no flux of n_i from the surface, this constant can be taken as zero, and the distribution of n_i is obtained by equating the term within braces in (2.25) to zero. This is identical to (2.19), and the result implies that the distribution approaches complete mixing if the chemical reactions are slow.

Equation (2.24) includes effects of both chemistry and dynamics. Which of these two effects is more dominant is determined by the relative magnitude of the characteristic times (or time constants) for chemical and dynamical processes. The time constant for chemistry (or photochemical lifetime) will be discussed in detail in Chapter 6 but it may be represented roughly by

$$\tau_c = 1/L_i \tag{2.26}$$

whereas the characteristic time for vertical eddy diffusion may be given by

$$\tau_v = \frac{H^2}{K}. \tag{2.27}$$

The distribution of n_i is determined mainly by chemistry if $\tau_c < \tau_v$ or $L_i > K/H^2$ and by dynamics if $\tau_v < t_c$ or $K > L_i H^2$.

Chapter 3

SOLAR RADIATION

Solar UV radiation is the ultimate agent producing most of the minor constituents in the middle atmosphere. We discuss in this chapter the spectrum of the incoming solar radiation and its attenuation due to molecular absorption and scattering as it penetrates through the atmosphere. Multiple scattering and surface albedo may also affect to an appreciable extent the intensity of UV radiation in the middle atmosphere.

3.1 Solar Spectrum and Units for Radiation

The electromagnetic waves emitted from the sun are composed of various types of radiation as listed in Table 3.1. The approximate range of wavelengths for each type of radiation is also given. Boundaries are not clearly defined in most cases and there is some overlap. Since the overall range from longest to shortest wavelengths is so large, it is convenient to use different units in each category; usually angström Å (10^{-8} cm) and

TABLE 3.1 The spectrum of electromagnetic waves.

Type of radiation	Range of wavelength (cm)	(various units)
Gamma ray	$<10^{-8}$	<1 Å
X-ray	$10^{-8}-10^{-6}$	1–100 Å
Extreme ultraviolet (EUV)	$10^{-7}-10^{-5}$	10–1000 Å
Ultraviolet (UV)	$10^{-6}-3.8\times10^{-5}$	10–380 nm
Visible	3.8×10^{-5}–8.1×10^{-5}	380–810 nm
Near Infrared	$8.1\times10^{-5}-10^{-4}$	0.81–1 μm
Infrared (IR)	$10^{-4}-2\times10^{-3}$	1–20 μm
Far Infrared	$2\times10^{-3}-10^{-2}$	0.02–0.1 mm
Microwave	$10^{-2}-10^{3}$	0.1 mm–10 m
Radio Wave	$>10^{3}$	>10 m

nanometer nm (10^{-7} cm) are used for the ultraviolet (UV) and visible ranges, and micrometer μ (10^{-4} cm) for the infrared (IR) range. In some cases, wavenumber, the reciprocal of wavelength, is also used for the IR range. The notation XUV is employed to represent X-ray and EUV radiation.

Nearly all radiation of wavelength less than 100 nm is absorbed in the thermosphere by nitrogen and oxygen. We will not consider them except in the ionization of N_2 and O_2 which is the main source of mesospheric NO. The latter will be discussed in Chapter 13. On some special occasions, X-ray of wavelength less than 1 nm and energetic particle radiation can penetrate into the region below 100 km. Our main concern here, however, lies in the range between ~135 and 400 nm, and the hydrogen Lyman-α radiation at ~121.6 nm; these are the most important components of solar radiation in the middle atmosphere.

The solar *energy flux* is generally measured in units of erg per cm^2 per sec over the wavelength range nm or Å; this unit is convenient for calculating atmospheric heating rates by solar radiation. However, the usage of the solar *photon flux* is more convenient for calculations of the photodissociation rates of molecules. A photon has an energy hν, where h is the Planck constant and ν is the frequency of the wave, i.e. the velocity of light divided by the wavelength. These two units are related through the following conversion relationships:

$$\begin{aligned} 1 \text{ erg cm}^{-2} \text{ sec}^{-1} &= 1 \text{ mwm}^{-2} \\ &= 5.0324 \times 10^{15}\lambda \text{ photons cm}^{-2} \text{ sec}^{-1} \end{aligned}$$

or

$$1 \text{ photon cm}^{-2} \text{ sec}^{-1} = 1.9871 \times 10^{-16}/\lambda \text{ erg cm}^{-2} \text{ sec}^{-1}$$

where λ represents wavelength in cm. The energy flux mw (milliwatt) m^{-2} in the MKS system of units is equivalent to erg cm^{-2} sec^{-1} in the cgs system. The relative value of the energy represented by photon flux to that represented by energy flux increases as λ increases.

The solar constant, the intensity of solar radiation in free space at the earth's mean solar distance, is usually expressd in cal cm^{-2} min^{-1}. The conversion relationship between this unit and the cgs unit is given by

$$1 \text{ cal cm}^{-2} \text{ min}^{-1} = 6.9758 \times 10^5 \text{ erg cm}^{-2} \text{ sec}^{-1}$$

or

$$1 \text{ erg cm}^{-2} \text{ sec}^{-1} = 1.4335 \times 10^{-6} \text{ cal cm}^{-2} \text{ min}^{-1}.$$

The value of 1 cal cm^{-2} is also known as a ''langley'' and is denoted by ly.

Recent studies of the solar UV and IR spectrum suggest a value of ~2.00 cal cm^{-2} min^{-1} for the solar constant. The actual intensity varies depending upon the sun-earth distance, and ranges from about 1.93 at aphelion (early July) to 2.07 at perhelion (early January). The long term variation in the solar constant is important in terms of climate studies, but our knowledge in this respect is very limited at this time.

3.2 Intensity of the Incoming Solar Radiation

Figure 3.1 illustrates the energy flux of solar radiation in 5 nm intervals for the spectral range 150 to 900 nm as observed at the top of the atmosphere. The intensity is a maximum at visible blue near 480 nm, and it decreases gradually toward the longer wavelengths, but rather rapidly toward the shorter wavelengths. In the visible and IR ranges, the intensity is close to that of a 6000 K black body, but it becomes smaller than the 6000 K black body radiation in the UV range. The equivalent temperature of the black body radiation approaches 4750 K in the UV range below 200 nm (see Appendix B for black body radiation).

The change in the equivalent temperature of the black body radiation with wavelength is caused by the difference in the regions of the solar atmosphere where the radiation of that particular wavelength is emitted. the solar temperature decreases from ~2×10^7 K near the center to ~4200 K at

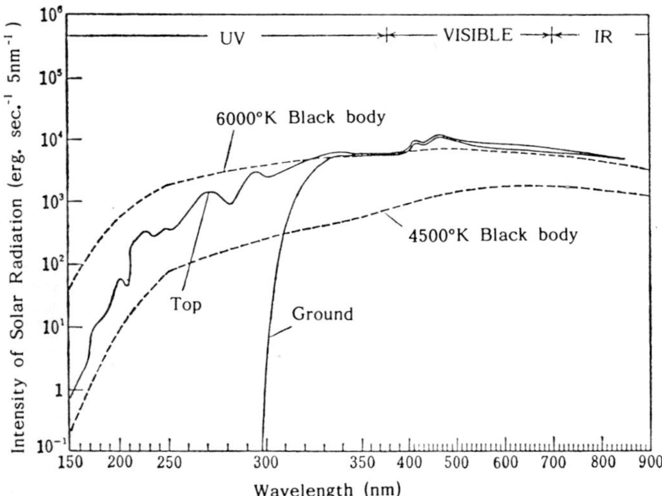

FIG. 3.1. Intensity of solar radiation at wavelengths between 150 and 900 nm observed on the ground surface and at the top of the atmosphere. Dashed curves indicate temperatures of blackbody radiation.

the surface of the photosphere. The temperature then increases in the chromosphere; it is approximately 6000 – 7000 K in the lower atmosphere and reaches 10^6 K in the solar corona.

Figure 3.2 compares various measurements of solar irradiance for the spectral range 120–180 nm (SIMON, 1978). The largest irradiance was measured by DETWILER *et al.* (1961) by using a photographic detection technique, but later measurements have shown much smaller values. This difference could be partly due to the actual variation of the solar radiation flux. In fact, there is some solar flux variability with the 11 year solar cycle in radiation in the Schumann-Runge continuum (TORR *et al.*, 1980). However, a large portion of the difference among observations may be attributed to the difference in measuring techniques. For instance, CARVER *et al.* (1972a) used ionization chambers, ACKERMAN and SIMON (1973), ROTTMAN (1974) and HEROUX and SWIRBALUS (1976) used photoelectric detectors and SAMAIN and SIMON (1976) deduced the solar radiance from photographic stigmatic spectra. The measurements are also affected by which portions of the solar disc have been measured and by the existence or non-existence of large sunspots or other major activity on the sun. The equivalent black body temperature takes a minimum value (~4400 K) at ~ 160 nm; it increases considerably at smaller wavelengths particularly below

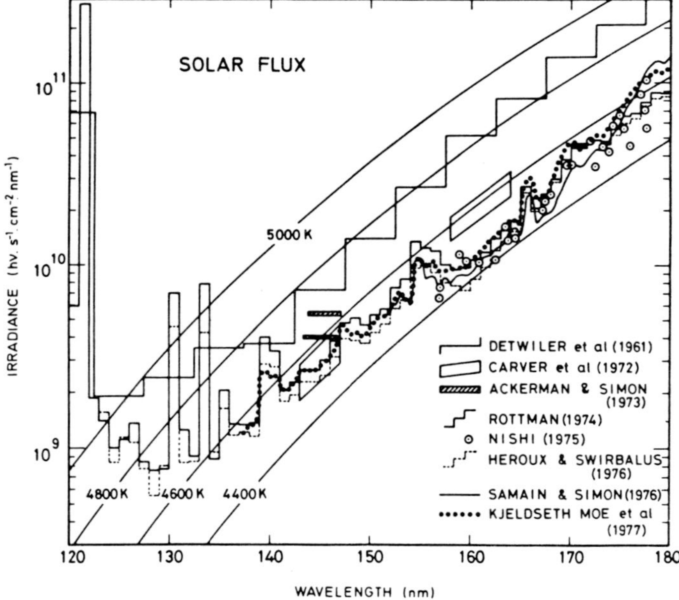

FIG. 3.2. Comparison of various measurements of solar UV radiation from 120 to 180 nm (SIMON, 1978).

$\sim$140 nm, whrereas it increases gradually towards larger wavelengths.

NICOLET (1980) has reviewed the existing data on solar UV radiation and concluded that there is no clear evidence to indicate an eleven year solar cycle variation for the radiation above 175 nm, although HEATH (1973) and HEATH and THEKAEKARA (1976) have obtained a factor of $\sim$2 variation at 200 nm and of not less than 4% variation at 300 nm.

The Lyman-α flux variation with solar activity probably represents the maximum possible difference that can occur in the solar flux above 100 nm. Figure 3.3 shows the frequency distribution of the number of observations of the solar Lyman-α flux for every 0.5 erg cm^{-2} see^{-1} interval (WEEKS, 1967). From this data we can derive the range of the Lyman-α flux intensity between the solar maximum and minimum as follows:

$$
\begin{aligned}
(I_\infty)_{\text{Lyman}-\alpha} &\doteqdot 5 \pm 1.5 \text{ erg cm}^{-2} \text{ sec}^{-1} \\
&\doteqdot (3 \pm 1) \times 10^{11} \text{ photons cm}^{-2} \text{ sec}^{-1}.
\end{aligned}
$$

Thus the flux can change by a factor of 2 from 4×10^{11} photons cm^{-2} sec^{-1} at the solar maximum to 2×10^{11} photons cm^{-2} sec^{-1} at the solar minimum. The solar Lyman-α flux can also change by a few percent during solar flare (LINDSAY, 1963), and a short term fluctuation may occur due to the 27 day solar rotation (VIDAL-MADJAR, 1975; ROTTMAN *et al.* 1982).

The solar flux in each band of the Schumann-Runge band system (175–205 nm) is compared in Fig. 3.4 between two data sources; one is compiled by ACKERMAN (1971) and the other observed by SAMAIN and SIMON (1976). Both indicate an increase of the flux with wavelength, but the 1976

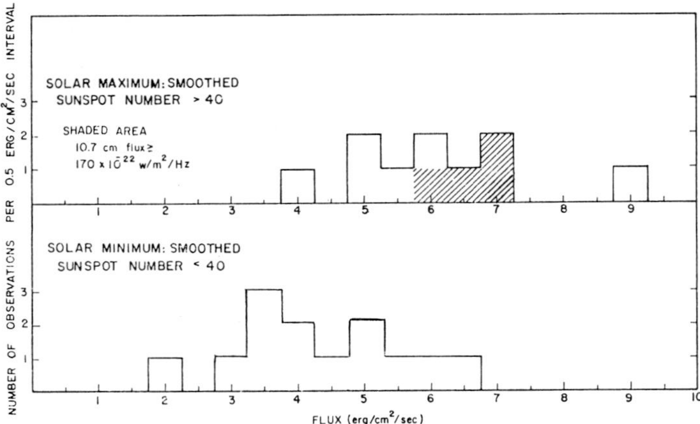

FIG. 3.3. Rocket measurements of Lyman-α by ion chambers for solar maximum and solar minimum (WEEKS, 1967).

values are systematically smaller than the 1971 values. This difference is probably caused by the different measuring techniques, rather than an actual variation during this time period. Even among recent observational data there is an estimated precision of ~20% in the flux measurements (NICOLET, 1980).

The solar fluxes reported by various experimenters are compared in Figs. 3.5 (a) - (c) for the wavelength range 170 - 210 nm, 210 - 280 nm, and 280 - 350 nm, respectively (SAMAIN and SIMON, 1976; SIMON, 1978). Again, the measurement by DETWILER *et al.* (1961) shows much larger irradiance than other observations below 210 nm. Other than that, there is general agreement among observations including some fine structure except that the data by ACKERMAN (1971) indicate appreciably larger values than any other observation. Again, this difference may not be due to an actual variation in the solar flux, but may imply the effects of different measuring techniques. Most stratospheric photochemical models before 1974 used the solar flux data compiled by ACKERMAN (1971), but most models after 1974 have reduced the solar flux according to the comparison shown in Figs. 3.2 and 3.5. The solar photon fluxes are tabulated in Appendix C for 70 subdivisions of the spectral range 135–400 nm and for the Lyman-α; these are used

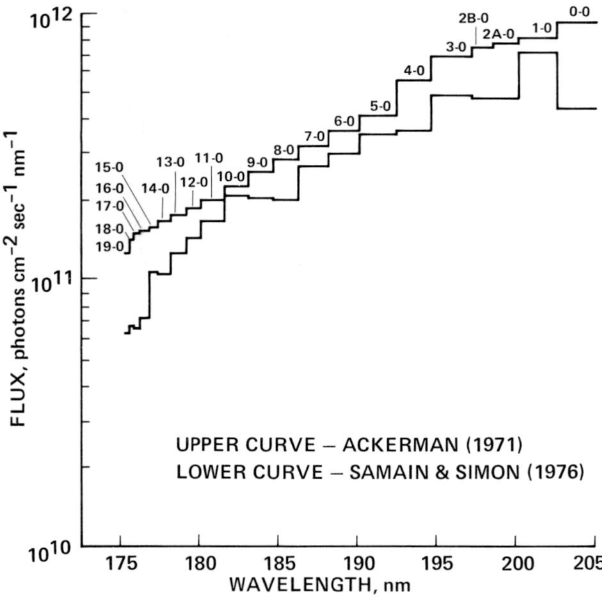

FIG. 3.4. Solar fluxes over each band of the O_2 Schumann-Runge band system.

in model calculations in Chapter 7.

3.3 Attenuation of Solar Radiation Intensity

The intensity of solar radiation changes as it penetrates through the atmosphere due to various processes of interaction between matter and the radiation field. If the radiation energy is converted to kinetic energy, the process is called *absorption*. The reverse process is called *thermal emission*. If the absorbed energy of solar photons is used to raise the internal energy of molecules without changing the kinetic energy, the *excitation* of molecules occurs. The excited molecules then re-emit the energy by transitions to lower energy levels; this phenomenum of induced emission is called *scattering*. The frequency of emitted radiation can generally be different from the frequency of the incident radiation, but we assume that *coherent scattering* of the identical frequency is most intense. We also assume a *simple scattering* in which there is no transfer of energy among various (electronic, vibrational and rotational) modes of internal energy.

Absorption and scattering can occur due to both atmospheric molecules and aerosol dust particles. The effect of atmospheric thermal emission is negligibly small in the UV range, since the temperature of the atmosphere is much lower than the radiative temperature of solar radiation. Thermal emission, however, becomes dominant in the radiative transfer of IR radiation (see Chapter 12). Here we will discuss mainly the effects of molecular absorption and scattering on the solar UV radiation.

3.3.1 Molecular absorption

Molecular absorption is important for attenuating the solar UV radiation in the middle atmosphere. The spectrum of solar radiation observed on the earth's surface shows a very sharp decrease in intensity below ~320 nm, and this radiation disappears completely below ~300 nm, as is seen in Fig. 3.1. This is mainly due to the absorption of radiation in this spectral range by atmospheric ozone. The ozone absorption below ~310 nm is indeed very large, and in 1881 Hartley proposed that the reason why there is no solar radiation below ~310 nm observed on the ground is the absorption by O_3 in the upper atmosphere. In particular, it was thought that at around 200 nm, where the O_3 absorption cross section is relatively small (see Fig. 4.1), solar radiation could be observed at high altitudes by balloon experiments.

All attempts by balloons to detect the solar radiation below 310 nm at high altitudes, however, have failed. It was only after World War II, when it became possible to carry out measurements at high altitudes using rockets, that solar radiation below 310 nm was actually observed in the up-

SOLAR FLUX
IRRADIANCE (hV.sec⁻¹.cm⁻².nm⁻¹)
WAVELENGTH (nm)
5000 K
4800 K
4600 K
4400 K
DETWILER et al (1961)
ACKERMAN et al (1971)
ACKERMAN & SIMON (1973)
SIMON (1974)
ACKERMAN (1974)
ROTTMAN (1974)
NISHI (1975)
HEROUX & SWIRBALUS (1976)
BRUECKNER et al (1976)
SAMAIN & SIMON (1976)

(a)

SOLAR FLUX
IRRADIANCE (hv. s⁻¹.cm⁻².nm⁻¹)
WAVELENGTH (nm)
5400 K
5200 K
5000 K
4800 K
DETWILER et al (1961)
ACKERMAN et al (1971)
BROADFOOT (1972)
HEATH (1973)
SIMON (1974)

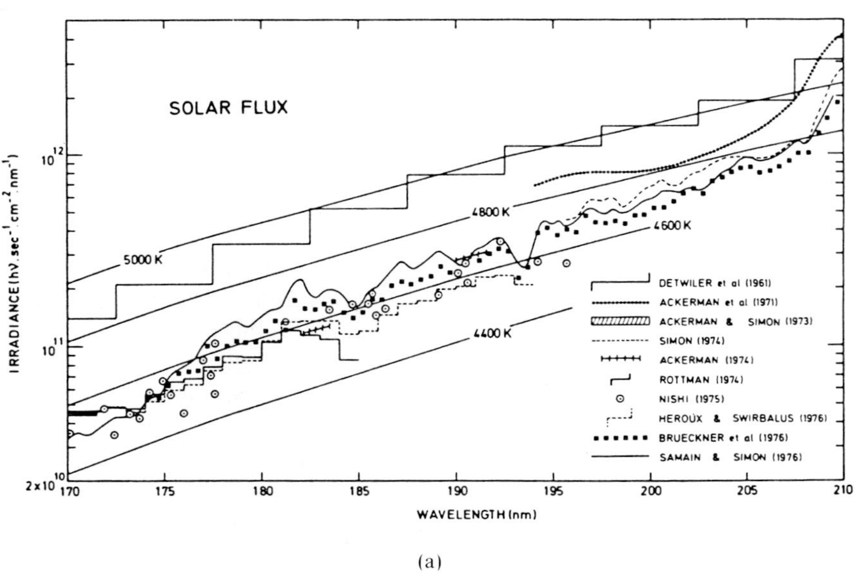

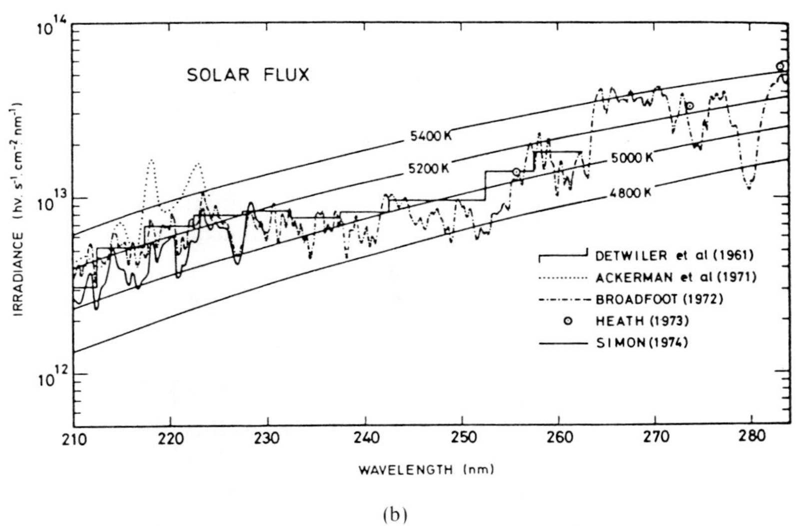

(b)

FIG. 3.5

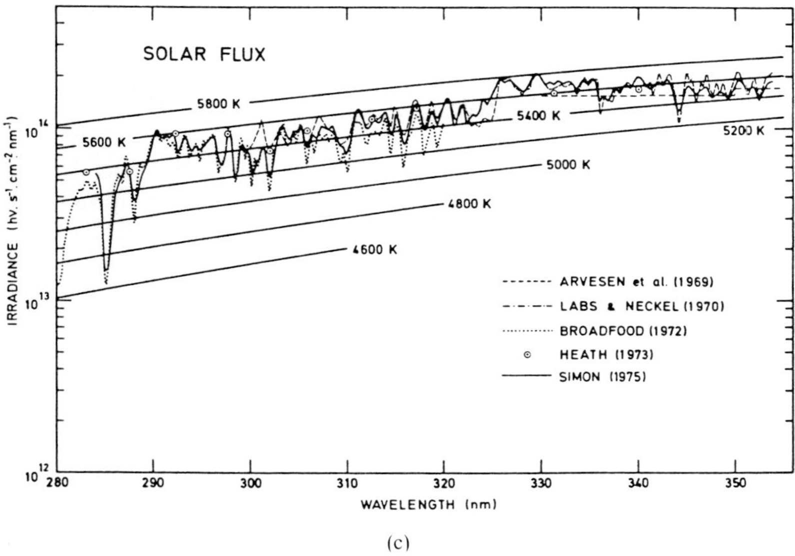

(c)

FIG. 3.5. (a) Comparison of various measurements of solar UV radiation from 170 to 210 nm (SIMON, 1978). (b) The same as Fig. 3.5 (a) but for 210–280 nm. (c) The same as Fig. 3.5 (a) but for 280–350 nm.

per atmosphere. The early balloons could not reach altitudes above 30 km. The change in the spectral range observed at various heights by early rocket observations is shown in Fig. 3.6 (BAUM *et al.*, 1946).

The decrease in the intensity of monochromatic radiation by molecular absorption is assumed by Lambert's law to be proportional to the intensity of radiation I, the number density of the absorber n_i, and the path length ds; it is expressed by

$$dI = -I\sum_i \sigma_i n_i \, ds \tag{3.1}$$

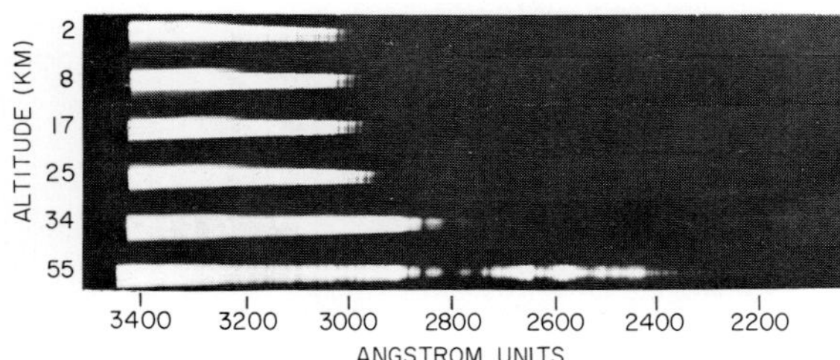

FIG. 3.6. Photograph of solar UV-spectra observed by rockets at various heights on October 10, 1946 (BAUM *et al.*, 1946).

where σ_i is the absorption cross-section of the ith constituent (absorber). For the stratified atmosphere, the path length ds is related to the vertical displacement dz by

$$\mathrm{d}s = \mathrm{d}z \sec \chi \tag{3.2}$$

where χ is the solar zenigh angle. This relationship is valid only for a plane parallel atmosphere (see Fig. 3.7) but gives a good approximation when χ is smaller than $\sim 80°$. When χ approaches or exceeds $90°$, the effect of the earth's curvature must be taken into account in the conversion factor between ds and dz; this factor is sometimes called *air mass* (see Appendix D).

(3.1) becomes by virtue of (3.2)

$$\frac{\partial I}{\partial z} = - \sum_i \sigma_i n_i I \sec \chi \tag{3.3}$$

whose solution is given by

$$I(z, \chi) = I_\infty \cdot \mathrm{e}^{-\tau \cdot \sec \chi} \tag{3.4}$$

where τ is the *optical depth* of the atmosphere as defined by

$$\tau = \sum_i \sigma_i \int_z^\infty n_i(z)\, \mathrm{d}z \tag{3.5}$$

and I_∞ is the intensity of the incoming solar radiation at the top of the atmosphere. The optical path passing through the atmosphere with an angle χ to the vertical direction is given by $\tau \cdot \sec \chi$.

The integral in (3.5) indicates the total number of absorbing molecules in the vertical column above the height z. If the distribution of n_i is given by (2.14), this integral can be calculated as

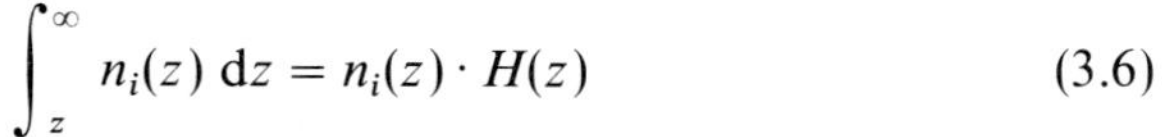

$$\int_z^\infty n_i(z)\, \mathrm{d}z = n_i(z) \cdot H(z) \tag{3.6}$$

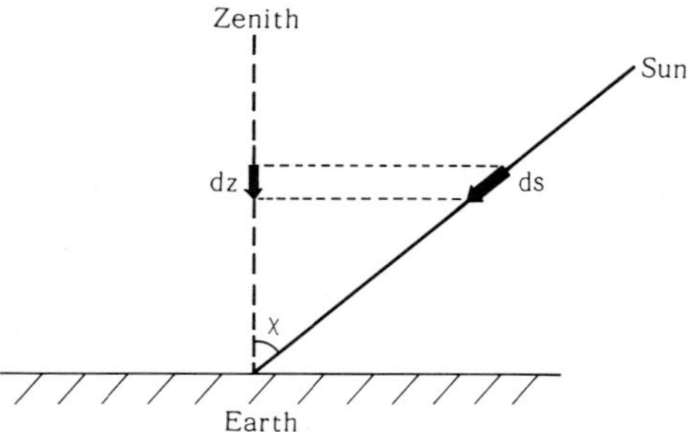

FIG. 3.7. Diagram illustrating the relationship between the slant optical path and vertical path in a plain parallel atmosphere.

Thus, the scale height $H(z)$ may be interpreted as the thickness of the uniform atmosphere in which all molecules above the height z are reduced to the physical condition (temperature and pressure) of that height. It must be noted that the relationship (3.6) is valid for any temperature distribution. From (3.5) and (3.6) we can write

$$\tau = \sum_i \sigma_i n_i(z) H(z) \tag{3.7}$$

It is usually enough to take the summation of i just for O_3 and O_2 in the middle atmosphere. In the troposphere, where there is no O_3 absorption except for a weak absorption in the Chappuis band (450 – 800 nm), the absorption by NO_2 must be included. For other molecules at least one of σ_i or n_i is extremely small, and their contribution to the total optical depth is negligibly small.

The total absorbed energy by all kinds of molecules can be calculated by

$$A = \sum_j \sigma_j n_j(z) I_\infty e^{-\tau(z)\cdot\sec\chi} \tag{3.8}$$

A has its maximum value under the condition $\partial A/\partial z = 0$, which gives

$$\sum_j \frac{1}{n_j}\frac{\partial n_j}{\partial z} = \frac{\partial \tau}{\partial z}\sec\chi. \tag{3.9}$$

We now consider the absorption of solar radiation by a single kind of molecule. This situation occurs approximately in cases of O_3 absorption in the stratosphere and O_2 absorption in the thermosphere, since these molecules are the main single absorbers in each case. Omitting the subscripts on n and σ, we can write (3.9) by virtue of (2.13) and (3.5) as

$$\frac{1}{T}\frac{\partial T}{\partial z} + \frac{1}{H} = \sigma \cdot n \cdot \sec\chi \tag{3.10}$$

which becomes, using (3.7)

$$\tau = \left(1 + \frac{\partial H}{\partial z}\right) \cdot \cos\chi. \tag{3.11}$$

Equation (3.11) indicates that the absorption of solar radiation has a maximum value at the height where the optical path ($\tau \cdot \sec\chi$) is equal to $1 + \partial H/\partial z$. For an isothermal atmosphere in which T, and therefore H, is constant with height, the absorption is maximum at the height where the optical depth is unity.

The height of maximum absorption in an isothermal atmosphere can be written explicitly from (3.10) using (2.15) as

$$z_m = H \ln (\sigma H n_0 \sec \chi). \tag{3.12}$$

We will call z_m the *penetration height*. The intensity of radiation at z_m is e^{-1} times the value of the incident radiation (I_∞) and most of the energy should be absorbed at the height range where the optical depth is close to but less than 1, i.e. $\tau \simeq 0.2$–1.0. Below this height, the intensity decreases exponentially, at a very rapid rate.

According to (3.12) the penetration height is lowest at vertical incidence ($\chi = 0$) and increases as χ increases. Thus, solar radiation can penetrate deeper in the atmosphere in summer than in winter, at lower latitudes than at higher latitudes, and near local noon than near sunrise or sunset.

Since σ changes with wavelength, the penetration height also depends upon the wavelength range over which the molecule absorbs solar radiation. Figure 3.8 illustrates the variation in the optical depth with height at various wavelengths, taking into account the effects of absorption by both O_3 and O_2. It can be seen that for $\lambda \geq 310$ nm the optical depth is always smaller than unity; therefore, no maximum absorption occurs and the attenuation of radiation is very small for the entire atmosphere. Below ~310 nm the optical depth becomes unity at different heights for different wavelengths.

Figure 3.9 shows the variation in the penetration height with wavelength. It can be seen that the radiation between ~200 and ~310 nm can penetrate to the stratosphere, and most of the energy is absorbed there. The radiation between ~130 and ~200 nm is absorbed mainly in the mesosphere, and below ~100 nm the radiation is absorbed mainly in the thermosphere. Figure 3.9 also shows the main absorbing molecules over various wavelength ranges. O_3 is the main absorber at ~240–320 nm, O_2 at ~100–200 nm, and both O_3 and O_2 absorb radiation at ~200 – 240 nm. Absorption below ~100 nm can occur by various molecules and atoms (N_2, O_2, N, O).

The spectrum of solar radiation intensity calculated for various heights is illustrated in Fig. 3.10, from which it is seen that radiation between ~220 and 260 nm attenuates markedly in the upper stratosphere and cannot reach the lower stratosphere. At around 200 nm, about 1% of the incoming radiation reaches the lower stratosphere. This is due to the fact that the absorption cross section of the O_3 Hartley continuum has a minimum around 200 nm (see Fig. 4.1).

3.3.2 Scattering and albedo

The intensity of solar radiation decreases due to scattering from any particular region of the atmosphere, and increases due to the supply of energy

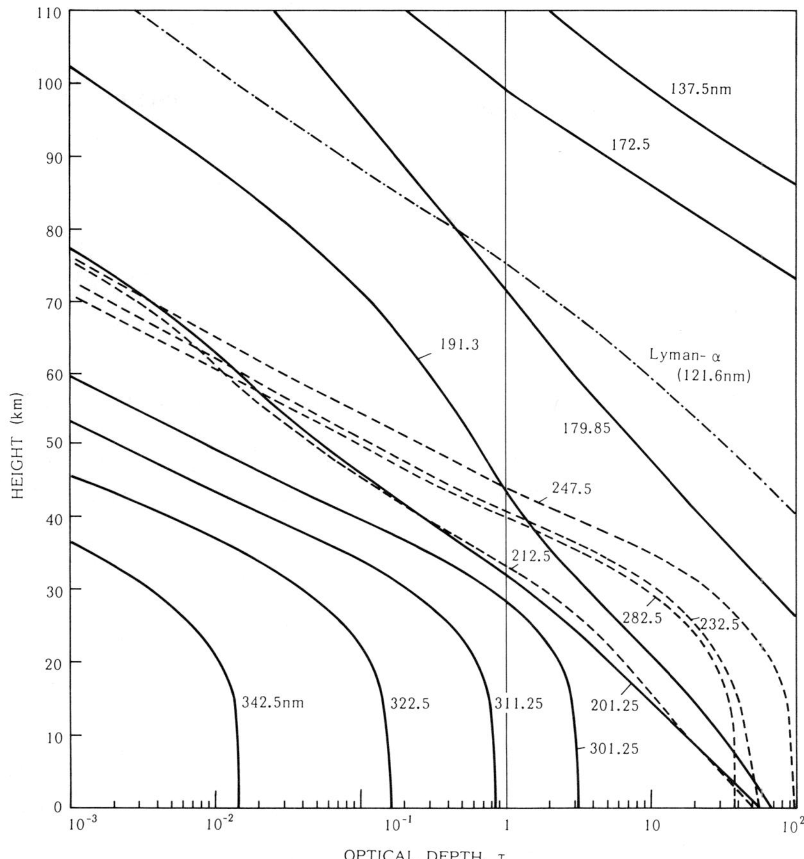

FIG. 3.8. Height profiles of optical depth due to absorption by O_3 and O_2 at various wavelengths.

to that region as a result of multiple scattering from the surrounding region.

The first effect is expressed by Lambert's law as

$$\mathrm{d}I = -\sigma_s n I \,\mathrm{d}s \tag{3.13}$$

where σ_s is the scattering cross section of the scatterers and n is the atmospheric number density. If the size of the scatterers (molecules or the inhomogeneity of the atmosphere) is less than a tenth of the wavelength of the radiation, the scattering is called *Rayleigh scattering*. The molecular scattering is Rayleigh scattering for most wavelengths under consideration. The scattering cross section for Rayleigh scattering is inversely propor-

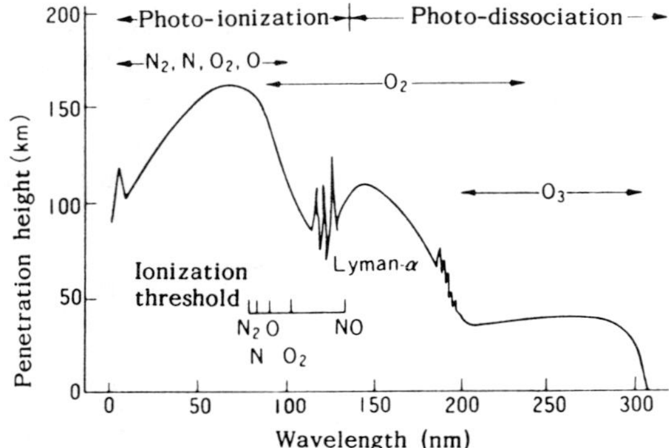

FIG. 3.9. Penetration height for vertical incidence of solar radiation at various wavelengths.

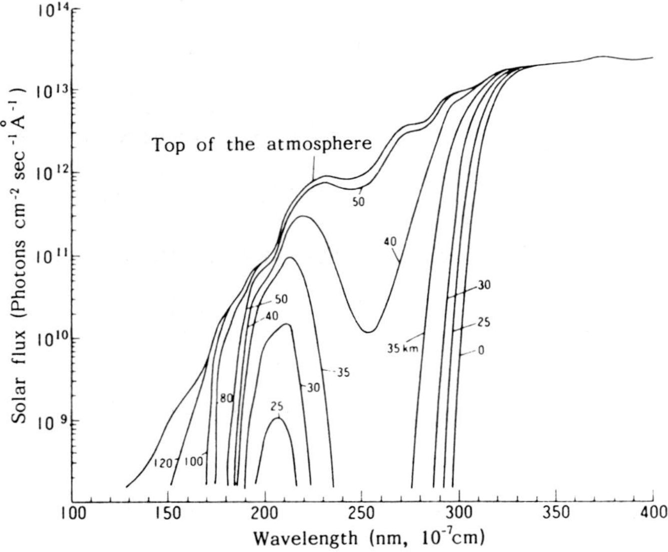

FIG. 3.10. Spectrum of the intensity of solar radiation predicted for various heights.

tional to the 4th power of wavelength (see (3.28) below).

The increase of the radiation intensity in a solid angle dΩ by multiple scattering may be expressed by

$$\mathrm{d}I = \frac{\sigma_s n}{4\pi} P(z, \theta, \varphi) I'(z, \theta, \varphi)\, \mathrm{d}\Omega\, \mathrm{d}s \tag{3.14}$$

where I' (z, θ, φ) is the intensity of the scattered radiation received at the height z from the direction θ and φ (spherical coordinates), and the scattering phase function $P(z, \theta, \varphi)$ characterizes the portion of radiation scattered in the direction (θ, φ) relative to the entire scattered radiation; P is normalized so that the integral over the entire solid angle is unity, i.e.

$$\frac{1}{4\pi} \int_{\Omega} P(z, \theta, \varphi) \, \mathrm{d}\Omega = \frac{1}{4\pi} \int_0^{2\pi} \int_{-\frac{\pi}{2}}^{\frac{\pi}{2}} P(z, \theta, \varphi) \sin \theta \cdot \mathrm{d}\theta \cdot \mathrm{d}\varphi = 1. \quad (3.15)$$

We now assume that the scattering is isotropic (independent of height and direction); therefore, P is unity. Taking into account the effects of both absorption and scattering, we combine (3.3) and (3.14) into

$$\cos \chi \cdot \frac{\partial I(z, \chi)}{\partial z} = -\left(\sum_i \sigma_i n_i + \sigma_s n \right) \cdot I(z, \chi) + \frac{\sigma_s n}{4\pi} \int_0^{2\pi} \int_{-\frac{\pi}{2}}^{\frac{\pi}{2}} I'(z, \theta, \varphi) \sin \theta \cdot \mathrm{d}\theta \cdot \mathrm{d}\varphi. \quad (3.16)$$

Introducing the *extinction cross-section* α defined by

$$\alpha \cdot n = \sum_i \sigma_i n_i + \sigma_s n \quad (3.17)$$

and the total optical depth for absorption and scattering

$$\tau = \tau_a + \tau_s = \int_z^{\infty} \alpha n \, \mathrm{d}z \quad (3.18)$$

and the scattering albedo

$$\beta = \frac{\tau_s}{\tau} = \frac{\sigma_s}{\alpha} \quad (3.19)$$

Equation (3.16) can be rewritten in the form

$$\cos \chi \cdot \frac{\partial I(\tau, \chi)}{\partial \tau} = I(\tau, \chi) - J(\tau) \quad (3.20)$$

with

$$J(\tau) = \frac{\beta}{4\pi} \int_0^{2\pi} \int_{-\frac{\pi}{2}}^{\frac{\pi}{2}} I'(\tau, \theta, \varphi) \sin \theta \cdot \mathrm{d}\theta \cdot \mathrm{d}\varphi \quad (3.21)$$

(3.20) is *Schwarzchild's equation*, the fundamental equation of radiative transfer, and (3.21) gives the source function due to scattering. The function $J(\tau)$ is composed of two terms; one is the scattering from the parallel beam of the direct (solar) radiation and the other is the contribution of the scattering from the surrounding atmospheric medium. Assuming a horizontally stratified atmosphere, in which I' does not depend upon φ, we can write

$$J(\tau) = \frac{\beta}{4\pi} I_{\infty} e^{-\tau \cdot \sec\chi} + \frac{\beta}{2} \int_0^{\frac{\pi}{2}} I'(\tau, \theta) \sin\theta \cdot d\theta \qquad (3.22)$$

for the downward flux ($\theta > 0$).

$I'(\tau, \theta)$ in (3.22) can be obtained by solving a differential equation which is similar to (3.20) except that χ is replaced by θ, i.e.

$$\cos\theta \cdot \frac{\partial I'(\tau, \theta)}{\partial \tau} = I'(\tau, \theta) - J(\tau). \qquad (3.23)$$

Inserting into (3.22) the solution of (3.23) obtained with the upper boundary condition that there can be no scattered radiation from outer space, we get

$$J(\tau) = \frac{\beta}{4\pi} I_{\infty} \cdot e^{-\tau \cdot \sec\chi} + \frac{\beta}{2} \int_0^{\tau} J(\tau') E_1(|\tau - \tau'|)\, d\tau' \qquad (3.24)$$

with

$$E_1(x) = \int_1^{\infty} e^{-xt}/t \, dt \qquad (3.25)$$

The numerical values of this exponential integral function (3.25) can be found, for instance, in Appendix 9 of Atmospheric Radiation (GOODY, 1964).

The solution for the reflected diffuse radiation or upward flux ($\theta < 0$) can be obtained by a method similar to the direct solar radiation or downward flux ($\theta > 0$) by using the air mass factor 1.66 instead of sec χ. This value of air mass corresponds to averaging sec χ by angular integration (see Appendix I). The lower boundary condition is given by the reflected flux corresponding to the fraction of the downward radiation arriving at the surface, i.e.

$$I_0 = A \cdot I(\tau_0, \chi) \qquad (3.26)$$

where τ_0 is the optical depth of the entire atmosphere and A represents a surface albedo.

In order to solve (3.24), it is useful to use the method of successive approximation. First, assuming that J is represented by the first term of the right-hand side of the equation, we calculate a new J via (3.24) by substituting the old J into the second term of the right-hand side. Repeating the same procedure, we can obtain the second, third and higher order approximations. This series of solutions is equivalent to the consecutive account of multiple scattering of the first, second, and higher orders (KONDRATYEV, 1969). Various simplified methods have been worked out by many workers (SHETTLE and WEIMAN (1970), LUTHER and GELINAS (1976), CALLIS *et al.*, (1976), YUNG (1976), SZE (1976), ISAKSEN *et al.* (1977), WOFSY (1978), MUGNAI *et al.* (1979), PETRONELLI *et al.* (1980), VISCONTI *et al.* (1980), and MEIER *et al.* (1982)) in order to include the effects of scattering on photo-dissociation rates of molecules in photochemical models.

The effects of multiple scattering may be evaluated using the *scattering albedo* β, which can be written using (3.17) and (3.19) as

$$\beta = \frac{\sigma_s}{\sigma_s + \sum_i \sigma_i f_i} \tag{3.27}$$

where f_i is the mixing ratio of the absorbing molecules. For isotropic Rayleigh scattering, σ_s is expressed by

$$\sigma_s = \frac{8\pi^3}{3} \frac{(r^2 - 1)^2}{n\lambda^4 L} \tag{3.28}$$

where L is the Loshmidt number (2.687×10^{19} cm^{-3}) and r is the refractive index of air given by Lorentz-Lorentz relation

$$\frac{1}{\rho} \frac{r^2 - 1}{r^2 + 2} = \frac{1}{\rho_0} \frac{r_0^2 - 1}{r_0^2 + 2} = \text{const.} \tag{3.29}$$

where ρ is the air density and the suffix $_0$ indicates the value at normal temperature and pressure on the ground surface. When both r and r_0 are not too different from unity we obtain from (3.29)

$$r^2 \doteqdot 1 + 2(r_0 - 1)\frac{\rho}{\rho_0}. \tag{3.30}$$

r_0 is a function of wavelength and its approximation value for the range of $\lambda = 0.2$ - 1.35 μm is given by the formula

$$r_0 = 1 + 6.4328 \times 10^{-5} + \frac{2.94981 \times 10^{-2}}{146 - 1/\lambda^2} + \frac{2.554 \times 10^{-4}}{41 - 1/\lambda^2}. \tag{3.31}$$

It must be noted that λ in (3.31) is in units of μm, whereas λ in (3.28) is in cgs units (cm).

The height variation of β is shown in Fig. 3.11 for various wavelengths; β changes between 0 and 1 depending upon the relative magnitude of σ_s to $\Sigma_i \sigma_i f_i$. It is generally large in the troposphere and lower stratosphere, where scattering molecules are abundant, and in the spectral range above 300 nm, where σ_i is small: in particular, β is almost unity below ~15 km for wavelengths above ~300 nm.

Molecular absorption above 300 nm is mainly due to O_3 and NO_2 in the height region below 20 km. Since the abundance of these molecules is relatively small below ~20 km, the optical depth τ_a $(=\Sigma_i \sigma_i \cdot n_i)$ does not change much at these heights as is seen in Fig. 3.12. The relatively large values of τ_a below ~360 nm and above ~400 nm are primarily due to O_3

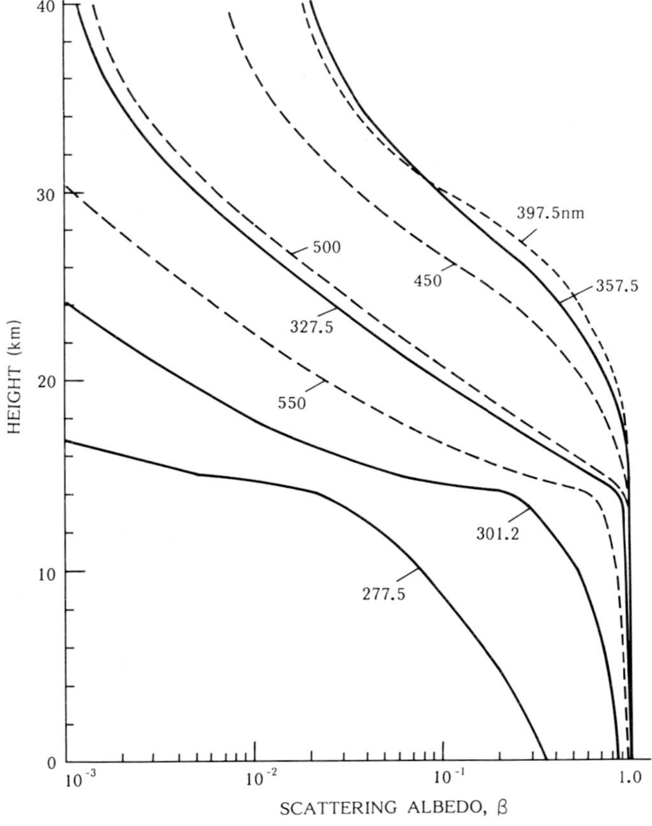

FIG. 3.11. Height profiles of scattering albedo at various wavelengths.

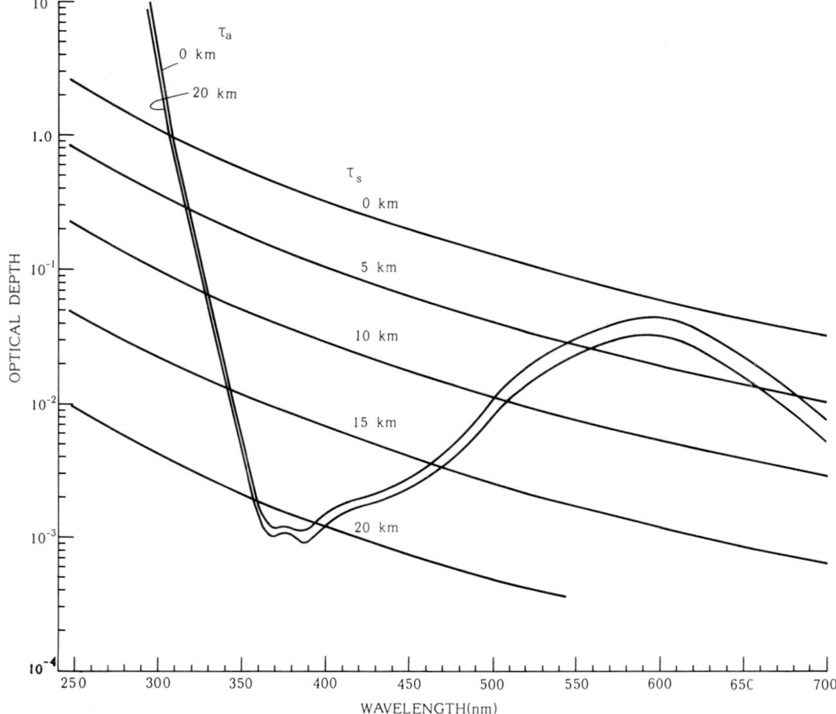

FIG. 3.12. Comparison of optical depths due to molecular absorption (τ_a) and to scattering (τ_s) at wavelengths 250–700 nm.

absorption in the Huggins and the Chappuis bands, respectively. Between 360 and 400 nm, NO_2 is the single main absorber.

The optical depth for scattering τ_s exceeds τ_a in a certain wavelength range above 310 nm depending on the altitude; in these spectral and height ranges, the effect of multiple scattering should become appreciable. The maximum effect is expected around 330 nm in the troposphere since the total optical depth is appreciable ($\gtrsim 0.2$) but not larger than unity and the scattering albedo is high ($\tau_s \gg \tau_a$). The results by calculations confirm this expectation (LUTHER and GELINAS, 1976; ISAKSEN *et al.*, 1977; NICOLET *et al.*, 1982). The solar flux increases by as much as 50% due to multiple scattering at these wavelengths in the troposphere.

The effects of surface reflection also occur above 300 nm, while below this wavelength only small amount of solar radiation can reach the surface. In most cases the total optical depth is small and the reflected solar flux can reach the upper stratosphere without significant attenuation. This is particularly true above 400 nm, if the solar zenith angle is not too large. Calculations using an average surface albedo of 0.25 indicate that the solar flux increases by as much as 50% because of surface reflection (WOFSY, 1978; NICOLET *et al.*, 1982).

Chapter 4

PHOTOABSORPTION AND PHOTODISSOCIATION

Upon absorbing solar radiation, atmospheric molecules can be decomposed into two or more fragments. This process is called *photodissociation* or *photolysis* if the products are all neutral atoms or molecules, and is called *photoionization* if the products include a molecular ion and one or more electrons. The energy required for photoionization is generally higher than that for photodissociation (see Fig. 3.9).

Photodissociation occurs mainly by absorption of radiation of wavelengths above ~130 nm (or photons of energy less than ~9.53 eV*); the process is essential for photochemistry in the middle atmosphere. The Lyman-α emission at 121.57 nm (or 10.19 eV) penetrates through the mesosphere and can dissociate some molecules such as O_2, H_2O, CO_2 and CH_4. Photoionization occurs mainly at wavelengths shorter than ~100 nm (or photon energies larger than ~12.4 eV) at altitudes above the mesopause, and plays important roles for the chemistries of NO molecule and water cluster ions in the mesosphere as well as ion chemistry in the thermosphere.

In this chapter, we will discuss photoabsorption of various molecules important to the photochemistry of the middle atmosphere and calculate the absorption and dissociation coefficients at various heights for each molecule. Photoionization processes will be discussed in Chapter 13.

4.1 Units for Absorption

Various units are used in the literature to express laboratory data on the intensity of molecular absorption of radiation. It is sometimes necessary to

*The conversion relation between energy E and wavelength λ is given by

$$\lambda(\text{nm}) = 1.239 \times 10^3 / E(\text{eV})$$

convert them to the appropriate unit for the specific problem.

The most commonly used unit in theoretical studies is the absorption cross-section σ (cm^2), which is defined in the Lambert-Beer law

$$I = I_\infty e^{-\tau} I_\infty e^{-\sigma \cdot N} \tag{4.1}$$

where I is the intensity of *in situ* radiation, I_∞ is the intensity of the incident radiation at the top of the atmosphere, and N is the total number of absorbing molecules per cm^2 in the slant column along the line of sight.

Another unit widely used, particularly by experimenters is the absorption coefficient k (cm^{-1}), which is the fractional decrease in intensity per unit distance traversed. Sometimes the molar absorption coefficient α ($cm^{-1}mole^{-1}$) is also used by chemists. The Lambert-Beer law can be expressed in terms of these coefficients

$$I = I_\infty e^{-k \cdot u} = I_\infty e^{-\alpha \cdot c \cdot u} \tag{4.2}$$

where u is the thickness in cm of the absorbing medium and c is the concentration of the absorbing species in moles per liter. Since u is equal to N divided by the Loschmidt number L, the following conversion relations hold among the three units:

$$k = \sigma \cdot L = \alpha \cdot c. \tag{4.3}$$

Inserting the numerical value for an ideal gas ($L = 2.687 \times 10^{19} cm^{-3}$ and $c = 1/22.4$ gr $liter^{-1}$), we obtain the following conversion relationships:

$$k = 2.687 \times 10^{19} \sigma = 0.0446\alpha \tag{4.4a}$$

$$\sigma = 3.7216 \times 10^{-20} k = 1.6614 \times 10^{-21} \alpha \tag{4.4b}$$

and

$$\alpha = 22.4k = 6.019 \times 10^{20} \sigma. \tag{4.4c}$$

In some cases the Lambert-Beer law is expressed using the base of 10 instead of the natural logarithmic base

$$I = I_\infty \cdot 10^{-k_{10} \cdot u} \tag{4.5}$$

Then, the absorption cross-section can be calculated by

$$\sigma = 2.3026 k_{10}/L = 8.56935 \times 10^{-20} k_{10}. \tag{4.6}$$

Nuclear physicists sometimes express the absorption cross-section in megabarn (Mbn), one of which is equivalent to 10^{-18} cm^2.

4.2 Photodissociation Coefficient

The rate of dissociation per molecule is called the *dissociation coefficient*. For the *i*th constituent it can be calculated by

$$J_i = \int_0^\infty \eta_i(\lambda) \cdot \sigma_i(\lambda) \cdot I_\infty e^{-\tau \sec \chi} \, d\lambda \tag{4.7}$$

where σ_i is the absorption cross-section of the *i*th molecule, and η_i is the quantum yield of photodissociation defined by

$$\eta_i = \frac{\text{number of particles produced per second}}{\text{number of photons absorbed per second}}. \tag{4.8}$$

τ in (4.7) is the *total* optical depth of the atmosphere due to absorption by *all kinds of molecules*. It is usually sufficient for the middle atmosphere to include only absorptions by O_2 and O_3, but the contribution of NO_2 may not be negligible, particularly in the lower stratosphere and troposphere.

The value of η_i is not well known in most cases and is usually assumed to be unity. It is very important, however, to learn precisely the quantum yield for production of excited atomic oxygen $O(^1D)$ in ozone photolysis near the end of the Hartley contumuum, since the reactions of $O(^1D)$ with N_2O and H_2O are the important sources of NO_x and HO_x in the middle and lower atmospheres.

It is generally sufficient to integrate (4.7) only for the wavelength range where σ_i is appreciable. That range depends strongly upon the molecule, and within that range, σ_i should change with λ in a different fashion for each molecule. If σ_i and τ change gradually with λ, J_i can be calculated with sufficient accuracy by

$$J_i = \sum_{\Delta\lambda} \bar{\eta}_i \bar{\sigma}_i \cdot \bar{I}_\infty e^{-\tau \sec \chi} \, d\lambda \tag{4.9}$$

where the upper bar indicates the average of the assigned quantity over the subdivided wavelength range $\Delta\lambda$, and the summation is taken for all subdivisions. Usually $\bar{I}_\infty \cdot \Delta\lambda$ is given by the integrated solar flux over the range $\Delta\lambda$.

When the variations of σ_i and τ with λ are large, it is time consuming to perform the calculation of J_i, since we then need many subdivisions in order to obtain meaningful average values for each quantities. If the variations include very rapid oscillations, it is formidable and practically impossible to perform such calculations. This situation occurs, for instance, in calculations of photodissociation rates for O_2 and NO in the Schuman-Runge bands (175–200 nm). In such cases, statistical band models will be used to simulate the profiles of the oscillatory variations of σ_i and τ. The absorption cross-sections of various gases have been reviewed, amongst others, by WATANABE *et al.* (1953), WATANABE (1958), SCHULTZ *et al.* (1962), THOMPSON *et al.* (1963), VOLMAN (1963), MCNESBY and OKABE (1964), CALVERT and PITTS (1966), MARR (1967), HUDSON (1971) and TURCO (1975).

4.3 Absorption Cross-Sections

We will now discuss the main characteristics of various photolytic processes relevant to middle atmosphere photochemistry.

4.3.1 Molecular oxygen (O_2)

The photolysis of O_2 is fundamentally important for photochemistry of the middle atmosphere; it is the main natural source of atomic oxygen (O), which is converted quickly to ozone (O_3) by three body recombinations with O_2. The O_2 absorption occurs mainly at wavelengths below ~242 nm. The spectrum of σ_{O_2} is shown in Fig. 4.1. The absorption data are taken from WATANABE (1958) for wavelengths below 175 nm, from HUDSON and MAHLE (1972a) and KOCKARTS (1972) for wavelength range between 175

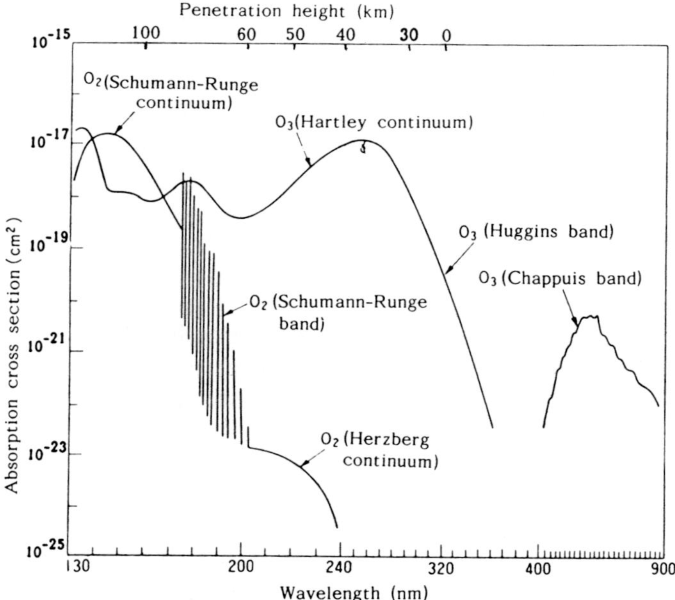

FIG. 4.1. Absorption spectrum of O_2 and O_3 at wavelengths between 130 and 900 nm.

and 205 nm, and from OGAWA (1971) and HASSEN and NICHOLLS (1971) for the wavelength range between 205 and 242 nm. The O_2 photodissociation in the mesosphere also occurs near the Lyman-α, whose absorption cross section is shown in Fig. 4.2 (MCNESBY and OKABE, 1964).

The potential energy diagram of the O_2 molecule is illustrated in Fig. 4.3. Each curve represents a potential energy curve for different levels of electronic energy. It has a minimum at the equilibrium internuclear distance; for smaller separation the internal energy of the molecule increases by repulsive forces between the two atoms and for larger separation it increases by forces of attraction. Each energy level splits into many discrete levels corresponding to the vibrational and rotational energies of the molecule. Thus, for example, the internuclear distance of the ground state $O_2(X^3\Sigma_g^-)$, having a certain amount of vibrational and rotational energies, should oscillate between a and b on the potential energy curve of O_2 $(X^3\Sigma_g^-)$.

If a molecule in the ground state gains energy by absorbing solar photons, it can make a transition to a higher energy level (excited state). It is most probable that the transition will occur along the vertical lines near the end points of the oscillation, a or b (the Franck Condon principle). If the transition occurs from a to a' or a'' the energy of the excited molecule is high enough to make the distance to the other end of the oscillation on the potential energy curves of $B^3\Sigma_u^-$ or $A^3\Sigma_u^-$ infinite and eventually separate

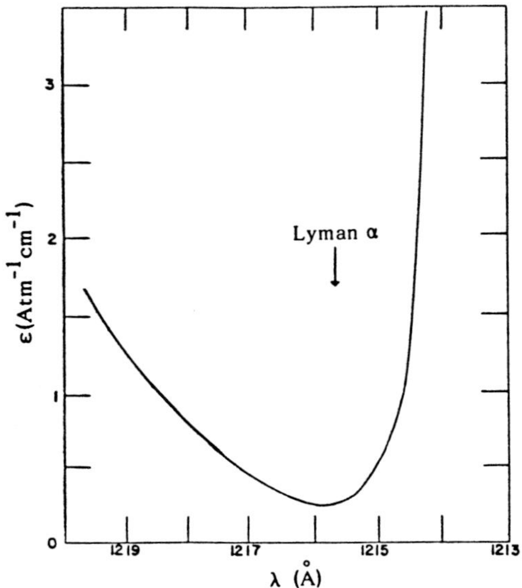

FIG. 4.2. High resolution absorption spectrum of O_2 near Lyman-α (MCNESBY and OKABE, 1964).

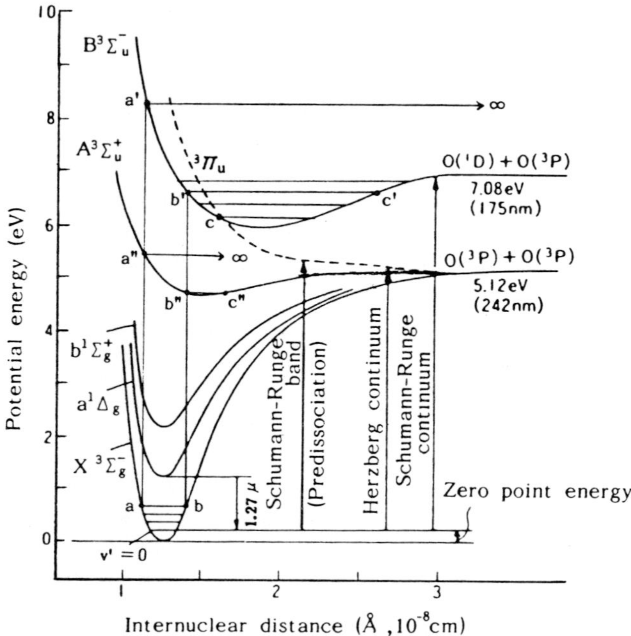

FIG. 4.3. Potential energy curves of O_2 molecule.

the molecule into two atoms. This is the mechanism of photodissociation. In order to make this happen, the molecule must absorb an energy higher than a certain critical value, which is called the *dissociation energy*. If the absorbed energy is lower than this critical value, the molecule may be excited to the state b' or b''; in this case, however, the oscillation of internuclear distance remains finite between b' and c' or between b'' and c'' and the molecule is not dissociated.

O_2 has two dissociation energies; one is ~5.12 eV and the other is ~ 7.08 eV. These two energies correspond to UV radiation of wavelengths ~ 242 nm and ~175 nm, respectively. Absorbing radiation between 205 and 242 nm, O_2 can be dissociated into two ground state O atoms by

$$(J_{1a})\quad O_2(X\,^3\Sigma_g^-) + h\nu(205 < \lambda < 242\text{ nm}) \rightarrow O_2(A\,^3\Sigma_u^+) \rightarrow O(^3P) + O(^3P).$$

In what follows, we will use the notation (J_i) to indicate the process of photodissociation and the notation J_i to indicate the dissociation coefficient. The absorption by (J_{1a}) is called the *Herzberg continuum absorption*. Since the transition from X to A is forbidden, the absorption cross-section of (J_{1a}) is very small ($< 10^{-23}$ cm^2). However, in the stratosphere where O_2 is abundant, photodissociation in the Herzberg continuum becomes the principal O_2 dissociation, since most of the radiation at shorter wavelengths is absorbed above the stratosphere and is not available in the stratosphere. The dissociation coefficients of O_2 calculated for various continuum and band absorptions are compared in Fig. 4.4.

The absorption of radiation below ~175 nm causes O_2 to be dissociated by

$$(J_{1b})\quad O_2(X^3\Sigma_g^-) + h\nu(\lambda < 175\ \text{nm}) \rightarrow O_2(B^3\Sigma_u^-) \rightarrow O(^3P) + O(^1D).$$

In this case one of the O atoms produced is in the excited state (^1D). This absorption is called the *Schumann-Runge continuum absorption*. Since the transition from X to B is allowed, the absorption cross-section is large ($>10^{-18}$ cm^2). The maximum absorption occurs near 142–145 nm and the absorption cross-section is $\sim 1{\cdot}4 \times 10^{-17}$ cm^2. O_2 dissociation in the S-R continuum is most important in the thermosphere (see Fig. 4.4).

When O_2 is excited to the state b' by absorption of radiation between 175 and 205 nm, the excited $O_2(B^3\Sigma_u^-)$ usually can not be dissociated as was discussed above (see Fig. 4.3). During the oscillation between b' and c', however, the internuclear distance sometimes starts to increase from c

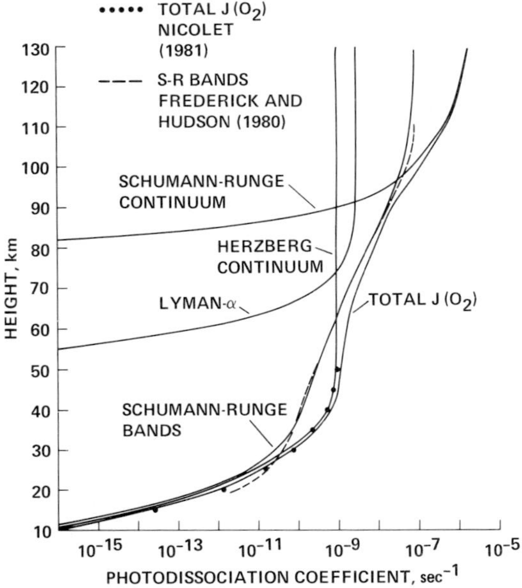

FIG. 4.4. Altitude profiles of O_2 photodissociation coefficient at various absorption continua and bands for an overhead sun. A dashed curve shows a S-R band calculated by FREDERICH and HUDSON (1980b) without including the effect of O_3 absorption. Dotted curve indicates the total dissociation coefficient calculated by NICOLET (1981a).

along the dashed curve representing a potential energy curve for $O_2(^3\Pi_u)$, which intersects at c with the curve of $O_2(B^3\Sigma_u^-)$. Thus, O_2 can be dissociated into two ground state O by

$$(J_{1c})\quad O_2(X^3\Sigma_g^-)+h\nu(175\leqq\lambda\leqq 205\text{ nm})\rightarrow O_2(B^3\Sigma_u^-)\rightarrow O_2(^3\Pi_u)\rightarrow O(^3P)+O(^3P).$$

This absorption is called the *Schumann-Runge band absorption*, and the mechanism is called *predissociation*. Predissociation is the main dissociation of O_2 in the mesosphere (see Fig. 4.4).

The transition leading to predissociation of O_2 occurs between the vibrational levels of $B^3\Sigma_u^-$ ($v'=0, 1, 2, \ldots$ or 19) and of $X^3\Sigma_g^-$ ($v''=0$ or 1). Each band also has fine structure due to various rotational lines. Thus, the spectrum of σ_{O_2} in the Schumann-Runge band system is very complicated, as is illustrated for example, in Fig. 4.5 for the 5-0 to 12-0 transitions at 200 K. σ_{O_2} changes more than two orders of magnitude over a very short range of wavelength.

Because of a very large oscillatory variation of σ_{O_2}, the exact calculation of J_{O_2} in the Schumann-Runge band is complicated and time consuming. σ_{O_2} can be evaluated only as an average over the band width (or resolution) of the measuring instrument. Most theories assume a Lorentzian line shape

$$\sigma_{O_2}(\nu)\cdot\Delta\nu=\frac{2k_0/\pi}{1+\left[\dfrac{2(\nu-\nu_i)}{\Delta\nu}\right]^2} \tag{4.10}$$

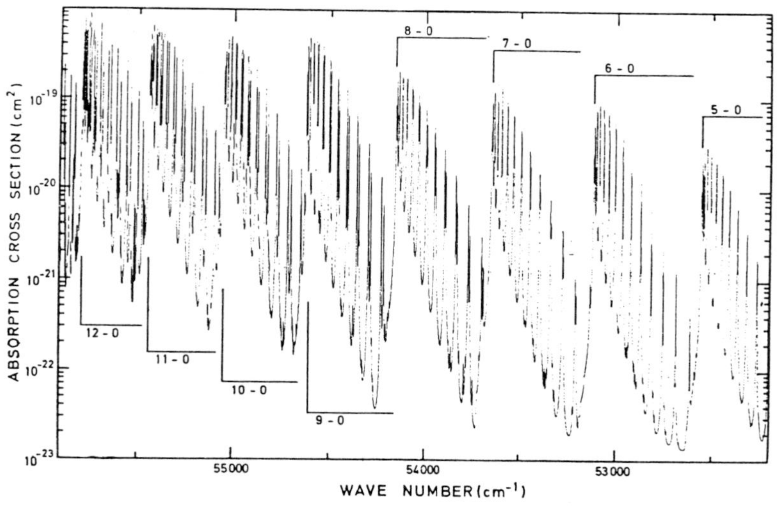

FIG. 4.5. Structure of (5–0) to (12–0) bands of the O_2 Schumann-Runge band system at 200 K temperature (ACKERMAN *et al.*, 1970).

where k_0 is the line integrated absorption cross- section, ν is the wavenumber in cm^{-1}, ν_i is the wavenumber at the center of a line and $\Delta\nu$ is the line width. k_0 is determined so that the calculated integral of (4.10) over the band width agrees with the experimental value. Actually, k_0 is related to the measured oscillator strength (f-value) by

$$k_0 = \frac{\pi e^2}{mc^2} f \doteqdot 8.838 \times 10^{-13} f \tag{4.11}$$

where m is the electron mass, c the velocity of light, and e the electron charge in e.s.u. It must be noted that the solar flux over wavenumber, but not over wavelength, should be multiplied by k_0 in calculating the dissociation coefficient.

The f-values have been obtained for each band by various workers (BETHKE, 1959b: HALMANN, 1966; FARMER *et al.*, 1968; ACKERMAN and BIAUME, 1970; ACKERMAN *et al.*, 1970; HASSEN *et al.*, 1970; ALLISON *et al.*, 1971; HUDSON and MAHLE, 1972a; HUEBNER *et al.*, 1975a; JULIENNE, 1976; LEWIS *et al.*, 1978 and 1979; FREDERICK and HUDSON, 1979a; BLAKE, 1979). The values deduced from various laboratory experiments are compared in NICOLET and PEETERMANS (1980). Although the values from various sources are of the same order of magnitude for each band, there are still differences which may lead to significant discrepancies in determining the absolute values for J_{O_2} and the atmospheric transmission at some wavelength range in the Schumann-Runge band system. However, these differences are generally not systematic; they occur in one direction for some bands and in the other direction for other bands. Therefore, the impact of using different sets of experimental data on the calculations of the total O_2 dissociation coefficient and the atmospheric transmission are limited (NICOLET and PEETERMANS, 1980).

The difficulty of performing the wavelength integral within a reasonable computation time has forced modelers to work out various conventional methods. For this purpose we need to evaluate two integrals for each subdivision; one is for the O_2 dissociation cross-section and is written

$$R(\lambda_k, z) \cdot \Delta\lambda_k = \int_{\lambda_k}^{\lambda_{k+1}} \sigma_{O_2}(\lambda) e^{-\tau_{O_2}(\lambda, z)} \, d\lambda \tag{4.12}$$

and the other is for the atmospheric transmission function given by

$$P(\lambda_k, z) \cdot \Delta\lambda_k = \int_{\lambda_k}^{\lambda_{k+1}} e^{-\tau_{O_2}(\lambda, z)} \, d\lambda \tag{4.13}$$

where $\Delta\lambda_k = \lambda_{k+1} - \lambda_k$ represents the wavelength range of the k-th subdivision. Then, the O_2 dissociation coefficient can be calculated by

$$J_{O_2}(z)=\sum_k \overline{I_\infty(\lambda_k)}\cdot\Delta\lambda_k\cdot R(\lambda_k, z)e^{\overline{-\tau_{0_2}(\lambda_k,z)}} \tag{4.14}$$

and the dissociation coefficients for the molecules other than O_2 by

$$J_i(z)=\sum_k \overline{I_\infty(\lambda_k)}\cdot\Delta\lambda_k\sigma_i(\lambda_k)P(\lambda_k, z)e^{\overline{-\tau_{0_2}(\lambda_k,z)}} \tag{4.15}$$

Based on their experimental data, HUDSON and MAHLE (1972b) have expressed variations in R and P as a second order polynomial function of temperature. The coefficients of the polynomials for 300 K are tabulated for some chosen values of the O_2 column density, $N(O_2)$, in the range of $10^{17} - 7\times10^{23}$ cm^{-2}, and for 19 subdivisions of wavelength, in Table E.1 of Appendix E. On the other hand, KOCKARTS (1976) has constructed sixth order polynomials of $N(O_2)$ to represent R and P, using the result obtained from elaborate calculations on the basis of his own data. The results of Hudson and of Kockarts are compared with each other in Figs. 4.6 (a) and (b) for each band. The abscissa indicates $N(O_2)$ but also shows a scale of approximate heights for the situation of overhead sun.

The two models are in general agreement, but a significant difference appears at large $N(O_2)$ in each band. According to Kockarts, the fine resolution used in his model generally leads to a larger dissociation coefficient than the value calculated with the average absorption cross-section, since the radiation can penetrate through the small windows between the absorption lines. It is more likely, however, that the very small O_2 dissociation and transmission in the Hudson and Mahle model is caused by absorption by the underlying Schumann-Runge continuum. At smaller $N(O_2)$ there is a slight but systematic difference of the O_2 dissociation cross section between the two models; the Hudson and Mahle model tends to give a larger value than the Kockarts model for the bands of $v' \geqq 15$, whereas the reverse is the case for the bands $v' \leqq 14$. Part of the reason for this might be a small but systematic difference of the oscillator strength measured in their experiments, although data are not available in the literature to confirm this.

In an attempt to simplify the numerical procedure and to facilitate numerical calculations of R and P, SHIMAZAKI and OGAWA (1974b) have assumed 7-th order polynomials in the form

$$\log R = \sum_{m=1}^{7} R_m\cdot(\log x)^m \tag{4.16}$$

and

$$\log P = \sum_{m=1}^{7} P_m\cdot(\log x)^m \tag{4.17}$$

where x represents the thickness in cm of the O_2 column density. The coeffi-

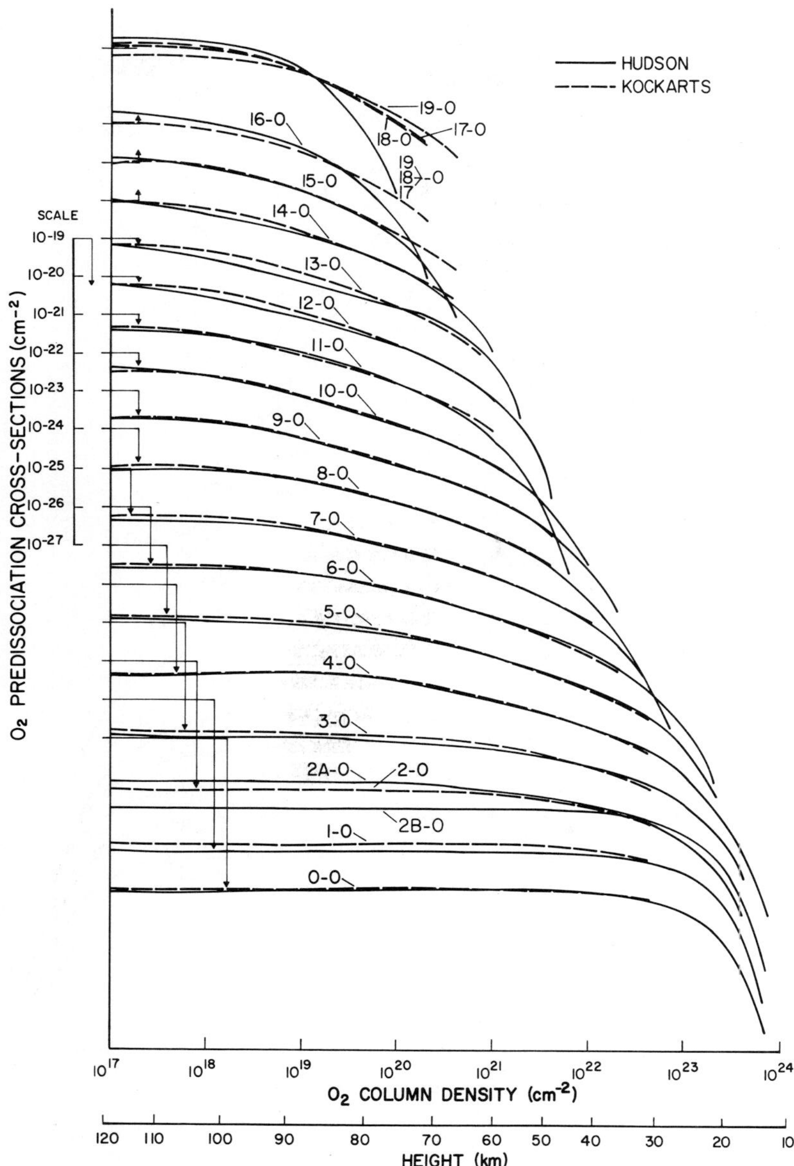

FIG. 4.6. (a) O_2 predissociation cross-section vs. O_2 column density (or height) for each of the Schumann-Runge bands. Results from HUDSON and MAHLE (1972b) and KOCKARTS (1976) are compared. The height scale corresponds to the case of an overhead sun.

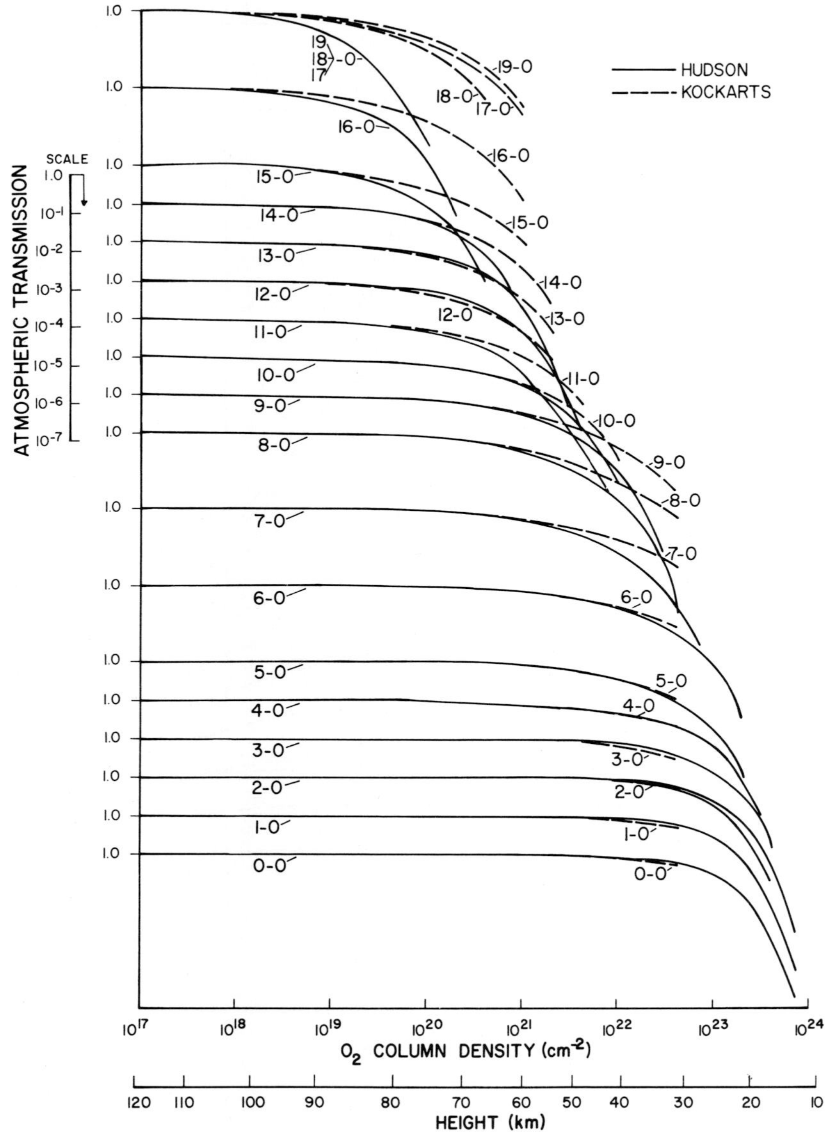

FIG. 4.6. (b) Atmospheric transmission vs. O_2 column density (or height) for each of the Schumann-Runge bands. Results from HUDSON and MAHLE (1972b) and KOCKARTS (1976) are compared. The height scale corresponds to the case of an overhead sun.

cients R_m's and P_m's have been determined by curve-fitting the formulae to the data of Hudson and Mahle by the least square method. The effect of temperature is included implicitly when the input data are calculated from the table of Hudson and Mahle, since $N(O_2)$ is actually a function of altitude which specifies the temperature (see also SHIMAZAKI, OGAWA, and FARRELL, 1977). This method should give essentially the same result as that of HUDSON and MAHLE (1972b) but the computer storage requirement in computer is reduced tremendously and the interpolation process needed to calculate R and P for any value of $N(O_2)$ from the tables in the Hudson and Mahle method can be entirely eliminated.

NICOLET and PEETERMANS (1980) have shown that a smaller order polynomial is acceptable, if the formulae

$$\log P = a + be^s \tag{4.18}$$

and

$$\log R = c + d/(1 + e^d) \tag{4.19}$$

are used and s and d are approximated by the polynomial functions of log $N(O_2)$. However, the expressions (4.16) and (4.17) have the advantage, since we will develop in Chapter 12 the codes for calculating the atmospheric heating and cooling rates by using the empirical formula of the same type of polynomials; this treatment could greatly simplify and facilitate the calculations of photodissociation rates and atmospheric heating and cooling rates in a model of coupled radiation and chemistry.

In order to improve the accuracy of polynomial approximation, the author recalculated coefficients R_m's and P_m's for two different ranges of $N(O_2)$; the results are tabulated in Appendix E. Table E.2 lists the coefficients applicable to the range of larger $N(O_2)$; this range covers mainly the stratosphere. For $N(O_2)$ larger than this range, the dissociation in the Schumann-Runge bands is very small and may be assumed to be essentially zero. The coefficients applicable to the range of smaller $N(O_2)$ are given in Table E.3; this range covers mainly the mesosphere and lower thermosphere. For $N(O_2)$ smaller than this range the dissociation coefficients are almost constant with height and they may be calculated by using the smallest value of $N(O_2)$ in the range.

It is evident from Fig. 4.4 that the dissociation in the Herzberg continuum (204–242 nm) is dominant in the stratosphere and lower mesosphere, whereas the Schumann-Runge bands (175–205 nm) gives the largest dissociation coefficient in the mesosphere and lower thermosphere. Our result for the total dissociation coefficient agrees well with the calculation by NICOLET (1981a). We also compare the result of our calculation for the Schumann-Runge bands with that of FREDERICK and HUDSON (1980b).

The agreement between the two results is excellent except below ~30 km, where the results of Frederick and Hudson show much larger O_2 dissociation coefficient than our result; this is because their calculations do not include the effect of attenuation of solar radiation by ozone absorption.

Recently, FREDERICK and MENTALL (1982) and HERMAN and MENTALL (1982) have measured the solar radiation flux in the stratosphere by the balloon-board double monochromater. They have observed that the solar irradiance attenuates more slowly between the two heights located in the height range 30 – 40 km than calculated in Fig. 4.4. They have suggested that the O_2 absorption cross-section in the 200 – 210 nm spectral region must be smaller by a factor of 0.52 – 0.68 than the values measured by HASSEN and NICHOLLS (1971). Such a decreased $\sigma(O_2)$ should allow more radiation to penetrate into the lower stratosphere, and the effect is favorable in resolving discrepancies between model predictions and observations on some long-lived source species including HNO_3, $CFCl_3$ and CF_2Cl_2 (FROIDERAUX and YUNG, 1982; BRASSEUR *et al.*, 1983a; see also p. 213). However, a reduced $\sigma(O_2)$ in the Herzberg continuum decreases the predicted O_3 concentration in the upper stratosphere and enhances the discrepancy between the model result and observations for O_3 at these heights (KO and SZE, 1983; see also p. 114 and p. 158).

The O_2 absorption cross section at Lyman-α is less than 10^{-20} cm^2 (see Fig. 4.2), and the dissociation coefficient can be approximated by (note that $I_\infty \approx 3\times 10^{11}$ cm^{-2} sec^{-1} from Section 3.3)

$$J_{O_2}(Ly-\alpha) = 3\times 10^{-9} e^{-10^{-20}N(O_2)}\,sec^{-1}. \tag{4.20}$$

if the atmosphere is thin ($N(O_2) \lesssim 10^{19}$ cm^{-2}). For $N(O_2) > 10^{19}$ cm^{-2} NICOLET and PEETERMANS (1980) have developed a formula

$$J_{O_2}(Ly-\alpha) = \frac{1.25\times 10^{-7}}{N(O_2)^{0.083}} e^{-4.17\times 10^{-19}N(O_2)^{0.917}}\,sec^{-1} \tag{4.21}$$

based on experimental data by CARVER *et al.* (1977). The calculated values by these formulae are included in Fig. 4.4.

4.3.2 Ozone (O_3)

Ozone, at ordinary temperature, is a blue unstable gas with a characteristic pungent odor. It absorbs radiation in two distinctive spectral regions; one is the ultraviolet (UV) centered around 255 nm and the other in the visible centered around 610 nm (see Fig. 4.1). The absorption cross-section has been well measured and there are no major discrepancies among the various measurements. Data in Fig. 4.1 have been taken from INN and TANAKA (1953).

The products of O_3 photolysis are atomic oxygen and molecular oxygen; both can be in different energy levels depending upon the energy of the absorbed photon (or wavelength of radiation). Table 4.1 shows the possible products of the primary fission and gives the longest (threshold) wavelengths at which their formation is energetically possible (JONES and WAYNE, 1970). The energy level diagram for atomic oxygen is illustrated in Fig. 4.7.

The strong O_3 absorption in the UV below ~310 nm was discovered by an Irish chemist W. N. Hartley in 1881, and is called the *Hartley continuum absorption*. Atomic oxygen is produced is in the excited ^1D state, and molecular oxygen is a singlet in either $O_2(^1\Delta_g)$ or $O_2(^1\Sigma_g^-)$. There is no evidence, however, that $O_2(^1\Sigma_g^+)$ is a primary product, although its production is energetically possible since the threshold wavelength is 266 nm.

Ozone photolysis plays an essential role in the photochemistry of the stratosphere. It is thought to be the primary process leading to production of several important energy-rich species. In particular, production of $O(^1D)$ is essential for producing nitrogen-oxygen compounds (NO_x) and hydrogen-oxygen compounds (HO_x) in the stratosphere. Even in the

TABLE 4.1 Threshold wavelength in nm for the production of various electronic states of O_2 and O in ozone photolysis.

		Molecular Oxygen (O_2)				
		$(X^3\Sigma_g^-)$	$(a^1\Delta_g)$	$(b^1\Sigma_g^+)$	$(A^3\Sigma_u^+)$	$(B^3\Sigma_u^-)$
Atomic	(^3P)	1180	611	463	230	170
Oxygen (O)	(^1D)	411	310	266	167	150
	(^1S)	234	196	179	129	108

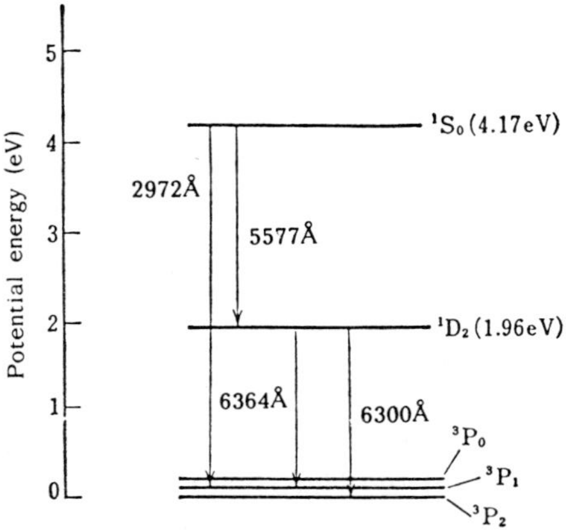

FIG. 4.7. Energy level diagram of O atom.

troposphere, production of HO_x can occur significantly by the reaction of $O(^1D)$ with the abundant water vapor. OH in the lower atmosphere is an important reactant in chemical loss reactions for many species.

Penetration of solar radiation between 300 and 320 nm through the lower atmosphere depends upon the amount of stratospheric ozone as well as the absorption cross-section of O_3. It has been argued recently that the stratospheric ozone density may be reduced significantly by various human activities (see Chapter 11). Production of $O(^1D)$ in the lower atmosphere may be affected by these changes in the stratospheric ozone density. Thus, it is important to know the quantum yield of $O(^1D)$ production at these wavelengths in order to calculate its production rate and to be able to assess the impact of the changes in the stratospheric ozone density on the tropospheric chemistry.

Ozone photolysis above 300 nm can produce $O(^1D)$ in the lower atmosphere, although the efficiency against the production of the ground state $O(^3P)$ drops rapidly as wavelength increases from 300 nm. The effect of the reduced absorption cross-section is partially offset by the increased solar flux above 300 nm. An analytical formula has been developed by MOORTGAT and KUDSZUS (1978) for the calculation of the efficiency (quantum yield) of $O(^1D)$ production as a function of wavelength (295–320 nm) and temperature (230–320 K); it is given by

$$\eta(\lambda, T) = A(\tau)\tan^{-1}\{B(\tau)\cdot(\lambda - \lambda_0(\tau)\} + C(\tau) \tag{4.22}$$

where $\tau = T - 230$, and λ is the wavelength expressed in nm. A, B, λ_0 and C are functions of τ and are given by the following interpolation polynomials of the third order:

$$A(\tau) = 0.369 + 2.85 \times 10^{-4}\cdot\tau + 1.28 \times 10^{-5}\cdot\tau^2 + 2.57 \times 10^{-8}\cdot\tau^3 \tag{4.23a}$$

$$B(\tau) = -0.575 + 5.59 \times 10^{-3}\cdot\tau - 1.439 \times 10^{-5}\cdot\tau^2 - 3.27 \times 10^{-8}\cdot\tau^3 \tag{4.23b}$$

$$\lambda_0(\tau) = 308.20 + 4.4871 \times 10^{-2}\cdot\tau + 6.9380 \times 10^{-5}\cdot\tau^2 - 2.5452 \times 10^{-6}\cdot\tau^3 \tag{4.23c}$$

and

$$C(\tau) = 0.518 + 9.87 \times 10^{-4}\cdot\tau - 3.94 \times 10^{-5}\cdot\tau^2 + 3.91 \times 10^{-7}\cdot\tau^3. \tag{4.23d}$$

The calculated η's for various temperatures are illustrated in Fig. 4.8.

The longest wavelength of the Hartley continuum is usually defined as 300 or 310 nm. Since the production of $O(^1D)$ is possible up to ~320 nm, however, we define for convenience in model calculations that the dissocia-

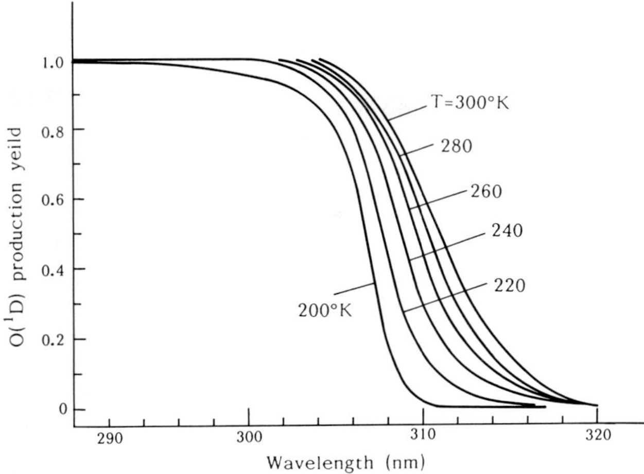

FIG. 4.8. Quantum yield of $O(^1D)$ production from O_3 photolysis as functions of wavelength and temperature.

tion coefficient in the Hartley continuum includes all contributions of $O(^1D)$ production at wavelengths up to 320 nm. Thus, for wavelengths shorter than 295–320 nm the ozone photolysis occurs by

$$(J_{2a})\quad O_3 + h\nu(\lambda < (295\text{–}320\text{ nm})) \rightarrow O(^1D) + O_2(^1\Delta_g)$$

but between 295 and 320 nm only η times the dissociation coefficient should be included in J_{2a}; the rest should be included in J_{2b}, which is discussed below.

The band absorption at 300 – 350 nm was first discovered in 1890 by an English astronomer W. Huggins in the spectrum of Sirius, and is named the *Huggins absorption band*. The photolysis products are $O(^3P)$ and $O_2(^1\Delta_g)$ and the process is written as

$$(J_{2b})\quad O_3 + h\nu((295\text{–}320\text{ nm}) < \lambda < 350\text{ nm}) \rightarrow O(^3P) + O_2(^1\Delta_g)\,.$$

It must be noted that only a portion $1-\eta$ of this O_3 dissociation coefficient should be included in J_{2b}; the rest should be included in J_{2a}.

Energetically $O_2(^1\Delta_g)$ can be formed at λ<611 nm and $O_2(^1\Sigma_g^+)$ at λ<463 nm, but neither can be formed without spin violation at λ>310 nm, Assuming that $O_2(^1\Delta_g)$ is formed only below 310 nm, however, we have difficulty with the quantitative interpretation of the observed airglow intensities on the basis of $O_2(^1\Delta_g)$ production by ozone photolysis. Thus, it seems to be clear that $O_2(^1\Delta_g)$ can be produced even above 310 nm.

A French physicist, J. Chappuis, discovered in 1880 the O_3 absorption in the visible range 450-800 nm; this is called the *Chappuis band absorption*.

The process is

$$(J_{2c})\quad O_3 + h\nu(450 < \lambda < 800\ \text{nm}) \rightarrow O(^3P) + O_2(X\,^3\Sigma_g^-).$$

Both products are in the ground state. Since the radiation at these wavelengths can penetrate to the ground without significant loss, and since the shorter wavelengths have been completely absorbed in the upper regions, the absorption in the Chappuis band should provide a major means of O_3 dissociation in the lower stratosphere and the troposphere, even if the absorption cross-section is small.

The profiles of J_{2a}, J_{2b} and J_{2c} calculated for various solar zenith angles are shown in Fig. 4.9. If the angle approaches or exceeds 90° the effects of grazing incidence must be taken into account (see Appendix D). J_{2c} is essen-

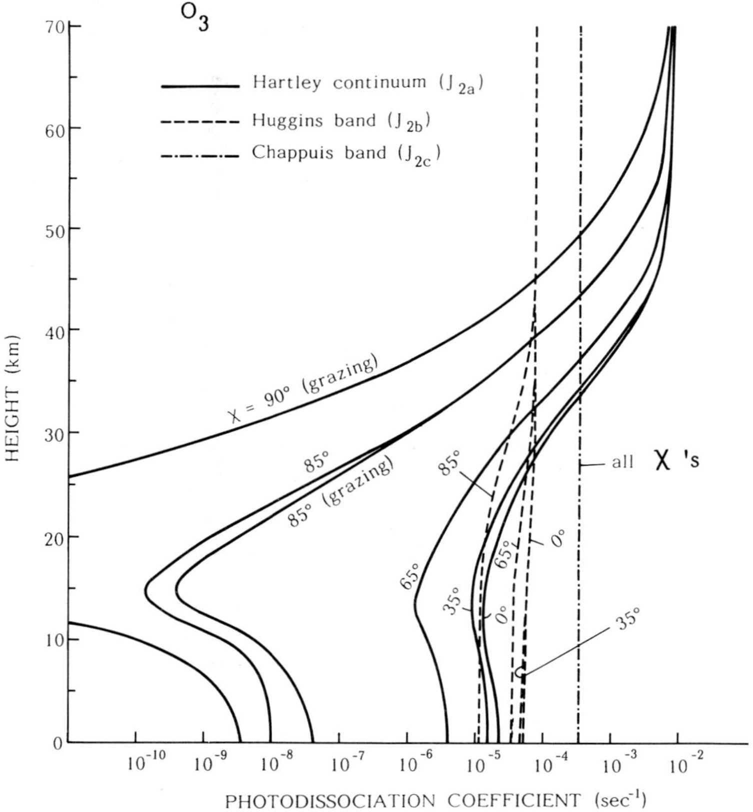

FIG. 4.9. Altitude profiles of O_3 photodissociation coefficient at various continua and absorption bands for different solar zenith angles. Grazing incidence calculations are also shown for $\chi=85°$ and 90°.

tially constant with height, since the radiation above 350 nm is completely transparent throughout the atmosphere.

It can be seen in Fig. 4.9 that the O_3 dissociation in and above the upper stratosphere (> ~30–35 km) occurs mainly in the Hartley continuum J_{2a}. Below this height the dissociation due to Chappuis band absorption J_{2c} becomes the dominant dissociation process. J_{2a}, and therefore $O(^1D)$ production, increases slightly in the troposphere; it has a minimum at the tropopause (assumed to be 14 km in these calculations), where the temperature is a minimum. It is evident that the increased $O(^1D)$ production rate with decreasing height in the troposphere is due to the strong temperature dependence of the quantum yield of $O(^1D)$ production (see Fig. 4.8).

4.3.3 *Other molecules*

There are many molecules in the middle atmosphere that are photochemically active and could influence the ozone density eighter directly or indirectly. Figures 4.10 (a), (b) and (c) illustrate the spectrum of absorption cross section of the molecules which are thought to be important in the photochemistry of the middle atmosphere, particularly the stratosphere. Photolyses of the molecules shown in Figs. 4.10 (a) and (b) are taken into account in most photochemical models, but those in Fig. 4.10 (c) include the molecules that could potentially be important but their impact on the stratospheric photochemistry has yet to be determined. The absorption cross sections are also tabulated in Appendix C.

In what follows, we will discuss briefly the photolysis of each molecule. The dissociation products, the threshold wavelength λ_m, the dissociation energy E, and the equivalent reaction heat ΔH are given whenever they are available. The dissociation products sometimes are different for different wavelength ranges, and in some cases the products have not been identified.

4.3.3.1 *Nitrogen-oxygen compounds*

N_2O (nitroud oxide)

$$(J_3)\quad N_2O + h\nu(\lambda < 240\,\text{nm}) \rightarrow N_2 + O(^1D)$$
$$\lambda_m = 341\,\text{nm}, E = 3.63\,\text{eV}, \Delta H = 83.7\,\text{Kcal/mol}.$$

The main products are N_2 and $O(^1D)$, but they may be NO and N with a quantum yield less than 0.01 (PRESTON and BARR, 1971). The threshold wavelength is 341 nm, but the actual photolysis occurs at wavelengths below ~240 nm, The earlier measurements of σ_{N_2O} by BATES and HAYS (1967) above 260 nm have not been confirmed by the later measurements of JOHNSTON and SELWYN (1975).

σ_{N_2O} increases rapidly with increasing temperature. HUDSON (1974)

thermosphere
mesosphere
stratosphere
troposphere
ABSORPTION CROSS SECTION (cm²)
10^{-16}
10^{-17}
10^{-18}
10^{-19}
10^{-20}
10^{-21}
10^{-22}
10^{-23}
10^{-24}
NO₂
HNO₃
HNO₄
HO₂
NO₃
HNO₂
NO₂
CH₄
H₂O₂
N₂O₅
H₂O
CO₂
N₂O
HNO₃
130 150 200 240 280 320 360 400 600 800
WAVELENGTH (nm)
(a)
thermosphere
mesosphere
stratosphere
troposphere
ABSORPTION CROSS SECTION (cm²)
10^{-16}
10^{-17}
10^{-18}
10^{-19}
10^{-20}
10^{-21}
10^{-22}
10^{-23}
CFCl₃
SO₂
CCl₄
ClONO₂
ClO
SO₂
Cl₂
HCl
CFCl₃
COS
CH₂O
HOCl
ClONO₂
CH₃Cl
CF₂Cl₂
130 150 200 240 280 320 360 400 440
WAVELENGTH (nm)
(b)

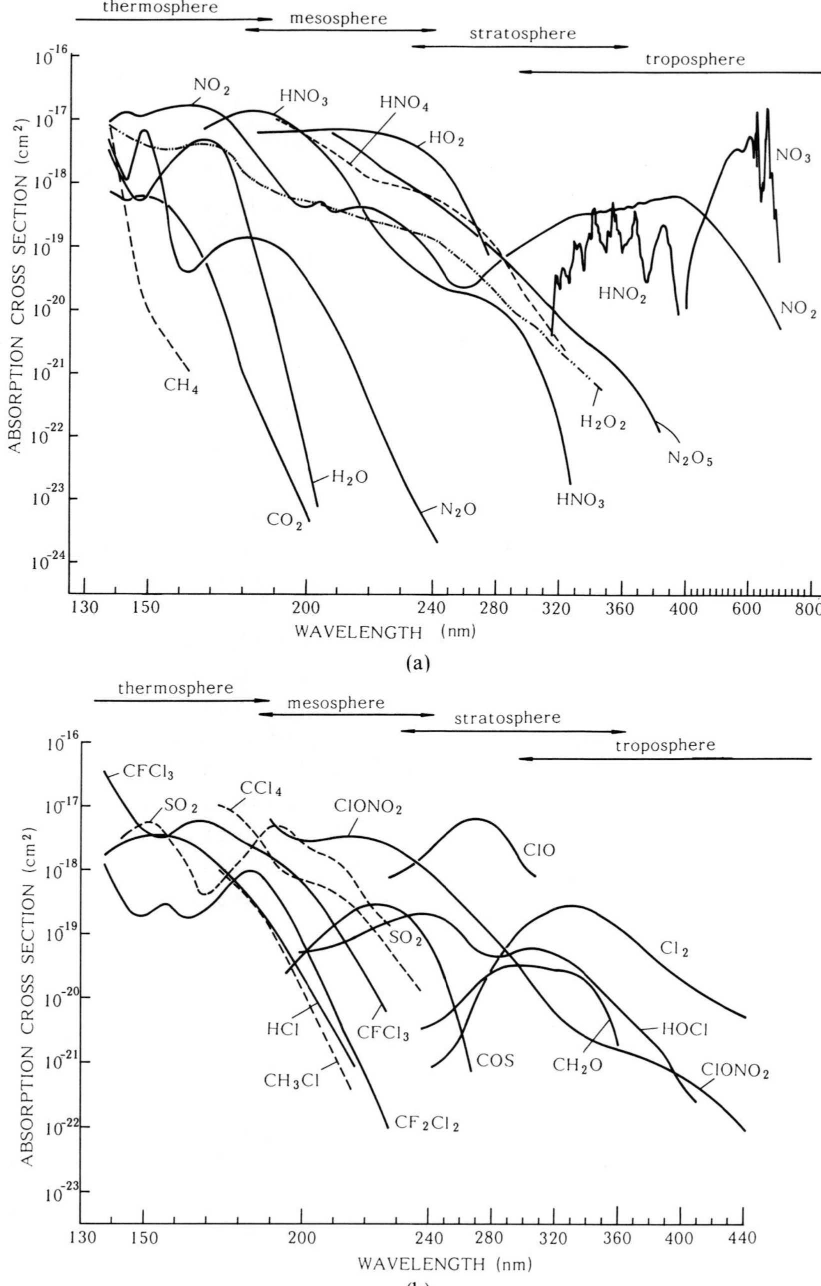

FIG. 4.10.

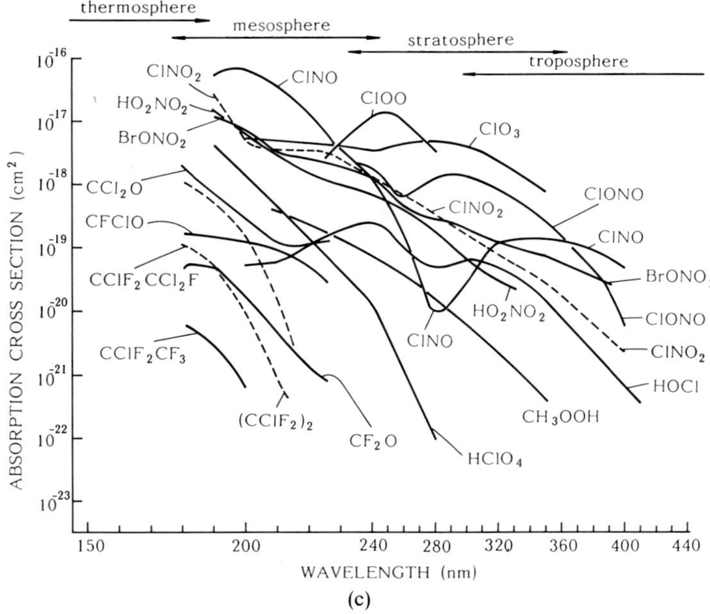

(c)

FIG. 4.10. (a)-(c) Spectra of absorption cross-sections for various molecules.

estimated that the maximum cross-section for 373 K was about 40% larger than that for 183 K. Those maxima occur near 185 and 182 nm, respectively. SELWYN *et al.* (1977) have determined the absorption spectrum at five temperatures as is shown in Fig. 4.11, and have formulated these data as a function of wavelength (173–240 nm) and temperature (194–302 K) as a nine-term polynomial in the following expression:

$$\ln \sigma(\lambda, T) = A_1 + A_2 \cdot \lambda + A_3 \cdot \lambda^2 + A_4 \cdot \lambda^3 + A_5 \cdot \lambda^4 + (T - 300) \exp(B_1 + B_2 \cdot \lambda + B_3 \cdot \lambda^2 + B_4 \cdot \lambda^3) \quad (4.24)$$

with

$A_1 = 68.21023$ $\qquad B_1 = 123.4014$
$A_2 = -4.071805$ $\qquad B_2 = -2.116255$
$A_3 = 4.301146 \times 10^{-2}$ $\qquad B_3 = 1.111572 \times 10^{-2}$
$A_4 = -1.777846 \times 10^{-4}$ $\qquad B_4 = -1.881058 \times 10^{-5}$
$A_5 = 2.520672 \times 10^{-7}$.

The absorption cross-section below 170 nm shown in Fig. 4.10 (a) is taken from MARR (1967).

NO (nitric oxide)

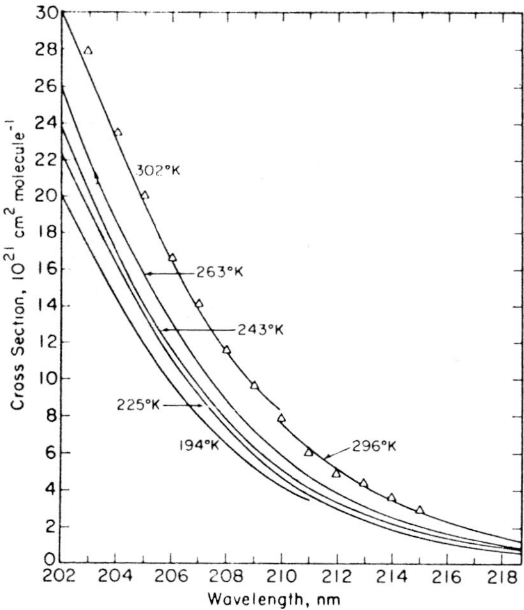

FIG. 4.11. Temperature dependance of the N_2O absorption cross-section (curves are from data by SELWYN *et al.*, 1977 and Δ indicates data from ZELIKOFF *et al.*, 1953).

(J_4) $NO + h\nu(\lambda < 190\,\text{nm}) \rightarrow N + O$
$\lambda_m = 190.8\,\text{nm}, E = 6.50\,\text{eV}, \Delta H = 149.6\,\text{Kcal/mol}.$

The predissociation of NO occurs by transitions between the ground state X^2 and various excited states $A^2\Sigma$, $B^2\Pi$, $C^2\Sigma$ and $D^2\Sigma$; they are called γ, β, δ, and ε band absorption, respectively (see Fig. 4.12). Of these the $\delta(0,0)$, $\delta(1,0)$, $\delta(2,0)$ and $\beta(12,0)$ bands are the most significant for dissociation in the middle atmosphere. Oscillator strengths (f-values) of these bands measured by various workers are compared in Table 4.2. The absorption in the ε band could be larger than the other bands, but a larger part of the absorbed energy will be lost by emission and may not go to dissociation (CIESLIK and NICOLET, 1973).

The precise calculation of J_{NO} is very complicated since the spectral range of NO predissociation overlaps with that of the O_2 predissociation in the Schumann-Runge band system. As an example, the band structures of the NO and O_2 absorption near the $\delta(1,0)$ band are illustrated in Fig. 4.13.

Figure 4.14 compares the profiles of J_{NO} calculated by two groups; one by CIESLIK and NICOLET (1973) and the other by FREDERICK and HUDSON (1979b). The two results agree well in the lower regions, and indicate that

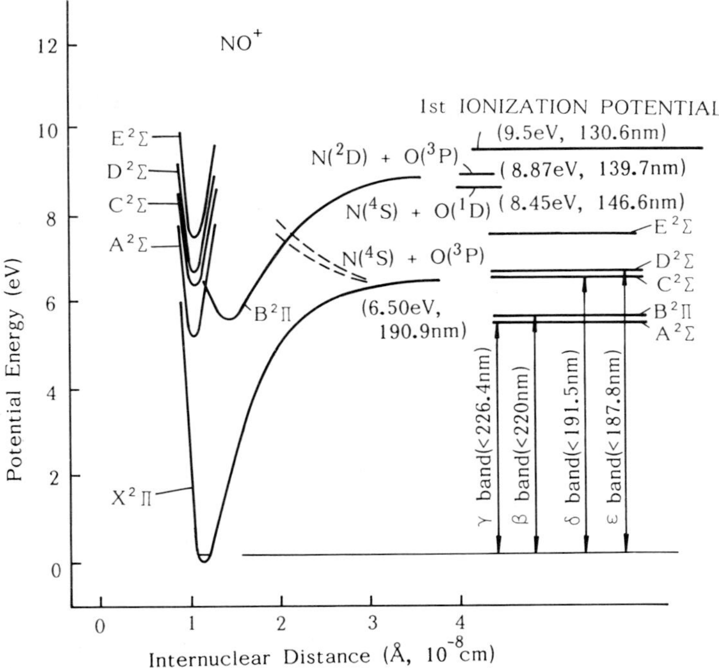

FIG. 4.12. Potential energy curves and energy level diagrams of NO.

the NO dissociation at the $\delta(0,0)$ band could occur even in the lower stratosphere for the overhead sun condition. There is a significant difference, however, between the two results at higher altitudes where the maximum dissociation occurs. The difference in the absolute value is due mainly to the larger solar flux values available at the time of the calculations of CIESLIK and NICOLET (1973), and the relatively large value for $\delta(0,0)$ calculated by Cieslik and Nicolet is mainly due to a larger oscillator strength assumed on the basis of experimental data by CALLEAR and PILLING (1970).

NICOLET (1979) has formulated the result of his most recent calculations on the total NO dissociation coefficient into the folllowing formula:

$$J_{NO} = 4.75 \times 10^{-6} e^{-AN(O_2)^{0.38}} \quad (4.25)$$

where A changes from 0.825×10^{-8} to 1.125×10^{-8} over the temperature range 175 – 275 K. The value of J_∞ (4.75×10^{-6} sec^{-1}) depends upon $N(O_2)$ and would change by $\pm 10\%$ in the entire mesosphere. This formula gives a smaller J_{NO} than those obtained in their earlier calculations and it agrees better with the results of FREDERICK and HUDSON (1979b).

TABLE 4.2 Data for major bands of NO predissociation.

band	Threshold wavelength (nm)	solar flux (ϕ) ($cm^{-2}sec^{-1}$ $wavenumber^{-1}$)	oscillator strength (f-value)	Dissociation coefficient (J_∞) (sec^{-1})
$\delta(0, 0)$	191.5	9.12×10^8	2.49×10^{-3}(BETHKE, 1959a)	2.01×10^{-6}
			5.60×10^{-3}(CALLEAR and PILLING, 1970)	4.03×10^{-6}
			2.80×10^{-3}(CIESLIK, 1977)	2.26×10^{-6}
$\delta(1, 0)$	182.8	4.34×10^8	5.78×10^{-3}(BETHKE, 1959a)	2.22×10^{-6}
			5.61×10^{-3}(CIESLIK, 1977)	2.16×10^{-6}
$\delta(2, 0)$	175.1	1.75×10^8	2.73×10^{-3}(BETHKE, 1959a)	4.22×10^{-7}
$\delta(4, 0)$	187.8	9.52×10^8	1.20×10^{-4}(BETHKE, 1959a)	1.01×10^{-7}
$\beta(12, 0)$	176.2	1.89×10^8	2.31×10^{-3}(BETHKE, 1959a)	3.86×10^{-7}

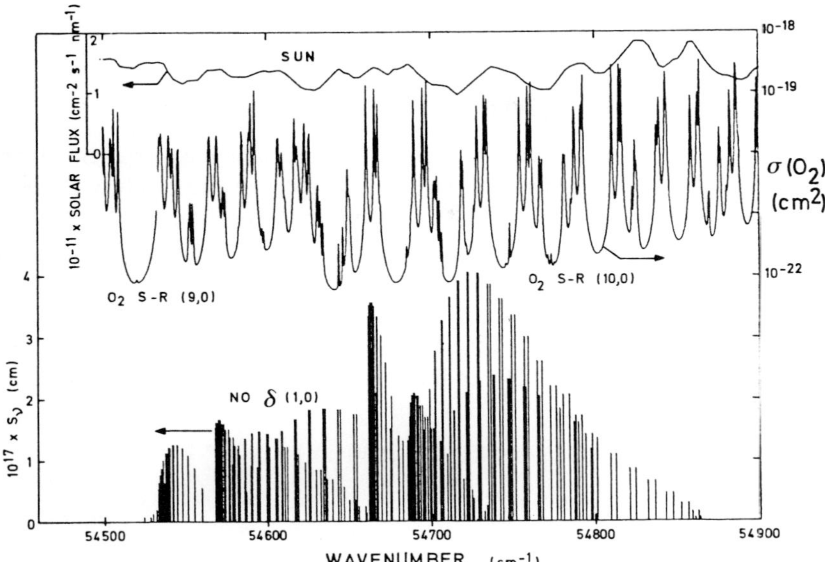

FIG. 4.13. Illustrating overlapping of NO absorption lines in the δ(1–0) band and O_2 rotational lines in the (9–0) and (10–0) bands of the Schumann-Runge band system. Top illustration is the solar flux at mean solar-earth distance (NICOLET and CIESLIK, 1980).

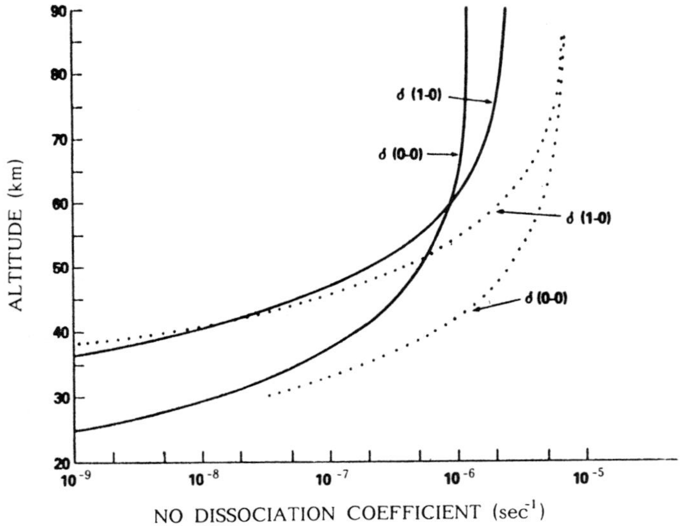

FIG. 4.14. Altitude profiles of the NO dissociation coefficients in the δ(0–0) and δ(1–0) bands for the overhead sun. Solid curves are results from FREDERICK and HUDSON (1979b): dotted curves from CIESLIK and NICOLET (1973).

NO_2 (nitrogen dioxide)

$$(J_5) \quad NO_2 + h\nu(190 < \lambda < 400\,\text{nm}) \rightarrow NO + O$$

$$\lambda_m = 398\,\text{nm}, E = 3.11\,\text{eV}, \Delta H = 71.7\,\text{Kcal/mol}.$$

NO_2 absorbs radiation in a wide spectral range. The measurements from 108–270 nm by NAKAYAMA *et al.* (1959) indicate that the product O could be in the excited state ^1D for $\lambda < 245$ nm, and in the even higher level ^1S for $\lambda < 170$ nm. The quantum yield decreases significantly between 380 and 390 nm (HARKER *et al.*, 1977) and there is a possibility that NO_2 may not be dissociated above this wavelength. However, NO_2 can absorb radiation at much larger wavelengths (see Fig. 4.10 (a)). The measurements by CREEL and ROSS (1976) at 457.9 – 630 nm indicate that NO_2 may be dissociated into $NO+O_2$ by excitation

$$NO_2 + h\nu(460 < \lambda < 630\,\text{nm}) \rightarrow NO_2{}^*$$

followed by the reaction

$$NO_2{}^* + NO_2 \rightarrow 2NO + O_2.$$

The absorption of solar radiation by NO_2 at visible wavelengths is very important for forming photochemical smog in the troposphere; it is also effective for reducing the intensity of solar radiation arriving at the ground surface.

NO_3 (nitrogen trioxide)

$$(J_6) \quad NO_3 + h\nu(500 < \lambda < 680\,\text{nm}) \rightarrow NO_2 + O \qquad (1)$$

$$\rightarrow NO + O_2 \qquad (2)$$

for

$$(1) \quad \lambda_m = 578\,\text{nm}, E = 2.14\,\text{eV}, \Delta H = 49.4\,\text{Kcal/mol}$$

$$(2) \quad \lambda_m = 900\,\text{nm}, E = 1.38\,\text{eV}, \Delta H = 31.7\,\text{Kcal/mol}.$$

Absorption by the free radical NO_3 occurs mainly above ~450 nm; therefore, photodissociation takes place in the troposphere as well as in the stratosphere. The process (1) requires a wavelength shorter than ~578 nm, but is is not known well whether the products for $\lambda < 578$ nm are (1) NO_2+O or (2) $NO+O_2$. MAGNOTTA and JOHNSTON (1980) recently calculated J_{NO_3} over 470 – 700 nm for the sun overhead and obtained the value at the surface to be 0.18 $\sec^{-1}$ for (1) and 0.022 $\sec^{-1}$ for (2).

The difference in the products between (1) and (2) is very important for stratospheric aeronomy, since the reaction (1) eventually produces two oxygen atoms by further dissociating NO_2, whereas (2) does not produce odd oxygen at all (note that photodissociation of NO is significant only in the mesosphere). The maximum σ_{NO_3} extends to wavelengths longer than 578 nm (GRAHAM and JOHNSTON, 1978; see also Fig. 4.10 (a)), but the electronic state of the product O_2 has not been determined.

N_2O_5 (dinitrogen pentoxide)

$$(J_7) \quad N_2O_5 + h\nu(210 < \lambda < 380\,\text{nm}) \rightarrow NO_2 + NO_3 \quad (1)$$
$$\rightarrow N_2O_4 + O \quad (2)$$

for

$$(1) \quad \lambda_m = 1340\,\text{nm}, E = 0.92\,\text{eV}, \Delta H = 21.3\,\text{Kcal/mol}$$
$$(2) \quad \lambda_m = 495\,\text{nm}, E = 2.50\,\text{eV}, \Delta H = 57.7\,\text{Kcal/mol}.$$

The dissociation of N_2O_5 is greatly complicated by branching between (1) and (2) and possible secondary chemical reactions. Reaction (1) is followed by the photolysis of NO_3, which itself has two branches as was discussed above. Reaction (2) is followed by the photolysis of N_2O_4, which has at least the following two paths (JOHNSTON and GRAHAM, 1974):

$$N_2O_4 + h\nu \rightarrow 2NO_2$$
$$\rightarrow N_2O_3 + O \rightarrow NO + NO_2 + O.$$

4.3.3.2 Nitrogen-hydrogen-oxygen compounds
HNO_2 (nitrous acid)

$$(J_8) \quad HNO_2 + h\nu(310 < \lambda < 395\,\text{nm}) \rightarrow OH + NO$$
$$\lambda_m = 590\,\text{nm}, E = 2.10\,\text{eV}, \Delta H = 48.4\,\text{Kcal/mol}.$$

The absorption cross-section of HNO_2 has been measured by STOCKWELL and CALVERT (1978) for the wavelength range 300 – 400 nm. Since the maximum absorption occurs near 360 nm, HNO_2 can be dissociated in the troposphere. The products could be $H+NO_2$, which would probably be followed by the reaction $H+NO_2 \rightarrow OH+NO$, and thus is indistinguishable from (J_8). The spectrum below 300 nm has been studied by COX and DERWENT (1976); σ_{HNO_2} has a maximum value of 4.25×10^{-18} cm^2 at ~215 nm, decreasing rapidly at longer wavelengths. The apparent larger cross-section near 200 – 220 nm, however, is not significant aeronomically, since at these wavelengths photolysis normally occurs in the mesosphere, where the abundance of HNO_2 is minimal and the solar flux is relatively small.

HNO_3 (nitric acid)

(J_9) $HNO_3 + h\nu(190 < \lambda < 320 \text{ nm}) \rightarrow OH + NO_2$

$\lambda_m = 598 \text{ nm}, E = 2.07 \text{ eV}, \Delta H = 47.7 \text{ Kcal/mol}.$

The UV absorption spectrum of HNO_3 vapor has been studied by JOHNSTON and GRAHAM (1973) for wavelengths 190 – 370 nm and by SCHMIDT *et al.* (1972) for 160 – 440 nm. They agree reasonably well with each other for the range 190 – 320 nm, but there is a major disagreement for the range 320 – 440 nm. Schmidt *et al.* have measured a strong absorption having a maximum cross-section of 3×10^{-18} cm^2 at ~410 nm, but Johnston and Graham obtained a value less than 10^{-22} cm^2 between 330 and 370 nm.

HNO_4 (peroxynitric acid)

(J_{10}) $HNO_4 + h\nu(190 < \lambda < 320 \text{ nm}) \rightarrow OH + NO_3.$

The UV absorption cross section of HNO_4 has been measured by GRAHAM *et al.* (1978) from 190 to 320 nm at 269 K and 20 torr total pressure. The region below 290 nm was also investigated by COX and PATRICK (1979). The two results agree well between 205 and 250 nm. Most earlier calculations assumed that the photolysis products were $HO_2 + NO_2$ (weaker bond), but the photolysis does not necessarily result in breaking the weaker bond (JESSON *et al.*, 1977). It seems to be more likely that the products are $OH + NO_3$ by analogy with $ClONO_2 + h\nu \rightarrow Cl + NO_3$ (see J_{24} at p. 72) since Cl and OH are both isoelectronic.

4.3.3.3 Hydrogen and carbon compounds

H_2O (water vapor), CH_4 (methane) and CO_2 (carbon dioxide)

(J_{11}) $H_2O + h\nu(135 < \lambda < 200 \text{ nm}) \rightarrow H + OH$

$\lambda_m = 242.4 \text{ nm}, E = 5.11 \text{ eV}, \Delta H = 117.8 \text{ Kcal/mol}$

(J_{12}) $CH_4 + h\nu(135 < \lambda < 160 \text{ nm}) \rightarrow CH_3 + H$

$\lambda_m = 276.7 \text{ nm}, E = 4.48 \text{ eV}, \Delta H = 103.2 \text{ Kcal/mol}$

(J_{13}) $CO_2 + h\nu(135 < \lambda < 200 \text{ nm}) \rightarrow CO + O$

$\lambda_m = 227.4 \text{ nm}, E = 5.45 \text{ eV}, \Delta H = 125.5 \text{ Kcal/mol}.$

These three molecules are dissociated mainly in the mesosphere and lower thermosphere. Dissociation of H_2O and CO_2 occurs at 135 – 200 nm including the O_2 Schuman-Runge band range. The absorption cross-sections have been measured by LAUFER and MCNESBY (1965) for H_2O and CH_4, by WATANABE and ZELIKOFF (1953) for H_2O, by THOMPSON *et al.* (1963) for H_2O, CO_2, and CH_4 and by INN *et al.* (1953), NAKATA *et al.* (1965), OGAWA (1971), SHEMANSKY (1972) and DEMORE and PATAPOFF

(1972) for CO_2. The measured CO_2 cross-section above ~200 nm may be mainly due to Rayleigh scattering (SHEMANSKY, 1972) and is not considered to lead to dissociation of CO_2.

These molecules are also dissociated at Lyman-α with the average cross-section 1.5×10^{-17} cm^2 for H_2O, 1.8×10^{-17} cm^2 for CH_4 and 5×10^{-20} cm^2 for CO_2. NICOLET (1981b) investigated the variation with the solar flux of the H_2O photodissociation at *Ly-α* at zero optical depth and obtained the formula

$$J_\infty(H_2O, Ly-\alpha) = 3.5\times10^{-6} + 1.2\times10^{-6}\frac{\Phi - 75}{25}\sec^{-1} \qquad (4.26)$$

where Φ is the solar flux at 10.7 cm (Wm^2 Hz^{-1}).

H_2O_2 (hydrogen peroxide) and HO_2 (perhydroxyl radical)

(J_{14}) $H_2O_2 + h\nu(135 < \lambda < 350\,\text{nm}) \rightarrow OH + OH$

$\lambda_m = 578\,\text{nm}, E = 2.14\,\text{eV}, \Delta H = 49.4\,\text{Kcal/mol}$

(J_{15}) $HO_2 + h\nu(180 < \lambda < 270\,\text{nm}) \rightarrow$ products.

Results of recent experiments on the photolysis of vapor phase of H_2O_2 by MOLINA *et al.* (1977) and LIN *et al.* (1978) are in excellent agreement for the spectral range 210 – 390 nm. They are also in fair agreement with earlier results by UREY *et al.* (1929) and HOLT *et al.* (1948); Urey's result shows slightly larger cross-section than Molina's result above 300 nm. The spectral range below 200 nm has been measured by SCHÜRGERS and WELGE (1968).

The measurement of HO_2 photolysis is rather difficult because of the rapid bimolecular combination reaction HO_2+HO_2. Two measurements by HOCHANADEL and GHORMLEY (1972) and PAUKERT and JOHNSTON (1972) show a broad maximum of the absorption cross section centering around 205–210 nm. The products of HO_2 photolysis have not been clearly determined, although there is a possibility that they are OH and O.

CH_2O (formaldehyde)

(J_{16}) $CH_2O + h\nu(240 < \lambda < 360\,\text{nm}) \rightarrow H + CHO$ (1)

$\rightarrow H_2 + CO$ (2)

for

(1) $\lambda_m = 335\,\text{nm}, E = 3.70\,\text{eV}, \Delta H = 85.2\,\text{Kcal/mol}$

(2) $\Delta H = -2.2\,\text{Kcal/mol}.$

The quantum yield in the photodecomposition of CH_2O has been measured by MOORTGAT and WARNECK (1979), CALVERT *et al.* (1972) and McQUIGG and CALVERT (1969). Process (1) produces atomic H and occurs below ~335 nm, the known dissociation energy. Above 335 nm only molecular products H_2 and CO can be formed. The quantum yields for each process are given in NASA (1979); the cross-sections for each branch are also given in Appendix C.

4.3.3.4 Chlorine containing molecules

HCl (*hydrochloric acid*)

$$(J_{17}) \quad HCl + h\nu(140 < \lambda < 220\,\text{nm}) \rightarrow H + Cl$$

$$\lambda_m = 280\,\text{nm}, E = 4.43\,\text{eV}, \Delta H = 102\,\text{Kcal/mol.}$$

The measured absorption coefficient by INN (1975) agrees well with the earlier results by ROMAND (1949) and MYER and SAMSON (1970) for the wavelength range 145 - 180 nm. Below 145 nm, the measurements by Inn and by Myer and Samson are in good agreement, but the result by Romand is about 35% lower. On the other hand, above 180 nm, Inn's measurement agrees well with those by Romand but the result by Myer and Samson is about 50 times larger than the other two results. It is difficult at this time to determine the cause of this large difference. The result of INN (1975) would probably be the best compromise for use in model calculations.

HOCl (*hypochlorous acid*)

$$(J_{18}) \quad HOCl + h\nu(200 < \lambda < 410\,\text{nm}) \rightarrow OH + Cl$$

$$\lambda_m = 513\,\text{nm}, E = 2.42\,\text{eV}, \Delta H = 55.6\,\text{Kcal/mol.}$$

The absorption cross-sections of HOCl have been measured by MOLINA and MOLINA (1978), KNAUTH *et al.* (1979), and MOLINA *et al.* (1980). All experimental data are in good agreement, but the theoretical study for the transition between the lowest singlet states by JAFFE and LANGHOFF (1978) indicates much smaller cross-sections than the observations; in particular, above 300 nm the theoretical values are negligibly small, whereas experimental data show absorption cross-sections of $\sim 6 \times 10^{-20}$ cm^2 around 310 nm and $\sim 5 \times 10^{-22}$ cm^2 at around 400 nm.

Cl_2 (*molecular chlorline*)

$$(J_{19}) \quad Cl_2 + h\nu(240 < \lambda < 450\,\text{nm}) \rightarrow Cl + Cl$$

$$\lambda_m = 500\,\text{nm}, E = 2.48\,\text{eV}, \Delta H = 57.1\,\text{Kcal/mol.}$$

Cl_2 absorbs UV radiation continuously between 240 and 450 nm. The absorption coefficients measured by SEERY and BRITTON (1964) at room temperature are in good agreement with those reported by GIBSON and BAYLISS (1933).

ClO (chlorine monoxide)

$$(J_{20}) \quad ClO + h\nu(230 < \lambda < 350\,\text{nm}) \rightarrow Cl + O$$

$$\lambda_m = 453\,\text{nm}, E = 2.74\,\text{eV}, \Delta H = 63.0\,\text{Kcal/mol}.$$

The cross-section of broad band absorption has been measured by JOHNSTON *et al.* (1969) for the spectral range 225–280 nm. The result agrees well with the cross-section measured at various band heads above ~263 nm by PORTER and WRIGHT (1953). The latter measurements extend up to ~ 303 nm. The recent measurements at band heads between 270 and 312 nm by JOURDAIN *et al.* (1978), however, give about 20% larger cross-sections, and the theoretical calculation by LANGHOFF *et al.* (1977) give even larger (by ~50%) values than the experimental values by Jourdrain *et al.* The theory also indicates that even with these larger cross-sections, the destruction of ClO by photolysis in the stratosphere is dominated by its loss due to chemical reactions with NO and O.

$CFCl_3$, CF_2Cl_2, CCl_4 (chlorofluoromethane, CFM)

$$(J_{21}) \quad CFCl_3 + h\nu(150 < \lambda < 226\,\text{nm}) \rightarrow CFCl + 2Cl$$

$$(J_{22}) \quad CF_2Cl_2 + h\nu(150 < \lambda < 226\,\text{nm}) \rightarrow CF_2 + 2Cl$$

$$(J_{23}) \quad CCl_4 + h\nu(174 < \lambda < 220\,\text{nm}) \rightarrow CCl_3 + Cl.$$

The molecules derived from saturated hydrocarbons such as methane (CH_4) or ethane (C_2H_6) by replacing a part or the entirety of H atoms by chlorine and/or fluorine are called *chlorofluorocarbons* (CFC). More generally, they are called *halocarbons* considering that other halogen atoms (bromine and iodine) can also be replacements for H atoms. If CFC are derived from CH_4, they are sometimes called *chlorofluoromethane* (CFM). The typical examples of CFM include $CFCl_3$ (triclorofluoromethane or Freon 11, abbreviated F_{11}). CF_2Cl_2 (dichlorodifluoromethane or Freon 12, abbreviated F_{12}) and CCl_4 (carbon tetrachloride).

CFC's release halogens by photolysis absorbing UV radiation in the stratosphere. Reviewing the UV absorption spectra of many CFC's, SANDORFY (1976) has described the general features of the location of the absorption bands as follows: compounds containing only one chlorine like CF_3Cl, CH_2HCl or C_2F_5Cl have their absorption maxima at 152 nm or shorter wavelengths, and the absorption is weakest among CFC. CFC's containing two or more chlorines absorb UV radiation at longer wavelengths (near 175–180 nm for $CFCl_3$ and CF_2Cl_2). These molecules absorb UV radiation more strongly and will be decomposed in the upper stratosphere. Bromine containing halocarbons absorb at even longer wavelengths and are subject to stronger photochemical decomposition in the stratosphere. Molecules such as CH_3Cl and CH_2Cl_2, which retain some of the four H atoms in methane, are generally more vulnerable photochemically than the molecules such as $CFCl_3$ and CF_2Cl_2, which replace all of the H atoms by halogens.

The absorption cross sections for the spectral range above 174 nm have been reviewed by WATSON (1977) for measurements made before 1977 and by HUDSON *et al.* (NASA Reference Pub., 1979) after 1977. Below 174 nm, the absorption cross sections of F_{11} and F_{12} have been measured by HUEBNER *et al.* (1975b). In most cases the products of the photolyses are not known well, although it is suggested by REBBERT and AUSLOOS (1975, 1976/1977) that there is a rapidly increasing possibility that the absorption of a photon will lead to the release of two chlorine atoms from $CFCl_3$ or CF_2Cl_2 and one chlorine from CCl_4.

$ClONO_2$ (chlorine nitrate)

$$(J_{24})\quad ClONO_2 + h\nu(190 < \lambda < 450\text{ nm}) \rightarrow Cl + NO_3.$$

The recent measurements by MOLINA and MOLINA (1979) give about 15% smaller cross-sections than the earlier measurements by ROWLAND *et al.* (1976). The cross-section of nitryl chloride ($ClNO_2$) is about 3 times larger, whereas that of bromine nitrate ($BrONO_2$) is about 20 times larger than that of $ClONO_2$ (ILLIES and TAKACS, 1976/1977; SPENCER and ROWLAND, 1978). The products of $ClONO_2$ photolysis is most probably Cl and NO_3 (CHANG *et al.*, 1979b), but it is not ruled out that they are ClO and NO_2 (ROWLAND *et al.*, 1976) or O and ClONO (SMITH W. S. *et al.*, 1977). In the latter case, however, the photolysis of ClONO occurs very rapidly in the stratosphere, producing either $Cl+NO_2$ or $ClO+NO$.

CH_3Cl (methyl chloride)

$$(J_{25})\quad CH_3Cl + h\nu(174 < \lambda < 220\text{ nm}) \rightarrow CH_3 + Cl.$$

The spectrum of absorption cross section here is very similar to that of HCl. The cross-sections measured by ROBBINS (1976) are about two times greater than those measured by HERZBERG and SCHEIBE (1930). Note that the values tabulated in Table 1 of ROBBINS (1976) do not match with the plotted values in the figure of the same paper; the plotted values are probably correct (see also, the table at p. 903 of the review paper by WATSON, 1977). The cross-section of methyl bromide (CH_3Br) is about 2 – 3 times larger than those of CH_3Cl and extends up to ~270 nm (ROBBINS, 1976).

4.3.3.5 Sulfur containing molecules

SO_2 (sulfur dioxide)

$$(J_{26}) \quad SO_2 + h\nu(135 < \lambda < 218 \text{ nm}) \rightarrow SO + O$$

$$\lambda_m = 218 \text{ nm}, E = 5.69 \text{ eV}, \Delta H = 130.9 \text{ Kcal/mol.}$$

The absorption of UV radiation in the spectral range 135 – 218 nm would lead to dissociation of SO_2. SO_2 also absorbs radiation at shorter wavelengths causing ionization (GOLOMB *et al.*;, 1962) and radiation of longer wavelengths leading to production of non-dissociated excited SO_2 in the troposphere (CALVERT *et al.*, 1978).

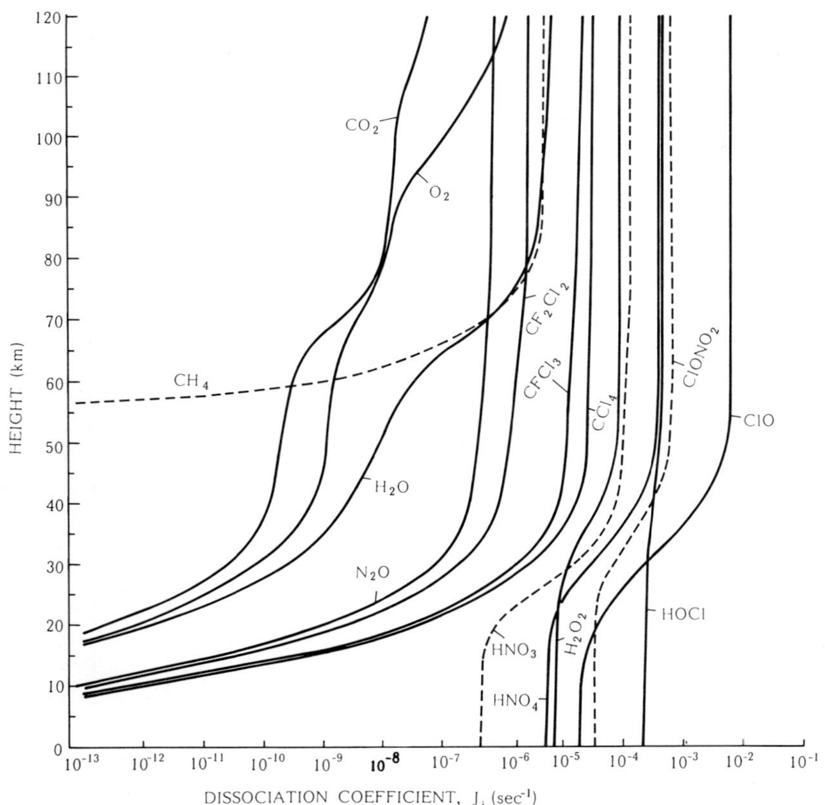

FIG. 4.15. (a) Altitude profiles of photodissociation coefficients calculated for various molecules for an overhead sun, Part 1.

COS (carbonyl sulfide)

$$(J_{27})\quad COS + h\nu(186 < \lambda < 260\,\text{nm}) \rightarrow CO + S$$

$$\lambda_m = 395\,\text{nm}, E = 3.14\,\text{eV}, \Delta H = 72.3\,\text{Kcal/mol.}$$

The absorption cross-section has a broad peak at ~220–225 nm, and the measured value by BRECKENBRIDGE and TANABE (1970) is $\sim 1.67 \times 10^{-19}$ cm^2, which is about one-half the values recently measured by RUDOLPH and INN (1981). Absorption is negligibly small above ~270 nm (MOLINA *et al.*, 1981). The threshold wavelengths for production of CS and O are smaller than 200 nm (KLEMM *et al.*, 1975).

4.4 Height Variations in Dissociation Coefficients of Various Molecules

Height variations of photodissociation coefficients calculated for vertical solar incidence ($\chi = 0$) are illustrated in Figs. 4.15(a) and (b) for various molecules. The values at some chosen heights are also shown in Table 4.3. J_{O_2} and J_{O_3} in Figs. 4.15(a) and (b) are the total dissociation coefficients for the entire spectrum range, i.e.

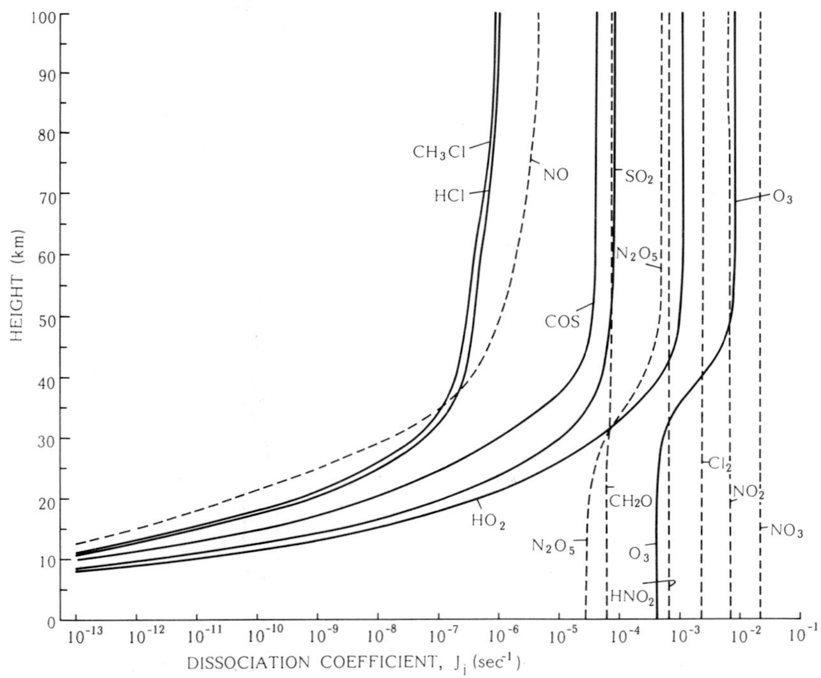

FIG. 4.15. (b) The same as Fig. 4.15(a) but Part 2.

TABLE 4.3 Photodissociation coefficients (in sec^{-1}) of various molecules predicted for overhead sun. The dissociation products are shown in parentheses.

	Photolyses	Photodissociation coefficients at various heights for overhead sun. 10 km	15 cm	20 km	25 km	30 km	45 km	60 km	80 km	100 km
J_{1a}	O_2(Herzberg, O, O)	5.07(−18)	8.21(−15)	5.25(−13)	7.42(−12)	4.93(−11)	7.48(−10)	9.02(−10)	9.13(−10)	9.14(−10)
J_{1b}	O_2(SR-cont., O, O(^1D))	0	0	0	0	0	0	0	3.67(−20)	7.40(−8)
J_{1c}	O_2(SR-band, O, O)	1.78(−18)	5.60(−15)	4.14(−13)	5.04(−12)	2.47(−12)	1.85(−10)	6.48(−10)	5.05(−9)	4.07(−8)
	O_2(Ly-α, O, O(^1D))	0	0	0	0	0	0	2.98(−13)	1.94(−9)	2.71(−9)
J_{2a}	O_3(Hartley, O_2, O(^1D))	1.44(−5)	1.22(−5)	1.96(−5)	4.26(−5)	1.26(−4)	4.44(−3)	7.52(−3)	7.66(−3)	7.67(−3)
J_{2b}	O_3(Huggins, O_2, O)	5.58(−5)	5.96(−5)	6.42(−5)	7.16(−5)	7.83(−5)	7.47(−5)	7.82(−5)	1.13(−4)	1.02(−4)
J_{2c}	O_3(Chappuis, O_2, O)	3.04(−4)	3.40(−4)	3.40(−4)	3.40(−4)	3.40(−4)	3.40(−4)	3.40(−4)	3.40(−4)	3.40(−4)
J_3	N_2O (N_2, O(^1D))	5.08(−15)	1.44(−11)	1.07(−9)	1.37(−8)	6.90(−8)	3.64(−7)	4.94(−7)	6.26(−7)	6.87(−7)
J_4	NO (N, O)	2.27(−14)	1.04(−12)	4.12(−11)	1.17(−8)	1.78(−8)	7.01(−8)	1.93(−6)	3.64(−6)	4.50(−6)
J_5	NO_2 (NO, O)	6.51(−3)	6.52(−3)	6.55(−3)	6.60(−3)	6.66(−3)	6.86(−3)	6.89(−3)	6.90(−3)	6.90(−3)
J_6	NO_3 (NO_2, O)	2.20(−2)	2.20(−2)	2.20(−2)	2.20(−2)	2.20(−2)	2.20(−2)	2.20(−2)	2.20(−2)	2.20(−2)

TABLE 4.3 (continued)

	Photolyses	Photodissociation coefficients at various heights for overhead sun. 10 km	15 km	20 km	25 km	30 km	45 km	60 km	80 km	100 km
J_7	N_2O_5 (NO_2, NO_3)	3.03(−5)	3.05(−5)	3.23(−5)	3.85(−5)	6.21(−5)	3.82(−4)	4.95(−4)	5.02(−4)	5.03(−4)
J_8	HNO_2 (OH, NO)	6.60(−4)	6.60(−4)	6.60(−4)	6.60(−4)	6.60(−4)	6.60(−4)	6.60(−4)	6.60(−4)	6.60(−4)
J_9	HNO_3 (OH, NO_2)	4.61(−7)	4.77(−7)	7.88(−7)	3.55(−6)	1.53(−5)	9.11(−5)	1.18(−4)	1.33(−4)	1.39(−4)
J_{10}	HNO_4 (OH, NO_3)	5.60(−6)	5.70(−6)	6.71(−6)	1.15(−5)	3.00(−5)	2.92(−4)	4.12(−4)	4.23(−4)	4.24(−4)
J_{11}	H_2O (H, OH)	3.71(−18)	1.71(−14)	1.77(−12)	3.26(−11)	2.56(−10)	4.74(−9)	3.19(−8)	3.17(−6)	5.02(−6)
J_{12}	CH_4 (H, CH_3)	0	0	0	0	0	0	5.37(−10)	3.50(−6)	4.89(−6)
J_{13}	CO_2 (CO, O)	1.92(−18)	4.20(−15)	2.99(−13)	4.04(−12)	2.34(−12)	1.66(−10)	2.96(−10)	1.05(−8)	1.67(−8)
J_{14}	H_2O_2 (OH, OH)	7.21(−6)	7.27(−6)	7.74(−6)	9.02(−6)	1.26(−5)	6.61(−5)	9.00(−5)	9.26(−5)	9.35(−5)
J_{15}	HO_2 (O, OH)	4.13(−12)	8.02(−9)	5.38(−7)	7.16(−6)	4.34(−5)	8.01(−4)	1.09(−3)	1.11(−3)	1.11(−3)
J_{16a}	CH_2O (H, CHO)	3.16(−5)	3.19(−5)	3.43(−5)	3.89(−5)	4.58(−5)	7.68(−5)	8.22(−5)	8.24(−5)	8.24(−5)
J_{16b}	CH_2O (H_2, CO)	5.80(−5)	5.82(−5)	5.95(−5)	6.16(−5)	6.44(−5)	7.44(−5)	7.54(−5)	7.54(−5)	7.54(−5)
J_{17}	HCl (H, Cl)	3.97(−15)	1.13(−11)	8.53(−10)	1.12(−8)	5.83(−8)	3.37(−7)	5.26(−7)	8.52(−7)	1.20(−6)
J_{18}	HOCl (OH, Cl)	2.22(−4)	2.23(−4)	2.30(−4)	2.42(−4)	2.63(−4)	3.77(−4)	4.19(−4)	4.21(−4)	4.21(−4)

J_{19}	Cl_2 (Cl, Cl)	2.12(−3)	2.12(−3)	2.15(13)	2.20(−3)	2.25(−3)	2.40(−3)	2.41(−3)	2.41(−3)	2.41(−3)
J_{20}	ClO (Cl, O)	2.10(−5)	2.27(−5)	3.76(−5)	8.48(−5)	2.41(−4)	4.03(−3)	5.96(−3)	6.07(−3)	6.07(−3)
J_{21}	$CFCl_3$ (2Cl)	1.39(−13)	3.68(−10)	2.66(−8)	3.35(−7)	1.69(−6)	1.07(−5)	1.39(−5)	1.67(−5)	1.87(−5)
J_{22}	CF_2Cl_2 (2Cl)	1.04(−14)	3.26(−11)	2.58(−9)	3.46(−8)	1.84(−8)	1.11(−6)	1.76(−6)	2.56(−6)	2.90(−6)
J_{23}	CCl_4 (Cl)	2.84(−13)	6.10(−10)	4.14(−8)	5.29(−7)	2.87(−6)	2.15(−5)	2.64(−5)	3.07(−5)	3.34(−5)
J_{24}	$ClONO_2$ (Cl, NO_3)	3.44(−5)	3.46(−5)	3.62(−5)	4.22(−5)	6.55(−5)	5.14(−4)	6.86(−4)	6.99(−4)	7.00(−4)
J_{25}	CH_3Cl (CH_3, Cl)	2.34(−15)	6.95(−12)	5.41(−10)	7.25(−9)	3.90(−8)	2.44(−7)	4.15(−7)	7.34(−7)	9.56(−7)

The numeral $A(n)$ should reas as $A \times 10^n$.

$$J_{O_2} = J_{1a} + J_{1b} + J_{1c} \tag{4.27}$$

and

$$J_{O_3} = J_{2a} + J_{2b} + J_{2c}. \tag{4.28}$$

For many molecules J_i decreases rapidly in the stratosphere; we learn from Fig. 4.10 that these molecules absorb solar radiation in the spectral range 200–320 nm, where there is a strong O_3 absorption (see Fig. 4.1). The exact height where the large decrease of J_i occurs depends upon the wavelength range where the molecule absorbs radiation most strongly. In general, the decrease occurs at lower heights as that wavelength increases.

It is interesting to note that dissociation of freons ($CFCl_3$, CF_2Cl_2) decreases sharply below ~40 km; the reason for this is that these molecules are dissociated mainly below ~230 nm (see Fig. 10(b)). Thus, freons are stable for photochemical decomposition until they reach above ~40 km, where they produce chlorine atoms by photolysis. Since this region coincides with the heights where O_3 is produced most effectively, the active reaction of chlorine with odd oxygen would reduce the formation of O_3 just in the region where it is produced.

When molecules have large absorption cross-sections above 320 nm, J_i is almost constant with height, because the radiation at these wavelengths can penetrate to the earth's surface without significant loss. Examples include NO_2, NO_3, HNO_2, Cl_2 and O_3 in the Chappuis band (J_{2c}).

For molecules absorbing radiation mainly below 200 nm, dissociation occurs mostly above the stratosphere; examples include CH_4, CO_2 and O_2 in the Schumann-Runge continuum (J_{1b}). Most dissociations of CH_4, CO_2 and H_2O are due to absorption in the Lyman-α.

For some molecules such as HNO_3, HO_2, N_2O_5 and O_3 in the Hartley continuum (J_{2a}), J_i first tends to decrease with decreasing height, but instead of decreasing indefinitely it approaches a constant value in the troposphere. Absorption by these molecules occurs in a wide spectral range covering both below and above 320 nm.

Chapter 5

OZONE AND NEUTRAL CHEMISTRY

A great variety of atoms and molecules is produced in the middle atmosphere from the primary constituents by photochemical processes. The most significant product is ozone, but other chemically active species are also important since they could sometimes affect ozone density seriously. In this chapter we will discuss first a formation theory of the ozone layer and then, the photochemistry of those constituents that affect ozone density in the middle atmosphere. Finally, a review will be given of rate constants for significant reactions in the photochemistry of the middle atmosphere, particularly the stratosphere.

5.1 Formation Theory of an Ozone Layer in a Pure Oxygen Atmosphere

For simplicity we consider here an atmosphere which is composed of oxygen only; no other constituents such as water vapor and nitrogen oxides are included. The chemically inactive N_2 could be included as a third body that promotes three body reactions. This atmosphere is called the *pure oxygen atmosphere*, and a formation theory of the ozone layer in such an atmosphere was first presented by CHAPMAN (1930); therefore, it is called the *Chapman theory* or *Chapman mechanism.*

When O_2 is dissociated, various forms of excited atomic oxygen such as $O(^1D)$ are produced, but they are quickly deactivated by collisions with major species and make transitions to the ground state O by

$$(R_1) \qquad O(^1D) + M \rightarrow O(^3P) + M + 45.4 \text{ Kcal/moke}$$

where M represents any molecule, but mostly N_2 and O_2 in the middle and lower atmosphere. The heat of reaction is given for N_2; all heats of reaction in this book are for the temperature 298 K. In what follows the notation (R_i) is used to represent the chemical reaction, and k_i to represent the reaction rate constant for the reaction (R_i). The expressions for k_i's are discussed in Section 5.6. The reaction (R_1) is called *deactivation* or *quenching* of the excited atomic oxygen $O(^1D)$. Thus, the O_2 photolysis eventually produces two ground state O atoms by

$$(J_{O_2}) \qquad O_2 + h\nu \rightarrow O + O.$$

The total dissociation coefficient J_{O_2} is given by (4.27).

From these oxygen atoms, ozone is formed by the three body reaction

$$(R_2) \qquad O + O_2 + M \rightarrow O_3 + M + 25.3 \text{ Kcal/mole}.$$

Ozone, on the other hand, is destroyed either by photolysis

$$(J_{O_3}) \qquad O_3 + h\nu \rightarrow O + O_2$$

or by the reaction

$$(R_3) \qquad O + O_3 \rightarrow 2O_2 + 93.7 \text{ Kcal/mole}.$$

The total dissociation coefficient J_{O_3} is given by (4.28). Although (R_3) is much slower than (J_{O_3}), it is nonetheless important since it results in a net loss of O_3. By contrast, the faster reaction (J_{O_3}) simply transforms O_3 into O, which easily reproduces O_3 by the equally fast reaction (R_2); therefore, (J_{O_3}) does not in effect destroy O_3.

Above the mesopause where O atoms are abundant, O will be converted to O_2 by

$$(R_4) \qquad O + O + M \rightarrow O_2 + M + 118 \text{ Kcal/mole}.$$

The reactions (J_{O_2}), (J_{O_3}), (R_2), (R_3), (R_4) are the basic reactions in the Chapman mechanism. The relationships among various forms of oxygen are illustrated in Fig. 5.1.

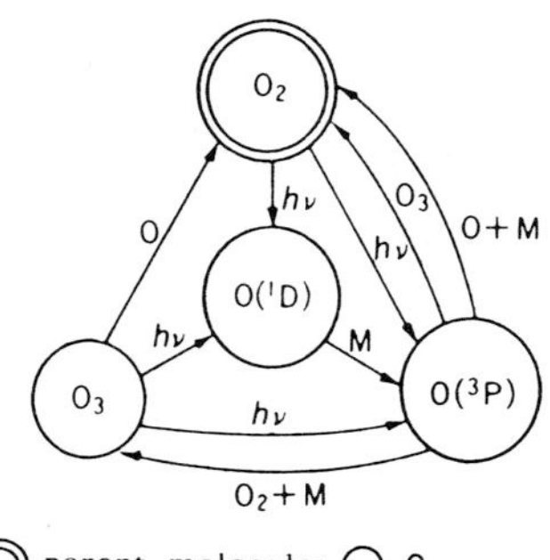

FIG. 5.1. Diagram for chemical reactions of odd oxygens (O_x).

Chemical rate equations for concentrations of O and O_3 are now written as*

$$\frac{\partial[O]}{\partial t} = 2J_{O_2}[O_2] + J_{O_3}[O_3] - k_2[O][O_2][M] - k_3[O][O_3] - 2k_4[O]^2[M] \tag{5.1}$$

40km	5.81×10^6	4.390×10^8	-4.434×10^8	-9.28×10^5	-4.1×10^2
25km	2.75×10^5	9.454×10^8	-9.456×10^8	-1.62×10^4	-1.05

$$\frac{\partial[O_3]}{\partial t} = k_2[O][O_2][M] - J_{O_3}[O_3] - k_3[O][O_3] \tag{5.2}$$

40km	4.434×10^8	-4.390×10^8	-9.28×10^5
25km	9.456×10^8	-9.454×10^8	-1.62×10^4.

Written below each term are the reaction rates in $cm^{-3}\ sec^{-1}$ at 40 km and 25 km. For these evaluations, we use photolysis rates of O_2 and O_3 and concentrations of O, O_2, O_3 and M given in Tables 4.3 and 7.3, respectively; these are taken from the result of a diurnally averaged photochemical model that will be discussed in Chapter 7 (note that the model includes all chemistries of HO_x, NO_x, and Cl_x as well as the pure oxygen chemistry).

There is a large difference of magnitude among various terms in (5.1) and (5.2); terms due to (J_{O_3}) and (R_2) are much larger than other terms and these two terms are almost equal in magnitude but of opposite sign. Because of these two large competing terms it is difficult to evaluate their difference accurately and to calculate [O] and $[O_3]$ by solving (5.1) and (5.2) by a successive iteration method as is discussed below.

If the initial value of $[O_3]$ is assumed to be slightly larger than the value determined by the equilibrium between (J_{O_3}) and (R_2), the right hand side of (5.1) is positive and therefore [O] would increase. Unless the time step is taken to be much smaller than the photochemical time constant for O, the increase of [O] calculated by (5.1) would be so large that it makes the right hand side of (5.2) positive; this causes the $[O_3]$ predicted by (5.2) to increase. The increased $[O_3]$ would further increase [O] calculated by (5.1). Thus, repeating these processes, both [O] and $[O_3]$ increase steadily and rapidly step by step; we will call this behavior the *snowball effect*. Since the photochemical time constant for O at 25 – 40 km is less than a second (see Fig. 6.3(a)) calculations of reasonable [O] and $[O_3]$ values by the iteration method would require a very small time step and therefore enormous time steps and computational time.

A convenient method to avoid this annoying problem of numerical in-

*Brackets [ ] indicate the concentration of the asigned species.

stability is to consider the *odd oxygen* O_x defined by the sum of O and O_3.* We can derive an equation for $[O_x]$ by adding the two equations (5.1) and (5.2) as follows:

$$\frac{\partial [O_x]}{\partial t} = \frac{\partial([O] + [O_3])}{\partial t} = 2\, J_{O_2}[O_2] - 2k_3[O][O_3] - 2k_4[O]^2[M]. \quad (5.3)$$

It must be noted that the large terms produced by the fast reactions (J_{O_3}) and (R_2) in (5.1) and (5.2) now cancelled each other out and do not appear in (5.3).

Since the terms due to (J_{O_3}) and (R_2) must be almost completely in balance, the ratio of [O] to $[O_3]$ should be maintained as

$$r = \frac{[O]}{[O_3]} = \frac{J_{O_3}}{k_2[O_2][M]} \quad (5.4)$$

from which expressions for $[O_3]$ and [O] are obtained in terms of r and $[O_x]$ as follows:

$$[O_3] = \frac{1}{1+r}[O_x], \quad [O] = \frac{r}{1+r}[O_x]. \quad (5.5)$$

Inserting these relations into (5.3) we obtain the equation for $[O_x]$

$$\frac{\partial [O_x]}{\partial t} = 2\, J_{O_2}[O_2] - 2\alpha[O_x]^2 \quad (5.6)$$

with

$$\alpha = \frac{k_3 r + k_4[M]r^2}{(1+r)^2}. \quad (5.7)$$

The values of $[O_3]$ and [O] are now given by (5.5) through the solution of (5.6).

In a steady state ($\partial/\partial t = 0$) the equilibrium density for O_x is given by $(J_{O_2}[O_2]/\alpha)^{1/2}$. The corresponding equilibrium densities for O_3 and O are now calculated from (5.5) as follows:

$$[O_3]_e = \sqrt{\frac{J_{O_2}[O_2]}{k_3 r + k_4[M]r^2}}, \quad [O]_e = \sqrt{\frac{J_{O_2}[O_2]}{k_3 r + k_4[M]r^2}} \cdot r. \quad (5.8)$$

If the reaction (R_4) is neglected as is justified in the stratosphere and mesosphere, equations (5.8) are reduced to

*This method is called the *family method*. See Chapter 6 for more details.

$$[O_3]_e = \sqrt{\frac{J_{O_2}k_2[M]}{J_{O_3}k_3}}[O_2], \quad [O]_e = \sqrt{\frac{J_{O_2}J_{O_3}}{k_2k_3[M]}}. \tag{5.9}$$

Steady state solution for $[O_3]$ and [O] may also be obtained by the following method: we first represent $[O_3]$ in terms of [O] from the equation obtained by setting to zero the right hand side of (5.3)

$$[O_3] = \frac{J_{O_2}[O_2]}{k_3[O]} - \frac{k_4[M]}{k_3}[O]. \tag{5.10}$$

Inserting (5.10) into (5.2) we have a cubic equation for [O]

$$k_3k_4[M][O]^3 + (J_{O_3}k_4[M] + k_2k_3[O_2][M])[O]^2 - k_3 J_{O_2}[O_2][O] - J_{O_2} J_{O_3}[O_2] = 0. \tag{5.11}$$

Neglecting k_4, (5.11) is reduced to a quadratic equation, whose solution is given by

$$[O] = \frac{J_{O_2} + \sqrt{J_{O_2} + 4 J_{O_2} J_{O_3}k_2/k_3[M]}}{2k_2[M]}. \tag{5.12}$$

Taking account of $J_{O_2} \ll J_{O_3}$ in the middle atmosphere, (5.12) is reduced to the equation equivalent to the equation for $[O]_e$ in (5.9). The equation equivalent to $[O_3]_e$ is readily obtained by (5.10) again by neglecting k_4.

In order to calculate $[O_3]_e$ and $[O]_e$ at a certain height using (5.8) or (5.9) we need the values of $[O_3]_e$ at higher altitudes to calculate the transmission τ_{O_3}, which is required for calculations of J_{O_3} and J_{O_2}. Thus, the equations (5.8) and (5.9) are not the explicit solution for $[O_3]_e$. However, the problem can be solved by first calculating $[O_3]_e$ at an altitude high enough so that τ_{O_3} may be assumed to be zero. We then use the result to calculate τ_{O_3} at the next lower height (grid point in the model) and calculate $[O_3]_e$ at that height. Repeating the same procedure step by step to the lower heights we can calculate $[O_3]_e$ and $[O]_e$ at all heights.

The right-hand side of the equation for $[O_3]_e$ in (5.9) is a function of height; it depends most strongly on J_{O_2}, $[M]$ and $[O_2]$. Since J_{O_2} increases with height (see Fig. 4.4), whereas both $[M]$ and $[O_2]$ decrease with height, the quantity proportional to $J_{O_2}{}^{1/2}[M]^{1/2}[O_2]$ should have a maximum at a certain height. An analysis similar to that which led to (3.11) would tell us that such a maximum would occur at the height where the O_2 optical depth is ~3, since $[M]$ is proportional to $[O_2]$. From Fig. 3.8 we learn that such a condition of the maximum ozone density occurs at ~25 - 40 km for the wavelengths 200 - 240 nm.

The profiles of $[O]_e$ and $[O_3]_e$ predicted for a solar incidence angle 35°

are shown by long-dashed curves in Fig. 5.2. Comparing the profile of $[O_3]_e$ with the average curve of the observed ozone density in the middle latitudes taken from the U. S. Standard Atmosphere 1976 (Appendix A, see also KRUEGER and MINZNER, 1976), it is evident that the theoretical value is appreciably (~2 times) larger than the observed values above ~25 km, and that the theoretical values are extremely small below that height. The large discrepancy between the theoretical and experimental values at lower heights is nearly removed by taking into account the effects of vertical mixing as is seen from a short-dashed curve. However, the discrepancy at higher altitudes is increased by this effect; it can be reduced and the theoretical curve and the observations may become compatible if we introduce the effects of photochemical reactions due to nitrogen-oxygen and hydrogen-ox-

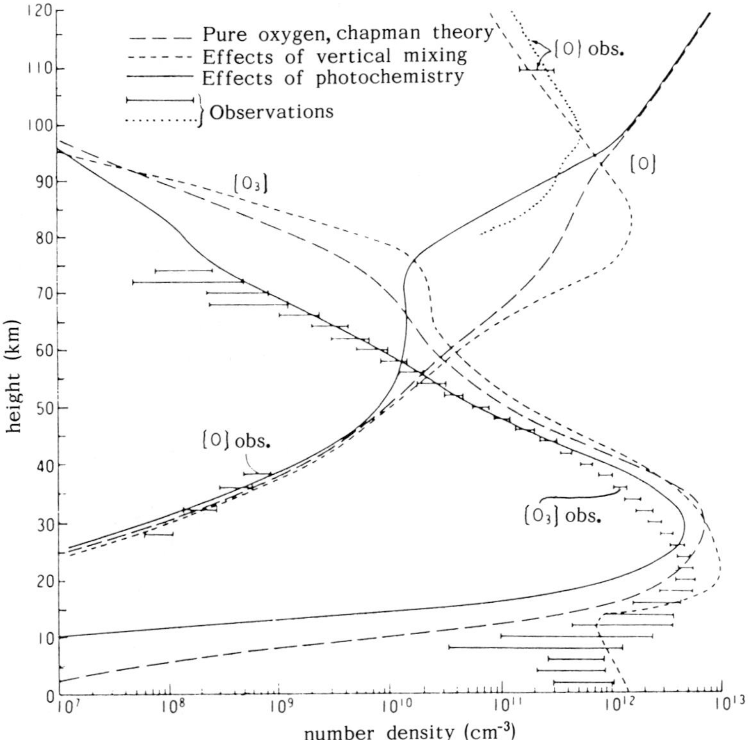

FIG. 5.2. Height distributions of [O] and $[O_3]$ predicted by Chapman theory (long-dashed curves) compared with observations. Effects of dynamics (vertical eddy transport) and chemistry including HO_x, NO_x and ClO_x reactions are shown by short-dashed and solid curves, respectively. Observational data are taken from KRUEGER and MINZNER (1976) for O_3, from DICKINSON *et al.* (1974) for thermospheric O, and from ANDERSON (1975) for stratospheric O.

ygen compounds as the solid curve shows.

As for atomic oxygen, the equilibrium profile agrees well with observations by the chemiluminescent method (ANDERSON, 1975) below ~40 km. The equilibrium $[O]_e$ increases steadily with height, and at ~120 km the value becomes almost two orders of magnitude larger than the rocket measurements shown by a dotted curve (DICKINSON *et al.*, 1974). This is caused by the fact that the production of O by O_2 dissociation in the Schumann-Runge continuum becomes very large above ~90 km (see Fig. 4.4). Vertical transport by molecular and eddy diffusion can carry this O downward and reduce the predicted [O] above ~100 km to values close to the observations. However, the O transported to lower heights then increases [O] at ~80 km, and it is the role of chemical reactions of hydrogen-oxygen compounds to reduce [O] at these heights, making them compatible with observations.

5.2 *Odd Oxygen Losses by Catalytic Reactions*

As was discussed in the previous section, the equilibrium ozone density predicted by the Chapman mechanism is larger than observation by a factor of ~2 in the middle and upper stratosphere and a factor of ~10 in the mesosphere. Since ozone is much more abundant in the stratosphere, the overprediction in the stratosphere is much greater in absolute amount and is thus more influential in the attenuation of solar UV radiation arriving at the surface.

These discrepancies between the theory and observations can be resolved if we consider the effects of photochemical reactions not considered in the Chapman mechanism. Since the reaction (R_2) is the only possible reaction producing O_3, and the strength of the natural source of O in the stratosphere (O_2 photodissociation) is very well known, there is no possibility of reducing the production rate of O_3. Only an increase in the O_3 loss rate is a possibility to modify the O_3 density photochemically. The stratospheric photochemistry has been discussed, amongst others, by CRUTZEN (1970), DÜTSCH (1971), JOHNSTON (1974), NICOLET (1975), ROWLAND and MOLINA (1975) and CICERONE (1981). These indicate that different constituents and different chemical reactions contribute to the additional O_3 loss mechanisms at different heights; we will now discuss photochemistries of important compounds in the middle atmosphere.

5.2.1 *Hydrogen-oxygen compounds (HO_x)*

The primary molecules for the production of hydrogen-oxygen compounds HO_x (H, OH and HO_2) in the middle atmosphere are H_2O and CH_4. The discussion here will be limited primarily to the direct production of HO_x from these molecules; the methane chemistry including hydrocarbons will be discussed later.

Since H_2O and CH_4 absorb solar UV radiation below 200 nm, they can be dissociated mainly in the mesosphere, producing H and OH. The product H can be converted to HO_2 quickly by the fast reaction

$$(R_5) \qquad H + O_2 + M \rightarrow HO_2 + M + 51.6 \text{ Kcal/mole.}$$

Both HO_2 and OH react with O by

$$(R_6) \qquad HO_2 + O \rightarrow OH + O_2 + 50.7 \text{ Kcal/mole}$$

and

$$(R_7) \qquad OH + O \rightarrow H + O_2 + 16.7 \text{ Kcal/mole.}$$

While these two reactions convert O atoms to molecular O_2, they also reproduce OH and H, respectively. These H and OH molecules, again by reactions (R_5) – (R_7), can convert O into O_2. The net effect can readily be seen if three chemical equations are added and the constituents common to both sides are eliminated, i.e.

$$\begin{array}{rll} & H + O_2 + M & \rightarrow HO_2 + M \\ & HO_2 + O & \rightarrow OH + O_2 \\ +) & OH + O & \rightarrow H + O_2 \\ \hline \text{(I)} & O + O + M & \rightarrow O_2 + M. \end{array}$$

The net reaction (I) is identical to (R_4). Thus, in the mesosphere the O loss by (R_4) is apparently intensified by the reactions of hydrogen compounds HO_x. Hydrogen changes its chemical form among the three components of HO_x (i.e. H, OH and HO_2), while destroying two O's in each cycle. During these cycles, the concentration of hydrogen compounds remains unchanged. This mechanism is the so-called *catalytic process* (*reaction*) of HO_x in destroying the odd oxygen densities. The mechanism is illustrated schematically in Fig. 5.3. Travelling clockwise from H along the thick arrows, we come back to the original H (H → HO_2 → OH → H). During this cycle the reactions (R_6) and (R_7) destroy O, but there is no change in HO_x.

It is seen in Fig. 5.3 that H can proceed counter-clockwise to OH by the reaction

$$(R_8) \qquad H + O_3 \rightarrow OH + O_2 + 76.9 \text{ Kcal/mole.}$$

The product OH then comes back clockwise to H by the reaction (R_7). The net effect is

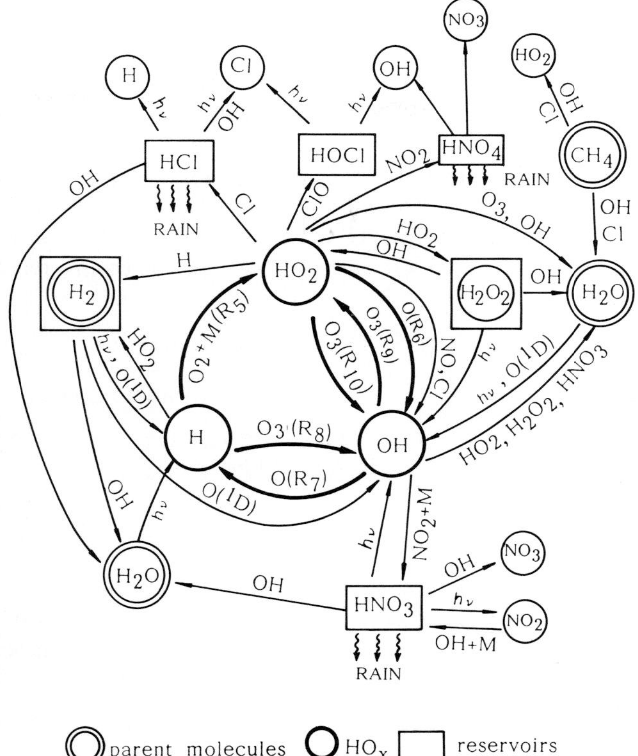

FIG. 5.3. Diagram for main chemical reactions of hydrogen-oxygen compounds (HO_x).

$$\begin{array}{lll} & H + O_3 & \rightarrow OH + O_2 \\ +) & OH + O & \rightarrow H + O_2 \\ \hline (II) & O + O_3 & \rightarrow 2O_2. \end{array}$$

The net reaction (II) is identical to (R_3), and constitutes a catalytic cycle by HO_x for the net loss of odd oxygen. This cycle is also important in the mesosphere where H is still abundant.

In the stratosphere, the above catalytic cycles (I) and (II) are both ineffective since H is less abundant. Instead, the reaction of OH and O_3 would become important; it changes OH into HO_2 by

$$(R_9) \qquad OH + O_3 \rightarrow HO_2 + O_2 + 43.0 \text{ Kcal/mole}.$$

This reaction proceeds counter-clockwise in Fig. 5.3; combining this with a clockwise reaction (R_6) results in another catalytic cycle

$$\begin{array}{lll} & OH + O_3 & \rightarrow HO_2 + O_2 \\ +) & HO_2 + O & \rightarrow OH + O_2 \\ \hline \text{(III)} & O + O_3 & \rightarrow 2O_2. \end{array}$$

The catalytic cycle (III) destroys odd oxygen efficiently in the upper and middle stratosphere, but it becomes less effective in the lower stratosphere and troposphere, since atomic O becomes increasingly scarce. In these regions a larger amount of O_3 would make the following reaction more effective:

$$(R_{10}) \quad HO_2 + O_3 \rightarrow OH + 2O_2 + 25.3 \text{ Kcal/mole}.$$

Combining (R_9) and (R_{10}) would constitute a catalytic cycle

$$\begin{array}{lll} & OH + O_3 & \rightarrow HO_2 + O_2 \\ +) & HO_2 + O_3 & \rightarrow OH + 2O_2 \\ \hline \text{(IV)} & 2O_3 & \rightarrow 3O_2. \end{array}$$

This cycle converts two ozone to three oxygen molecules.

We have discusseed the fact that the production of HO_x by the H_2O photolysis (J_{11}) is limited to regions in and above the mesosphere, but HO_x can also be produced in the lower regions by

$$(R_{11}) \quad H_2O + O(^1D) \rightarrow 2OH + 28.4 \text{ Kcal/mole}.$$

The excited $O(^1D)$ is produced by O_3 photolysis in the Hartley continuum. Since the tail of the Hartley continuum near 320 nm can penetrate through the troposphere, $O(^1D)$ can be produced at levels all the way through the stratosphere and troposphere. In particular, the production of OH in the troposphere could be surprisingly large because of the large amount of H_2O. Thus, the catalytic cycle (IV) can effectively destroy odd oxygens even in the troposphere. The production of OH in the troposphere is important since it reacts actively with many constituents such as CH_4, CO, HCl and various hydrocarbons, and destroys them (ATKINSON *et al.*, 1976).

The reaction rates in the catalytic cycles (I) – (IV) are tabulated in Table 5.1 for selected heights, along with the reaction rates of the O_x loss reactions in pure oxygen chemistry. In each cycle the slowest reaction should be *rate determining**, i.e. it controls the speed of the net effect of the catalytic

*The idea of a *rate determining* (or *rate limiting*) reaction may be understood if we consider a flow of vehicles on the highway. The speed of the entire flow is determined by the speed at the place where the vehicles proceed at the slowest speed due to, for instance, congestion. The effect may also be called the *bottleneck effect*.

TABLE 5.1 Chemical reaction rates in various HO_x catalytic cycles compared with those in pure oxygen reactions. ◎ shows the rate determining reaction in each cycle and boldface indicates that the reaction is faster than or comparable to the pure oxygen reactions at the same height.

		HO_x catalytic cycle	Reaction rate (in cm^{-3} sec^{-1}) at 80 km	60 km	40 km	20 km	10 km
I	(R_5)	$H+O_2+M \rightarrow HO_2+M$	**2.67×10^5**	**5.69×10^5**	◎1.56×10^5	**4.19×10^3**	**1.68×10^5**
	(R_6)	$HO_2+O \rightarrow OH+O_2$	◎**2.65×10^5**	◎**5.50×10^5**	2.71×10^5	1.26×10^2	**2.07×10**
	(R_7)	$OH+O \rightarrow H+O_2$	**5.73×10^5**	**6.75×10^5**	1.87×10^5	◎2.89×10	◎2.95×10^{-1}
II	(R_8)	$H+O_3 \rightarrow OH+O_2$	◎**3.05×10^5**	◎**1.23×10^5**	◎3.43×10^4	◎5.53	1.50
	(R_7)	$OH+O \rightarrow H+O_2$	**5.78×10^5**	**6.75×10^5**	1.87×10^5	1.92×10	◎2.95×10^{-1}
III	(R_9)	$OH+O_3 \rightarrow HO_2+O_2$	◎4.28	◎4.47×10^2	◎1.86×10^5	**2.30×10^4**	**9.76×10^3**
	(R_6)	$HO_2+O \rightarrow OH+O_2$	**2.65×10^5**	**5.50×10^5**	2.71×10^5	◎1.26×10^2	◎**2.07×10**
IV	(R_9)	$OH+O_3 \rightarrow HO_2+O_2$	4.28	4.47×10^2	1.86×10^5	**2.30×10^4**	◎**9.76×10^3**
	(R_{10})	$HO_2+O_3 \rightarrow OH+2O_2$	◎1.11×10^{-1}	◎2.78×10	◎1.07×10^4	◎**8.01×10^3**	◎**3.16×10^4**
		Pure oxygen chemistry					
	(R_4)	$O+O+M \rightarrow O_2+M$	3.39×10^3	1.42×10^3	2.05×10^2	4.48×10^{-2}	6.05×10^{-6}
	(R_3)	$O+O_3 \rightarrow 2O_2$	7.30×10^2	5.30×10^4	9.30×10^5	2.60×10^3	3.52

cycle. Those rate determining reactions are indicated by ◎.

The boldfaced numbers in Table 5.1 indicate that the reactions are faster than or comparable to the O_x loss reactions in pure oxygen chemistry. If some reactions in a cycle are faster than the pure oxygen reactions but the rate determining reaction of that cycle is slower than the pure oxygen reactions, then that cycle does not greatly affect the O_x loss of the Chapman mechanism. It can be seen in Table 5.1 that the cycles (I) and (II) are most effective in destroying O_x at 80 and 60 km, the cycle (III) at 40 km, and the cycle (IV) at 20 and 40 km. In particular, the rate determining reactions of the cycles (I) and (IV) give much larger O_x loss than the pure oxygen chemistry at 80 km (mesosphere) and 10 km (troposphere), respectively.

Figure 5.3 illustrates the mechanism of producing HO_x from the primary molecules (H_2O, CH_4 and H_2) and the catalytic reactions of HO_x. Other important features illustrated are the strong interactive reactions among members of HO_x; these include

(R_{12a}) $\quad H + HO_2 \rightarrow H_2 + O_2 + 52.6$ Kcal/mole

(R_{12b}) $\quad H + HO_2 \rightarrow H_2O + O + 50.9$ Kcal/mole

(R_{12c}) $\quad H + HO_2 \rightarrow 2OH + 33.9$ Kcal/mole

(R_{13}) $\quad OH + OH \rightarrow H_2O + O + 17.0$ Kcal/mole

(R_{14}) $\quad OH + OH + M \rightarrow H_2O_2 + M + 51.1$ Kcal/mole

(R_{15}) $\quad OH + HO_2 \rightarrow H_2O + O_2 + 67.6$ Kcal/mole

and

(R_{16}) $\quad HO_2 + HO_2 \rightarrow H_2O_2 + O_2 + 33.4$ Kcal/mole.

Reactions (R_{12b}), (R_{13}) and (R_{15}) reproduce the parent molecule H_2O, but reactions (R_{12a}), (R_{14}) and (R_{16}) produce new molecules H_2 and H_2O_2. These two molecules are sinks for HO_x, but at the same time they work as sources by the photolysis of H_2O_2 (photolysis of H_2 is a very slow process)

(J_{13}) $\quad H_2O_2 + h\nu \rightarrow 2OH$

and the reactions

(R_{17}) $\quad H_2 + O(^1D) \rightarrow H + OH + 43.5$ Kcal/mole

(R_{18}) $\quad H_2 + O \rightarrow H + OH - 1.9$ Kcal/mole

and

(R_{19}) $\quad H_2O_2 + O \rightarrow HO_2 + OH + 17.2$ Kcal/mole.

A molecule that works as both sink and source is called a *reservoir*· H_2 and H_2O_2 are reservoirs for HO_x. H_2 can also be a parent gas since it is injected in the atmosphere from the earth's interior as a component of volcanic gas.

Both H_2 and H_2O_2 also react with OH by

$$(R_{20}) \qquad H_2 + OH \rightarrow H_2O + H + 15.0 \text{ Kcal/mole}$$

and

$$(R_{21}) \qquad H_2O_2 + OH \rightarrow H_2O + HO_2 + 34.2 \text{ Kcal/mole.}$$

These reactions reproduce H and HO_2, while destroying OH.

5.2.2 Nitrogen-oxygen compounds (NO_x)

The most abundant molecule in the atmosphere, N_2 is photochemically inactive and will not be oxidized under normal conditions. Except by the ionic reactions in the thermosphere, N_2 can not be the source for nitrogen-oxygen compounds in the middle atmosphere.

N_2O is introduced in the atmosphere from the earth's soil by the process of denitrification (see Section 11.1.3). Transported by vertical motion into the stratosphere, N_2O produces NO by the reaction

$$(R_{22a}) \qquad N_2O + O(^1D) \rightarrow 2\,NO + 81.2 \text{ Kcal/mole.}$$

It must be noted, however, that the reaction is equally capable of producing N_2 by

$$(R_{22b}) \qquad N_2O + O(^1D) \rightarrow N_2 + O_2 + 124.5 \text{ Kcal/mole.}$$

NO can quickly react with O_x and produce NO_2 by the reactions

$$(R_{23}) \qquad NO + O_3 \rightarrow NO_2 + O_2 + 47.8 \text{ Kcal/mole}$$

and

$$(R_{24}) \qquad NO + O + M \rightarrow NO_2 + M + 73.1 \text{ Kcal/mole.}$$

The reaction (R_{23}) is much more important than (R_{24}) in the stratosphere, but in and above the mesosphere (R_{24}) becomes more important for converting NO to NO_2 because of the increasing amount of O.

The product NO_2 will return to NO either by the photolysis

$$(J_5) \qquad NO_2 + h\nu \rightarrow NO + O$$

or by reaction with O

$$(R_{25}) \qquad NO_2 + O \rightarrow NO + O_2 + 45.6 \text{ Kcal/mole.}$$

During the cycle NO → NO_2 → NO, odd oxygen is destroyed by (R_{23}) and (R_{25}).* Thus, NO and NO_2 constitute a catalytic cycle destroying O_x.

*In the mesosphere (R_{24}) and (R_{25}) may constitute a catalytic cycle to destroy atomic oxygen, but the effect is small compared with the catalytic cycles by HO_x(I) and (II).

The net effect is found by adding the three reactions (R_{23}), (J_5), and (R_{25}) after multiplying (R_{23}) by 2 as follows:

$$\begin{array}{rll} & 2(NO + O_3) & \rightarrow 2(NO_2 + O_2) \\ & NO_2 + h\nu & \rightarrow NO + O \\ +) & NO_2 + O & \rightarrow NO + O_2 \\ \hline (V) & 2O_3 & \rightarrow 3O_2. \end{array}$$

It must be noted here that in the troposphere and lower stratosphere where atomic oxygen is scarce, the fast chain reaction occurs through (R_{23}), (J_5) and (R_2), whose net effect on O_x is null.

The catalytic cycle of NO and NO_2 is similar to that of OH and HO_2 (cycle III), but they differ in that the dissociation of NO_2 is a strong source for O, whereas the dissociation of HO_2 is small in the stratosphere and its products are not well known.

The photochemistry of NO_x is illustrated in Fig. 5.4. The reaction of N and O_2 produces NO rather than NO_2 by*

(R_{26}) $\quad N + O_2 \rightarrow NO + O + 31.8$ Kcal/mole.

This reaction is usually slow, but if N is in an excited state the reaction

(R_{26a}) $\quad N(^2D) + O_2 \rightarrow NO + O + 87$ Kcal/mole

is a fast reaction for producing NO (see Section 13.2).

The reaction

(R_{27}) $\quad N + O_3 \rightarrow NO + O_2 + 125.4$ Kcal/mole

may destroy O_3, but the product NO would produce O by photolysis rather than destroy O by the reaction NO+O. Thus, N and NO do not constitute a catalytic cycle, in contrast to the catalytic cycle (II) due to H and OH. N and NO actually react with each other by

(R_{28}) $\quad N + NO \rightarrow N_2 + O + 75.0$ Kcal/mole.

This reaction is an important net loss mechanism for NO_x in the lower thermosphere. There are some additional reactions among members of NO_x such as

(R_{29}) $\quad N + NO_2 \rightarrow N_2O + 41.8$ Kcal/mole

(R_{30}) $\quad NO + NO_3 \rightarrow 2NO_2 + 22.7$ Kcal/mole.

*Note that the reaction of H and O_2 is a very fast three body reaction producing HO_2 (see R_5).

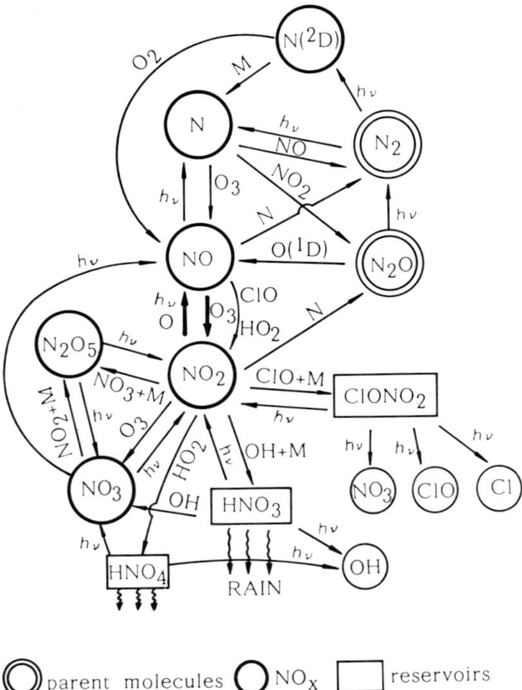

FIG. 5.4. Diagram for main chemical reactions of nitrogen-oxygen compounds (NO_x).

The above discussion leads to the conclusion that the reaction (V) is the only important catalytic cycle destroying O_x by NO_x in the middle atmosphere. This is a remarkable simplification in contrast with the four different catalytic cycles by HO_x affecting O_x at different heights. The O_x loss by (V) is dominant in the middle stratosphere, where the effects of (I) – (IV) by HO_x are relatively small. The catalytic cycle (V) by NO_x is important since it occurs in the region where a high ozone density is predicted by the Chapman mechanism.

The rate of change in $[O_x]$ due to the three reactions in (V) is given by

$$\frac{\partial[O_x]}{\partial t} = J_5[NO_2] - k_{23}[NO][O_3] - k_{25}[NO_2][O] \qquad (5.13)$$

30 km	6.10×10^6	-6.06×10^6	-5.07×10^5
20 km	6.33×10^6	-6.25×10^6	-2.25×10^4.

Written below each term on the right hand side are the reaction rates in $cm^{-3}\ sec^{-1}$ at 30 km and 20 km. It is evident that the reaction (R_{25}) is the slowest reaction, and therefore the rate determining reaction. The other two larger terms are almost equal in magnitude but of opposite sign. In fact, the photochemical time constants for NO and NO_2 are very short at these heights, and equilibrium is always maintained for them. The equilibrium condition for NO is written as

$$O = \frac{\partial[NO]}{\partial t} = J_5[NO_2] + k_{25}[NO_2][O] - k_{23}[NO][O_3]. \qquad (5.14)$$

Equation (5.13) can be transformed by virtue of (5.14) as follows:*

$$\frac{\partial[O_x]}{\partial t} = -2k_{25}[NO_2][O]. \qquad (5.15)$$

The result is consistent with the earlier statement that the net effect of the catalytic cycle (V) is given by the rate determining reaction (R_{25}). Comparing (5.15) with the second term of the right-hand side of (5.3), we learn that the O_x loss rate due to the catalytic cycle (V) should become comparable to the O_x loss rate in the Chapman mechanism, if the condition

$$k_{25}[NO_2] \sim k_3[O_3]$$

is satisfied. Inserting $k_{25} \sim 9.3 \times 10^{-12}$ cm^3 sec^{-1}, $k_3 \sim 7.7 \times 10^{-16}$ cm^3 sec^{-1}, and $[O_3] \sim 4 \times 10^{12}$ cm^{-3} at 30 km, we find that $[NO_2] \sim 3.3 \times 10^8$ cm^{-3}. The observed NO_2 concentration in the stratosphere is in the order of 10^9 cm^{-3}; therefore, it is established that the catalytic loss of O_x by (V) greatly exceeds the O_x loss by the Chapman mechanism in the stratosphere.

Since there is no fast reaction between NO and NO_2, the photochemistry of NO_x is much simpler than that of HO_x. However, the reactions

(R_{31}) $\qquad NO_2 + O_3 \rightarrow NO_3 + O_2 + 25.1$ Kcal/mole

and

(R_{32}) $\qquad NO_2 + O + M \rightarrow NO_3 + M + 50.2$ Kcal/mole.

produce a radical NO_3 (there is no reaction corresponding to R_{31} in HO_x chemistry; the reaction of HO_2 with O_3 produces OH rather than HO_3). The product NO_3 from (R_{31}) or (R_{32}) further reacts with NO_2 by the three body reaction

(R_{33}) $\qquad NO_3 + NO_2 + M \rightarrow N_2O_5 + M + 22.2$ Kcal/mole.

The product N_2O_5 is decomposed either by photolysis, i.e.

*The same result can be obtained without assuming (5.14) if we consider that NO_2 is a member of the O_x family (see Chapter 6).

(J_7) $N_2O_5 + h\nu \rightarrow NO_2 + NO_3$

or by the reaction

(R_{34}) $N_2O_5 + M \rightarrow NO_2 + NO_3 + M - 22.2$ Kcal/mole.

At night when atomic oxygen disappears quickly from the stratosphere, the production of NO by (J_5) and (R_{25}) ceases, whereas conversion of NO to NO_2 by (R_{23}) continues. Part of the increased NO_2 then goes into the production of NO_3 by (R_{31}) and then N_2O_5 by (R_{33}). Since there are no effective chemical losses at night, these molecules increase throughout the nighttime period. After sunrise, photolysis starts to destroy NO_3 and N_2O_5. Thus, no instantaneous photochemical equilibrium is maintained for NO_3 and N_2O_5, but the nighttime production and the daytime loss must be balanced for these constituents in order to maintain the long term stability.

5.2.3 Chlorine-oxygen compounds (ClO_x)

Like HO_x and NO_x, chlorine-oxygen compounds (Cl and ClO) can destroy O_x by the chain reaction of

(R_{35}) $Cl + O_3 \rightarrow ClO + O_2 + 38.7$ Kcal/mole

and

(R_{36}) $ClO + O \rightarrow Cl + O_2 + 56.0$ Kcal/mole.

The net effect is

$$
\begin{array}{lll}
 & Cl + O_3 & \rightarrow ClO + O_2 \\
 +) & ClO + O & \rightarrow Cl + O_2 \\
\hline
\text{(VI)} & O + O_3 & \rightarrow 2O_2.
\end{array}
$$

This catalytic cycle occurs mainly in the same height range as the catalytic cycle of NO_x(V). There are several additional catalytic cycles of ClO_x which also include reactions of species from HO_x and NO_x families; they occur slightly at lower heights and are generally less significant than the cycle (VI) (WUEBBLES and CHANG, 1981).

The natural sources for atmospheric chlorine may be HCl in volcanic gases and NaCl in the ocean, but freon gases (CFM) injected in the atmosphere by human activities may be an important source for stratospheric chlorine (see Chapter 11). Cl is released by photolyses of CFM or their reactions with $O(^1D)$.

Photochemistry of chlorine compounds is illustrated in Fig. 5.5. Other halogen compounds (Bromine and Fluorine compounds) may also play catalytic reactions for O_x losses. The reaction rate constant for the reaction

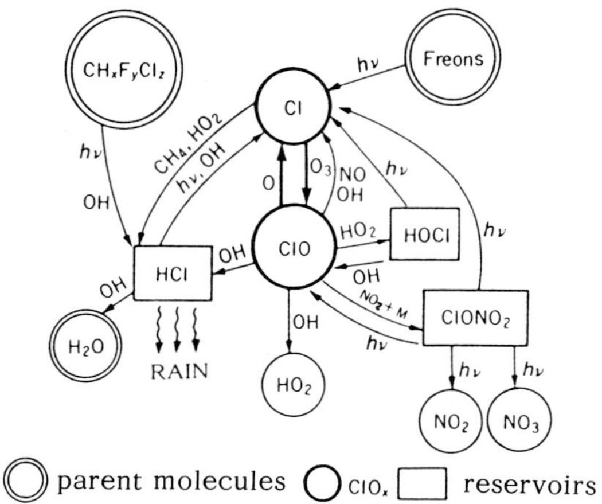

FIG. 5.5. Diagram for main chemical reactions of chlorine-oxygen compounds (ClO_x).

of BrO and O is indeed about twice large as that of ClO and O. However, the current abundance of the atmospheric Bromide is far less than the chloride and its impacts on $[O_x]$ may be limited. The photochemistry of Bromine and its effects on ozone has been discussed by YUNG *et al.* (1980), and that of Fluorine has been examined by SZE (1978).

The relative importance of various catalytic cycles (I) – (VI) depends upon the concentrations of reactants as well as the reaction rate constants. Based on the result of model calculations (see Chapter 7) the reaction rates of the rate determining reactions are calculated for each cycle and compared in Fig. 5.6. Also, the relative magnitude of the O_x losses by catalytic reactions of NO_x, HO_x and ClO_x as well as by pure oxygen chemistry are illustrated in Fig. 5.7 as a percentage diagram. The pure oxygen chemistry gives the largest O_x loss at ~42 – 52 km, the maximum of ~46% occurring at ~45 km. Below these heights the catalytic cycle (V) of NO_x is the dominant O_x loss having the maximum loss (~69%) at ~30 km, whereas above these heights catalytic cycles (I) and (II) of HO_x are the dominant O_x losses. In the lower stratosphere and troposphere the catalytic cycle (IV) of HO_x is the most important loss for O_x. The HO_x catalysis provides almost 100% of the O_x loss in the troposphere and also in the mesosphere above ~65 km. The O_x loss by the catalytic cycle (VI) of ClO_x has a maximum at 30 – 45 km, but its effect is about an order of magnitude less than that of the NO_x cycle (V). It must be noted, however, that the model simulates the status before the injection of a large amount of CFM's takes place (see p. 152). It is possible that the impact of the ClO catalytic cycle increases and will become comparable to that of NO_x catalytic cycle as the CFM injection to

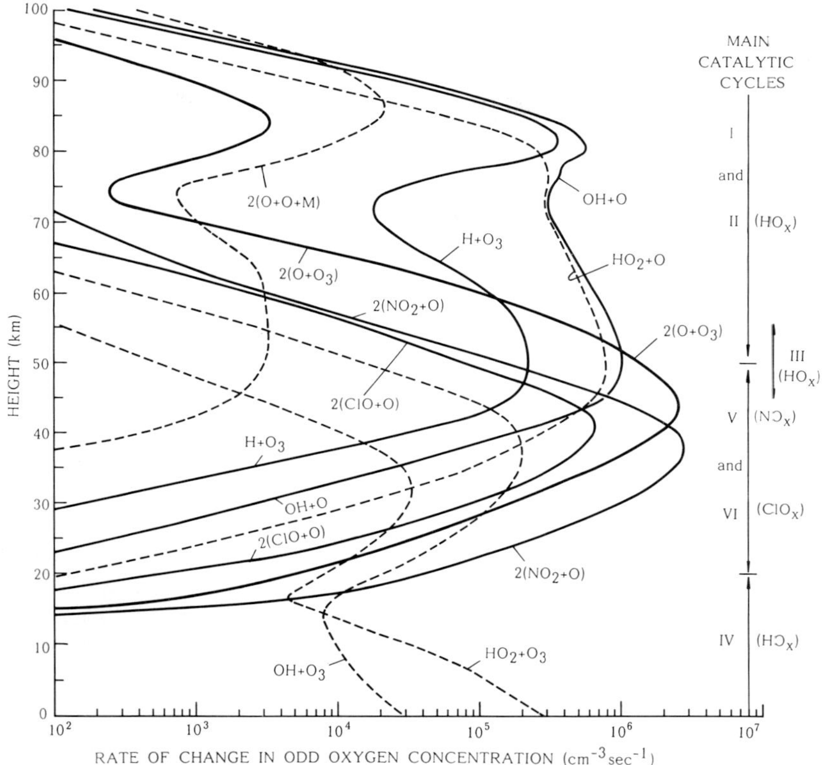

FIG. 5.6. Comparison of the O_x loss rates by various catalytic cycles by HO_x, NO_x, and ClO_x.

the atmosphere continues.

5.3 *Interactions among HO_x, NO_x and ClO_x*

If both HO_x and NO_x exist in the atmosphere at the same time, mutual interaction should occur by

$$(R_{37}) \qquad OH + NO_2 + M \rightarrow HNO_3 + M + 49.5 \text{ Kcal/mole}.$$

While HNO_3 is a sink for both HO_x and NO_x by (R_{37}), it can also be a source for them by photolysis (J_9); therefore, HNO_3 is a reservoir for HO_x and NO_x. HNO_3 also reacts with OH by

$$(R_{38}) \qquad HNO_3 + OH \rightarrow H_2O + NO_3 + 17.9 \text{ Kcal/mole}$$

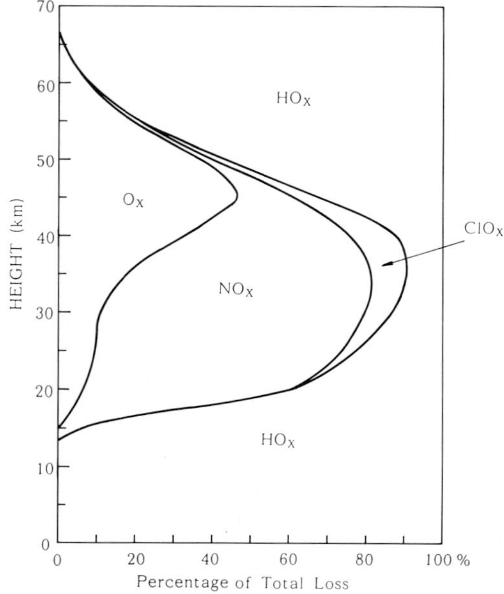

FIG. 5.7. Relative magnitude (in %) of odd oxygen loss rates by pure oxygen chemistry (O_x) and various catalytic cycles of NO_x, HO_x and ClO_x. Freon injection rates in 1955 are assumed.

and acts as a sink for OH and a source for NO_3, which eventually produces NO and NO_2.

HNO_3 falls down constantly into the troposphere and will be lost from the atmosphere by *rainout* or *washout* effects*. Thus, HNO_3 eventually acts as a sink for HO_x and NO_x in the stratosphere for a long time period.

HNO_2 and HNO_4 are other reservoirs for HO_x and NO_x; they are produced by

$$(R_{39}) \qquad NO + OH + M \rightarrow HNO_2 + M + 49.9 \text{ Kcal/mole}$$

or

$$(R_{40}) \qquad NO_2 + HO_2 + M \rightarrow HNO_4 + M + 21.3 \text{ Kcal/mole.}$$

and they are lost mainly by photolyses (J_8), (J_{10}) and chemical reactions

$$(R_{41}) \qquad HNO_2 + OH \rightarrow NO_2 + H_2O + 40.4 \text{ Kcal/mole}$$

or

**Rainout* is the process in which the molecule is used as a condensation nucleus for rain droplet formation, whereas *washout* is the process in which the molecule is dissolved into the rain droplet.

(R_{42}) $HNO_4 + OH \rightarrow H_2O + O_2 + NO_2$.

Another important reaction between HO_x and NO_x is

(R_{43}) $HO_2 + NO \rightarrow OH + NO_2 + 4.8$ Kcal/mole.

The significance of this reaction will be discussed in Sections 5.4 and 11.1.1.

HCl works as a reservoir for HO_x and ClO_x. It is a sink for Cl and HO_2 by the reaction

(R_{44}) $Cl + HO_2 \rightarrow HCl + O_2 + 51.6$ Kcal/mole

and for Cl by the reaction

(R_{45}) $Cl + CH_4 \rightarrow HCl + CH_3 - 1.7$ Kcal/mole

or

(R_{46}) $Cl + H_2 \rightarrow HCl + H - 0.96$ Kcal/mole.

On the other hand, HCl is a source for H and Cl by photolysis (J_{17}) and for Cl by the reactions

(R_{47}) $HCl + OH \rightarrow H_2O + Cl + 16.0$ Kcal/mole

or

(R_{48}) $HCl + O \rightarrow OH + Cl + 7.4$ Kcal/mole.

However, HCl will be lost from the stratosphere by falling down continuously into the troposphere, where it will eventually be washed out.

ClO and NO_2 may be lost in the production of $ClONO_2$ by the reaction

(R_{49}) $ClO + NO_2 + M \rightarrow ClONO_2 + M + 26.0$ Kcal/mole.

Thus, $ClONO_2$ is a sink for NO_2 and ClO. Since these two molecules are the reactants of the rate determining reactions of catalytic cycles (V) and (VI), respectively, the reaction (R_{49}) reduces the O_x loss by these two catalytic reaction cycles. How much the O_x loss will be diminished depends upon the processes of reproducing NO_x and ClO_x from $ClONO_2$.

The photolysis of $ClONO_2$ would produce either (a) $Cl+NO_3$ or (b) $ClO+NO_2$. In the former case, NO_3 is further dissociated and the products may be either (1) $NO+O_2$ or (2) NO_2+O. Since route (2) produces O_x while route (1) does not, there is an ambiguity for the net effect of the decomposition of $ClONO_2$ on the O_x loss.

$ClONO_2$ also can react with OH and O by

(R_{50}) $ClONO_2 + OH \rightarrow HOCl + NO_3$

and

$(R_{51})\qquad ClONO_2 + O \rightarrow ClO + NO_3.$

HOCl is another reservoir for HO_x and ClO_x; in addition to (R_{50}) it can be produced by the reaction

$(R_{52})\qquad ClO + HO_2 \rightarrow HOCl + O_2 + 43.5$ Kcal/mole.

The main loss of HOCl is due to the photolysis (J_{18}) producing OH and Cl, and the reaction

$(R_{53})\qquad HOCl + OH \rightarrow H_2O + ClO + 24.1$ Kcal/mole.

The fast reaction between NO and ClO

$(R_{54})\qquad NO + ClO \rightarrow NO_2 + Cl + 9.1$ Kcal/mole

is significant since it controls the relative abundance of NO_2 and ClO, the important reactants of the rate determining reactions in the catalytic cycles (V) and (VI), respectively. Thus, for instance, if there is a large amount of background NO, an increased ClO_x due to freon injection may increase the O_x loss through the catalytic cycle (V) of NO_x rather than that by the cycle (VI) of ClO_x, because most ClO is converted to NO_2 by (R_{54}).

5.4 Methane Chemistry and Smog Reactions

Methane may be a source of water vapor in the middle atmosphere. The direct transport of tropospheric water vapor into the stratosphere is limited by cold temperatures near the tropopause; water vapor will be condensed out by saturation and a large part of it will be lost before entering the stratosphere. This phenomenon is called the *cold trap*. Since the dew point of CH_4 is much lower (~111.4 K) than that of water vapor (273 K), there is no cold trap effect on CH_4; it can enter the stratosphere easily. Then, the reaction

$(R_{55})\qquad CH_4 + OH \rightarrow CH_3 + H_2O + 14.3$ Kcal/mole

may produce H_2O in the stratosphere.

CH_4 also reacts with $O(^1D)$ by

$(R_{56a})\qquad CH_4 + O(^1D) \rightarrow CH_3 + OH + 43.0$ Kcal/mole

or

$(R_{56b})\qquad CH_4 + O(^1D) \rightarrow H_2 + CH_2O + 113.0$ Kcal/mole.

A product OH from (R_{56a}) can be used to produce another H_2O molecule from CH_4 by the reaction (R_{55}). From CH_3, a product from both (R_{55}) and (R_{56a}), various forms of hydrocarbons (CH_nO_m) are produced by the following reactions:

(R_{57}) $\quad CH_3 + O_2 + M \rightarrow CH_3O_2 + M + 33.2$ Kcal/mole

(R_{58}) $\quad CH_3 + O \rightarrow CH_2O + H + 68.1$ Kcal/mole

(R_{59}) $\quad CH_3O_2 + NO \rightarrow CH_3O + NO_2 + 11.9$ Kcal/mole

(R_{60}) $\quad CH_3O + O_2 \rightarrow CH_2O + HO_2 + 28.9$ Kcal/mole

(J_{16}) $\quad CH_2O + h\nu \rightarrow H + CHO$

(R_{61}) $\quad CH_2O + OH \rightarrow CHO + H_2O + 32.3$ Kcal/mole

(R_{62}) $\quad CH_2O + O \rightarrow CHO + OH + 15.0$ Kcal/mole

and

(R_{63}) $\quad CHO + O_2 \rightarrow CO + HO_2 + 34.9$ Kcal/mole.

These reaction schemes are illustrated in Fig. 5.8. During the course of these reaction series, H_2O may again be produced by (R_{61}). The reaction (R_{59}) converts NO to NO_2, from which O will be released by photolysis (J_5). Since O_3 is easily produced from O by the reaction (R_2), the reaction (R_{59})

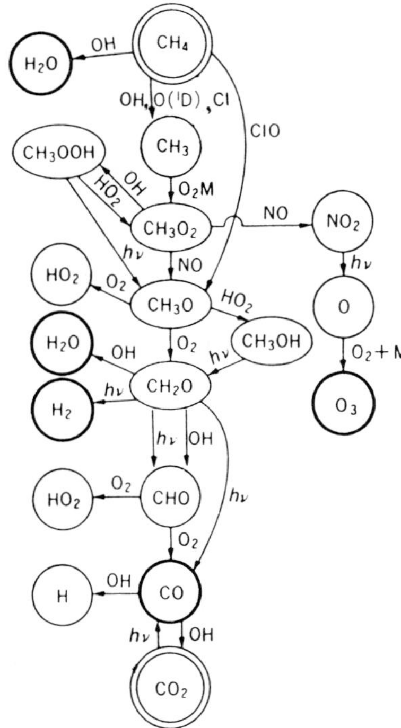

FIG. 5.8. Diagram for methane and hydrocarbon chemistry.

eventually produces O_3. The end product of the above series includes CO. Thus, the CH_4 chemistry can produce H_2O, O_3 and CO in the middle atmosphere.

The reaction (R_{59}) is a mechanism producing *photochemical smog* in the troposphere. The photolysis of the reaction product NO_2 can occur efficiently in the troposphere because of a large absorption cross section of NO_2 above 320 nm including the wavelengths of visible light (see Fig. 4.10(a)), and produces O_x. Plenty of hydrocarbons and NO are produced from automobiles and factories. Thus, the type of the reaction similar to (R_{59}) may occur in the polluted atmosphere and produce a large amount of O_3 near the surface by interaction with visible solar radiation.

CO will be converted to CO_2 by

$$(R_{64}) \qquad CO + OH \rightarrow CO_2 + H + 24.8\ \text{Kcal/mole.}$$

This reaction can initiate the following series of reactions:

$$\begin{array}{lll} (R_{64}) & CO + OH & \rightarrow CO_2 + H \\ (R_5) & H + O_2 + M & \rightarrow HO_2 + M \\ (R_{43}) & HO_2 + NO & \rightarrow OH + NO_2 \\ (J_5) & NO_2 + h\nu & \rightarrow NO + O \\ +)\ (R_2) & O + O_2 + M & \rightarrow O_3 + M \\ \hline & CO + 2O_2 + h\nu & \rightarrow CO_2 + O_3. \end{array}$$

The net effect is the production of O_3; these are *smog reactions* producing O_3 in the lower stratosphere and troposphere. They are effective only if the reaction (R_{43}) is fast enough. If the reaction of HO_2 and O_3 is faster than (R_{43}), the following reactions should occur:

$$\begin{array}{lll} (R_{64}) & CO + OH & \rightarrow CO_2 + H \\ (R_5) & H + O_2 + M & \rightarrow HO_2 + M \\ (R_{10}) & HO_2 + O_3 & \rightarrow OH + 2O_2 \\ (J_{2a}) & O_3 + h\nu & \rightarrow O_2 + O(^1D) \\ +)\ (R_{11}) & O(^1D) + H_2O & \rightarrow 2OH \\ \hline & CO + H_2O + 2O_3 + h\nu & \rightarrow CO_2 + 2O_2 + 2OH. \end{array}$$

The net effect now is the loss of O_3. Thus, reaction (R_{43}) is very important in determining which series of reactions actually occur. If the reaction (R_{43}) is faster than (R_{10}), O_3 will be produced, whereas if (R_{43}) is slower than (R_{10}), O_3 will be lost in the lower stratosphere and troposphere.

5.5 Rate Constants of Chemical Reactions

5.5.1 Bimolecular reactions

Most chemical reactions in the middle atmosphere are second-order (bimolecular) or third-order (three body) reactions. In bimolecular reactions, two molecules (or atoms) encounter one another. If we consider a bimolecular reaction involving an atom and a diatomic molecule,

$$X + YZ \underset{(r)}{\overset{(f)}{\rightleftarrows}} XY + Z \tag{5.17}$$

where (f) and (r) represent the forward and reverse reactions, respectively, the reaction rate for the forward reaction is given by

$$\frac{\partial[X]}{\partial t} = \frac{\partial[YZ]}{\partial t} = -k[X[YZ] \tag{5.18}$$

in which k is the *reaction rate constant* or *specific reaction rate*. k is usually in units of cm^3 $molecule^{-1}$ sec^{-1}, but sometimes it is expressed in units of liter $mole^{-1}$ sec^{-1}, which is about 1.667×10^{-21} cm^3 $molecule^{-1}$ sec^{-1}. It is customary in physical science to omit naming the unit of particles; therefore, cm^3 $molecule^{-1}$ sec^{-1} is often expressed simply as cm^3 sec^{-1}.

The relationships between the potential energies of reactant molecules and product molecules are illustrated schematically in Fig. 5.9. The reaction coordinate (abscissa) denotes some measure of the change in molecular configurations between the two states and the transient state. E_f and E_r represent the *activation energies* for the forward and reverse reactions, respectively. In order to drive the the reaction forward, i.e. from left to right, the system needs to acquire energy E_f, whereas the reverse reaction needs energy E_r. The difference between the two energies

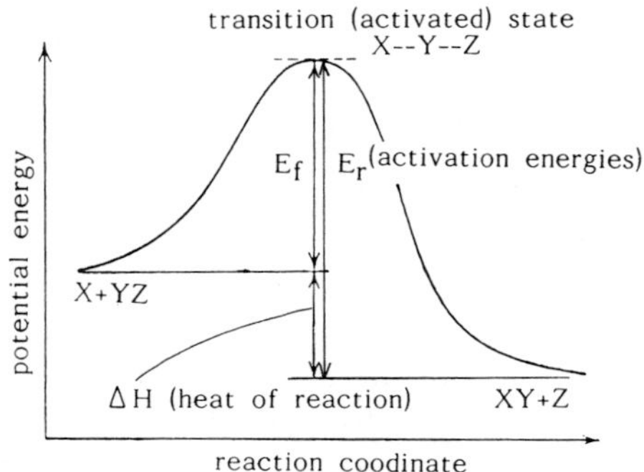

FIG. 5.9. Variation of potential energy accompanying the bimolecular reaction $X+YZ \rightarrow XY+Z$.

$$\Delta H = E_r - E_f \tag{5.19}$$

is the *heat of reaction*. If $\Delta H > 0$ the forward reaction produces heat and the reaction is called an *exothermic reaction*, whereas if $\Delta H < 0$ the reaction results in energy absorption and is called an *endothermic reaction*. Endothermic reactions are rare unless some external energy is supplied.

The temperature of the gas indicates the mean kinetic energy per molecule. Each molecule has a different energy (velocity) and its distribution is usually given by the Maxwellian distribution law. There are always some molecules that have much larger velocities (energies) than the average. If a molecule has energy greater than the activation energy, the chemical reaction can occur for that molecule. If the temperature of the entire gas increases, the number of molecules having energy greater than the activation energy increases, and more molecules can react per unit time; therefore, the reaction would proceed faster, and the specific reaction rate would be larger. Thus the reaction rate constant depends upon the temperature and the relationship between k and T is expressed by the *Arrhenuis formula*

$$k = A\,\mathrm{e}^{-\frac{E}{RT}} \tag{5.20}$$

where E is the activation energy, R is the gas constant, and A is the *Arrhenuis constant* (*A factor*). A and E can be determined by measuring k's at various temperatures.

The reaction rate constant for bimolecular reactions also depends upon the kinds of molecules involved in the reaction; it is in the wide range 10^{-10} - 10^{-20} $\mathrm{cm^3\ sec^{-1}}$. The upper limit of k may be evaluated by the following simple consideration of collision theory: According to the kinetic theory of gases, the mean velocity of molecules of temperature T and mass m is given by

$$\langle v \rangle = \sqrt{\frac{8kT}{\pi m}} \tag{5.21}$$

where k is the Boltzmann constant. For N_2 molecules and $T = 300$ K, the mean velocity is calculated to be $\sim 4.8 \times 10^4$ $\mathrm{cm\ sec^{-1}}$. If the collisional cross section is written as

$$\sigma = \pi R^2 \tag{5.22}$$

the value of R for most atmospheric molecules is several times larger than the radius of the orbit of H atoms ($\sim 0.53 \times 10^{-8}$ cm). Therefore, σ is $\sim 2 \times 10^{-15}$ $\mathrm{cm^2}$, and the rate of elastic collision, which is given by the product of σ and $\langle v \rangle$, is $\sim 10^{-10}$ $\mathrm{cm^3\ sec^{-1}}$. This is called a *collision theory rate constant*. Since not all molecular collisions result in a chemical reaction, the actual reaction rate constant is smaller than the collision theory rate constant.

5.5.2 *Three-body reactions*

In some cases the reaction of two molecules (or atoms) can proceed more easily by the presence of a third body (molecule). The reaction is then called a *third-order*, *termolecular* or *three-body reaction*. A typical process is written as

$$A + B + M \rightarrow AB + M \tag{5.23}$$

in which the third body M does not change but plays a role to facilitate the reaction between A and B. The ozone production reaction (R_2) is an example of a three-body reaction.

The process of a three-body reaction may be characterized by the production of an intermediate molecule AB^* with sufficient energy to decompose to A and B. The entire process may occur via

$$A + B \underset{k_2}{\overset{k_1}{\rightleftarrows}} AB^* \tag{5.24}$$

$$AB^* + M \xrightarrow{k_3} AB + M \tag{5.25}$$

where the rate constants k_1, k_2 and k_3 refer to the above reactions for the present discussion only. (Do not confuse them with the rate constants of the reactions (R_1), (R_2) and (R_3)).

Since there is a possibility that AB^* returns to $A+B$ by the reverse reaction k_2, it is necessary for M to absorb that excess energy and to help the reaction k_3 to occur. By equating the production and loss rates of AB^* by the above reactions, we obtain

$$k_1[A][B] = k_2[AB^*] + k_3[AB^*][M] \tag{5.26}$$

The rate of production of AB is given by

$$\frac{d[AB]}{dt} = k_3[AB^*][M] \tag{5.27}$$

which, if $[AB^*]$ is substituted from (5.23), becomes

$$\frac{d[AB]}{dt} = \frac{k_1 k_3 [A][B][M]}{k_2 + k_3[M]}. \tag{5.28}$$

If $[M]$ is larger than k_2/k_3 (*high pressure limit*), the equation (5.28) is reduced to a form similar to that of a bimolecular reaction

$$\frac{\mathrm{d}[AB]}{\mathrm{d}t} = k_1[A][B] \tag{5.29}$$

and if $[M]$ is smaller than k_2/k_3 (*low pressure limit*) we have a form similar to that of a three-body reaction

$$\frac{\mathrm{d}[AB]}{\mathrm{d}t} = \frac{k_1 k_3}{k_2}[A][B][M]. \tag{5.30}$$

The collision theory rate constants k_1 and k_3 are both $\sim 10^{-10}$ cm^3 sec^{-1}, whereas the constant k_2 for a *first-order* or *unimolecular reaction* is given by dividing the mean velocity ($\sim 4.8 \times 10^4$ cm sec^{-1}) by R in (5.22) and is $\sim 1.8 \times 10^{12}$ sec^{-1}. Thus, the value k_1 k_3/k_2 in (5.30) based on a collision theory is $\sim 5.5 \times 10^{-33}$ cm^6 sec^{-1}. The unit for the three body reactions is sometimes given in liter^2 mole^{-2} sec^{-1}, which is equivalent to $\sim 2.77 \times 10^{-42}$ cm^6 sec^{-1}.

Putting

$$k_\infty = k_1, \quad k_0 = k_1 k_3/k_2 \tag{5.31}$$

one can write from (5.28) the rate constant for the three-body reaction as follows:

$$k = \frac{1}{[A][B][M]}\frac{\mathrm{d}[AB]}{\mathrm{d}t} = \frac{k_0}{1 + k_0[M]/k_\infty} \tag{5.32}$$

where k_0 and k_∞ are the low and high pressure limit rate constant, respectively. Both k_0 and k_∞, and therefore k are functions of temperature. If the temperature dependence of k is expressed by a formula similar to the Arrhenuis formula

$$k = A\left(\frac{T}{300}\right)^b \mathrm{e}^{-\frac{E}{RT}}. \tag{5.33}$$

E usually becomes negative. There is no physical meaning for a negative activation energy, but the reason why E becomes negative for the three-body reaction may be that k_0 decreases as temperature increases because of a rapid increase of k_2 with temperature.

We have seen that the reaction becomes the bimolecular type reaction when $[M]$ is large. This sounds strange since M is the third body essential for the three-body reactions. However, it could be understood by considering that if $[M]$ is large, the reaction k_3 proceeds faster and the slower reaction k_1 becomes the rate determining reaction. Thus, the net speed of the reaction is determined by the bimolecular reaction k_1.

5.5.3 *Results of laboratory measurements*

The reaction rate constants of most of the important chemical reactions in the middle atmosphere have been measured in laboratories during the period of CIAP's intensive studies and thereafter. Table 5.2 lists the measured rate constants for most reactions discussed in previous sections taken from reviews at several different times since CIAP (1975).

The values in the first column are those used in CIAP studies and the following columns are those that became available after CIAP. One can learn from the table how the measurement of each reaction has progressed since CIAP. The symbol S indicates that the rate constant has remained unchanged since the previous review, and therefore is the same as the value in the nearest left column. If S appears more often, it means either (1) the rate constant has been well determined or (2) the reaction is relatively unimportant and there is no necessity for more accurate measurement. On the other hand, if the value has changed often from review to review, it implies that the reaction needed more accurate measurement because it was an important reaction and/or its rate constant included a greater uncertainty. For example, the increased rate constant for (R_{43}) $NO+HO_2 \rightarrow NO_2+OH$ by a factor of 40 from the CIAP value (HOWARD and EVENSON, 1977; SIMONAITIS and HEICKLEN, 1978; HOWARD, 1979: LEU, 1979; THRUSH and WILKINSEN, 1981) has dramatically changed the assessment of the impact of SST injected NO_x on stratospheric ozone, since as was discussed in Section 5.4 the larger k_{43} would produce O_3 rather than destroy it in the lower stratosphere when NO_x is injected from SST's.

The relatively poor data on the Chlorine-included reactions in CIAP (1975) and NBS (1975) would imply that the importance of these reactions has been recognized only after CIAP, in connection with the sudden uprise of the controversial CFM problems. Laboratory data on chlorine reactions were very limited at the time of CIAP; most data have been obtained since CIAP. Actually, the reviews of NASA (1977) and NASA (1979) were primarily concerned with the problem of the impacts of CFM's on stratospheric ozone.

In NASA (1979), JPL (1981, 1982) and WMO (1981) the rate constants for three body reactions are given by a formula extended from (5.32) to

$$k = \left(\frac{k_0(T)}{1 + k_0(T)\cdot[M]/k_\infty(T)}\right)\cdot 0.6^{[1+\{\log_{10}(k_0(T)\cdot[M]/k_\infty(T)\}^2]^{-1}}. \quad (5.34)$$

The reaction rate constants have been determined and recommended by each panel after reviewing much laboratory data. There are too many laboratory data for each reaction to give the complete references here. The near complete list can be found in FAA (1981) and detailed discussions

TABLE 5.2 Measured chemical reaction rate constants taken from reviews at different times since CIAP. S indicates that the value is same as that in the nearest left column.

	Chemical Reactions	CIAP(1975) NBS(1975)	NASA(1977) NBS(1977)
$R_2^{@}$	$O+O_2+M \to O_3+M$	$1.06\times10^{-34}e^{510/T}$	S
R_4	$O+O+M \to O_2+M$	$3.8\times10^{-30}(1/T)e^{-170/T}$	$5.2\times10^{-35}e^{900/T}$
R_3	$O+O_3 \to 2O_2$	$1.9\times10^{-11}e^{-2300/T}$	S
R_1	$O(^1D)+N_2 \to O+N_2$	5.4×10^{-11}	$2.0\times10^{-11}e^{107/T}$
R_1	$O(^1D)+O_2 \to O+O_2$	7.4×10^{-11}	$2.9\times10^{-11}e^{67/T}$
	$O(^1D)+O_3 \to 2O_2$	2.7×10^{-10}	1.2×10^{-10}
	$O_2(^1\Delta_g)+O_2 \to 2O_2$	$2.2\times10^{-18}(T/300)^{0.8}$	S
	$O_2(^1\Delta_g)+O_3 \to 2O_2+O$	$6.6\times10^{-13}e^{-1560/T}$	$4.5\times10^{-11}e^{-2830/T}$
	$O_2(^1\Delta_g) \to O_2+h\nu$	2.6×10^{-4}	
R_5	$H+O_2+M \to HO_2+M$	$2.1\times10^{-32}e^{290/T}$	S
R_6	$HO_2+O \to OH+O_2$	$8.0\times10^{-11}e^{-500/T}$	3.5×10^{-11}
R_7	$OH+O \to H+O_2$	4.2×10^{-11}	4.0×10^{-11}
R_8	$H+O_3 \to OH+O_2$	2.6×10^{-11}	$1.0\times10^{-10}e^{-516/T}$
R_9	$OH+O_3 \to HO_2+O_2$	$1.6\times10^{-12}e^{-1000/T}$	$1.9\times10^{-12}e^{-1000/T}$
R_{10}	$HO_2+O_3 \to OH+2O_2$	$1.0\times10^{-13}e^{1250/T}$	$7.3\times10^{-14}e^{-1275/T}$
R_{11}	$H_2O+O(^1D) \to 2OH$	3.5×10^{-10}	2.3×10^{-10}
R_{18}	$H_2+O \to OH+H$	$3.0\times10^{-14}Te^{-4480/T}$	$8.8\times10^{-12}e^{-4200/T}$
R_{17}	$H_2+O(^1D) \to OH+H$	2.9×10^{-10}	9.9×10^{-11}
R_{19}	$H_2O_2+O \to HO_2+OH$	$2.75\times10^{-12}e^{-2125/T}$	S
R_{12a}	$H+HO_2 \to H_2+O_2$	$4.2\times10^{-11}e^{-350/T}$	S
R_{12b}	$H+HO_2 \to H_2O+O$	$8.3\times10^{-11}e^{-500/T}$	S
R_{12c}	$H+HO_2 \to 2OH$	$4.2\times10^{-10}e^{-950/T}$	S
R_{13}	$OH+OH \to H_2O+O$	$1.0\times10^{-11}e^{-550/T}$	S
R_{14}	$OH+OH+M \to H_2O_2+M$	$2.5\times10^{-33}e^{-2550/T}$	$1.25\times10^{-32}e^{900/T}$
R_{15}	$OH+HO_2 \to H_2O_2+O_2$	$2\times10^{-11}-2\times10^{-10}$	3×10^{-11}
R_{16}	$HO_2+HO_2 \to H_2O_2+O_2$	$3.0\times10^{-11}e^{-500/T}$	2.5×10^{-12}
R_{20}	$H_2+OH \to H_2O+H$	$3.6\times10^{-11}e^{-2590/T}$	
R_{21}	$H_2O_2+OH \to H_2O+HO_2$	$1.7\times10^{-11}e^{-910/T}$	$1.0\times10^{-11}e^{-750/T}$
R_{22a}	$N_2O+O(^1D) \to 2NO$	1.1×10^{-10}	5.5×10^{-11}
R_{22b}	$N_2O+O(^1D) \to N_2+O_2$	1.1×10^{-10}	5.5×10^{-11}
R_{23}	$NO+O_3 \to NO_2+O_2$	$9.0\times10^{-13}e^{-1200/T}$	$2.3\times10^{-12}e^{-1450/T}$
R_{24}	$NO+O+M \to NO_2+M$	$4.2\times10^{-33}e^{940/T}$	$1.55\times10^{-32}e^{584/T}$
R_{25}	$NO_2+O \to NO+O_2$	9.1×10^{-12}	9.3×10^{-12}
R_{26}	$N+O_2 \to NO+O$	$1.1\times10^{-14}Te^{-3150/T}$	$5.5\times10^{-12}e^{-3220/T}$
R_{27}	$N+O_3 \to NO+O_2$	$3.0\times10^{-11}e^{-1200/T}$	
R_{28}	$N+NO \to N_2+O$	2.7×10^{-11}	$8.2\times10^{-11}e^{-410/T}$
R_{29}	$N+NO_2 \to N_2O+O$	1.4×10^{-12}	$2.0\times10^{-11}e^{-800/T}$
R_{30}	$NO+NO_3 \to 2NO_2$	8.7×10^{-12}	2.0×10^{-11}

NASA (1979) FAA (1980)	JPL (1981) WMO (1981)	JPL (1982)
$6.2\times10^{-34}(T/300)^{-2}$	S	$6.0\times10^{-34}(T/300)^{-2.3}$
$4.8\times10^{-33}(T/300)^{-2}$		
$1.5\times10^{-11}e^{-2218/T}$	S	S
$1.8\times10^{-11}e^{107/T}$	S	S
$3.2\times10^{-11}e^{67/T}$	S	S
S	S	S
S		
$1.2\times10^{-11}e^{-2400/T}$		
$5.5\times10^{-32}(T/300)^{-1.4}$	S	S
S	4.0×10^{-11}	$3.0\times10^{-11}e^{200/T}$
S	$2.3\times10^{-11}e^{110/T}$	$2.2\times10^{-11}e^{117/T}$
$1.4\times10^{-10}e^{-470/T}$	S	S
$1.6\times10^{-12}e^{-940/T}$	S	S
$1.1\times10^{-14}e^{-580/T}$	$1.4\times10^{-14}e^{580/T}$	S
S	2.2×10^{-10}	S
$1.6\times10^{-11}e^{-4570/T}$		
S	1.1×10^{-10}	S
S	$1.0\times10^{-11}e^{-2500/T}$	S
1.36×10^{-11}		
9.4×10^{-13}		
3.24×10^{-11}		
$1.0\times10^{-11}e^{-500/T}$	$4.5\times10^{-12}e^{-275/T}$	$4.2\times10^{-12}e^{-242/T}$
$k_0=2.5\times10^{-31}(T/300)^{-0.8}$	S	$6.9\times10^{-31}(T/300)^{-1.0}$
$k_\infty=3.0\times10^{-11}(T/300)^{-1.0}$	S	$1.0\times10^{-11}(T/300)^{-1.0}$
4×10^{-11}	8×10^{-11}	$(7+4P_{atm})\times10^{-11}$
S	3×10^{-12}	$(3.4+2.5P_{atm})\times10^{-14}e^{1150/T}$
$1.2\times10^{-11}e^{-2200/T}$	$7.7\times10^{-12}e^{-2100/T}$	$6.1\times10^{-12}e^{-2030/T}$
S	$2.9\times10^{-12}e^{-160/T}$	$3.1\times10^{-12}e^{-187/T}$
6.2×10^{-11}	7.2×10^{-11}	6.7×10^{-11}
4.8×10^{-11}	4.4×10^{-11}	4.9×10^{-11}
S	$3.8\times10^{-12}e^{-1580/T}$	$2.2\times10^{-12}e^{-1430/T}$
$k_0=1.2\times10^{-31}(T/300)^{-1.8}$	S	S
$k_\infty=3.0\times10^{-11}(T/300)^{0.3}$	3.0×10^{-11}	S
S	S	S
$4.4\times10^{-12}e^{-3220/T}$	S	S
$<10^{-15}$	S	S
3.4×10^{-11}	3.7×10^{-11}	3.4×10^{-11}
S	1.4×10^{-12}	S
S	S	S

TABLE 5.2 (continued)

	Chemical Reactions	CIAP(1975) NBS(1975)	NASA(1977) NBS(1977)
R_{31}	$NO_2+O_3 \rightarrow NO_3+O_2$	$1.1\times10^{-13}e^{-2450/T}$	$1.2\times10^{-13}e^{-2450/T}$
R_{32}	$NO_2+O+M \rightarrow NO_3+M$	1.0×10^{-31}	
R_{33}	$NO_3+NO_2+M \rightarrow N_2O_5+M$	$4.8\times10^{-31}-1.78\times10^{-30}$	
R_{34}	$N_2O_5+M \rightarrow NO_2+NO_3+M$	$1.26\times10^{-25}-1.1\times10^{-21}$	
R_{43}	$NO+HO_2 \rightarrow NO_2+OH$	2.0×10^{-13}	8.0×10^{-12}
R_{37}	$OH+NO_2+M \rightarrow HNO_3+M$	$8.04\times10^{-31}-3.5\times10^{-30}$	$5.5\times10^{-31}-3.5\times10^{-30}$
R_{38}	$HNO_3+OH \rightarrow NO_3+H_2O$	1.3×10^{-13}	8.9×10^{-14}
R_{39}	$OH+NO+M \rightarrow HNO_2+M$	$2.2\times10^{-32}e^{1110/T}$	
R_{41}	$HNO_2+OH \rightarrow NO_2+H_2O$	1.3×10^{-13}	6.6×10^{-12}
R_{40}	$HO_2+NO_2+M \rightarrow HNO_4+M$		
R_{42}	$OH+HNO_4 \rightarrow NO_2+H_2O+O_2$		
R_{35}	$Cl+O_3 \rightarrow ClO+O_2$	1.85×10^{-11}	$2.7\times10^{-11}e^{-257/T}$
R_{36}	$ClO+O \rightarrow Cl+O_2$	5.3×10^{-11}	$7.7\times10^{-11}e^{-130/T}$
R_{44}	$Cl+HO_2 \rightarrow HCl+O_2$		3.0×10^{-11}
	$\rightarrow OH+ClO$		
R_{45}	$Cl+CH_4 \rightarrow HCl+CH_3$	$5.6\times10^{-11}e^{-1790/T}$	$7.3\times10^{-12}e^{-1260/T}$
R_{46}	$Cl+H_2 \rightarrow HCl+H$	$<8\times10^{-16}$	$3.5\times10^{-11}e^{-2290/T}$
R_{47}	$HCl+OH \rightarrow H_2O+Cl$	$2.8\times10^{-12}e^{-400/T}$	$3.0\times10^{-12}e^{-425/T}$
R_{48}	$HCl+O \rightarrow OH+Cl$	$1.74\times10^{-12}e^{-2260/T}$	$1.14\times10^{-11}e^{-3370/T}$
	$HCl+O(^1D) \rightarrow OH+Cl$		1.4×10^{-10}
	$Cl+O_2+M \rightarrow ClOO+M$		1.7×10^{-33}
	$Cl+ClOO \rightarrow Cl_2+O_2$		1.6×10^{-10}
	$Cl+ClOO \rightarrow ClO+ClO$		1.1×10^{-11}
	$Cl+OClO \rightarrow ClO+ClO$		5.9×10^{-11}
	$ClO+ClO \rightarrow Cl+ClOO$	$<2.2\times10^{-15}$	$1.0\times10^{-12}e^{-1238/T}$
	$ClO+ClO \rightarrow Cl_2+O_2$	2.4×10^{-14}	$5.0\times10^{-13}e^{-1238/T}$
R_{49}	$ClO+NO_2+M \rightarrow ClONO_2+M$		$\dfrac{3.3\times10^{-23}T^{-3.34}}{1+8.7\times10^{-9}T^{-0.6}\sqrt{M}}$
R_{50}	$ClONO_2+OH \rightarrow HOCl+NO_3$		$1.2\times10^{-12}e^{-333/T}$
R_{51}	$ClONO_2+O \rightarrow ClO+NO_3$		$3.0\times10^{-12}e^{-808/T}$
	$CH_3Cl+OH \rightarrow CH_2Cl+H_2O$		$2.2\times10^{-12}e^{-1142/T}$
R_{52}	$ClO+HO_2 \rightarrow HOCl+O_2$		2.0×10^{-13}
R_{53}	$HOCl+OH \rightarrow H_2O+ClO$		
	$HOCl+O \rightarrow OH+ClO$		

NASA (1979) FAA (1980)	JPL (1981) WMO (1981)	JPL (1982)
S	S	S
$k_0=9.0\times10^{-32}(T/300)^{-2.0}$	S	S
$k_\infty=2.2\times10^{-11}$	S	S
$k_0=1.4\times10^{-30}(T/300)^{-2.8}$	S	$2.2\times10^{-30}(T/300)^{-2.8}$
$k_\infty=9.0\times10^{-13}(T/300)^{0.7}$	8.0×10^{-13}	1.0×10^{-12}
$8.8\times10^{-6}e^{-9700/T}$		
$4.3\times10^{-12}e^{200/T}$	$3.7\times10^{-12}e^{240/T}$	S
$k_0=2.6\times10^{-30}(T/300)^{-2.9}$	S	S
$k_\infty=2.4\times10^{-11}(T/300)^{-1.3}$	S	S
8.5×10^{-14}	$1.1\times10^{-14}e^{650/T}$	$9.4\times10^{-15}e^{778/T}$
$k_0=6.7\times10^{-31}(T/300)^{-3.3}$	S	$6.7\times10^{-31}(T/300)^{-2.5}$
$k_\infty=3.0\times10^{-11}(T/300)^{-1.0}$	S	1.5×10^{-11}
S		
$k_0=2.1\times10^{-31}(T/300)^{-5.0}$	S	$2.3\times10^{-31}(T/300)^{-4.6}$
$k_\infty=6.5\times10^{-12}(T/300)^{-5.0}$	$6.5\times10^{-12}(T/300)^{-2.0}$	4.2×10^{-12}
5×10^{-13}	4×10^{-12}	$1.3\times10^{-12}e^{380/T}$
$2.8\times10^{-11}e^{-257/T}$	S	S
S	S	S
4.5×10^{-11}	4.8×10^{-11}	$1.8\times10^{-11}e^{170/T}$
		$4.1\times10^{-11}e^{-450/T}$
$9.9\times10^{-12}e^{-1359/T}$	$9.6\times10^{-12}e^{-1350/T}$	S
S	S	$3.7\times10^{-11}e^{-2300/T}$
$2.8\times10^{-12}e^{-425/T}$	S	S
S	S	$1.0\times10^{-11}e^{-3340/T}$
S	S	S
$2.0\times10^{-33}(T/300)^{-1.3}$	S	S
1.0×10^{-10}	1.4×10^{-10}	S
5.0×10^{-12}	8.0×10^{-12}	S
S	S	S
no recommendation	S	S
no recommendation	S	S
$k_0^*=1.6\times10^{-31}(T/300)^{-3.4}$		$1.8\times10^{-31}(T/300)^{-3.4}$
or $3.5\times10^{-32}(T/300)^{-3.8}$	$4.5\times10^{-32}(T/300)^{-3.8}$	S
$k_\infty=1.5\times10^{-11}(T/300)^{-1.9}$	S	S
S	S	S
S	S	S
S	$1.8\times10^{-12}e^{-1112/T}$	S
5.2×10^{-12}	$4.6\times10^{-13}e^{710/T}$	S
$3.0\times10^{-12}e^{-800/T}$	$3.0\times10^{-12}e^{-150/T}$	S
$1.0\times10^{-11}e^{-2200/T}$	S	S

TABLE 5.2 (continued)

	Chemical Reactions	CIAP(1975) NBS(1975)	NASA(1977) NBS(1977)
R_{54}	$ClO+NO \rightarrow NO_2+Cl$		$1.0\times10^{-11}e^{200/T}$
	$ClO+OH \rightarrow Cl+HO_2$		
	$CFCl_3+O(^1D) \rightarrow (2\cdot ClO_x)$		2.3×10^{-10}
	$CF_2Cl_2+O(^1D) \rightarrow (2\cdot ClO_x)$		2.0×10^{-10}
R_{55}	$CH_4+OH \rightarrow CH_3+H_2O$	$2.95\times10^{-12}e^{-1770/T}$	$2.36\times10^{-12}e^{-1710/T}$
R_{56a}	$CH_4+O(^1D) \rightarrow CH_3+OH$	4×10^{-10}	1.3×10^{-10}
R_{56b}	$CH_4+O(^1D) \rightarrow H_2+CH_2O$	4.0×10^{-11}	1.4×10^{-11}
R_{57}	$CH_3+O_2+M \rightarrow CH_3O_2+M$	2.6×10^{-31}	
R_{58}	$CH_3+O \rightarrow CH_2O+H$		
R_{59}	$CH_3O_2+NO \rightarrow CH_3O+NO_2$	$6.6\times10^{-14}e^{-500/T}$	$3.3\times10^{-12}e^{-500/T}$
	$CH_3O_2+NO_2+M \rightarrow CH_3O_2NO_2+M$		
	$CH_3O_2+CH_3O_2 \rightarrow CH_3O_2CH_3+O_2$		4.6×10^{-13}
	$2CH_3O+O_2$		
	$CH_2O+CH_3OH+O_2$		
R_{60}	$CH_3O+O_2 \rightarrow CH_2O+HO_2$	$1.6\times10^{-13}e^{-3300/T}$	S
R_{61}	$CH_2O+OH \rightarrow CHO+H_2O$	1.4×10^{-11}	$3.0\times10^{-11}e^{-250/T}$
R_{62}	$CH_2O+O \rightarrow CHO+OH$	1.6×10^{-13}	$2.0\times10^{-11}e^{-1450/T}$
R_{63}	$CHO+O_2 \rightarrow CO+HO_2$	5.7×10^{-12}	6.0×10^{-12}
R_{64}	$CO+OH \rightarrow CO_2+H$	1.4×10^{-13}	S

@The attached numbers correspond to the reaction numbers in the text.

*Two values are recommended for the low pressure limit rate constant k_0 for R_{49}. This reflects the uncertainty in the status of the molecular structure (isomer) for the product of the reaction.

leading to the recommended values in NASA (1979), JPL (1981, 1982) and WMO (1981). In some cases the recommended value is a simple average or a weighted mean of several qualified experimental data.

Most of the reaction rate constants in the review of WMO (1981) are about the same as those recommended in earlier reviews (NASA, 1979; JPL, 1981). However, for some reactions the recommended values in WMO (1981) are significantly different from previous values; these include faster rates for the reaction R_{15}: $OH+HO_2$ (HOCHANADEL *et al.*, 1980; LII *et al.*, 1980; THRUSH and WILKINSEN, 1981; COX *et al.*, 1981; KEYSER, 1981; SRIDHARAN *et al.*, 1981; KURYLO *et al.*, 1981; DEMORE, 1982; TEMPS and WAGNER, 1982), the reaction R_{38}: $OH+HNO_3$ (WINE *et al.*, 1981; NELSON *et al.*, 1981; KURYLO *et al.*, 1982) and the reaction R_{42}: $OH+HNO_4$ (LITTLEJOHN and JOHNSTON, 1980; BARNES *et al.*, 1981; TREROR *et al.*, 1982), as well as a slower rate for the reaction R_{49}:

NASA (1979) FAA (1980)	JPL (1981) WMO (1981)	JPL (1982)
$7.8\times10^{-12}e^{250/T}$	$6.2\times10^{-12}e^{294/T}$	S
9.1×10^{-12}	S	$5.1\times10^{-12}e^{180/T}$
2.2×10^{-10}	S	2.3×10^{-10}
1.4×10^{-10}	S	S
$2.4\times10^{-12}e^{-1710/T}$	S	S
S	1.4×10^{-10}	S
S	1.5×10^{-11}	1.4×10^{-11}
$k_0=2.2\times10^{-31}(T/300)^{-2.2}$	S	S
$k_\infty=2.0\times10^{-12}(T/300)^{-1.7}$	S	S
1.0×10^{-10}	1.3×10^{-10}	1.4×10^{-10}
7.0×10^{-12}	7.4×10^{-12}	$4.2\times10^{-12}e^{180/T}$
$k_0=2.0\times10^{-30}(T/300)^{-4.0}$	$1.5\times10^{-30}(T/300)^{-4.0}$	S
$k_\infty=5.0\times10^{-13}(T/300)^{-4.0}$	$6.5\times10^{-12}(T/300)^{-2.0}$	S
3.6×10^{-13}	S	$1.6\times10^{-13}e^{220/T}$
$5.0\times10^{-13}e^{-2000/T}$	$1.0\times10^{-12}e^{-2050/T}$	$1.2\times10^{-13}e^{-1350/T}$
1.0×10^{-11}	S	S
$3.2\times10^{-11}e^{-1550/T}$	$3.0\times10^{-11}e^{-1550/T}$	S
5.0×10^{-12}	S	$3.5\times10^{-12}e^{140/T}$
$1.35\times10^{-13}(1+\text{Patm})$	S	S

$ClO+NO_2+M$(KNAUTH, 1978). Two values had been recommended for k_{49}; one, the higher rate constant, assumes all products go to $ClONO_2$; the other, the lower rate constant, assumes about 2/3 of the overall reaction forms an unknown *isomer* (having the same constituent elements but different molecular structure). WMO (1981) recommends using the lower value for the purpose of model calculations. However, the error limit is significantly large to encompass the higher value.

The reaction rate constants used in the model calculations in this book were taken mostly from WMO (1981) and JPL (1981). For some reactions newer values have become available since then as are given as JPL (1982) in Table 5.2. One significant difference in the rate constants in this new set from previous values is a faster rate of a factor of two for the reaction R_6: $O+HO_2$ at stratospheric temperatures. Adopting the faster rate constant for this reaction, however, KO and SZE (1983) have found that the discrepancy between the predicted ozone density and observations in the upper stratosphere becomes even larger; the earlier model predicted smaller ozone densities than observed values at 45–50 km (see Fig. 5.2).

Chapter 6

PHOTOCHEMICAL TIME CONSTANTS

As was discussed in Section 2.3.2 the distribution of minor constituent is controlled mainly by photochemistry, if the photochemical life time is shorter than the characteristic transport time. The photochemical time constant is essential for determining whether a constituent is controlled mainly by photochemistry or by dynamics.

There are many photochemically active species in the middle atmosphere that are strongly coupled with each other by rapid interchange reactions. They usually behave as a set called a *family*. While each member of the family can change rapidly to other forms of constituents within the same family, the family as a whole will change with a much slower rate, since the fast exchange reactions will not affect the total concentration of the family. Thus, photochemical time constants for families are much longer than those for individual constituents within the family; therefore families are usually more strongly subjected to transport effects.

Photochemical time constants are often used explicitly or implicitly in discussions of variations in middle atmosphere minor constituents. For instance, all of the following occur with different time scales: variations at sunrise, sunset, and during solar eclipses; and changes in ozone concentration caused by nuclear explosions, solar proton events, and human activities such as NO_x emission from supersonic transports (SST) and Cl_x emission from chlorofluoromethanes (CFM). A knowledge of the photochemical time constants of the relevant constituents and their families is essential for analysing these variations.

6.1 Mathematical Formulations

A simple but most commonly used measure for the chemical time constant is a reciprocal of the loss coefficient (see (2.26)). If the chemical loss term is expressed by a linear relation to the concentration and the proportional constant is assumed to be β, the time constant may be expressed by

$$\tau_{e-\text{fold}} = 1/\beta. \tag{6.1}$$

Since the number density decreases by a factor of e after this time, it is called the *e-fold time constant*. Another widely used measure of the time constant is the *half life time*, which is the time required for the number density to be reduced to one half of the initial value; it is given by

$$\tau_{\text{half}-\text{life}} = \ln(2)/\beta \doteqdot 0.693/\beta. \tag{6.2}$$

In general, however, the change in concentration is governed by the chemical rate equation, which includes the production and nonlinear loss terms as well as the linear loss terms. Thus, the temporal variation in concentration n is given by

$$\frac{\partial n}{\partial t} = Q - \beta n - \alpha n^2 \tag{6.3}$$

where Q represents the total production rate, and the second and third terms of the right-hand side are the total linear and quadratic loss terms, respectively. Most individual constituents have linear loss terms (β-type), but for families, the significant loss appears as terms in quadratic form (α-type) in most cases as is discussed in Section 6.3.

The general solution of Eq. (6.3) is given by (see Appendix F)

$$n = -\frac{\beta}{2\alpha} + \left(n_e + \frac{\beta}{2\alpha}\right) \frac{n_0 + \dfrac{\beta}{2\alpha} + \left(n_e + \dfrac{\beta}{2\alpha}\right)\tanh \gamma t}{n_e + \dfrac{\beta}{2\alpha} + \left(n_0 + \dfrac{\beta}{2\alpha}\right)\tanh \gamma t} \tag{6.4}$$

where n_0 is the initial value, n_e is the equilibrium (steady state) value obtained by setting the right-hand side of (6.3) to zero:

$$n_e = \frac{-\beta + \sqrt{\beta^2 + 4\alpha Q}}{2\alpha} \tag{6.5}$$

and γ in (6.4) is given by

$$\gamma = \sqrt{\beta^2 + 4\alpha Q/2.} \tag{6.6}$$

Figure 6.1 illustrates the variation of n/n_e against γt calculated from (6.4) for various initial values of n_0/n_e between 0 and 2 and for different values of the parameter

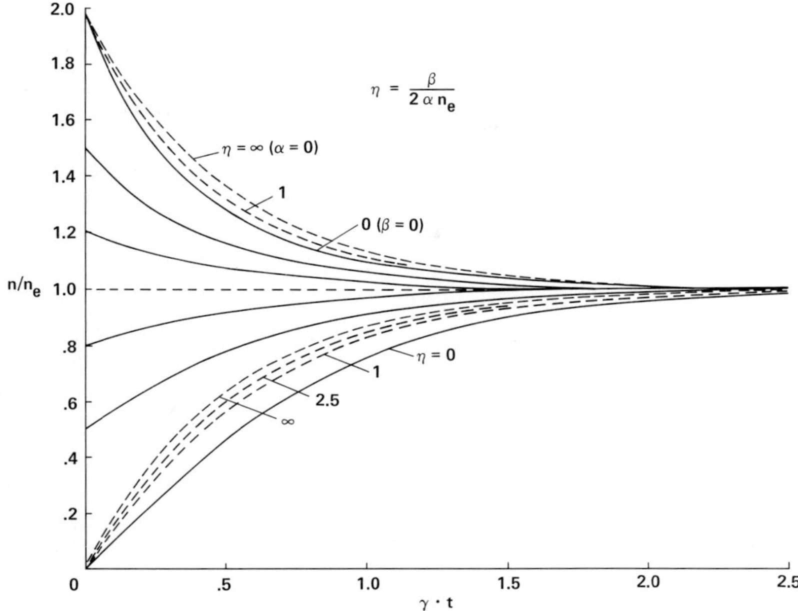

FIG. 6.1. Variation in n/n_e against γt calculated from Eq. (6.4).

$$\eta = \frac{\beta}{2\alpha n_e}. \tag{6.7}$$

In case of $\alpha=0$, the above Eqs. (6.4) – (6.6) reduce to

$$n = n_e + (n_0 - n_e)\,\mathrm{e}^{-\beta t} \tag{6.8}$$

$$n_e = Q/\beta, \quad \gamma = \beta/2 \quad \text{and} \quad \eta = \infty \tag{6.9}$$

and if $\beta=0$, we obtain the following equations:

$$n = n_e \frac{n_e + n_0 - (n_e - n_0)\,\mathrm{e}^{-2\sqrt{\alpha Q}\,t}}{n_e + n_0 + (n_e - n_0)\,\mathrm{e}^{-2\sqrt{\alpha Q}\,t}} \tag{6.10}$$

$$n_e = \sqrt{\frac{Q}{\alpha}}, \quad \gamma = \sqrt{\alpha Q} \quad \text{and} \quad \eta = 0. \tag{6.11}$$

The curves $\eta=\infty$ and $\eta=0$ in Fig. 6.1 indicate these two extreme cases. The curve $\eta=1$ corresponds to the case of 3 $\beta^2 \sim 4\alpha Q$.

From (6.4) one can calculate the time required for n to change from the initial value $\varepsilon_1 \cdot n_e$ to the end value $\varepsilon_2 \cdot n_e$ as follows:

$$\tau(\varepsilon_1,\varepsilon_2) = 2f_\varepsilon/\gamma$$
$$= \tanh^{-1}\left[\frac{\varepsilon_2-\varepsilon_1}{1-\varepsilon_1\cdot\varepsilon_2+\dfrac{\eta}{1+\eta}(1-\varepsilon_1)(1-\varepsilon_2)}\right]\Big/\gamma. \tag{6.12}$$

The factor f_ε is plotted in Fig. 6.2 as functions of ε_1 and ε_2. The intersection of each curve with the abscissa gives the value of ε_1 (initial value) for that curve. Figure 6.2(a) illustrates cases of $\varepsilon_1=0$ and 0.5, whereas Fig. 6.2(b) illustrates cases of $\varepsilon_1=1.5$ and 2.0.

NICOLET (1972) and BANKS and KOCKARTS (1973) have chosen the value $f_\varepsilon=0.275$ to define the chemical time constant. If $\eta=0$ (or $\beta=0$) this value of f_ε gives the time for n to change from 0 to 50% of n_e, from 50% to 80% of n_e, or from twice to 1.25 times n_e. It is evident, however, that the value of f_ε giving the change in n from 0 to 50% of n_e depends upon the value of η; in fact, if $\eta=\infty$ (or $\alpha=0$) the corresponding value of f_ε should be $\sim$ 0.173.

We now extend the definition of the chemical time constant to cases with any value of η as follows:

$$\tau_{1/2} = \tau(0,0.5) = 2\tanh^{-1}\left(\frac{2\gamma}{4\gamma+\beta}\right)\Big/\sqrt{\beta^2+4\alpha Q}$$
$$= \ln\left(\frac{\beta+6\gamma}{\beta+2\gamma}\right)\Big/\sqrt{\beta^2+4\alpha Q}. \tag{6.13}$$

In the special case of $\eta=0$ (or $\beta=0$), (6.13) is reduced to

$$\tau_{1/2}(\eta=0) = \tanh^{-1}\left(\frac{1}{2}\right)\Big/\sqrt{\alpha Q} \cong \frac{0.55}{\sqrt{\alpha Q}} \tag{6.14}$$

which agrees with the formula of NICOLET (1972) and BANKS and KOCKARTS (1973). If $\eta=\infty$ (or $\alpha=0$), (6.13) is reduced to a formula consistent with (6.2).

If $Q=0$ there is no equilibrium value for n. In this case the following form can be derived from (6.4) as a limiting case of $n_e \to 0$, or by solving (6.3) directly, uner the condition of $Q=0$:

$$n = n_0\,\mathrm{e}^{-\beta t}\Big/\left[1+\frac{\alpha n_0}{\beta}(1-\mathrm{e}^{-\beta t})\right]. \tag{6.15}$$

The time required for n to decrease to one-half the initial value n_0 is calculated by

$$\tau_{1/2}(Q=0) = \ln\left(\frac{2\beta+\alpha n_0}{\beta+\alpha n_0}\right)\Big/\beta. \tag{6.16}$$

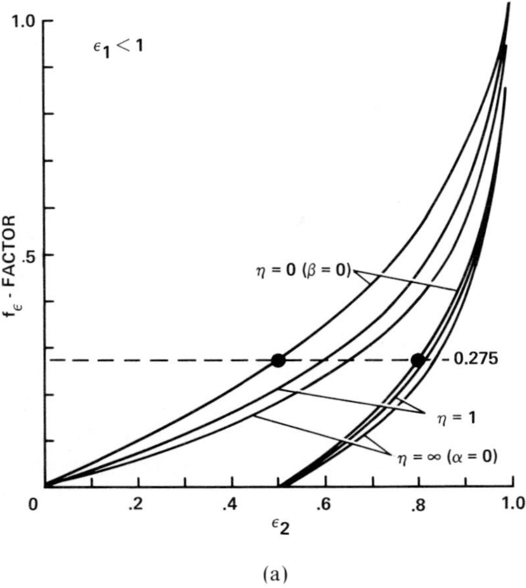

(a)

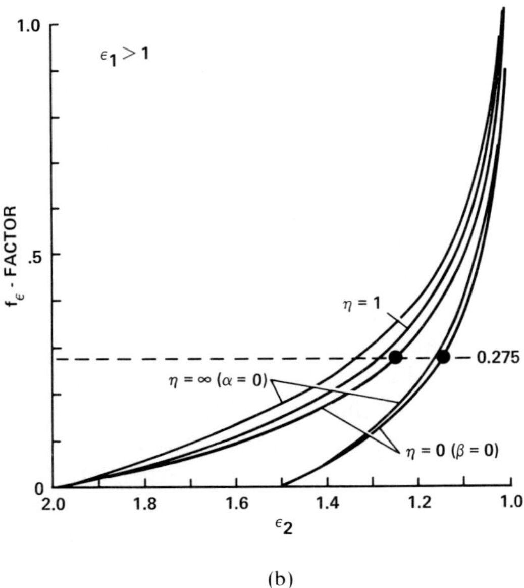

(b)

FIG. 6.2. Variations in f_ε against ε_2: (a) for the cases of $\varepsilon_1 = 0$ and 0.5 and (b) for the cases of $\varepsilon_1 = 1.5$ and 2.0.

If $\alpha=0$, (6.16) again gives a result consistent with (6.2).

If both Q and β are equal to zero, (6.16) still gives the time constant as a limiting case of $\beta \to 0$, but it is more appropriate to consider the direct solution of (6.3) given by

$$n = \frac{n_0}{1 + \alpha n_0 t} \tag{6.17}$$

from which we calculate the time constant

$$\tau_{1/2}(Q = 0, \beta = 0) = 1/\alpha n_0. \tag{6.18}$$

It may be concluded from the above discussion that it is appropriate to define the photochemical time constant by (6.13) in most general cases, by (6.16) if $Q=0$, and by (6.18) if both Q and β are equal to zero. The time constant defined here gives the time required for n to increase from 0 to one-half the equilibrium value. If $Q=0$, however, the time constant is the time for n to be reduced to half of the initial value; in case of $Q=0$ and $\alpha \neq 0$ the time constant depends upon the initial value of n.

6.2 *Time Constants for Individual Constituents*

Photochemical time constants depend upon concentrations of the reactants of the chemical reactions as well as the reaction rate constants and photolysis coefficients. Observational data on concentrations of the middle atmosphere constituents are limited, and complete height profiles are not available for most constituents. Thus, for the purpose of present study we use concentration profiles at noon and midnight calculated in a time-dependent model using the reaction rate constants of WMO (1981) and JPL (1981) (see Chapter 7; concentrations are given in Table 7.3).

Once dissociation coefficients, reaction rate constants, and concentrations of the reactants are known, calculations of the production term and loss coefficients (α, β), and therefore all quantities in (6.13) are straightforward for individual constituents. Strictly speaking, the time constants defined here indicate the speed of change in n for the short time period for which Q, α, and β do not change significantly. Special care must be taken for those periods such as near sunrise and sunset, when large and rapid changes occur for these parameters. However, the time constants of larger than around a day can be calculated properly by considering the representative values for daytime and at night.

The time constants for the pure oxygen constituents are illustrated in Fig. 6.3(a). The daytime time constant for $O(^1D)$ is very short ($\ll 10^{-3}$ sec) at all heights; therefore, $O(^1D)$ is always in photochemical equilibrium. The nighttime time constant is undetermined since there is essentially no $O(^1D)$

at night. The time constant of $O(^3P)$ is about the same during the daytime and at night; it is very small ($\sim 10^{-5}$ sec) near the surface but increases rapidly with height reaching the order of a month at 100 km. This increase is caused by a rapid decrease in concentrations of O_2 and M (mostly N_2), which serve as reactants in the main loss reaction for $O(^3P)$. The concentration of $O(^3P)$ decreases very rapidly after sunset below $\sim$80 km, where the time constant becomes less than $\sim$1 hour. Above this height, the $O(^3P)$ concentration remains almost constant throughout the day because of a long photochemical time constant.

The daytime time constant of O_3 is of the order of one hour in the stratosphere and of the order of a minute in the mesosphere, whereas the nighttime time constant is of the order of a year below $\sim$70 km, above which it decreases rapidly. It must be noted, however, that the small O_3 time constant during the daytime represents the time scale for O_3 to decompose by photolysis; actually, O_3 can be reproduced very quickly from the photolysis products, atomic oxygen, as would be expected from the very short time constant of $O(^3P)$ in the stratosphere and mesosphere. Thus, the time constant of O_3 shown in Fig. 6.3(a), particularly those during the daytime do not represent the time scale by which the ozone density actually changes. Later, we will discuss the importance of considering the odd-oxygen family in this respect.

Figures 6.3(b), (c) and (d) illustrate noon and midnight time constants for nitrogen-oxygen compounds, hydrogen-oxygen and hydrocarbon compounds, and chlorine compounds, respectively. The time constant changes from constituent to constituent and also changes with altitude in a very wide range indeed, that is, from a very small fraction of a second to a period much longer than 100 years. The concentrations of constituents having time constants much longer than a day do not change appreciably throughout the day, whereas the constituents having time constants much smaller than an hour generally keep almost constant concentrations, although they could be different between day and night. Some constituents have time constants of the order of an hour, and concentrations of these constituents undergo large diurnal variations; they include N_2O_5, H_2O_2, and $ClONO_2$.

The time constants for parent (or primary) molecules such as H_2O, N_2O, CO_2, and chlorofluoromethanes ($CFCl_3$, CF_2Cl_2) are particularly large (> 100 years) in the troposphere and lower stratosphere (nighttime time constants are essentially infinite), but they decrease markedly in the middle and upper stratosphere. This implies that these molecules can be decomposed effectively only in and above the middle stratosphere.

As one approaches the surfact the time constants of HNO_3, HCl, HNO_4 (night), and H_2O_2 (night) decreases rapidly to the values of 10^5–10^6 sec.

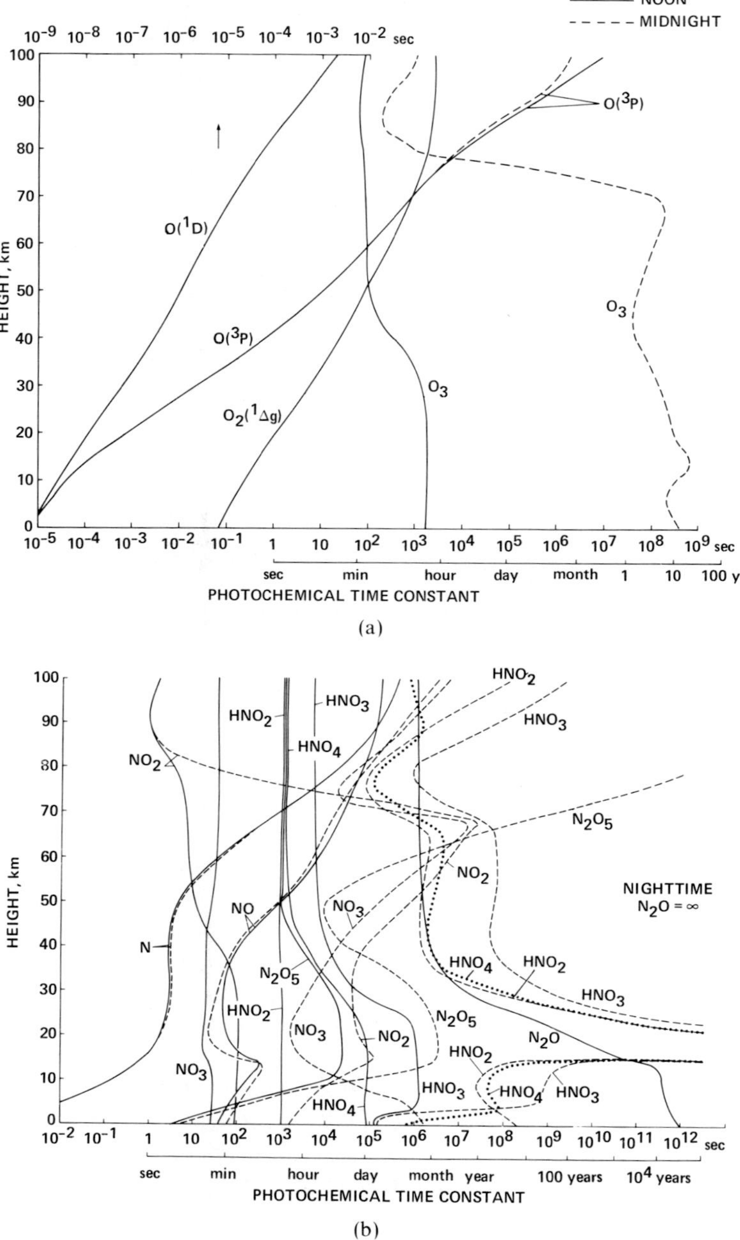

FIG. 6.3. (a) Height profiles of photochemical time constants at noon and midnight for pure oxygen constituents. (b) Height profiles of photochemical time constants at noon and midnight for nitrogen-oxygen compounds.

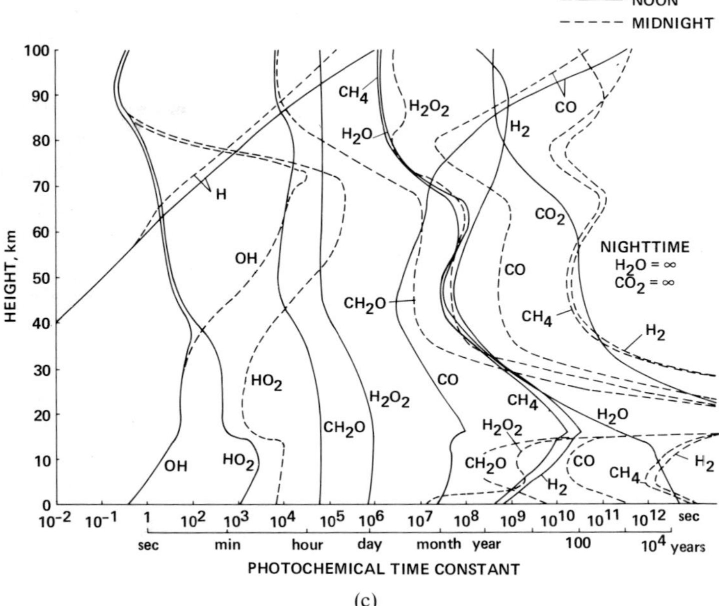

(c)

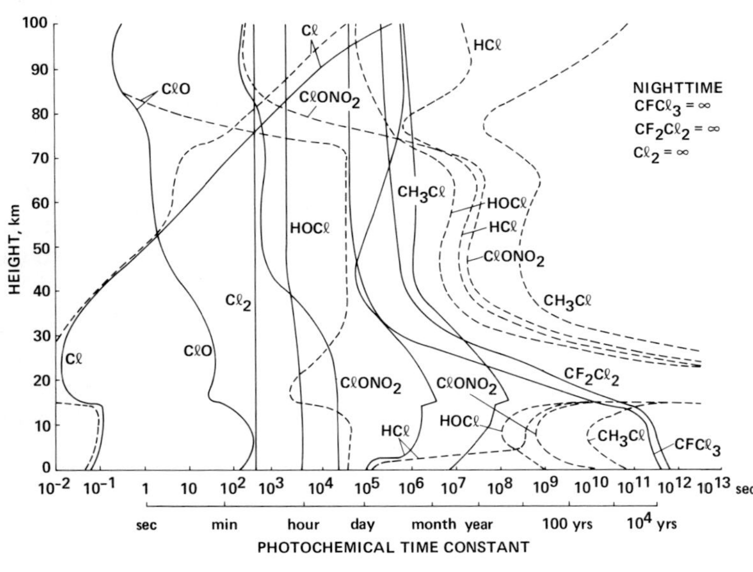

(d)

FIG. 6.3. (c) The same as Fig. 6.3(a), but for hydrogen-oxygen compounds and hydrocarbons. (d) The same as Fig. 6.3(a), but for chlorine-oxygen compounds and chlorofluoromethanes.

This reflects the scavenging effects in the troposphere. The model assumes the washout and rainout coefficients to be 5×10^{-6} sec^{-1} for HNO_3 and HCl, 10^{-6} sec^{-1} for HNO_4, and 5×10^{-7} sec^{-1} for H_2O_2 below 3 km (CRUTZEN *et al.*, 1978). Above 3 km the coefficient is assumed to decrease with height up to the tropopause with a scale height of 3.3 km. The model also assumed scavenging coefficients for other molecules including NO, NO_2, and ClO, but the effects on these species and on H_2O_2 (day) and HNO_4 (day) are not visible in Fig. 6.3(b)–(d), since they have much smaller photochemical time constants than the time constants due to scavenging.

Time constants for some constituents decrease with decreasing height over the entire troposphere; they include N, NO, NO_2 (night), NO_3 (night), N_2O_5, OH, HO_2, Cl, CO, and CH_3Cl. A detailed inspection reveals that the main loss reactions for these constituents at this height range have rather large activation energies. The substantial increase of the atmospheric temperature from tropopause to surface should cause a large increase in the reaction rate constants for these reactions and, therefore, a decrease in the photochemical time constants.

6.3 *Time Constants for Families*

If there are fast exchange photochemical reactions among some constituents, they can be transported as a group without major changes in their concentrations. Within that group (family) relative abundances among constituents are kept at almost constant values determined by their exchange rates. Thus, it is the time constant of the family that really controls the changes in concentration of an individual constituent in the family.

A special technique is required for calculating loss coefficients for the family. Since a member constituent (n_i) and the main constituent (n_m) of the family is exchanged by fast reactions, the ratio between the two constituents

$$r_i = n_i/n_m \tag{6.19}$$

can be defined as functions of the rate constants and concentrations of the reactants of the fast exchange reactions. The total concentration of the family is then given by

$$n_0 = n_m + \sum_s n_s = n_m\left(1 + \sum_s r_s\right) \tag{6.20}$$

where the summation of s should be taken for all members of the family which have exchange reactions with n_m. n_0 should change little by exchange reactions, but would change largely by slower nonexchange reactions that cause an exit from or an entrance into the family.

From (6.19) and (6.20) we have a relationship between n_i and n_0 as follows:

$$n_i = r_i \cdot n_m = \frac{r_i}{1 + \sum_s r_s} n_0 = f_i \cdot n_0. \tag{6.21}$$

Thus, the reaction rate of a nonexchange reaction between the ith and jth constituents can be written in terms of n_0 either by

$$k_{ij} n_i n_j = (k_{ij} f_i n_j) \cdot n_0 \tag{6.22}$$

if the jth constituent is not a member of the family, or by

$$k_{ij} n_i n_j = (k_{ij} f_i f_j) \cdot n_0^2 \tag{6.23}$$

if the jth constituent is also a member of the family. The coefficients of n_0 and n_0^2 in (6.22) and (6.23) give β and α, respectively, for the loss coefficients of the family concentration.

We have shown that a loss reaction for n_i can be either of the quadratic (α-type) or linear (β-type) form depending upon the assumption of whether the reactant (n_j) is a member of the family or not. For example, the reaction NO_2+O, which is an important reaction for odd-oxygen loss in the stratosphere, is a β-type loss process for odd oxygen ($O+O_3$), but it becomes an α-type loss mechanism if NO_2 is also a member of odd oxygen ($O+O_3+NO_2$), since both species in the reaction are now members of the family. For convenience, in later discussions we will use the notation NO_x-(β) and $NO_x(\alpha)$ to represent the afore-mentioned two cases, respectively, for the effect of reactions between NO_x and O_x on the loss of odd-oxygen concentration.

We now examine how the family time constant will be changed by including various candidate species in the family. Since the family time constants are determined by slower non-exchange reactions, they should be much larger than the time constants of exchange reactions. If inclusion of a constituent in a family does not make the family time constant larger than the exchange time between that constituent and other members of the family, that constituent is not qualified to be a member of the family. In the following we will show and discuss the effects of candidate constituents on the family time constants of various families:

6.3.1 O_x family

Let us first consider pure oxygen chemistry and define the odd-oxygen family as the sum of O and O_3. The predicted time constant for O_x at noon is shown by the solid curve (a) in Fig. 6.4. It is much larger than the time constants of the individual constituents O and O_3, particularly in the stratosphere and troposphere. The O_x time constant is less than a day in the upper stratosphere, increases with decreasing height reaching ~10 years in the lower stratosphere, and continues to increase descending into the troposphere.

Photochemistry of nitrogen-oxygen (NO_x) compounds is known to have a serious effect on pure oxygen chemistry. Consider the effects of the three most significant reactions (NO_2 photolysis and reactions (R_{23}) and (R_{25})). We predict the time constant for O_x ($=O+O_3$) to be shown by a curve NO_x (β) in Fig. 6.4, since the effect appears as the linear loss term in $[O_x]$ in this case. Whereas the $NO_x(\beta)$ chemistry does not significantly change the O_x time constant from the value for the pure oxygen chemistry in and above the middle stratosphere, it reduces the O_x time constant significantly in the lower stratosphere and troposphere. In particular, it is of the order of 10^5 sec in the lower stratosphere and is much smaller than the characteristic time of transport by vertical eddy mixing, which is shown in Fig. 6.4 by a heavy dashed curve labeled $1/2(H^2/k)$, where k is the eddy diffusion coeffi-

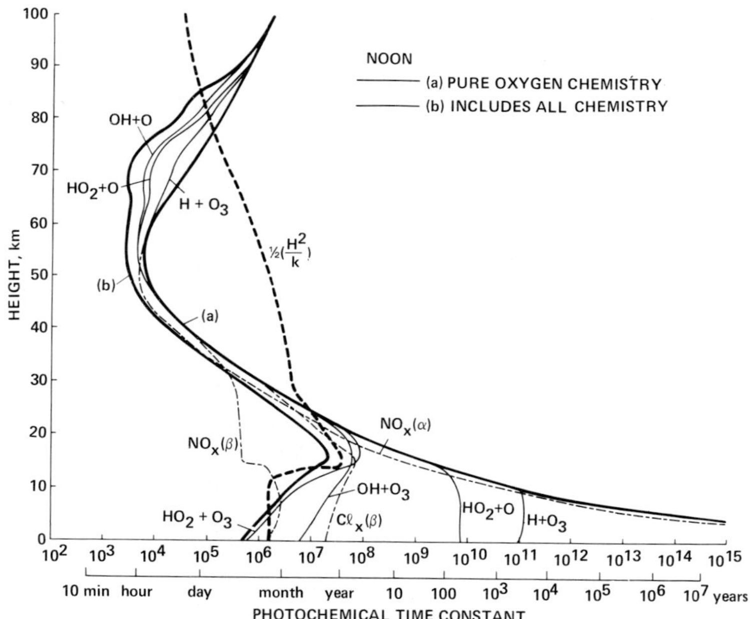

FIG. 6.4. Height profiles of photochemical time constants for odd oxgen (O_x) families at noon, due to major reactions affecting ozone densities. Heavy broken curve indicates the dynamical time constant due to the vertical eddy mixing.

cient and H is the atmospheric scale height. The factor 1/2 is introduced in order to compare the diurnally averaged transport time with the photochemical time constant at noon. The latter is about half the diurnal average at the equinoxes; however, the exact reduction factor depends upon the season (the relative length of day and night) and is also affected by detailed diurnal variations.

The time constant of transport described by vertical eddy diffusion may not represent the time constant of actual transport processes, particularly in the lower stratosphere and troposphere. A simple consideration of large scale hemispheric motion from equator to poles with a mean horizontal velocity of 1 m/sec would give us a transport time of $\sim 10^7$ sec ($\sim$4 months). Using the latitudinal profile of the observed meriodional velocity (OORT and RASMUSSEN, 1970), one may calculate a global transport time of about 1–3 months in the lower stratosphere. The characteristic time scale for planetary motion system in the troposphere would be of the order of at least of a week (VAN MIEGHEM, 1978). Thus, the O_x time constant calculated by the $NO_x(\beta)$ chemistry is generally small compared with the transport time of global scale motions in the lower stratosphere. Then, the ozone concentration should have been controlled by chemistry rather than dynamics. This is in conflict with the well-established knowledge that the distribution of ozone in the lower stratosphere is controlled mainly by dynamics.

The problem is solved by considering NO_2 as a member of the O_x family. This is reasonable since NO_2 has rapid exchange reactions with O and O_3 by

$$NO_2 \xrightarrow{h\nu} O \xrightarrow{O_2+M} O_3 \overset{NO}{\dashrightarrow} NO_2.$$

A dashed arrow indicates the rate-determining reaction, which is the slowest reaction of the series; however, it should be the fastest reaction among parallel reactions if there is more than one route for the same conversion reaction (converting O_3 to NO_2 in this case).

The time constant predicted for the family O_x $(=O+O_3+NO_2)$ is shown by the notation $NO_x(\alpha)$ in Fig. 6.4. The effect of the $NO_x(\alpha)$ chemistry occurs mainly through a nonexchange reaction (R_{25}) $NO_2+O \rightarrow NO+O_2$, which is much slower than the exchange reaction (R_{23}) $NO+O_3 \rightarrow NO_2+O_2$. The latter reaction played an essential role in the $NO_x(\beta)$ chemistry, but now it does not affect the family since it removes O_3 but produces another family member: NO_2. Thus, the family time constant due to $NO_x(\alpha)$ chemistry is much larger than that due to $NO_x(\beta)$ chemistry in the lower stratosphere and troposphere. It is only slightly less than the time constant for pure oxygen chemistry (curve (a)) throughout the stratosphere and troposphere. A difference of a factor of $\sim$2 in the middle and upper stratosphere corresponds to the effect of the catalytic destruction of O_3 by the NO_x chemistry. This effect reduces the ozone concentration at these heights from the larger values predicted in the pure oxygen chemistry to values closer to those observed. It must be noted that such an effect cannot be predicted by the $NO_x(\beta)$ chemistry. Moreover, the $NO_x(\beta)$ chemistry would make the O_x time constant too small in the lower stratosphere and troposphere. Thus, it is essential to include NO_2 in the O_x family in order to correctly predict the effects of NO_x chemistry on the ozone distribution. This is also reasonable if one regards NO_2 as an oxygen atom attached to an NO molecule, just as O_3 can be thought of as an oxygen atom attached to an O_2 molecule.

Figure 6.4 also illustrates the effects of other chemical reactions that influence ozone. The Cl_x chemistry has an effect similar to the NO_x chemistry, but the magnitude of the effect is much smaller. The effect of $Cl_x(\alpha)$ is negligibly small; the curve $Cl_x(\alpha)$ is too close to the curve (a) to show the difference in the figure, whereas the effect of $Cl_x(\beta)$ is more than an order of magnitude less than that of $NO_x(\beta)$. The effect of Cl_x chemistry, however, should be enhanced if chlorine-bearing species such as freons increase in the atmosphere.

Hydroperoxyl (HO_2) has exchange reactions with O_3 through the sequence

$$HO_2 \xrightarrow{NO} NO_2 \xrightarrow{h\nu} O \xrightarrow{O_2 + M} O_3 \xrightarrow{OH} HO_2.$$

Thus, HO_2 has some odd-oxygen character, which might be denoted by the symbol OH–O. However, fast nonexchange reactions such as (R_{15}) $HO_2 + OH \rightarrow H_2O + O_2$ emphasize the even-oxygen character which might be denoted by the symbol $H{-}O_2$. Since the reactions for which $H{-}O_2$ dominates are much faster than those associated with OH–O, one can ignore the O_x character of HO_2. Actually, if HO_2 is included in O_x, the family time constant becomes too small ($\sim 1.5 \times 10^4$ sec at 20 km), due to the fast reaction (R_{15}) previously discussed.

It is evident from Fig. 6.4 that the reactions $OH + O_3$ and $HO_2 + O_3$ reduce the O_x time constant markedly below ~20 km, whereas in and above the mesosphere the reactions $OH + O$, $HO_2 + O$, and $H + O_3$ reduce the O_x time constant significantly. The general feature of Fig. 6.4 indicates that the O_x time constant is much smaller than the transport time constant between ~25 and ~80 km, and the O_x density is controlled mainly by photochemistry at these heights. Above ~80 km the transport effects, including those of molecular diffusion, dominate the chemical effects, and below ~25 km the transport effects *only slightly* exceed the chemical effects because of the fast reactions of HO_x with O_3.

We now investigate more extensively the effect of the definition of family members on the O_x time constant by taking into account *all* chemical reactions in Table 5.2. A solid curve ① in Fig. 6.5 indicates the time constant for $O+O_3$ (for simplicity, notation O is used here to indicate both $O(^3P)$ and $O(^1D)$). This family time constant is always larger than the exchange times between O and O_3, which are shown by dashed curves labeled as $O(^3P)$ or $O(^1D)$.

The family time constant for $O_x(=O+O_3+NO_2)$ is shown by curve ②, which indicates much larger values than curve ① in the middle and lower stratosphere. Inclusion of ClO in O_x does not greatly change the O_x time constant, as is seen by curve ③. This is mainly because of reaction (R_{54}) $ClO+NO \rightarrow NO_2+Cl$, which is faster than the reaction ClO+O below ~ 40 km and causes a rapid loss of ClO. If both NO_2 and ClO are members of O_x, however, reaction (R_{54}) does not affect the O_x loss, since it removes ClO but produces NO_2; therefore, the O_x time constant becomes slightly larger (compare curves ② and ④ in Fig. 6.5).

The exchange time of NO_2 with O_3 is shown by a dash-dot curve labeled NO_2 in Fig. 6.5. The time constant for $O_x(=O+O_3+NO_2)$ is larger than this exchange time at an altitude range of ~10–30 km. Thus, inclusion of NO_2 in the O_x family is justified in the middle and lower stratosphere. Other possible candidate constituents for the O_x family and their exchange

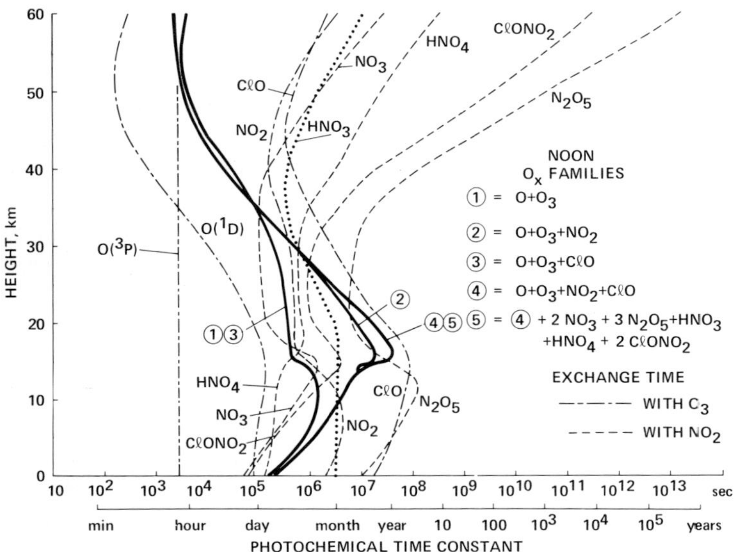

FIG. 6.5. Height profiles of photochemical time constants for various O_x families at noon, compared with exchange times among family members.

reactions with O_3 or NO_2 include the following:

$$NO_3: \quad NO_3 \xrightarrow{h\nu} NO_2 \left(\xrightarrow{h\nu} O \xrightarrow{O_2 + M} O_3 \overset{NO}{\underset{<40\ \mathrm{km}}{\dashrightarrow}} NO_2 \right) \overset{O_3}{\underset{>40\ \mathrm{km}}{\dashrightarrow}} NO_3;$$

$$N_2O_5: \quad N_2O_5 \xrightarrow{h\nu,\, M} NO_2 \left(\xrightarrow{h\nu} O \xrightarrow{O_2 + M} O_3 \xrightarrow{NO} NO_2 \right) \overset{NO_3 + M}{\dashrightarrow} N_2O_5;$$

$$HNO_3: \quad HNO_3 \overset{h\nu}{\underset{<20\ \mathrm{km}}{\dashrightarrow}} NO_2 \left(\xrightarrow{h\nu} O \xrightarrow{O_2 + M} O_3 \overset{NO}{\underset{20-35\ \mathrm{km}}{\dashrightarrow}} NO_2 \right)$$

$$\overset{OH + M}{\underset{>35\ \mathrm{km}}{\dashrightarrow}} HNO_3;$$

$$HNO_4: \quad HNO_4 \xrightarrow{h\nu} NO_3 \xrightarrow{h\nu} NO_2 \left(\xrightarrow{h\nu} O \xrightarrow{O_2 + M} O_3 \overset{NO}{\underset{<20\ \mathrm{km}}{\dashrightarrow}} NO_2 \right)$$

$$\overset{HO_2 + M}{\underset{>20\ \mathrm{km}}{\dashrightarrow}} HNO_4;$$

$$ClO: \quad ClO \xrightarrow{NO,\, h\nu} O \xrightarrow{O_2 + M} O_3 \overset{Cl}{\dashrightarrow} ClO; \quad \text{and}$$

$$ClONO_2: \quad ClONO_2 \xrightarrow{h\nu} NO_3 \xrightarrow{h\nu} NO_2 \left(\xrightarrow{h\nu} O \xrightarrow{O_2 + M} O_3 \overset{NO}{\underset{<15\ \mathrm{km}}{\dashrightarrow}} NO_2 \right)$$

$$\overset{ClO + M}{\underset{>15\ \mathrm{km}}{\dashrightarrow}} ClONO_2.$$

If all candidate constituents are included in O_x, we find that the family time constant ⑤ is about the same as curve ④.

Comparing the family time constant with the exchange time of each constituent with O_3 or NO_2 (illustrated by dash or dash-dot curves), we may argue that it is appropriate to include NO_2, NO_3, HNO_3, HNO_4, and $ClONO_2$ in the O_x family at least in the middle and lower stratosphere. The exchange time of N_2O_5 with NO_3 should be much smaller than with NO_2; therefore, N_2O_5 could be a family member if NO_3 is already in the O_x family. Similarly, if NO_2 and $ClONO_2$ are already in the family, the exchange time of ClO with family members by

$$ClO \xrightarrow{NO} NO_2 \xrightarrow{ClO + M} ClONO_2 \xrightarrow{h\nu} Cl \xrightarrow{O_3} ClO$$

is much smaller and it is reasonable to include ClO in the family. The above chemical reaction series seems to destroy two family members (ClO and O_3), but the photolysis of $ClONO_2$ produces NO_3 from which two family members (NO_2 and O) can be produced. Thus, there is no change in the total concentration of the family by the above exchange reactions.

The foregoing discussion should be influenced by the time scale of the phenomena under consideration. The conclusion may be justified for variations on a time scale longer than the order of a month. For calculations with a steady state model, however, it is appropriate to include all candidate constituents in the O_x family, since essentially infinite time is required to reach the steady state. For shorter time scale variations such as diurnal variation, it may be sufficient to consider O_x as the sum of only O and O_3 since the exchange times of other constituents with O_3 are all much larger than a day.

6.3.2 *NO_x family*

Photochemical time constants of NO and NO_2 are both of the order of a minute in the stratosphere, as is seen in Fig. 6.3(b). These time constants represent the exchange times between the two constituents. Taking into account the effects of other slower reactions, we calculate the family time constant for NO_x ($=NO+NO_2$), as is shown by curve ① in Fig. 6.6. Since this time constant is much larger than the exchange time between NO and NO_2 it is necessary to consider the sum of NO and NO_2 for variations of species on a time scale greater than the order of a minute.

The exchange reactions of other candidate constituents for NO_x with NO or NO_2 are as follows:

$$NO_3: \quad NO_3 \xrightarrow{h\nu} NO_2 \xrightarrow{O_3} NO_3;$$

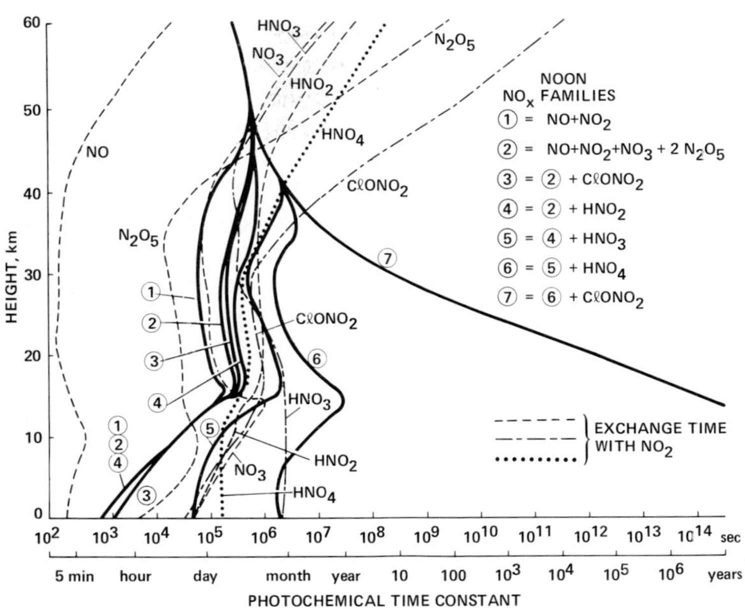

FIG. 6.6. The same as Fig. 6.5 but for NO_x families.

$$N_2O_5: \quad N_2O_5 \underset{<15\,\text{km}}{\overset{h\nu,\,M}{\dashrightarrow}} NO_2 \underset{>15\,\text{km}}{\overset{NO_3+M}{\dashrightarrow}} N_2O_5;$$

$$HNO_2: \quad HNO_2 \overset{h\nu}{\longrightarrow} NO \overset{OH+M}{\dashrightarrow} HNO_2;$$

$$HNO_3: \quad HNO_3 \underset{<25\,\text{km}}{\overset{h\nu}{\dashrightarrow}} NO_2 \underset{>25\,\text{km}}{\overset{OH+M}{\dashrightarrow}} HNO_3;$$

$$HNO_4: \quad HNO_4 \underset{<15\,\text{km}}{\overset{h\nu}{\dashrightarrow}} NO_3 \overset{h\nu}{\longrightarrow} NO_2 \underset{>15\,\text{km}}{\overset{HO_2+M}{\dashrightarrow}} HNO_4;$$

$$ClONO_2: \quad ClONO_2 \overset{h\nu,\,O,\,OH}{\longrightarrow} NO_3 \overset{h\nu}{\longrightarrow} NO_2 \overset{ClO+M}{\dashrightarrow} ClONO_2; \quad \text{and}$$

$$N: \quad N \overset{O_2+M}{\longrightarrow} NO \overset{h\nu}{\dashrightarrow} N.$$

Inclusion of NO_3 and N_2O_5 increases the NO_x time constant at ~15–45 km, (curve ② in Fig. 6.6). The exchange times between these constituents and NO_2 are smaller than the family time constant ② at these heights. Further inclusion of $ClONO_2$ slightly increases the NO_x time constant (curve ③). In this case the main loss occurs via the reaction of NO_2 (or NO) with HO_x, which produces oxyacids of nitrogen (HNO_2, HNO_3, and HNO_4). If these oxyacids are included in the family, the reactions between NO_2 (or NO) and HO_x will be eliminated as loss processes for the family, and the family time constant will become larger accordingly (curves ④, ⑤, and ⑥). In particular, inclusion of HNO_3 and HNO_4 would make the family time constant large at all heights below ~45 km. If all oxyacids are members of the NO_x family, the family time constant will be of the order of a month; in this case the major NO_x loss occurs by the reaction of NO_2 with ClO.

If both $ClONO_2$ and all three oxyacids of nitrogen are included as family members of NO_x, all reactions between NO_2 and ClO and between NO_2 (or NO) and HO_x will be eliminated from the list of loss mechanisms of the family, since they always produce other family constituents. The main loss for NO_x in this case occurs through reactions (R_{27}) $N+NO \rightarrow N_2+O$ and (R_{28}) $N+NO_2 \rightarrow N_2O+O$, both of which are not very significant in the stratosphere because of the small amount of N present. Thus, the family time constant becomes extremely large (curve ⑦ in Fig. 6.6), and the family is *nearly* conserved. It should be noted that the sum $N+NO+NO_2+NO_3+2N_2O_5+HNO_3+HNO_4+ClONO_2$ is usually called "NO_y" as opposed to $NO+NO_2$, which is usually called "NO_x."

*6.3.3 Cl_x family**

Candidate constituents and their exchange reactions with Cl (or ClO) for

*Instead of ClO_x, a notation Cl_x is used here because the family could include non-oxygen carrying species such as HCl.

the Cl_x family include the following:

$$\text{Cl:}\quad Cl \xrightarrow[>50\ \text{km}]{O_3} ClO \xrightarrow[<50\ \text{km}]{NO,\ O,\ h\nu} Cl;$$

$$\text{HCl:}\quad HCl \xrightarrow{OH,\ O,\ h\nu} Cl \xrightarrow{H_2,\ CH_4,\ HO_2} HCl;$$

$$\text{HOCl:}\quad HOCl \left\{ \begin{array}{l} \xrightarrow{O,\ OH} \\ \xrightarrow[<15\ \text{km}]{h\nu} Cl \xrightarrow{O_3} \end{array} \right\} ClO \xrightarrow[>15\ \text{km}]{HO_2} HOCl;\quad \text{and}$$

$$\text{ClONO}_2\text{:}\quad ClONO_2 \xrightarrow[<30\ \text{km}]{h\nu} Cl \xrightarrow{O_3} ClO \xrightarrow[>30\ \text{km}]{NO_2+M} ClONO_2.$$

The time constant for $Cl_x(=Cl+ClO)$ is larger than the exchange time between these two species in the middle atmosphere, as is seen by comparing curve ① with the dashed curve labeled Cl in Fig. 6.7. Including $ClONO_2$ and HCl in the family increases the family time constants (curves ②and ③, respectively), but they are still smaller than the exchange times between the respective constituent and ClO (or Cl). HOCl has a short exchange time with ClO and its inclusion in the family enlarges the family time constant (curve ④).

The main loss for the family ② occurs via reactions (R_{44})-(R_{46}) and (R_{52}),

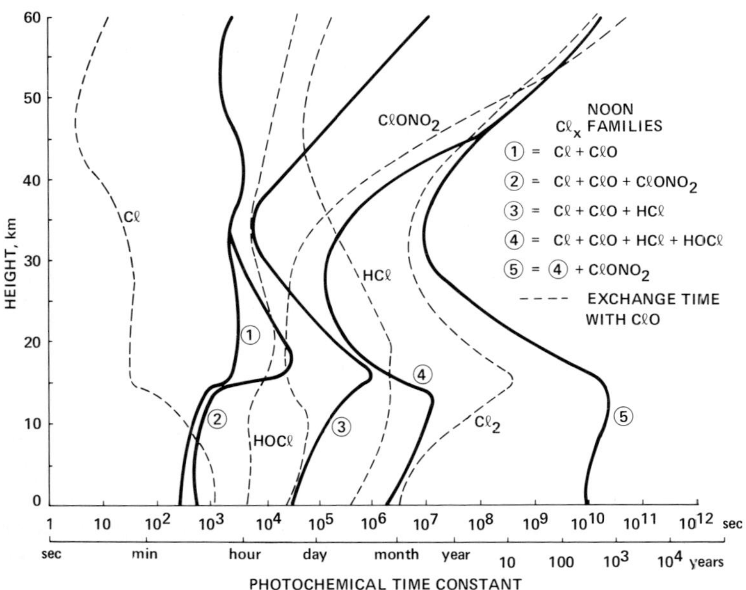

FIG. 6.7. The same as Fig. 6.5 but for ClO_x families.

the loss for ③ by reactions (R_{49}) and (R_{52}), and the loss for ④ by reaction (R_{49}). These reactions produce HCl, HOCl, or $ClONO_2$; therefore, if these three molecules are included in the family, the afore-mentioned reactions can no longer be the loss reactions for the family members. Thus, the family time constant becomes extremely large (curve ⑤). If Cl_2, ClOO, and OClO are further included in the family, the Cl_x family time constant becomes infinite, since *all* production and loss terms in the relevant chemical reactions are canceled out, and the Cl_x family is *completely* conserved.

6.3.4 HO_x family

Various hydrogen-containing constituents have exchange reactions with HO_2 (or OH) as follows:

$$\text{OH:}\quad \text{OH}\left\{\begin{array}{l}\xrightarrow{\text{CO}}\text{H}\xrightarrow{\text{O}_2+M} \\ \xrightarrow[>40\text{ km}]{\text{O}_3}\end{array}\right\}\text{HO}_2\xrightarrow[<40\text{ km}]{\text{O, O}_3}\text{OH};$$

$$\text{H:}\quad \text{H}\xrightarrow{\text{O}_2+M}\text{HO}_2\xrightarrow[<25\text{ km}]{\text{O, O}_3}\text{OH}\xrightarrow[>25\text{ km}]{\text{O, CO, H}_2}\text{H};$$

$$\text{H}_2\text{O}_2:\quad \text{H}_2\text{O}_2\xrightarrow[<50\text{ km}]{h\nu}\text{OH}\xrightarrow{\text{O}_3}\text{HO}_2\xrightarrow[>50\text{ km}]{\text{HO}_2}\text{H}_2\text{O}_2;$$

$$\text{HNO}_2:\quad \text{HNO}_2\xrightarrow{h\nu}\text{OH}\xrightarrow{\text{HO}+M}\text{HNO}_2;$$

$$\text{HNO}_3:\quad \text{HNO}_3\xrightarrow[<40\text{ km}]{h\nu}\text{OH}\xrightarrow[>40\text{ km}]{\text{NO}_2+M}\text{HNO}_3;$$

$$\text{HNO}_4:\quad \text{HNO}_4\xrightarrow[<30\text{ km}]{h\nu}\text{OH}\xrightarrow{\text{O}_3}\text{HO}_2\xrightarrow[>30\text{ km}]{\text{NO}_2+M}\text{HNO}_4;$$

$$\text{HCl:}\quad \text{HCl}\xrightarrow{h\nu}\text{H}\xrightarrow{\text{O}_2+M}\text{HO}_2\xrightarrow{\text{Cl}}\text{HCl};\quad\text{and}$$

$$\text{HOCl:}\quad \text{HOCl}\xrightarrow[20\text{–}35\text{ km}]{h\nu}\text{OH}\left\{\begin{array}{l}\xrightarrow{\qquad\text{ClONO}_2\qquad} \\ \xrightarrow{\text{O}_3}\text{HO}_2\xrightarrow[\text{other heights}]{\text{ClO}}\end{array}\right\}\text{HOCl}.$$

The family time constant for $HO_x(=H+OH+HO_2)$ is shown by curve ①in Fig. 6.8. It is larger than the exchange times among family members in the middle atmosphere, but the family time constant is still not very large; it is only about 10 minutes in most of the region. This small value is caused by fast nonexchange reactions such as (R_{15}), (R_{38}), (R_{42}) and (R_{55}).

Including HNO_2, HNO_3, HNO_4, and HCl increases the family time constant, as is seen by curves ②, ③, ④, and ⑤, respectively. However, exchange times between these constituents (except for HNO_2) and HO_2 are generally larger than the family time constant when the respective consti-

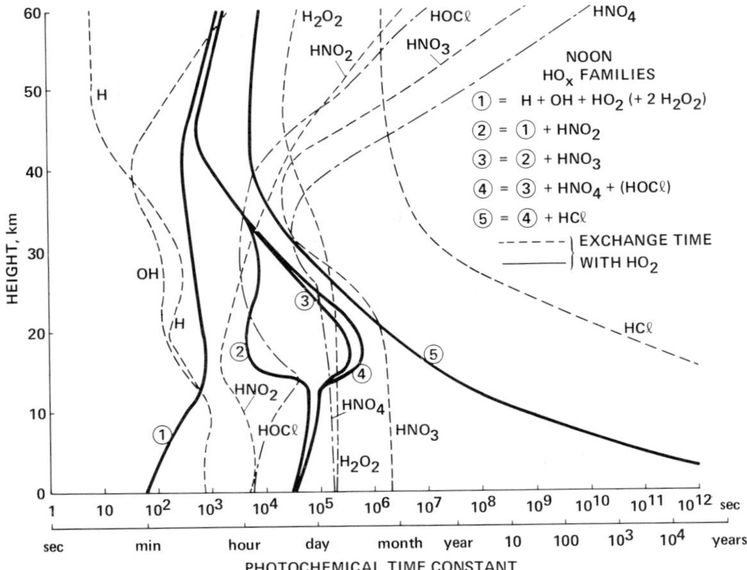

FIG. 6.8. The same as Fig. 6.5 but for HO_x families.

tuent is included in the family. Thus,it is not apppropriate to consider an HO_x family including HNO_3, HNO_4, and HCl, although inclusion can make the family time constant somewhat larger.

In model calculations it is appropriate to solve three chemical rate equations for H, OH, and HO_2 *simultaneously*; since that family has such a small time constant (< 10 min), it is not necessary to consider transport effects on the family. The method of solving simultaneous rate equations can guarantee conservation of the total concentration of family members just as in the family treatment (see Section 7.1.2.2).

Photochemical time constants presented in this chapter should be affected by reaction rate constants and species concentrations used. Omission of the effect of multiple scattering and albedo on photodissociation coefficients may have caused some error up to a factor of ~2 in the lower stratosphere and troposphere (see Section 3.3.2). Neglect of ion chemistry may have caused some error particularly for NO in and above the mesosphere (see Chap. 13). However, the essential features presented here should not be greatly influenced by them.

Chapter 7

PHOTOCHEMICAL MODELS

Photochemical modeling is a useful tool for gaining a better understanding of minor constituents. Measurements of minor constituents in the middle atmosphere are still limited both in space and time; in particular, simultaneous observations are very scarce. Some constituents are very hard to observe. By comparing a theoretical model with observations of some constituents for which data are available, one can determine the validity of the model and may be able to understand physical and chemical mechanisms occurring in the middle atmosphere. In return, the knowledge obtained from the model, particularly the predicted range of number densities, would greatly assist in designing field and laboratory measurements. One can also use the model for studies of potential impacts of various parameters including the NO_x emission from SST's, the increased ClO_x from CFM's and variabilities of the solar UV radiation.

7.1 One-Dimensional Models

A model which considers the variations in one direction only is a one-dimensional model. The direction is usually taken as vertical in most photochemical models, since variations of minor consitients are most significant in that direction. There are many one-dimensional photochemical models in the literature including COLEGROVE *et al.* (1965), HUNT (1966), KENESHEA (1967), SHIMAZAKI (1967, 1968), HESSTVEDT (1968, 1972), SHIMAZAKI and LAIRD (1970, 1972), NICOLET (1970b, 1974), CRUTZEN (1973, 1974), MCCONNEL and MCELROY (1973), SHIMAZAKI and WUEBBLES (1973), MCELROY *et al.* (1974), SHIMAZAKI and OGAWA (1974a and b), WHITTEN and TURCO (1974a and b), CHANG (1974), STEWART and HOFFERT (1975), WUEBBLES and CHANG (1975), KURJEZA (1975), ASHBY (1976), OGAWA (1976), TUCK (1977), TURCO and WHITTEN

(1977), LOGAN *et al.* (1978), RUNDEL *et al.* (1978), LUTHER *et al.* (1979) and others.

A comparison among models developed prior to 1975, including a comparison of the models with observations, was published by SHIMAZAKI and WHITTEN (1976). Since then, many new models have been published; models are compared with observations in each paper. Most models explain fairly well the general features of the observations. However, there are still discrepancies between the models and some observations and also among various models. Models do not necessarily agree with each other in detail, and observations sometimes indicate inconsistencies and include some uncertainties as well as experimental errors.

Photochemical models are sensitive to various imput parameters such as solar fluxes, photoabsorption cross-sections, chemical reaction schemes and reaction rate constants, and transport parameters (e.q. eddy diffusion coefficient) as well as to the method of numerical analysis and boundary conditions. Therefore, it is usually difficult to identify the causes of discrepancies among models; most published model papers do not describe the details sufficiently enough to allow a complete comparison. In this book we describe details of a particular 1-D model developed by the author, for a general interest and also to serve as a basis for comparison with other models and observations.

7.1.1 Effective eddy diffusion coefficients

The concept of eddy diffusion has been introduced in photochemical models in order to express in mathematical form the effects of atmospheric mixing on the distribution of n_i (COLGROVE *et al.*, 1965; SHIMAZAKI, 1967). The atmosphere below the turbopause is always more or less turbulent, and some form of wave motion (e.g. planetray waves, gravity waves) can propagate into the middle and upper atmosphere from the troposphere. Different sizes of eddy motion transport the air mass and exercise frictional viscous force on its motion. The dynamic viscous coefficient due to eddy motion is called the *eddy viscosity coefficient*. The eddy diffusion coefficient K in photochemical models, however, is a parameter determined so that the predicted variation in n_i may closely represent the observed variation. A chosen value of K by such a method of parameterization may be called the *effective eddy diffusion coefficient*. We will consider here only the vertical variation, since the horizontal variation in n_i is much smaller.

We will now try to derive the flux equation due to eddy diffusion (2.25) by the phenomenological approach without inquiring into the detailed physics of eddy diffusion processes and to calculate the effective eddy diffusion coefficient in some special cases; some detailes on the physical aspects of eddy transport due to planetary waves and gravity waves will be discussed in Chapter 9.

As was discussed in Section 2.2, if the ith constituent is mixed completely with the major constituents, the distribution of n_i is given by the condition that the mixing ratio is constant with height or $\partial f_i/\partial z=0$. We assume that if the distribution of n_i deviates from complete mixing, the motion for redistribution should occur so that the distribution of n_i may approach the state of complete mixing. The velocity of such a motion should be proportional to but opposite in sign to the gradient of f_i and may be expressed by Fick's law:

$$w_i = -K\frac{1}{f_i}\frac{\partial f_i}{\partial z}. \tag{7.1}$$

Here K is defined as a proportionality constant. Using (2.22) and (2.13) one can derive the form of the flux, $n_i w_i$, consistent with (2.25).

An equation of the type (7.1) is also suggested by *mixing length theory*, in which the turbulent fluctuation of n_i is given by

$$n' = -l\cdot\frac{\partial n_i}{\partial z} \tag{7.2}$$

where l is the mixing length. K is defined in this case by the time average over the mixing period of the product of l and the turbulent velocity w',

$$K = \overline{lw'}. \tag{7.3}$$

The actual distribution of n_i would sometimes deviate markedly from complete mixing and the height variation of n_i may be represented by

$$\frac{\partial n_i}{\partial z} = -\left(\frac{1}{T}\frac{\partial T}{\partial z} + \frac{1}{H_s}\right)n_i \tag{7.4}$$

where H_s is the scale height of the ith constituent, determined from its observed distribution as the distance by which n_i decreases by a factor of e^{-1}.

Thus the eddy flux (2.25) becomes by virtue of (7.4).

$$\phi_i = -Kn_i\left(\frac{1}{H} - \frac{1}{H_s}\right) \tag{7.5}$$

For upward transport ($\phi_i > 0$) to occur by eddy mixing, it is required that $H_s < H$. Integrating (2.24) from z to ∞ under the assumption of a steady state, we obtain

$$\int_z^\infty (Q_i - L_i n_i)\, \mathrm{d}z = \phi_\infty - \phi_i$$
$$= \phi_\infty + K n_i \left(\frac{1}{H} - \frac{1}{H_s}\right) \tag{7.6}$$

where ϕ_∞ is the flux at the top of the atmosphere and is negligibly small for most primary molecules. If we choose a primary constituent for which there is no photochemical production, (7.6) can be written as

$$-\int_z^\infty L_i n_i\, \mathrm{d}z = K n_i \left(\frac{1}{H} - \frac{1}{H_s}\right). \tag{7.7}$$

This equation tells us that in a steady state the number of molecules lost by photochemical reactions in the region above z must be supplied by eddy diffusion flux from below.

In order to evaluate K from (7.7), two primary molecules CH_4 and N_2O have been chosen. There is no essential chemical production of these constituents in the middle atmosphere, and their chemical loss processes are simple; CH_4 is lost mainly by reactions with OH and photolysis and N_2O by a reaction with $O(^1D)$. $O(^1D)$ is produced largely by O_3 photolysis in the middle atmosphere and can be calculated from the distribution of the observed O_3 density. The observations of OH density are limited; thus we use the profile of OH density predicted by the one dimensional photochemical model. By estimating the scale height H_s for CH_4 and N_2O from their observed distributions we can calculate the variation in K from (7.7); the result, along with the profile of K used in our one dimensional model, is shown in Fig. 7.1.

The two predicted variations in K shown in Fig. 7.1 are similar in the stratosphere, but they differ from each other in the troposphere and mesosphere. Since there are no observations of CH_4 and N_2O above ~45 km, we assumed the same scale height H_s as was determind for the lower regions; this should cause some uncertainties in the calculated K above ~45 km. K generally is larger in the region where the chemical loss rate is larger. A bulge around 50 km in the CH_4 curve corresponds to the region where the CH_4 loss rate through reaction with OH is large, and the increase in K above ~80 km is due to the increasing loss rate by CH_4 photolysis. The N_2O curve shows a K value monotonically increasing with height in the middle atmosphere. This is due to the fact that the loss mechanism for N_2O is simpler, i.e. it occurs by a single reaction with $O(^1D)$. The apparent agreement between the two K's in the stratosphere is rather striking when we recall that there is no *a priori* requirement that they must have the same variation. By applying a technique essentially the same as was used here,

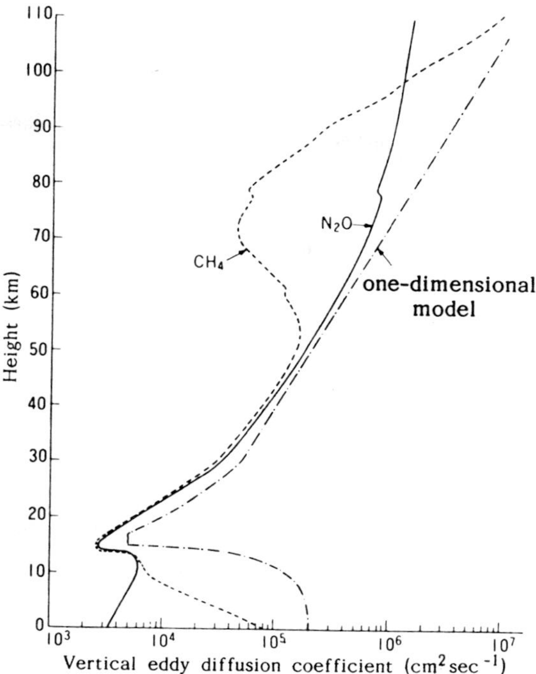

FIG. 7.1. Vertical profiles of effective eddy diffusion coefficients necessary for maintaining the observed densities of CH_4 and N_2O in the middle atmosphere and the profile used in our one-dimensional model.

SCHMELTOKOPF *et al.* (1977) have calculated the vertical distribution of K in the lower and middle stratosphere at various latitudes from the observed profiles of N_2O. Individual profiles of the predicted eddy diffusion coefficient are very diverse, but their average is close to the profile calculated in Fig. 7.1.

7.1.2 Methods of numerical analyses

7.1.2.1 A single continuity equation

We first consider a single component, whose concentration n changes with time by

$$\frac{\partial n_t}{\partial t} = \dot{n}_t = f(n_t, z, t). \tag{7.8}$$

If the transport terms are neglected, the function f does not depend upon height z, and $\dot{n}_t$ actually represents the local imbalance between the production and loss rates.

A simple numerical method to calculate n_t from (7.8) is

$$n_t = n_{t-\Delta t} + \dot{n}_{t-\Delta t} \cdot \Delta t. \tag{7.9}$$

Since the right-hand side can be evaluated from known quantities at $t - \Delta t$ explicitly, the method is called the *explicit method.* The equation indicates that a small imbalance between the production and loss terms initially assumed is amplified by the factor Δt; therefore, if Δt is large, n_t would take either large positive or negative values depending upon the sign of $\dot{n}_{t-\Delta t}$ (or upon the relative magnitude of the production and loss terms at $t-\Delta t$). Thus, the numerical solution becomes unstable unless Δt is chosen to be smaller than the photochemical time constant. Since the photochemical time constants are often very small ($\ll 1$ sec) as was discussed in Chapter 6, it is time consuming and impractical to use such a small Δt in model calculations.

The above problem may be solved by replacing $\dot{n}_{t-\Delta t}$ in (7.9) by $\dot{n}_t$. Since the right-hand side now includes the unknown value n_t implicitly through (7.8), the method is called the *implicit method.* If $\dot{n}_t$ is given by the rate equation (6.3), we have

$$n_t = n_{t-\Delta t} + \Delta t \cdot (Q - \beta n_t - \alpha n_t^2) \tag{7.10}$$

whose solution is given by

$$n_t = \frac{-(1+\beta \cdot \Delta t) + \sqrt{(1+\beta \cdot \Delta t)^2 + 4\alpha \Delta t (n_{t-\Delta t} + Q \cdot \Delta t)}}{2\alpha \cdot \Delta t}. \tag{7.11}$$

If $\alpha \to 0$, (7.11) reduces to

$$n_t = \frac{n_{t-\Delta t} + Q \cdot \Delta t}{1 + \beta \cdot \Delta t}. \tag{7.12}$$

It is evident that both (7.11) and (7.12) give a positive value for n_t regardless of the magnitude of Δt. Thus, we can obtain a stable solution by using a Δt larger than the photochemical time constant; however, a smaller Δt is still required for conservation of particles. The latter problem will be discussed further in Section 7.1.2.2.

If the effects of transport are included, (7.8) becomes a partial differential equation, which may be written for the ith constituent in the form

$$\frac{\partial n_i}{\partial t} = Q_i - L_i n_i - \frac{\partial \phi_i}{\partial z} \tag{7.13}$$

where L_i is a linearized loss coefficient* and ϕ_i is the total flux due to eddy

*A quadratic loss term may be linearized by the Taylor expansion $-\alpha n^2 = -\alpha (n_0 + \Delta n)^2 \doteqdot -\alpha n_0^2 - 2\alpha n_0 \cdot \Delta n = -2\alpha n_0\, n + \alpha n_0^2$ where n_0 is the first guess for n.

and molecular diffusion

$$\phi_i = -K\left\{\frac{\partial n_i}{\partial z} + \left(\frac{1}{T}\frac{\partial T}{\partial z} + \frac{1}{\bar{H}}\right)n_i\right\} - D_i\left\{\frac{\partial n_i}{\partial z} + \left(\frac{1}{T}\frac{\partial T}{\partial z} + \frac{1}{H_i}\right)\cdot n_i\right\} \tag{7.14}$$

where K and D_i are the coefficients of eddy and molecular diffusion, respectively, H_i is the scale height of the constituent under consideration, and $\bar{H}$ is the scale height of the atmosphere. The equation (7.14) may be simplified to

$$\phi_i = -(K+D_i)\left\{\frac{\partial n_i}{\partial z} + \left(\frac{1}{T}\frac{\partial T}{\partial z} + \frac{1}{H_A}\right)n_i\right\} \tag{7.15}$$

with

$$\frac{1}{H_A} = \frac{1}{K+D_i}\left(\frac{K}{\bar{H}} + \frac{D_i}{H_i}\right). \tag{7.16}$$

The flux divergence term in (7.13) includes second order derivatives of n_i with respect to z. We now represent the differential operations by their finite differences such as (the subscript i is omitted in the following:)

$$\frac{\partial n}{\partial t} = \frac{n^{l+1}(k) - n^l(k)}{\Delta t} \tag{7.17}$$

$$\frac{\partial n}{\partial z} = \frac{n^{l+1}(k+1) - n^{l+1}(k-1)}{2\cdot\Delta z} \tag{7.18}$$

$$\frac{\partial^2 n}{\partial z^2} = \frac{n^{l+1}(k+1) - 2\cdot n^{l+1}(k) + n^{l+1}(k-1)}{(\Delta z)^2} \tag{7.19}$$

where Δt and Δz are increments of time and height, respectively, the upper script l indicates the time step and k represents the height step by

$$z(k) = z_0 + (k-1)\cdot\Delta z \quad (k=1, 2, 3, \cdots, K). \tag{7.20}$$

Thus, z_0 is the lower boundary and the upper boundary is at $z_u = z_0 + (K-1)\cdot\Delta z$ (see Fig. 7.2).

Using the finite differences defined in (7.17) – (7.19) , the implicit scheme of (7.13) may be expressed in the form

$$-A^l(k)\cdot n^{l+1}(k+1) + B^l(k)n^{l+1}(k) - C^l(k)n^{l+1}(k-1) = D(k) \tag{7.21}$$

where A^l, B^l, C^l, and D^l are coefficients and can be evaluated from known quantities by

$$A^l(k) = \frac{(K+D)_k}{(\Delta z)^2} \tag{7.22}$$

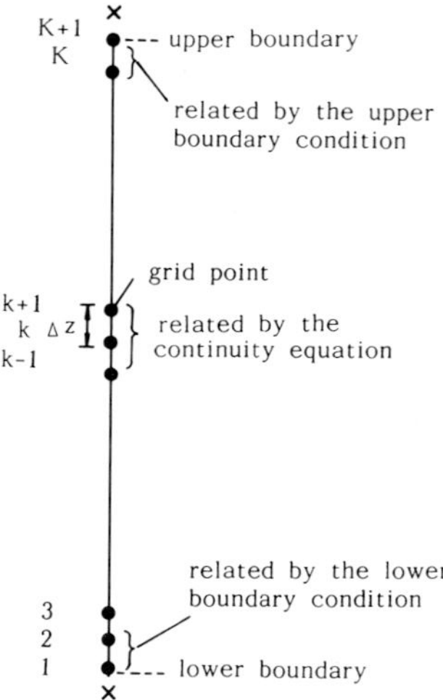

FIG. 7.2. Schematic illustration of grid points in one-dimensional model.

$$B^l(k) = \frac{1}{\Delta t} + \frac{1}{(\Delta z)^2}\left\{(K+D)_k\left(2 - \frac{T(k+1)}{T(k)} - \frac{\Delta z}{H_A(k)}\right) - (K+D)_{k-1}\right\} \tag{7.23}$$

$$C^l(k) = \frac{(K+D)_{k-1}}{(\Delta z)^2}\left\{2 - \frac{T(k)}{T(k-1)} - \frac{\Delta z}{H_A(k-1)}\right\} \tag{7.24}$$

and

$$D^l(k) = \frac{n^l(k)}{\Delta t} + Q^l(k). \tag{7.25}$$

(7.21) has three unknowns $n^{l+1}(k+1)$, $n^{l+1}(k)$, and $n^{l+1}(k-1)$, and the equations can be written for various k's between 2 and $K-1$; therefore, we have $K-2$ equations for K unknowns $n^{l+1}(k)$, $k=1,2,\ldots, K$. To solve the system of these simultaneous equations, we need two additional equations; they are supplied by the upper and lower boundary conditions. These conditions will be discussed in Section 7.1.2.3.

A convenient method which is suitable for automatic computation in solving simultaneous equations of the type (7.21) has been developed by many workers (e.g. RICHTMEYER and MORTON, 1967). Putting

$$n^{l+1}(k) = E(k)\cdot n^{l+1}(k+1) + F(k) \tag{7.26}$$

and substituting (7.26) and a similar equation for $n^{l+1}(k-1)$ into (7.21) gives the following recurrence formulae for $E(k)$ and $F(k)$

$$E(k)=\frac{A(k)}{B(k)-C(k)E(k-1)} \tag{7.27}$$

and

$$F(k)=\frac{D(k)+C(k)F(k-1)}{B(k)-C(k)E(k-1)} \tag{7.28}$$

where the upper script l for A, B, C and D have been omitted.

Consideration of (7.26) at the lower boundary $k=1$ leads to

$$E(1)=0, \quad F(1)=n^{l+1}(1). \tag{7.29}$$

If we know the lower boundary value $n^{l+1}(1)$, $E(k)$ and $F(k)$ can be calculated from equations (7.27) - (7.29) inductively in order of increasing k. Then, if we know the upper boundary value $n^{l+1}(K)$ we can calculate $n^{l+1}(k)$ from (7.26) inductively in order of decreasing k.

7.1.2.2 Coupled continuity equations

Considering the many constituents involved in the system, we have to solve continuity equations for all constituents. A method of solving these equations by simple successive iteration will face a difficulty of numerical instability unless we use a very small time step, as was discussed for the case of O and O_3 in Section 5. (snowball effect, see p. 81). Since photochemical time constants are widely different among constituents,* one must use a time step smaller than the shortest time constant in order to approach conservation of particles. This would require enormous amounts of computer time for numerical solutions.

There are several convenient methods to solve the above problem. One is the *family method* discussed in Section 6.3. The other is a method of *simultaneous solution*. The chemical production and loss terms in the continuity equation for the ith constituent usually include the concentration of other constituents n_j ($j \neq i$). In order to obtain the solution in which the total of any specific kind of atom is conserved after the time step Δt, all n_i and n_j appearing in the continuity equation for n_i must be considered as unknown variables in the implicit scheme equations. It is not sufficient to consider n_i as the only unknown variable in the continuity equation for n_i. This can be shown in the following example (SHIMAZAKI, 1972):

Consider the variations in [H] and [OH] by the chemical reactions

$$OH+O \rightarrow H+O_2 \tag{R$_7$}$$

and

*The system is then called *stiff*, and equations are the *stiff differential equations*.

$$H + O_3 \rightarrow OH + O_2. \tag{R$_8$}$$

First, if we consider that [H] and [OH] are the only variables in the chemical rate equations for [H] and [OH], respectively, the rate equations for [H] and [OH] can be written as follows:

$$[H]_1 = [H]_0 + (k_7[O][OH]_0 - k_8[O_3][H]_1) \cdot \Delta t \tag{7.30}$$

$$[OH]_1 = [OH]_0 + (k_8[O_3][H]_0 - k_7[O][OH]_1) \cdot \Delta t \tag{7.31}$$

where the suffix indicates the time step (0 for $t - \Delta t$ and 1 for t). Solving these equations we have

$$[H]_1 = ([H]_0 + k_7[O][OH]_0 \cdot \Delta t)/(1 + k_8[O_3] \cdot \Delta t) \tag{7.32}$$

and

$$[OH]_1 = ([OH]_0 + k_8[O_3][H]_0 \cdot \Delta t)/(1 + k_7[O] \cdot \Delta t) \tag{7.33}$$

Summing up (7.32) and (7.33) gives

$$[H]_1 + [OH]_1 = [H]_0 + [OH]_0 + \frac{(k_8[O_3] - k_7[O])(k_8[O_3][H]_0 - k_7[O][OH]_0)}{(1 + k_8[O_3] \cdot \Delta t)(1 + k_7[O] \cdot \Delta t)} (\Delta t)^2 \tag{7.34}$$

which indicates that [H]+[OH] are not conserved unless the third term of the right-hand side is equal to zero. This term is proportional to the square of Δt; therefore Δt must be very small in order to have conservation of hydrogen, unless the numerator is equal to zero or the condition of photochemical equilibrium prevails ($k_8 [O_3] [H]_0 = k_7 [O] [OH]_0$).

Next, considering that both [H] and [OH] are unknown variables in the rate equations for [H] and [OH], we have the following equations in place of (7.32) and (7.33):

$$[H]_1 = [H]_0 + (k_7[O][OH]_1 - k_8[O_3][H]_1) \cdot \Delta t \tag{7.35}$$

and

$$[OH]_1 = [OH]_0 + (k_8[O_3][H]_1 - k_7[O][OH]_1) \cdot \Delta t. \tag{7.36}$$

Solving the above two equations simultaneously, we obtain the solution

$$[H]_1 = \frac{[H]_0 + k_7[O]([H]_0 + [OH]_0) \cdot \Delta t}{1 + (k_7[O] + k_8[O_3]) \cdot \Delta t} \tag{7.37}$$

and

$$[OH]_1 = \frac{[OH]_0 + k_8[O_3]([H]_0 + [OH]_0) \cdot \Delta t}{1 + (k_7[O] + k_8[O_3]) \cdot \Delta t} \tag{7.38}$$

Adding these two equations produces

$$[H]_1+[OH]_1=[H]_0+[OH]_0 \tag{7.39}$$

which indicates that hydrogen is conserved regardless of Δt.

In the above example, [O] and $[O_3]$ are assumed to be unchanged with time, but if they change as is actually the case, we have to solve rate equations for [H], [OH], [O] and $[O_3]$ simultaneously by considering all four concentrations as unknown variables. In the most general case, we have to solve all continuity equations for the n_i's *simultaneously* by considering all n_i's as unknown variables. The non-linear term due to the product of n_i and n_j must be linearized by Taylor's expansion.* Various numerical techniques are available for facilitating matrix calculations to solve such large numbers of simultaneous equations.

The method of simultaneous equations is suitable for calculating concentrations of a group of constituents which are active photochemically with each other and for which the transport effects can be neglected. Example of such a group is the HO_x family composed of H, OH and HO_2. The method is also useful for partitioning the family concentration among members of the family, once the family concentration has been calculated by including the transport effect.

Still another method to overcome the numerical problems associated with stiff differential equations is a *multistep method*, which assumes that n_t depends not only on $n_{t-\Delta t}$ and $\dot{n}_{t-\Delta t}$, but also on those values at several earlier time steps. Thus, instead of (7.9) we write

$$n_t=\sum_{j=1}^{K_1}\alpha_j\cdot n_{t-j\Delta t}+\sum_{j=0}^{K_2}\beta_j\cdot\dot{n}_{t-j\Delta t}\Delta t. \tag{7.40}$$

If $\beta_0=0$, the right-hand side does not include quantities at t; therefore, the expression is explicit. If $\beta_0\neq 0$, however, (7.40) is an implicit expression for n_t.

In the implicit method (7.40) is used to calculate $\dot{n}_t$ for the initial guess for n_t denoted by $n_{t(0)}$. The pair of $n_{t(0)}$ and $\dot{n}_{t(0)}$ do not necessarily satisfy (7.8) and the error is defined by

$$\varepsilon_{(0)}=-\dot{n}_{t(0)}+f(n_{t(0)}). \tag{7.41}$$

We now write a second approximation for n_t by adding the correction term proportional to $\varepsilon_{(0)}$, i.e.

$$n_{t(1)}=n_{t(0)}+C_{(0)}\cdot\varepsilon_{(0)}. \tag{7.42}$$

*$n_i^{l+1}n_j^{l+1}=n_i^l n_j^l+n_j^l(n_i^{l+1}-n_i^l)+n_i^l(n_j^{l+1}-n_j^l)=n_j^l n_i^{l+1}+n_i^l n_j^{l+1}-n_i^l n_j^l$.

Repeating the same procedure we obtain the solution successively by

$$n_{t(m)}=n_{t(m-1)}-C_{(m)}\cdot\{\dot{n}_{t(m-1)}-f(n_{t(m-1)})\}. \tag{7.43}$$

Choosing the numerical values for α_i, β_i and $C_{(m)}$ as given in Table 7.1, GEAR (1971) was able to work out an effective method of solving the stiff differential equations.

The multistep method is simpler and more efficient for programming than the family method, but the family method has an advantage in studying the details of photochemical processes and dynamical effects on minor constituents.

7.1.2.3 Boundary conditions

As discussed in Section 7.1.2.1, boundary conditions at the lower and upper boundaries are necessary in order to solve the continuity equation (differential equation). For photochemically active species or constituents having very short chemical time constants, the boundary values for n_i are easily calculated by photochemical equlibrium (2.23). For long-lived species or constituents having longer chemical time constants, the boundary conditions may be given either by a given fixed value or a given flux value.

Upper boundary conditions

A simple method to give the upper boundary condition is to assume no flux through the upper boundary ($k=K$). From (7.15) the condition

TABLE 7.1 Coefficients of stiffly stable methods for the case of $K_2=0$ in Eq. (7.40). (GEAR, 1971)

K_1	2	3	4	5	6
β_0	2/3	6/11	12/25	60/137	60/147
α_1	4/3	18/11	48/25	300/137	360/147
α_2	−1/3	−9/11	−36/25	−300/137	−450/147
α_3		2/11	16/25	200/137	400/147
α_4			−3/25	−75/137	−225/147
α_5				12/137	72/147
α_6					−10/147
$C_{(0)}$	2/3	6/11	25/50	120/274	720/1764
$C_{(1)}$	3/3	11/11	50/50	274/274	1764/1764
$C_{(2)}$	1/3	6/11	35/50	225/274	1624/1764
$C_{(3)}$		1/11	10/50	85/274	735/1764
$C_{(4)}$			1/50	15/274	175/1764
$C_{(5)}$				1/274	31/1764
$C_{(6)}$					1/1764

$\phi(K)=0$ yields (the subscript i is omitted here)

$$n(K+1)=\gamma(K)\cdot n(K) \tag{7.44}$$

where

$$\gamma(K)=1-\left\{\frac{1}{T}\frac{\partial T}{\partial z}+\frac{1}{H_A}\right\}_K \Delta z=2-\frac{T(K+1)}{T(K)}-\frac{\Delta z}{H_A(K)}. \tag{7.45}$$

Inserting (7.44) into (7.26) at $k=K$, we obtain the upper boundary value

$$n(K)=\frac{F(K)}{1-\gamma(K)\cdot E(K)}. \tag{7.46}$$

The no flux condition may not be appropriate if the upper boundary is located at relatively low altitudes, and for some constituents such as H and H_2, the flux can never be zero because significant amounts of these light gases are always escaping from the top of the atmosphere. The upper boundary value in such cases is calculated by the following method:

Integrating (7.13) from $k=K$ to $k=\infty$ and assuming that the integrals of n, Q and Ln from K to ∞ are approximated by the product of these quantities and the scale height at the upper boundary, $H(K)$,* we obtain

$$\frac{n^{l+1}(K)-n^l(K)}{\Delta t}\cdot H(K)+n^{l+1}(K)\frac{\partial H(K)}{\partial t}=$$
$$Q(K)\cdot H(K)-\frac{L(K)H(K)}{n^l(K)}n^{l+1}(K)-\phi_\infty+\phi_K \tag{7.47}$$

where ϕ_∞ is the escape flux at the top of the atmosphere. The escape flux for H (atomic hydrogen) is $\sim 10^8$ cm^{-2} sec^{-1} (see Appendix G), whereas for all other species the escape flux is practically negligible.

The flux at the upper boundary ϕ_K is expressed from (7.15) by virtue of (7.45) as**

$$\phi_K=-(K+D)_K\frac{n^{l+1}(K+1)-\gamma(K)n^{l+1}(K)}{\Delta z}. \tag{7.48}$$

Inserting $n^{l+1}(K+1)$ obtained from (7.26) into (7.48), and substituting the result into (7.47), we obtain the upper boundary value

*Most models use kT/m_ig for the scale height for the three quantities (n_i, Q, and Ln_i), but they can be evaluated more properly from the predicted height variations of each quantity near the upper boundary.

**Note that the same notation K is used here for the eddy diffusion coefficient and the suffix to indicate the upper boundary.

$$n^{l+1}(K)=\frac{F(K)+a(K)}{1-\gamma'(K)\cdot E(K)} \tag{7.49}$$

where

$$a(K)=\frac{\Delta z}{(K+D)_K}E(K)\left\{Q(K)\cdot H(K)-\phi_\infty+\frac{n^l(K)\cdot H(K)}{\Delta t}\right\} \tag{7.50}$$

and

$$\gamma'(K)=\gamma(K)-\frac{\Delta z}{(K+D)_K}\left\{L(K)\cdot H(K)+\frac{H(K)}{\Delta t}+\left(\frac{\partial H}{\partial t}\right)_K\right\}. \tag{7.51}$$

The expression (7.49) is an extended version of (7.46).

Lower boundary conditions

A fixed lower boundary value may be justified for constituents whose concentrations are stable and well determined by observations. For some constituents, the flux condition may be physically more reasonable if the rate of influx from the ground soil or human activities is known. These two conditions, however, are equivalent to each other, i.e. we need a certain amount of flux to maintain the fixed value at the lower boundary. The relationship between the fixed lower boundary value and the equivalent flux is obtained by the following way:

Integrating (7.13) between the lower boundary ($k=1$) and the upper boundary ($k=K$), we obtain

$$\frac{\int_1^K n^{l+1}(k)\,\mathrm{d}k-\int_1^K n^l(k)\,\mathrm{d}k}{\Delta t}=\int_1^K Q^l(k)\,\mathrm{d}k-\int_1^K L^l(k)n^{l+1}(k)\,\mathrm{d}k-\phi_K+\phi_1. \tag{7.52}$$

To obtain the relationship between $n^{l+1}(k)$ and $n^{l+1}(1)$ we first put

$$F(k)=\kappa(k)+\mu(k)\cdot n^{l+1}(1). \tag{7.53}$$

Substituting (7.53) and a similar equation for $F(k-1)$ into (7.28) gives the following recurrence formulae for κ and μ

$$\kappa(k)=\frac{D(k)+C(k)\kappa(k-1)}{B(k)-C(k)E(k-1)} \tag{7.54}$$

and

$$\mu(k)=\frac{C(k)\cdot\mu(k-1)}{B(k)-C(k)E(k-1)}. \tag{7.55}$$

Putting $k=1$ in (7.53) and comparing the result with (7.29) we obtain

$$\kappa(1)=0, \quad \mu(1)=1. \tag{7.56}$$

From (7.54) – (7.56), the values of $\kappa(k)$ and $\mu(k)$ can be calculated inductively in order of increasing k.

Now we write the relationship between $n^{l+1}(k)$ and $n^{l+1}(1)$ in the form

$$n^{l+1}(k)=\lambda(k)+\nu(k)\cdot n^{l+1}(1). \tag{7.57}$$

Substitution of this and a similar equation for $n^{l+1}(k+1)$ as well as (7.53) into (7.26) yields

$$\lambda(k)=E(k)\lambda(k+1)+\kappa(k) \tag{7.58}$$

and

$$\nu(k)=E(k)\nu(k+1)+\mu(k). \tag{7.59}$$

Using (7.49), (7.53) and (7.57) we obtain the upper boundary values of λ and ν as follows:

$$\lambda(K)=\frac{\kappa(K)+a(K)}{1-\gamma'(K)E(K)} \tag{7.60}$$

and

$$\nu(K)=\frac{\mu(K)}{1-\gamma'(K)E(K)}. \tag{7.61}$$

From (7.58) and (7.59) together with (7.60) and (7.61) one can calculate $\lambda(k)$ and $\nu(k)$ inductively in order of decreasing k.

Finally, substituting the integral forms of (7.57) and $L^l(k)\ n^{l+1}(k)$ into (7.52) and neglecting ϕ_K compared with ϕ_1, we obtain an expression for the lower boundary value

$$n^{l+1}(1)=\frac{\dfrac{1}{\Delta t}\displaystyle\int_1^K \{n^l(k)-\lambda(k)\}\ \mathrm{d}k+\int_1^K \{Q(k)-L(k)\lambda(k)\}\ \mathrm{d}k+\phi_1}{\dfrac{1}{\Delta t}\displaystyle\int_l^K \nu(k)\ \mathrm{d}k+\int_1^K L(k)\nu(k)\ \mathrm{d}k}. \tag{7.62a}$$

If a deposition velocity v_0 instead of the flux ϕ_1 is given at the lower boundary, the lower boundary value of n may be calculated by

$$n^{l+1}(1)=\frac{\dfrac{1}{\Delta t}\displaystyle\int_1^K \{n^l(k)-\lambda(k)\}\ \mathrm{d}k+\int_1^K \{Q(k)-L(k)\lambda(k)\}\ \mathrm{d}k}{\dfrac{1}{\Delta t}\displaystyle\int_1^K \nu(k)\ \mathrm{d}k+\int_1^K L(k)\nu(k)\ \mathrm{d}k-v_0}. \tag{7.62b}$$

If the lower boundary value is prescribed as n_0, the boundary condition should be equivalent to the flux condition

$$\phi_L = n_0 \int_1^K v(k)\{\left\{L(k)+\frac{1}{\Delta t}\right\} dk - \int_1^K \{Q(k)-L(k)\lambda(k)\} dk$$
$$-\frac{1}{\Delta t}\int_1^K \{n^l(k)-\lambda(k)\} dk. \tag{7.63}$$

exposed at the lower boundary.

If the input flux at the lower boundary is larger (or smaller) than (7.63), the lower boundary value of n should be larger (or smaller) than n_0. The assumed lower boundary values and the corresponding boundary fluxes calculated with the steady state diurnally averaged model (see the next section) are tabulated in Table 7.2.

The global averages of the injection or deposition rates corresponding to the lower boundary fluxes predicted in the model are also tabulated in Table 7.2. The injection rates for $CFCl_3$ and CF_2Cl_2 are $\sim 3.85\times 10^4$ and $\sim 4.0\times 10^4$ metric tons per year, respectively, which are about equal to the values estimated for the middle of 1950's (see Fig. 11.2). Thus, our model has simulated, as far as the freon concentrations are concerned, a steady state that would be reached when the injection continued indefinitely at the rate of ~1955. This certainly represents the situation before any significant

TABLE 7.2 Fixed lower boundary values assumed for the long lived constituents and the equivalent fluxes necessary to maintain those values at the lower boundary (0 km except for H_2O, for which it is 14 km, the assumed tropopause). The global average injection or deposition rates are also shown.

	Lower bondary values, n_o		Equivalent flux ($cm^{-2}sec^{-1}$)	Global average rate (metric tons yr^{-1})
	Number density (cm^{-3})	Mixing Ratio		
H_2O	7.868×10^{12}	3.0×10^{-7}	1.089×10^{9}	1.29×10^{7}
H_2	2.623×10^{13}	1.0×10^{-6}	6.915×10^{8}	3.7×10^{5}
CH_4	4.196×10^{13}	1.6×10^{-6}	7.617×10^{10}	3.28×10^{8}
CO_2	8.654×10^{15}	3.3×10^{-4}	-4.517×10^{11}	-5.35×10^{9}
CO	7.868×10^{12}	3.0×10^{-7}	3.901×10^{11}	2.94×10^{9}
H_2O	1.790×10^{13}	3.5×10^{-6}	-6.482×10^{9}	-3.14×10^{7}
$CFCL_3$	3.934×10^{9}	1.5×10^{-10}	1.185×10^{6}	3.85×10^{4}
CF_2Cl_2	6.556×10^{9}	2.5×10^{-10}	1.225×10^{6}	4.00×10^{4}
CCl_4	3.409×10^{9}	1.3×10^{-10}	1.093×10^{6}	4.53×10^{4}
CH_3Cl	1.574×10^{10}	6.0×10^{-10}	1.865×10^{8}	2.53×10^{6}
O_x	6.0×10^{11}	2.288×10^{-8}	-2.002×10^{11}	-2.58×10^{9} (O_3)
NO_x	4.0×10^{10}	1.525×10^{-9}	2.096×10^{10}	1.69×10^{8} (NO)
Cl_x	2.6×10^{10}	9.914×10^{-10}	2.209×10^{10}	2.11×10^{8} (Cl)

injection of freons into the atmosphere has occurred.

The downward flux calculated for O_x corresponds to a global O_3 destruction rate of $\sim 2.58 \times 10^6$ metric tons per year; this is slightly larger than the observed values in the range of $0.5 - 1.5 \times 10^6$ metric tons per year (GALBALLY and ROY, 1980). The predicted flux of N_2O is equivalent to the global production rate of $\sim 1.29 \times 10^7$ metric tons per year, which is close to the value $\sim 1.4 \times 10^7$ metric tons per year estimated on the denitrification processes (see Section 11.1.3). Studies of CH_4 and H_2 fluxes from various solid types imply global production rates of $\sim 2.7 \times 10^8$ and $\sim 1.8 \times 10^4$ metric tons per year, respectively (KOYAMA, 1963), which are smaller than the model predicted values (3.28×10^8 and 3.7×10^5 metric tons per year, respectively), particularly for H_2. The difference may be accounted for by the fluxes due to volcanic gases and industrial waste gases. BLAKE *et al.* (1982) have estimated the yearly flux of $(4.0 \pm 1.3) \times 10^8$ metric tons needed to maintain the CH_4 concentration in steady state at its recent level of about 1.6 ppmv.

The negative flux of CO_2 at the surface may imply the dissolution of CO_2 into the ocean; this is the major sink of atmospheric CO_2. The CO emission rate is dependent on soil temperature and soil organic carbon content; CO is emitted from the surface under acid condition but emission turns into deposition when the soils are irrigated (CONRAD and SEILER, 1982). Thus, the CO flux can be either positive or negative depending upon locations and times. The downward (negative) flux of H_2O at the turbopause (the lower boundary set for H_2O) may imply that water vapor is produced in the stratosphere by methane chemistry (see Section 5.4).

7.1.3 Steady state models

Numerical method

A steady state model represents a condition to which the solution of the continuity equation approaches asymptotically. Technically, the same numerical method as was discussed in Section 7.1.2 can be used just be setting $\Delta t = \infty$ or eliminating the terms having Δt in the denominator in (7.25), (7.50), (7.51) and (7.62). The upper script l now represents the iteration number instead of the time step number. The solution sometimes changes greatly from iteration to iteration in the early stages of iteration, because the term relating $n^{l+1}(k)$ directly to $n^l(k)$, i.e. the first term of the right-hand side of (7.25) is omitted. This is particularly the case if unrealistic initial distributions are assumed. Problems arise, amongst others, when the initial ozone density distribution assumed is unrealistic or the distribution becomes unrealistic during the course of iteration; they give adverse effects on other minor constituents, which in turn adversely affect the ozone densities. Thus, the final result could become very unrealistic for many constituents. It is usually appropriate to assume as initial distributions for $[O_3]$ and [O] the photochemical equilibrium values of the pure oxygen atmosphere as given by (5.8) or (5.9).

In order to avoid the abovementioned problem during the course of iteration, it is useful to replace the value of n at the end of each iteration by a combination of the new value n^{l+1} and the old value n^l as follows:

$$n(k)=\theta n^{l+1}(k)+(1-\theta)n^l(k). \tag{7.64}$$

In most cases it is suitable to take θ as 0.5.

It may be most appropriate in the steady state model to assume that families of O_x, NO_x and Cl_x are composed of the *maximum* possible number of member constituents in each family, since we are considering the extreme case of $\Delta t \to \infty$. Thus, families are defined as

$$O_x=O(^3P)+O(^1D)+O_3+NO_2+HNO_3+2NO_3+3N_2O_5+2ClONO_2 \tag{7.65}$$

$$NO_x=N+NO+NO_2+NO_3+HNO_2+HNO_3+HNO_4+2N_2O_5 +ClONO_2 \tag{7.66}$$

and

$$Cl_x=Cl+ClO+2Cl_2+HCl+HOCl+ClONO_2+CH_3Cl. \tag{7.67}$$

The continuity equations including transport terms are solved for these families as well as other individual constituents having chemical time constants much longer than a day. Since the chemical time constants for these families are very long, chemical production and loss terms are so small that the transport effects should appear clearly on the density distributions of the families.

If a constituent appears in more than one family, its concentration is calculated by partitioning it within the family it is most strongly related to, and the result is used as an input in other families to calculate the remaining constituents. For example, concentrations of NO_2, NO_3, HNO_3, and N_2O_5 are calculated by partitioning within the NO_x family and that of $ClONO_2$ from the Cl_x family, and they are subtracted from O_x and NO_x in order to calculate the remaining constituents of these two families.

For hydrogen compounds (HO_x), the method of solution of simultaneous equations may be appropriate, since chemical time constants for H, OH and HO_2 are all very small except for H in the thermosphere. Continuity equations including transport terms should be solved for H (above the mesopause), H_2O (above the tropopause), H_2, H_2O_2 and various hydrocarbons (CH_6, CH_2O, CH_3O_2 etc....). The distribution of H_2O in the troposphere is prescribed in the model based on observations or assuming humidity (see Section 12.1.3 for the assumed profile for H_2O in the troposphere).

Calculations of the O_x system, the NO_x system, the HO_x system and the Cl_x systems are repeated by iteration until all constituents converge to stable values. Actually, our model repeats the calculation of the NO_x, HO_x and Cl_x systems five times in each system, and the coupling among the three systems is calculated by applying 10 iterations. Then, the result is coupled with the O_x system by 10 iterations, after the calculation of the O_x system has been repeated by 5 iterations within the system. The whole calculation is repeated five times, and we get a reasonably converged solution (SHIMAZAKI and OGAWA, 1974a).

Diurnal averaging

Steady state photochemical models normally simulate a daytime condition of a fixed sun, but sometimes it is useful to consider a model for the day-night average condition. For those constituents for which photodissociation is the dominant production (or loss) process and its time constant is much shorter than a day the former model should give a result comparable to observations. For those constituents whose chemical time constants are at least of the order of a day, however, the model for day-night average conditions seems to better predict the observational data (SHIMAZAKI and WHITTEN, 1976).

For a diurnally averaged model, the photodissociation rates must be averaged over an entire day. Since there is no photolysis at night, we actually need to calculate the integral of an exponential function in (3.8) over the daylight hours

$$\bar{J}=\frac{1}{24}J_\infty\int_{\substack{\text{day-light}\\ \text{hours}}} e^{-\tau/\mu(t)}\,dt \tag{7.68}$$

where J_∞ is the dissociation coefficient at the top of the atmosphere. We consider here monochromatic radiation (actually, $\bar{J}$, J_∞ and τ are functions of wavelength). Other notations in (7.68) are

$$\mu=\cos\chi=A+B\cdot\cos(\omega(t-12)) \tag{7.69}$$

with

$$A=\sin\phi\cdot\sin\delta,\quad B=\cos\phi\cdot\cos\delta \tag{7.70}$$

where ϕ is the latitude, δ is the solar declination, ω is the hour angle 15°/hour, and t is the local time.

The exponential integral function in (7.68) may be approximated by

$$I=2\int_{\text{sunrise}}^{\text{noon}} e^{-\tau/\mu(t)}\,dt \simeq 2\cdot a e^{-b\tau}. \tag{7.71}$$

Examining various matching techniques, COGLEY and BORUCKI (1976) have concluded that the following pair of a and b give a simple, one-term approximation that is fairly uniform over the height range 0–60 km with a maximum error of ~4% with no empirical correction,

$$a=\frac{\bar{\mu}}{2}\left(\frac{2t_0}{\overline{\mu^2}}\right)^{\frac{1}{2}},\quad b=\left(\frac{2t_0}{\overline{\mu^2}}\right)^{\frac{1}{2}} \tag{7.72}$$

where t_0 is the half-day length given by

$$\omega t_0=\cos^{-1}(-A/B) \tag{7.73}$$

and

$$\bar{\mu}=2\int_{12-t_0}^{12}\mu\,dt=2\left\{At_0+\frac{B}{\omega}\sin(\omega t_0)\right\} \tag{7.74}$$

$$\overline{\mu^2}=2\int_{12-t_0}^{12}\mu^2\,dt=2\left\{A^2t_0+\frac{2AB}{\omega}\sin(\omega t_0)+B^2\left(\frac{t_0}{2}+\frac{\sin(2\omega t_0)}{4\omega}\right)\right\}. \tag{7.75}$$

Alternative methods are also available for calculations of diurnally averaged photolytic rate coefficients in the literature including RUNDEL (1977) and KRAMER and WIDHOPF (1978).

Many constituents have different concentrations between day and night, and very often the concentrations are fairly constant for each of daytime and nighttime hours. TURCO and WHITTEN (1978) have considered a two level model, in which the daytime concentration n^D and the nighttime concentration n^N are constant and are represented by their averages in each case. A diurnally averaged concentration $\bar{n}$ is then defined by

$$\bar{n}=\frac{T^D n^D+T^N n^N}{T^D+T^N}=\frac{1+rt}{1+t}n^D \tag{7.76}$$

where T^D and T^N represent the light and dark hours, respectively. t and r indicate the ratio between the nighttime values and the daytime values, i.e.

$$t = T^N/T^D, \quad r = n^N/n^D. \tag{7.77}$$

The diurnally averaged photodissociation and chemical reaction rates are now expressed in terms of $\bar{n}$ as follows:

$$\overline{J_j n_j} = \frac{T^D J_j n_j^D}{T^D + T^N} = \bar{J}_j n_j^D = \alpha_j \bar{J}_j \bar{n}_j \tag{7.78}$$

and

$$\overline{k_{ij} n_i n_j} = k_{ij} \frac{T^D n_i^D n_j^D + T^N n_i^N n_j^N}{T^D + T^N} = \beta_{ij} k_{ij} \bar{n}_i \bar{n}_j. \tag{7.79}$$

Thus, the photodissociation coefficient and the chemical reaction constant in the continuity equation for $\bar{n}$ must be modified by the factors α_j and β_{ij} from $\bar{J}_j$ and k_{ij}, respectively. The two scaling factors are given by

$$\alpha_j = \frac{1+t}{1+r_j t} = 1 + t\frac{1-r_j}{1+r_j t} \tag{7.80}$$

and

$$\beta_{ij} = \frac{(1+r_i r_j t)(1+t)}{(1+r_i t)(1+r_j t)} = 1 + t\frac{(1-r_i)(1-r_j)}{(1+r_i t)(1+r_j t)}. \tag{7.81}$$

It is evident from (7.80) that if there is no diurnal variation in n_j (or $r_j = 1$), $\alpha_j = 1$; therefore, the α_j factor is necessary only for photolyses of constituents having large diurnal variations. It is interesting to note that α_j becomes very small if $n^N \gg n^D$ or $r_j \gg 1$. For instance, a very small α_j for the NO_3 photolysis makes the predicted diurnal average for $[NO_3]$ (and also for $[N_2O_5]$) much larger than the values calculated in a fixed sun steady state model. The result is reasonable since the time-dependent diurnal model predicts much larger $[NO_3]$ and $[N_2O_5]$ than the fixed sun steady model (SHIMAZAKI and OGAWA, 1974a; WHITTEN and TURCO, 1974a).

(7.81) indicates that $\beta_{ij} = 1$ if $r_i = 1$ or $r_j = 1$; therefore, the scaling factor β_{ij} is necessary only if *both* constituents involved in the chemical reaction have large diurnal variations. It is also noted that $\beta_{ij} > 1$ if both r_i *and* r_j are greater than unity or smaller than unity, and that $\beta_{ij} < 1$ if one of r_i and r_j is greater than unity and the other is smaller than unity.

It is interesting to note that $\beta_{ij} < 1$ for the reaction $NO_2 + O$ but $\beta_{ij} > 1$ for the reaction $ClO + O$ in the stratosphere. Thus, the scaling factor β_{ij} would decrease the O_x loss rates by the NO_x catalytic cycle, whereas it would increase those by the ClO_x catalytic cycle, from those values calculated using the diurnally averaged concentrations and the usual (unmodified) reaction rate constants. This is caused by the fact that the diurnal variation of ClO is in phase with that of O, whereas NO_2 is out of phase with O.

In the diurnally averaged model, ratios of concentration among family members must be specified using diurnal averages. For instance, if the daytime concentration ratio n_i^D/n_j^D is determined by the rapid exchange reactions composed of the photolysis of n_j and the chemical reaction between n_i and n_l, the ratio of diurnally averaged concentrations of n_i and n_j can be determined by

$$\frac{\bar{n}_i}{\bar{n}_j} = \frac{n_i^D}{n_j^D}\frac{1+r_i t}{1+r_j t} = \frac{J_j}{k_{il} n_l^D}\frac{1+r_i t}{1+r_j t} = \frac{\alpha_j \bar{J}_j}{\beta_{il} k_{il} \bar{n}_l}(1+r_i r_l t). \tag{7.82}$$

Scaling factors α_j and β_{ij} may be calculated from concentrations at noon and midnight of a diurnal model. It must be noted that each aeronomical model requires an independent evaluation of the scaling factors, depending upon the kind of constituents and their chemical reaction schemes assumed in the model. Once these scaling factors are determined for the adopted photochemical reaction schemes, however, they may be treated approximately as invarient parameters for model calculations of long term variations (TURCO and WHITTEN, 1978).

A diurnally averaged model has been calculated by the above discussed method for latitude 35° and equinox condition using the chemical reactions listed in Table 5.2 with the reaction rate constants of WMO (1981). The height profiles of various constituents predicted by the model are illustrated in Figs. 7.3(a) – (c).

The most significant feature of this model is the lower concentrations of OH and HO_2 and the higher concentrations of NO_2 than previous models, particularly in the lower stratosphere. This is a natural consequence of adopting faster loss rates for OH by reactions $OH+HO_2$, $OH+HNO_3$, and $OH+HNO_4$ as recommended in WMO (1981) and JPL (1981). The larger $[NO_2]$ is the result of the smaller [OH], since it reduces the rate of conversion of NO_2 to HNO_3 by the reaction $OH+NO_2+M$.

The recent evaluation of the reaction rate constants by JPL (1982) recommends a faster rate for the reaction $O+HO_2 \rightarrow OH+O_2$ (R_6; see Table 5.2). KO and SZE (1983) have reported that this new reaction rate would result in a factor of about 1.3 increase in [OH] and also a factor of about 1.4 increase in [Cl] and [ClO] above 40 km. These increased concentrations of OH and ClO_x should decrease the predicted O_3 concentration, enhancing the discrepancy between the model result and observations for $[O_3]$ above 40 km (see p. 114). A smaller O_2 absorption cross section in the Herzberg continuum as is recently suggested from observations of the solar irradiance in the upper stratosphere (HERMAN and MENTALL, 1982; FREDERICK and MENTALL, 1982) makes the predicted ozone concentration

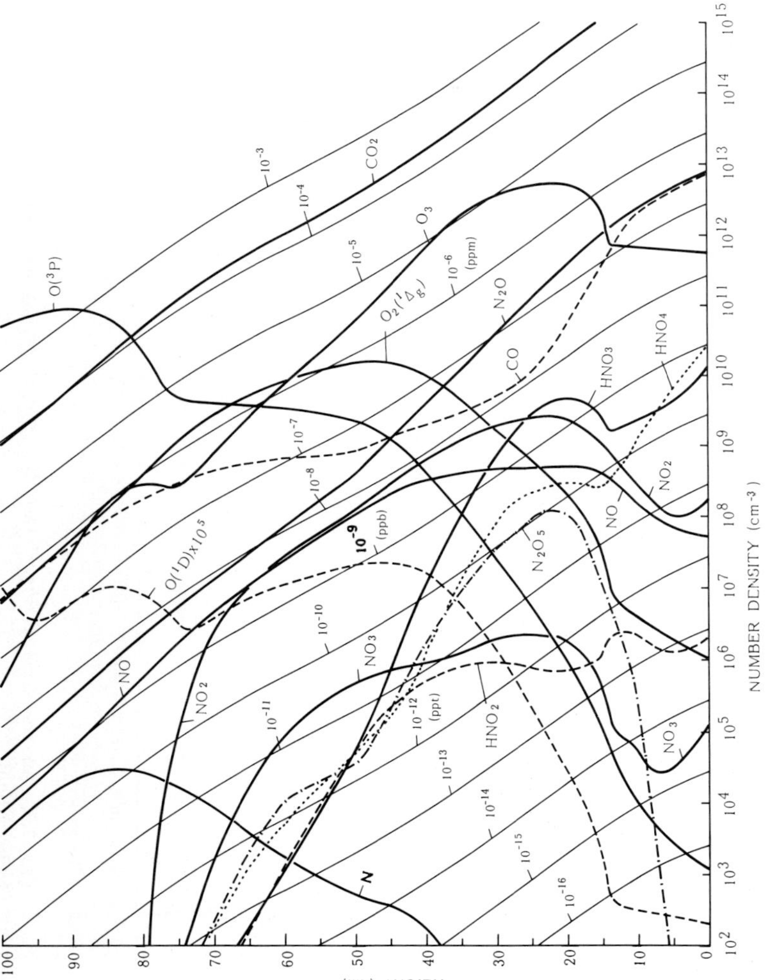

FIG. 7.3. (a) Height profiles of various minor constituents predicted in one-dimensional durnally averaged steady state model: I pure oxygen, nitrogen-oxygen and carbon-oxygen compounds.

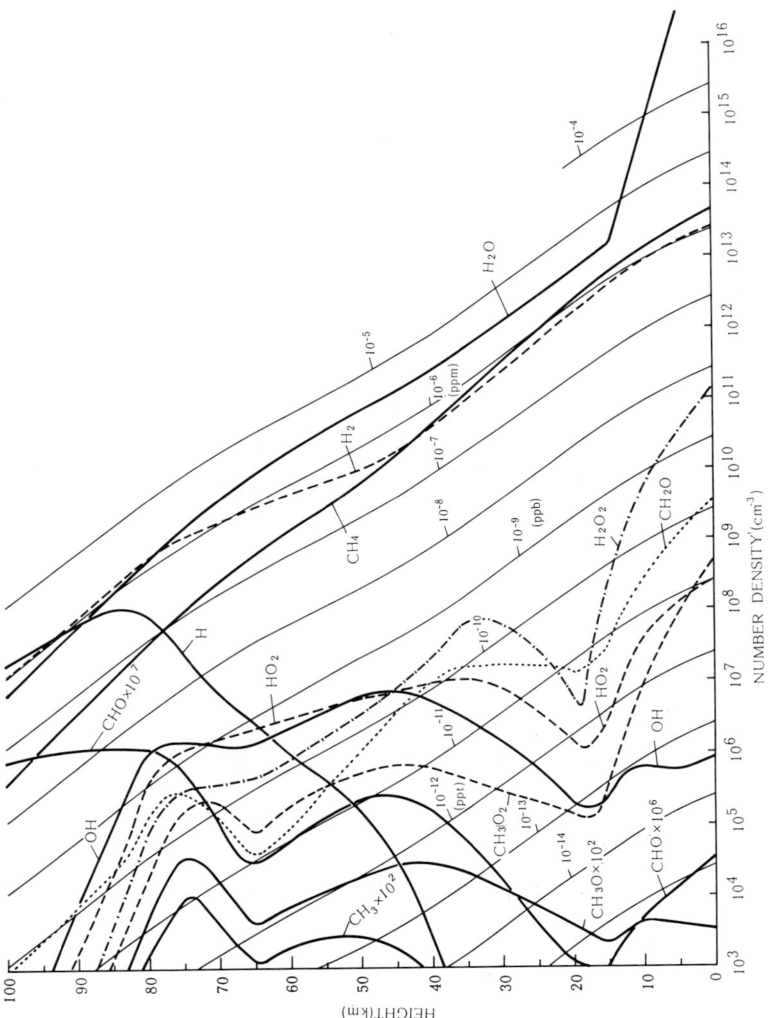

FIG. 7.3. (b) The same as Fig. 7.3(a): II hydrogen-oxygen compounds and hydrocarbons.

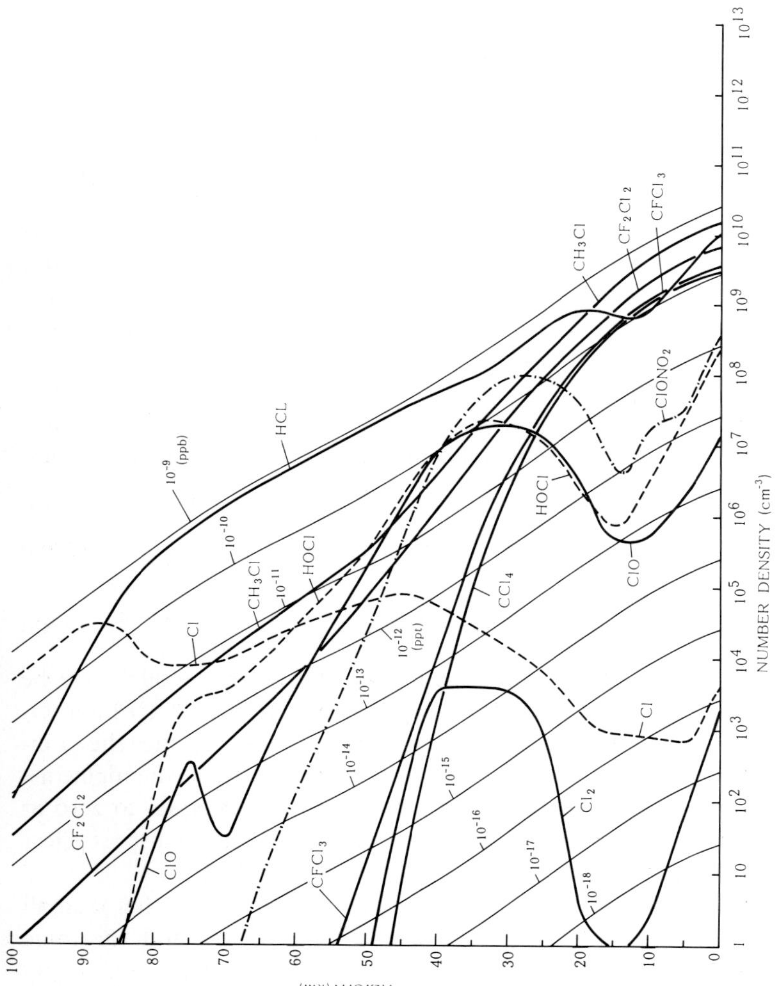

FIG. 7.3. (c) The same as Fig. 7.3(a): III chlorine-oxygen compounds and chlorofluoromethanes.

even lower, causing further deterioration of the agreement between model and observations for $[O_3]$ in the upper stratosphere, although it was reported by FROIDEVAUX and YUNG (1982) that such a reduced Herzberg cross section would be favorable for bringint the predicted stratospheric concentrations of HNO_3, $CFCl_3$ and CF_2Cl_2 into closer agreement with observations.

7.1.4 Diurnal models

A steady state model does not represent the actual state for the constituents whose concentrations change largely with a diurnal cycle. For some constituents a two-level diurnal model represents a good approximation, particularly if each of the daytime and nighttime values remains nearly constant at different levels (for some constituents the nighttime value becomes essentially zero). If there are large diurnal variations throughout the day and night, however, those diurnal variations could affect the levels of other constituents that have no diurnal variations themselves. Thus, it is essential to consider a fully diurnal model.

In our diurnal model the continuity equation containing the transport terms has been solved for the individual species that have photochemical time constants much larger than a day and also for the following families:

$$O_x = O(^3P) + O(^1D) + O_3 \tag{7.83}$$

$$NO_x = NO + NO_2 + NO_3 + 2 \cdot N_2O_5 + ClONO_2 \tag{7.84}$$

and

$$Cl_x = Cl + ClO + HOCl + ClONO_2. \tag{7.85}$$

The choice of these family constructure for the diurnal model may be justified by the discussion in Chapter 6. At night, however, $O(^3P)$ and O_3 are treated as individual species instead of a family member, since the family method always gives a null concentration for $O(^3P)$ under the nighttime condition (note that the ratio $[O]/[O_3]$ given by (5.4) is equal to zero at night). Because chemical reactions are less active at night, transport effect becomes more important for each of $O(^3P)$ and O_3 at night.

A numerical problem in the diurnal model lies in the fact that a small time step Δt ($\ll$min) is required, particularly at twilight and for some period after sunset when sudden and large changes occur for many chemical species. For instance, if the amount of O_3 produced by the reaction (R_2) after sunset is calculated by multiplying Δt to the instantaneous production rate, the result could greatly exceed the amount of O available for O_3 production unless Δt is taken to be small enough to make that product smaller than the amount of O. Since there is essentially no production of O at night, the above situation would give a large spurious addition to

O_3, which seriously affects various other constituents. Is is more appropriate for purposes of computer economy to solve this problem by setting an upper limit for the O_3 production rate based on the available amount of O rather than by reducing Δt (SHIMAZAKI, 1981). Following this procedure and also using the family method, the diurnal model was successfully calculated using $\Delta t = 12$ min throughout the entire day.

In Table 7.3 are given the concentrations of all species predicted in our diurnal model for noon and midnight at several chosen heights. The diurnal variations for [O] and [O_3] at various heights are also shown in Figs. 7.4(a) and (b), respectively. There is essentially no diurnal variation for [O_x] below ~40 km and above ~85 km, since the photochemical time constants exceed a day (see Fig. 6.4). Thus, no diurnal variation can be seen for [O_3] below ~40 km and for [O] above ~85 km.

Most O is converted to O_3 and disappears below ~85 km at night; therefore, [O] has a large day-night difference. Because the amount of daytime O is much less than O_3 below ~55 km, however, the ozone density does not show an appreciable difference between day and night. Above ~ 55 km the nighttime [O_3] is considerably larger than the daytime [O_3] because larger amounts of O change to O_3 at night. This change occurs in a short time period just after sunset, and since there is no more O production at night after all of the available O has been changed to O_3, the nighttime [O_3] remains nearly constant for the rest of the night until sunrise when photolysis starts to destroy it rather suddenly and to decrease its density.

Somewhat large variations can be seen in the daytime [O_3] at 70 - 80 km. While these variations are certainly caused by the fact that the photochemical time constant of O_x is in of the order of an hour at these heights (see Fig. 6.4), the chemical reactions affecting these O_3 variations are most likely related to the HO_x chemistry in the mesosphere (see Chapter 5). The diurnal variations in [OH] at various heights are illustrated in Fig. 7.5. It is evident that the daytime variation in [OH] at 70 km is inversely correlated to that in [O_3]. It is also noted that [OH] is larger during the daytime than at night below ~70 km, whereas the reverse is the case above ~80 km. This may be related to the fact that a considerable amount of H remains at night above ~80 km because of a longer chemical time constant (>1 day), whereas H is quickly lost at night in the lower region. H is a source for other forms of HO_x through the reaction (R_5): $H+O_2+M \rightarrow HO_2+M$.

The noon and midnight profiles of [O] and [O_3] and those constituents directly affecting [O_x] by catalytic reactions are illustrated in Fig. 7.6. Some details of the changes in [O] and [O_3] profiles near sunrise and sunset are shown in Fig. 7.7; these are taken from the model calculated by OGAWA and SHIMAZAKI (1975) including ionic reactions and using a smaller time

TABLE 7.3 Concentrations (cm^{-3}) of various constituents predicted in the time-dependent diurnal model*

z(km)	10	15	20	25	30	45	60	80	100
T(°K)	225.6	208.0	211.9	217.1	224.6	269.6	256.5	192.1	206.2
k($cm^2\ sec^{-1}$)	1.34(5)	5.00(3)	9.66(3)	2.20(4)	5.01(4)	1.43(5)	4.09(5)	1.66(6)	6.73(6)
N_2	7.11(18)	3.47(18)	1.50(18)	6.59(17)	2.93(17)	3.02(16)	4.79(15)	3.03(14)	9.12(12)
O_2	1.78(18)	8.68(17)	3.76(17)	1.65(17)	7.32(16)	7.56(15)	1.20(15)	7.57(13)	2.28(12)
$O(^3P)$	1.48(4)	1.85(5)	2.56(6)	1.37(7)	6.75(7)	3.95(9)	9.06(9)	2.12(10)	5.30(10)
	1.19(−9)	2.14(−16)	2.27(−19)	1.21(−12)	1.93(−7)	6.14(−2)	7.20(−1)	4.94(9)	5.23(10)
$O(^1D)$	1.83(−2)	1.08(−1)	1.13(0)	5.70(0)	3.01(1)	6.46(2)	2.76(2)	8.22(1)	4.07(2)
	0	0	0	0	0	0	0	0	0
O_3	6.47(11)	2.25(12)	5.46(12)	5.05(12)	3.96(12)	2.05(11)	6.99(9)	1.13(8)	2.21(5)
	6.43(11)	2.25(12)	5.45(12)	5.04(12)	3.96(12)	2.02(11)	1.48(10)	2.97(8)	2.97(6)
$O_2(^1\Delta_g)$	1.19(7)	9.32(7)	6.16(8)	1.74(9)	5.00(9)	4.94(10)	2.03(10)	1.99(9)	6.39(6)
	0	0	0	0	0	0	8.77(−5)	1.84(6)	3.20(4)
H_2O	5.13(14)	1.51(13)	6.37(12)	2.74(12)	1.19(12)	1.19(11)	1.59(10)	4.96(8)	5.52(6)
OH	9.52(5)	7.06(5)	8.34(5)	1.76(6)	3.92(6)	1.83(7)	4.98(6)	3.51(5)	1.26(2)
	3.64(3)	6.44(−1)	7.87(−3)	9.36(0)	1.82(3)	8.97(4)	5.47(4)	2.04(6)	9.01(2)
HO_2	1.25(8)	9.36(6)	7.00(6)	1.41(7)	2.58(7)	1.49(7)	3.79(6)	2.76(5)	5.44(1)
	1.75(6)	4.86	2.07(−1)	2.21(2)	4.35(4)	2.86(6)	2.13(6)	1.05(6)	5.44(1)
H	2.09(−1)	1.99(−1)	2.43(−1)	7.62(−1)	7.68(0)	1.08(5)	2.46(6)	7.00(7)	4.03(7)
	7.84(−4)	1.79(−7)	2.18(−9)	3.20(−6)	1.18(−3)	1.10(0)	7.71(0)	6.26(7)	4.17(7)
H_2O_2	7.82(9)	1.15(8)	1.09(7)	3.72(7)	8.33(7)	6.93(6)	4.80(5)	2.12(4)	1.73(0)
	8.00(9)	1.19(8)	1.12(7)	3.82(7)	8.63(7)	9.04(6)	8.93(5)	9.15(4)	2.07(0)
H_2	8.34(12)	3.84(12)	1.52(12)	6.06(11)	2.44(11)	1.93(10)	4.06(9)	4.97(8)	1.25(7)
N_2O	2.50(12)	1,13(12)	3.88(11)	1.30(11)	4.21(10)	1.22(9)	7.11(7)	2.32(6)	2.25(4)
NO	8.92(7)	7.34(8)	8.19(8)	7.16(8)	7.29(8)	4.64(8)	4.53(7)	6.83(5)	8.90(3)
	3.9(−10)	5.5(−16)	3.8(−19)	1.6(−12)	2.40(−7)	1.18(−1)	2.16(0)	6.81(5)	8.96(3)

NO_2	5.36(7)	5.03(8)	1.54(9)	1.53(9)	1.48(9)	2.56(7)	4.63(4)	1.30(1)	1.36(−3)
	8.43(7)	1.19(9)	2.15(9)	1.94(9)	1.77(9)	4.58(8)	4.52(7)	2.05(2)	1.48(−3)
NO_3	3.20(4)	1.24(5)	7.10(5)	8.89(5)	9.54(5)	1.32(4)	9.77(0)	4.34(−4)	3.33(−9)
	2.79(6)	9.10(6)	3.04(7)	4.48(7)	6.88(7)	2.46(7)	1.10(5)	7.56(−1)	6.10(−6)
N_2O_5	1.47(6)	9.26(7)	4.69(8)	4.42(8)	3.17(8)	1.29(4)	3.97(−1)	2.32(−8)	1.4(−16)
	1.45(6)	9.91(7)	5.48(8)	5.40(8)	4.43(8)	2.89(6)	3.55(2)	3.15(−8)	5.9(−17)
N	3.76(−8)	1.34(−4)	1.80(−2)	1.11(0)	3.18(1)	1.22(3)	4.06(3)	3.48(4)	4.38(3)
	0	0	0	0	0	0	4.4(−34)	1.77(4)	4.35(3)
HNO_3	2.02(9)	2.67(9)	4.80(9)	2.69(9)	5.14(8)	6.89(5)	5.48(2)	1.50(−1)	8.59(−7)
	1.98(9)	2.67(9)	4.80(9)	2.68(9)	5.24(8)	9.60(5)	3.88(3)	1.31(−1)	1.16(−6)
HNO_4	9.44(8)	2.76(8)	2.74(8)	1.81(8)	6.94(7)	1.48(4)	3.61(0)	3.95(−5)	1.7(−11)
	9.34(8)	2.82(8)	2.79(8)	1.86(8)	7.61(7)	4.33(5)	2.90(3)	3.23(−3)	2.9(−12)
HNO_2	3.45(5)	1.23(6)	8.52(5)	7.74(5)	8.26(5)	2.60(5)	1.26(3)	9.29(−2)	2.86(−8)
	3.54(5)	1.09(6)	4.17(5)	4.45(5)	5.20(5)	1.86(5)	2.88(3)	9.3(0)	1.77(−5)
CH_4	1.33(13)	6.03(12)	2.21(12)	8.10(11)	3.00(11)	1.56(10)	1.59(9)	4.19(7)	4.57(5)
CH_3	1.65(−2)	9.10(−3)	2.22(−2)	1.06(−1)	6.12(−1)	4.90(1)	3.38(1)	5.02(1)	3.32(−1)
	6.03(−5)	7.70(−9)	1.5(−10)	3.32(−7)	1.38(−4)	1.02(−1)	1.76(−1)	4.71(−2)	9.77(−7)
CH_3O_2	2.41(7)	5.44(5)	3.15(5)	4.14(5)	5.32(5)	8.26(5)	1.51(5)	2.82(4)	4.48(2)
	1.99(7)	8.39(5)	2.77(5)	4.39(5)	5.91(5)	5.12(5)	1.20(5)	2.76(4)	4.52(2)
CH_3O	7.93(1)	6.47(1)	8.11(1)	1.69(2)	3.64(2)	7.56(2)	1.25(2)	8.09(1)	2.72(−1)
	2.9(−16)	7.6(−23)	3.3(−26)	4.1(−19)	1.3(−13)	1.18(−7)	4.71(−6)	7.99(1)	2.68(−1)
CH_2O	1.69(8)	2.84(7)	1.53(7)	1.57(7)	1.69(7)	4.22(6)	1.14(5)	3.88(5)	2.20(3)
	1.91(8)	3.20(7)	1.55(7)	1.55(7)	1.51(7)	3.89(6)	1.78(5)	8.23(3)	4.28(−2)
CHO	7.18(−4)	2.33(−4)	3.19(−4)	1.01(−3)	3.81(−5)	7.06(−2)	1.47(−2)	2.94(−1)	1.84(−1)
	7.80(−7)	4.8(−11)	6.5(−13)	1.76(−9)	7.51(−7)	9.26(−5)	1.62(−5)	1.48(−3)	3.25(−6)
CO	2.03(12)	7.05(11)	1.26(11)	2.78(10)	9.75(9)	1.64(9)	7.96(8)	2.00(8)	7.47(6)

TABLE 7.3 (continued)

CO_2	2.75(15)	1.27(15)	5.08(14)	2.07(14)	8.52(13)	7.26(12)	1.02(12)	5.15(10)	1.05(9)
Cl	2.06(3)	1.44(3)	4.95(3)	1.60(4)	3.00(4)	2.29(5)	6.36(4)	1.13(4)	1.05(3)
	1.5(−17)	5.2(−30)	2.3(−34)	7.7(−24)	7.8(−16)	5.50(−8)	2.83(−7)	7.75(3)	1.03(3)
ClO	5.85(6)	1.41(6)	1.10(7)	3.87(7)	5.34(7)	2.53(6)	1.07(4)	1.12(11)	8.63(−4)
	1.40(3)	4.15(−3)	5.43(−4)	3.67(0)	2.00(3)	2.35(4)	5.69(2)	8.75(1)	1.13(−2)
Cl_2	3.41(1)	1.24(0)	8.41(1)	1.17(3)	2.59(3)	1.40(1)	1.92(−4)	4.2(−11)	4.7(−19)
	8.59(1)	2.02(0)	1.05(2)	2.18(3)	7.97(3)	5.33(1)	5.59(−3)	1.64(−6)	2.9(−14)
HCl	1.72(9)	6.36(8)	8.28(8)	5.13(8)	2.19(8)	3.38(7)	5.24(6)	2.58(5)	4.41(3)
	1.77(9)	6.38(8)	8.28(8)	5.12(8)	2.18(8)	3.38(7)	5.22(6)	2.62(5)	4.44(3)
HOCl	3.68(7)	8.64(5)	4.55(6)	2.76(7)	5.70(7)	5.89(5)	6.71(2)	1.35(−1)	1.56(−9)
	1.13(7)	8.68(5)	7.56(6)	1.72(7)	1.99(7)	2.90(6)	9.72(4)	5.11(2)	7.17(−7)
$ClONO_2$	3.03(6)	3.53(6)	3.65(7)	5.08(7)	1.96(7)	9.99(1)	7.05(−5)	1.5(−12)	1.6(−22)
	3.76(7)	5.23(6)	4.49(7)	1.00(8)	1.10(8)	5.65(5)	3.78(2)	1.36(−9)	2.7(−21)
$CFCl_3$	1.25(9)	5.47(8)	1.51(8)	3.23(7)	4.25(6)	4.63(2)	6.90(−2)	1.88(−5)	4.22(−8)
CF_2Cl_2	2.08(9)	9.31(8)	3.02(8)	9.15(7)	2.52(7)	2.89(5)	5.82(3)	7.56(1)	8.85(−1)
CCl_4	1.08(9)	4.73(8)	1.26(8)	2.45(7)	2.50(6)	2.19(1)	1.82(−4)	4.62(−9)	3.2(−12)
CH_3Cl	4.84(9)	2.12(9)	6.53(8)	1.96(8)	5.79(7)	1.09(6)	5.62(4)	1.81(3)	3.19(1)

*The upper line indicates noon values, whereas the bottom line indicates midnight values. If there is only one line, the constituent has little diurnal variation. Assumed temperature (T) and eddy diffusion coefficient (k) are also given. The numeral $A(n)$ should read as $A \times 10^n$.

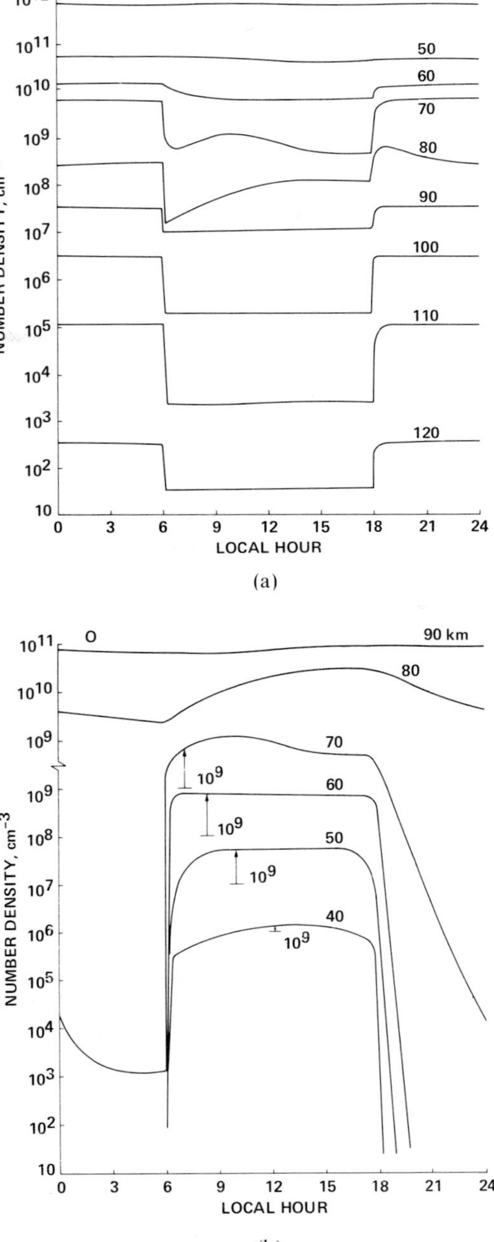

FIG. 7.4. (a) Diurnal variations in $[O_3]$ at various heights. (b) Diurnal variations in [O] at various heights.

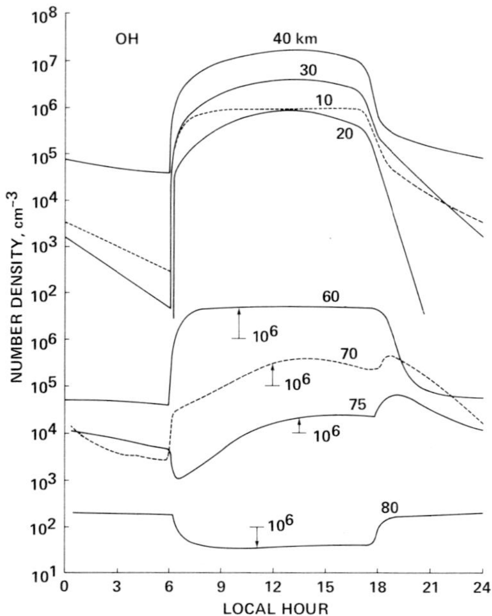

FIG. 7.5. Diurnal variations in [OH] at various heights.

step. Diurnal variations of some typical constituents at 25 km and 45 km are depicted in Fig. 7.8 (a) and (b), respectively. The predicted larger O_3 density at night than during the daytime above $\sim$50 km is consistent with the result of twilight ozone measurements by solar occultation from the AES satellite (GUENTHER *et al.*, 1977).

7.2 *Multi-Dimensional Models*

The greatest deficiency of a one-dimensional model is that the transport occurs only in vertical direction and its effects are simulated poorly by a single parameter, the eddy diffusion coefficient. Of course, in the real atmosphere the motion occurs in the three-dimensional domain, and the treatment of transport in a 3-D model certainly is the best approach for the simulation. The 3-D model can also treat fully the interaction between dynamics and radiation.

Some tropospheric 3-D global circulation models have been extended to include the stratosphere (e.g. KASAHARA and SASAMORI, 1974; MANABE and MAHLMAN, 1976; O'NEILL *et al.*, 1982). They are used particularly for transport of the stratospheric constituents by CUNNOLD *et al.*, (1975),

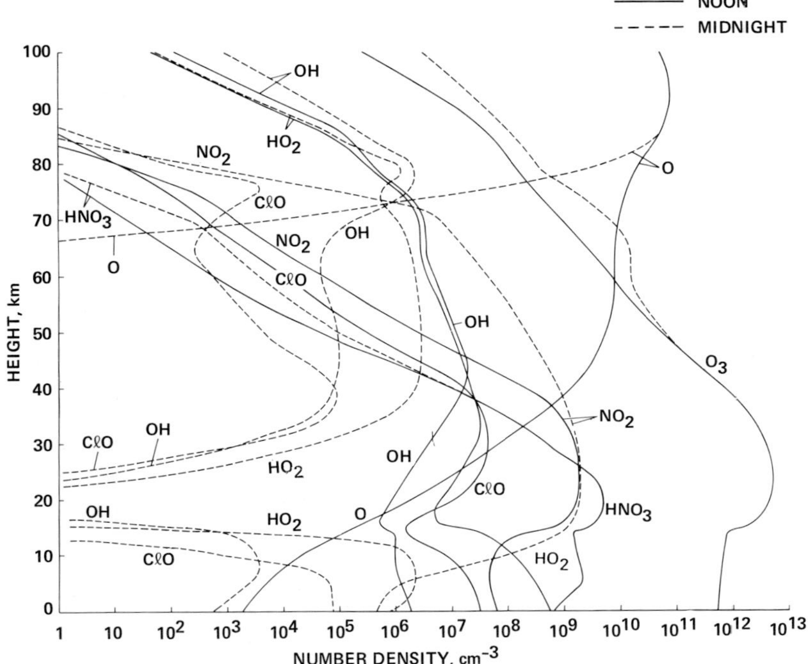

FIG. 7.6. Height profiles of concentrations of various constituents at noon and midnight predicted in our diurnal model.

MAHLMAN and MOXIM (1978), LEVY *et al.* (1979), ALLAN *et al.* (1981), and others. However, the chemistry that can be treated in 3-D models has a severe limitation because of the enormous computer time and memory sizes required, and all existing 3-D stratospheric models are far from comprehensive in terms of chemistry.

Two-dimensional models are perhaps the best compromise between one-dimensional models that have a poor representation for transport and three-dimensional models that have very limited chemistry. In 2-D models all variables are separated into two components; the zonal (longitudinal) mean and the deviation from it. The zonal mean velocities constitute the large scale meridional mean circulation, and the time average of the product between the components of fluctuation for velocity and particle concentration constitutes the eddy transport flux (see Eq. (9.3)). The magnitude of the eddy transport is generally comparable to transport by the mean motion.

Many 2-D models have been developed by various investigators, in-

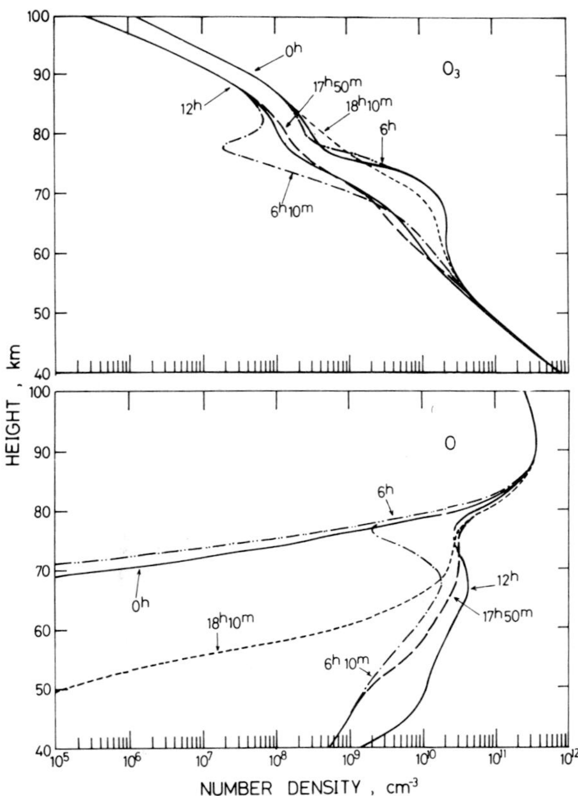

FIG. 7.7. Height profiles of $[O_3]$ and [O] near sunrise and sunset (OGAWA and SHIMAZAKI, 1975).

cluding SHIMAZAKI *et al.* (1973), SHIMAZAKI (1974), WIDHOPF and TAYLOR (1974), PRINN *et al.* (1975), HIDALGO and CRUTZEN (1977), BRASSEUR (1978), WHITTEN *et al.* (1977), BORUCKI *et al.* (1980), HYSON *et al.* (1980), HOLTON (1981), MILLER *et al.* (1981), GIDEL *et al.* (1983), and GARCIA and SOLOMON (1983). Most of them use the meridional circulation calculated, for instance, by LOUIS (1974) and the eddy transport represented by the eddy diffusion tensor (*K*-values). The *K*-values were first determined by REED and GERMAN (1965) using the data on heat transport by eddy motions. Later, they were reevaluated by GUDICKSEN *et al.* (1968) and LUTHER (1973). Some 2-D models use these *K*-values, but others determine their own *K*-values so that the predicted distributions of O_3 and other constituents may match observations satisfactorily. More discussions on *K*-values are given in Section 9.3. Equations and a numerical method of solv-

(a)

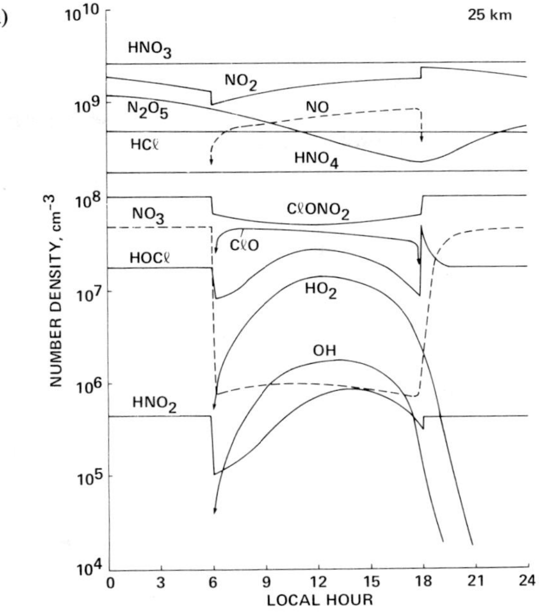

(b)

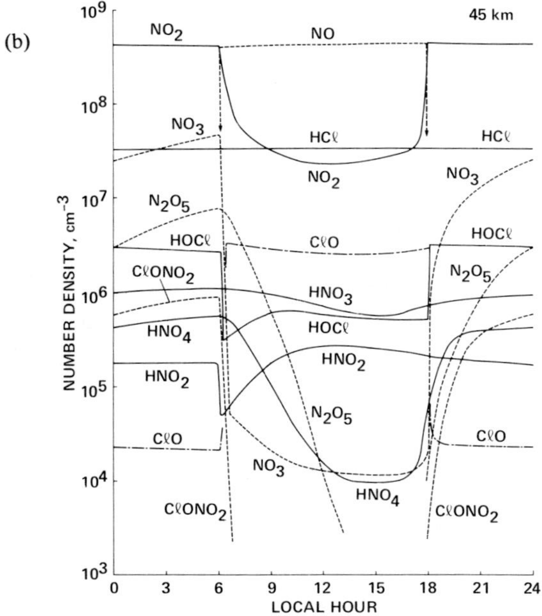

FIG. 7.8. (a) Diurnal variations in concentrations of various constituents at 25 km predicted in our diurnal model. (b) The same as (a) but at 45 km.

ing them in 2-D models are discussed in Appendix H.

Newer models are available for zonally averaged mean circulation (e.g. SCHOEBERL and STROBEL, 1978a; HOLTON and WEHRBEIN, 1981). Instead of specifying the mean circulation, some 2-D models can calculate it within the model so that it can be consistent with the radiation field of the model (RAO-VUPPUTURI, 1973, 1976; HARWOOD and PYLE, 1975, 1977). These models have advantages in that the feedback from ozone changes to the fields of radiation and dynamics can be calculated in the model.

Although 2-D models can treat the meridional transport as well as the vertical transport, these transports are still represented by poorly known parameters, the K-values. In addition, it is now well known that the flux by eddy motions almost completely compensates for the flux by mean motions, and that evaluations of a meaningful transport as the difference between two large quantities is a very difficult task. Recently, however, the latter problem seems to be partially solved by considering the Lagrangian mean transport (see Section 9.2.2 for details).

Chapter 8

OBSERVATIONS OF MIDDLE ATMOSPHERE COMPOSITION

Observations of middle atmosphere constituents can be made either by *in situ* measurements or remote sensing techniques. In *in situ* measurements the instruments may be designed to be most sensitive to each constituent and the measured signal is more easily interpreted than for remote sensing. Although finer height resolution can be achieved by *in situ* measurements, the region of observation is limited. Sample modification can occur either by contamination through outgasing or reactions of samples with surfaces of instruments and platforms. On the other hand, the technique of remote sensing may be applied to a larger area and the identification of measured constituens with very small abundances can be made accurately; thus, it is suitable for an exploratory search for unknown compounds. Generally speaking, *in situ* measurements are better suited to studies of local problems, whereas remote sensing is better suited for studying the global structure (EHHALT, 1980; HARRIES, 1980a). In the following, both techniques are briefly discussed, and the special aspects of observations and their results will be discussed for each constituent, particularly for O_3.

8.1 Observational Techniques

8.1.1 In situ Measurements

In situ measurements refer to observations made at the location of the instrument. They include *direct measurements* made by instruments taken into the middle atmosphere by means of balloons, aircrafts or rockets, and *indirect measurements* in which samples are taken from the middle atmosphere and are analyzed in laboratories on the ground. Balloons usually go up to ~30 km, but in some special cases experiments can be made up to 40–45 km. The altitude of aircracft is usually limited to lower than ~25 km, whereas rockets can go through the entire region of the middle at-

TABLE 8.1 *In situ* measurements by various direct methods.

Method	Constituent	Platform	References
1. Electrochemical	O_3	balloon	Brewer and Milford, 1960; Torres and Bandy, 1978; Thornton and Niazy, 1983
2. Mass spectrometer	CO_2	balloon	Mauersberger and Finstad, 1980
	O_3, Ar	balloon	Mauersberger *et al.*, 1981
(photoinonization)	NO	gondola	Maier *et al.*, 1978
	O_3	rocket	Trinks, 1975
	Ar, O, O_2, N_2	rocket	Scholtz and Offermann, 1974
(ion composition)	H_2SO_4	balloon	Arnold *et al.*, 1981 c
3. Chemiluminescence	NO	aircraft	Loewenstein *et al.*, 1974
		balloon	Ridley *et al.*, 1975; Drummond *et al.*, 1977; Evans *et al.*, 1977; Weiler *et al.*, 1980
		rocket	Horvath and Mason, 1978
	O_3	balloon	Regener, 1964; Hering and Dütsch, 1965
		rocket	Hilsenrath, 1980
4. Resonance fluorescence			
(resonance lamp)	O	balloon	Anderson, 1975
		rocket	Dickinson *et al.*, 1974 and 1980
	OH	balloon	Anderson, 1976
	HO_2	balloon	Anderson *et al.*, 1981
	Cl, ClO	balloon	Anderson *et al.*, 1980
(laser beam)	OH	aircraft	Davis *et al.*, 1976
	NO_2	ground	Tucker *et al.*, 1975
(Lyman-α)	H_2O	aircraft	Kley and Stone, 1978
(Lyman-β)	H_2	rocket	Feldman and Takacs, 1975
5. Hygrometers			
(aluminum oxide)	H_2O	aircraft	Hilsenrath *et al.*, 1977
		balloon	Vanderhoff, 1974
(frost point)	H_2O	balloon	Mastenbrook, 1971, and 1974
(α-radiation)	H_2O	balloon	Sissenwine *et al.*, 1968
6. Mercury oxide	H_2	aircraft	Schmidt, 1974
7. Silver film senser	O	rocket	Henderson and Schiff, 1970
8. Gas chromatography	F_{11}, F_{12}	ground	Lovelock, 1974
	F_{11}, F_{12}, CCl_4, N_2O	aircraft	Tyson *et al.*, 1978 b
	CH_3Cl	ship	Rasmussen *et al.*, 1980 a and b
	$CHClF_2$, CH_3CCl_3	ground	Rasmussen and Khalil, 1981
	CF_4, C_2F_6	ground	Penkett *et al.*, 1981

mosphere. However, rockets can stay in the middle atmosphere for only a very short period, and the parachute technique is sometimes used in order to make extended observations.

A large variety of techniques are available for *in situ* measurements. Table 8.1 lists various methods of direct measurements along with the constituents that have been measured by each method. The *chemiluminescence* method for measuring the NO density uses the fluorescence of wavelength 1.2 ± 0.6 μm emitted from NO_2 produced by the chemical reaction of the ambient NO and a known quantity of O_3 supplied by an ozone bombe (cylinder).

The *resonance fluorescence* technique measures the intensity of fluorescence emitted from atoms or molecules that are excited to higher energy levels by a microwave plasma discharge (resonance lamp) or a UV tunable laser beam. Atomic oxygen is detected by measuring the emission at the allowed transition $^3P - {}^3S$ at 130.4 nm, whereas OH is detected by the electronic transition $A^3\Sigma - X^2\Pi$ extending from 306.4 to ~312.0 nm. KLEY and STONE (1978) have used a Lyman-α photodissociation hygrometer detecting fluorescence emitted from OH, the products from the H_2O photolysis.

Atomic chlorine is measured by using the $^2D_{5/2} - {}^2P_{3/2}$ transition at 118.9 nm. The ClO molecule excited at the $A^2\Pi$ state predissociates more rapidly than it relaxes to the ground state ($X^2\Pi$). However, ClO can be measured by converting it to Cl via the rapid chemical reaction $ClO + NO \rightarrow Cl + NO_2$; the resulting Cl atoms can then be detected by resonance fluorescence. Similarly, the HO_2 measurement can be accomplished by converting it to OH by the rapid reaction $HO_2 + NO \rightarrow OH + NO_2$ followed by detection of OH by the resonance fluorescence technique.

There are three different methods for collecting samples. The simplest technique is *grab sampling*, and this method is useful for measuring relatively stable constituents such as N_2O, $CFCl_3$, CCl_4 etc. *Cryogenic sampling* can collect larger samples and can reduce the influence of contamination and wall effects. This method is suitable for simultaneous measurements of a number of trace gases. The *impregnated filter method* is a selective sampling technique and can measure acidic gases. Table 8.2 summarizes the constituents measured by various sampling techniques. The collected samples are usually analyzed by a gas chromatograph equipped with an electron capture detector in the laboratory.

8.1.2 Remote sensing

Remote sensing is a technique to detect and measure the constituents at a distance in space by measuring by spectroscopic methods the absorption or emission of radiation by molecules. Both natural and artificial radiation

TABLE 8.2 *In situ* measurements by various collection methods.

Method	Constituent	Sampled by	References
1. Grab sampling	N_2O	surface	Cicerone *et al.*, 1978; Roy, 1979; Goldan *et al.*, 1978
	CH_4	balloon	Bush *et al.*, 1978
	CF_4	antarctic	Rasmussen *et al.*, 1979
	H_2	aircraft	Ehhalt *et al.*, 1977
	Halogens	antarctic	Duce *et al.*, 1973
	F_{11}, F_{12}, N_2O	balloon	Schmeltokopf *et al.*, 1975; Goldan *et al.*, 1980
	F_{11}, F_{12}, CCl_4, CH_3Cl	aircraft	Robinson *et al.*, 1976
	CO, CCl_4, $CFCl_3$, CH_4	arctic	Wilkniss *et al.*, 1975
	Halogenated compounds	aircraft	Cronn *et al.*, 1977
	F_{11}, F_{12}, F_{22}, F_{113}, CH_4, CO, CO_2, N_2O, CH_3Cl, CCl_4, CH_3CCl_3, COS, SF_6	aircraft	Leifer *et al.*, 1981
2. Cryogenic Sampling	N_2O	aircraft	Ehhalt *et al.*, 1975 a; Krey *et al.*, 1977
	SO_2	aircraft	Georgii and Meixner, 1980; Inn *et al.*, 1981
	COS	aircraft	Inn *et al.*, 1979 and 1981
	C_2H_6, C_3H_8	balloon	Rudolph *et al.*, 1981
	F_{12}, N_2O	aircraft, balloon	Tyson *et al.*, 1978a
	F_{11}, F_{12}	balloon, rocket	Heidt *et al.*, 1975
	H_2O, CH_4, H_2	balloon	Pollock *et al.*, 1980
	CH_4, CO, CO_2, H_2, H_2O, N_2O	rocket	Ehhalt *et al.*, 1975 b
	H_2, CH_4, CO, N_2O, F_{11}, F_{12}	balloon	Fabian *et al.*, 1979 a, and 1981
3. Impregnated filter sampling	HNO_3	aircraft	Huebert and Lazrus, 1978
		ship	Huebert, 1980
		aircraft, balloon	Lazrus and Gandrud, 1974
	HCl	balloon, aircraft	Lazrus *et al.*, 1977
	HCl, HBr, Cl, Br	balloon, aircraft	Lazrus *et al.*, 1976
	SO_2, HCl, HF	volcanic, gases	Lazrus *et al.*, 1979

may be used as the source for absorption measurements. The wavelength of radiation used ranges from the ultraviolet (UV) through the infrared (IR) to the microwave. The observations may be made from platforms at ground level, from aircraft, balloons, rockets and orbiting satellites, and in a variety of observational geometries; the limb-sounding method is particularly useful since the long optical path makes the observations of absorption and emission easier.

Various types of instrument are being used in spectroscopic observations of middle atmosphere composition. The *grating spectrometer* is perhaps the most well known device and has been used extensively by Murcray and his group (e.g. MURCRAY *et al.*, 1967). The *grille spectrometer* developed by GIRARD (1963) is essentially a modified grating spectrometer and has been used by French and Belgian group (e.g. FONTANELLA *et al.*, 1975). The *Michelson interferometer* has been used by, amongst others, BALUTEAU and BUSSOLTTI (1973), HARRIES and BURROUGHS (1971), FARMER (1974), and CLARK and KENDALL (1976); this technique is useful in observing a wide range of wavelengths *simultaneously*. The far IR range is observed mainly by this method. The filter radiometer or photometer has been used by many investigators including SEELEY and HOUGHTON (1961), BREWER and THOMSON (1972), KUHN and STEARNS (1973), CHALONER *et al.* (1978), DRUMMOND and JARNET (1978) and ROSCOE *et al.* (1981); this is a simpler and more convenient instrument and is particularly useful for measurements when the absorption band being measured is clearly separated from the nearby absorption bands of other molecules.

Table 8.3 lists the constituents observed by absorption or scattering measurements in the UV range. For each observation, the wavelength range, platform and radiation source are given. BUV observes the solar UV radiation backscattered from the atmosphere. Remote sensing in the IR range is particularly useful since many molecules have vibrational and rotational bands in this range. Table 8.4 summarizes the IR absorption measurements, mainly for stratospheric molecules, whereas Table 8.5 gives the emission measurements mostly in the IR and far IR ranges. For some measurements the emissions in the UV range (e.g. for OH and H) and the millimeter range (e.g. for N_2O and O_3) have been used.

8.2 Observations of Ozone

8.2.1 Total ozone observations

8.2.1.1 Dobson spectrophotometers

DOBSON (1931) developed a photoelectric spectrophotometer for measuring the total amount of ozone (ΣO_3) from the ground surface. The instrument is the standard for ozone measurement and has been used extensively

TABLE 8.3 Remote sensing by UV absorption or scattering measurements.

Constituent	Wavelength (nm)	Measured from	Radiation source	References
O_3	see Table 8.6	ground	sun	Dobson, 1931
	200–400	aircraft	sun	Falconer and Holdeman, 1975
	290–330	aircraft	sun	Hanser *et al.*, 1978
	300–340, 600	balloon	sun	Paetzold, 1955
	240, 290, 450	rocket	sun	Nagata *et al.*, 1971
	265, 287.5, 300	rocket	sun	Krueger, 1973
	280.3, 279.6	satellite (OSO-8)	solar occ.	Miller F. *et al.*, 1979
	255–301	satellite (OGO-4)	BUV	London *et al.*, 1977
	250–340	satellite (Nimbus-4)	BUV	Heath *et al.*, 1973; Hilsenrath and Heath, 1979
		(Nimbus-7)		Heath *et al.*, 1975; Krueger, 1983; McPeters *et al.*, 1984
	246–298	satellite (OAO-2)	stellar occ.	Roble and Hays, 1974
	255–310	satellite (OAO-3)	stellar occ.	Riegler *et al.*, 1977
	255	satellite (AE-5)	BUV	Guenther *et al.*, 1977
	188–310, 223-340.4	satellite (SMO)	sun (limb scan.)	Rusch *et al.*, 1983
	308	ground	laser	Gibson and Thomas, 1975; Megie *et al.*, 1977; Uchino *et al.*, 1978; Pelon and Megie, 1982
NO	215	rocket	sun	Barth, 1966; Meira, 1971; Baker *et al.*, 1977; Iwagami and Ogawa 1981

NO_2	430–450	balloon	sun	Evans *et al.*, 1977; Goldman *et al.*, 1978; Ogawa *et al.*, 1981
	437–450	ground	sun	Kulkarni, 1975; Noxon *et al.*, 1979
NO_3	662	ground	moon	Noxon *et al.*, 1978
	662	ground	incandescent lamp	Noxon *et al.*, 1980
H	121.567 (Ly-α) 102.5 (Ly-β)	rocket	sun	Anderson D. E. *et al.*, 1980
OH	308	ground	sun	Burnett and Burnett, 1981
	307.995	ground	dye laser	Perner *et al.*, 1976
	307.995	balloon	lidar	Heaps and McGee, 1983
	306.4	rocket	sun	Anderson, 1971
Na	589	ground, rocket	dye laser	Gibson and Sanford, 1972; Blamont *et al.*, 1972
CH_2O	326, 329, 339	ground	sun	Platt and Perner, 1980
HNO_2	354, 368			
NO_3	623, 662			
NO_2	354, 357, 365			
O_3	328			
SO_2	~300			

TABLE 8.4 Remote sensing by IR absorption measurements.

Constituent	Wavelength (μm)	Wave number (cm^{-1})	Measured from	Reference
H_2O	6.27	1595	balloon	Murcray *et al.*, 1969; Ackerman, 1974
	1.88	5331	balloon	Houghton and Seeley, 1960; Murcray *et al.*, 1966; Zander, 1973
	2.7	3704	rocket	O'Brien and Evans, 1981
	5.2922*	1889.58	balloon	Patel *et al.*, 1974
CH_4	7.67	1304	balloon	Kyle *et al.*, 1969
	7.67**	1304	satellite	Prabhakara *et al.*, 1974
	3.31	3019	balloon	Cumming and Lowe, 1973
	3.31, 3.55	3019, 2818	balloon	Ackerman and Muller, 1973
CF_4	7.75–7.84	1290–1275	balloon	Goldman *et al.*, 1979
CH_2O	3.48, 3.56	2807, 2869	ground	Barbe *et al.*, 1979
CO	4.67	2143	balloon	Goldman *et al.*, 1973 a; Zander *et al.*, 1981
N_2O	4.50	2224	balloon	Goldman *et al.*, 1973 b
	8.44	1185	aircraft	Coffey *et al.*, 1981 a
NO	5.33	1876	balloon	Toth *et al.*, 1973; Ackerman *et al.*, 1973 and 1975
	5.2976*	1887.63	balloon	Patel *et al.*, 1974
NO_2	6.18	1618	balloon	Goldman *et al.*, 1970
	6.18, 3.44	1618, 2906	balloon	Ackerman and Muller, 1972 and 1973
	3.39–3.51	2950–2850	ground	Camy-Peyret *et al.*, 1983
HNO_3	7.55	1325	balloon	Murcray *et al.*, 1968
	5.81	1722	aircraft	Coffey *et al.*, 1981 a
HCl	3.47	2886	aircraft	Farmer *et al.*, 1976
	3.47	2886	balloon	Ackerman *et al.*, 1976; Williams *et al.*, 1976
	40–165	250–60	balloon	Chance *et al.*, 1980
HCN	3.04	3287	aircraft	Coffey *et al.*, 1981 b

HF	2.48	4039 }	balloon	BUIJS *et al.*, 1980; ZANDER, 1981
HCl	3.40	2945 }		
ClO	12.0	833	balloon	MENZIES, 1979; MENZIES *et al.*, 1981
$CFCl_3$	10.83	923 }	balloon	MURCRAY *et al.*, 1975 a
CF_2Cl_2	11.83	847 }		
F-11, F-12	8–16	1250–625	ground	KAGGAN *et al.*, 1983;
F-22				ZANDER *et al.*, 1983
$CHClF_2$	12.06	829	balloon	GOLDMAN *et al.*, 1981
$ClONO_2$	7.74	1292	balloon	MURCRAY *et al.*, 1979
	12.82	780	balloon	MURCRAY *et al.*, 1977
COS	4.85	2062	aircraft	MANKIN *et al.*, 1979
O_3, CH_4, CO, CO_2, N_2O, H_2O, HCl, HF	2.25–5.55	4500–1800	balloon	FARMER *et al.*, 1980

*use a spin-flip Raman laser;
**use the radiation from the earth's surface, and all others use the solar radiation as a radiation source.

TABLE 8.5 Remote sensing by emission measurements.

Constituent	Wavelength (various units)	Wave number (cm^{-1})	Measured from	References
O_3	9.6 μm	1042	rocket	Stair *et al.*, 1974
			satellite (Nimbus-4)	Prabhakara *et al.*, 1976
			satellite (Nimbus-6, -7)	Gille *et al.*, 1980 b
	2.95 mm (101.8 GHz)	3.39	ground*	Wilson and Schwartz, 1981
$O_2(^1\Delta_g)$	1.27 μm	787.4	baloon	Evans *et al.*, 1969
			rocket	Wood *et al.*, 1970; Han *et al.*, 1973; Baker *et al.*, 1974; Yamamoto *et al.*, 1983
			satellite (SMO)	Thomas *et al.*, 1983 a
H_2O	20–40 μm	500–250	aircraft	Harries, 1973; Kuhn *et al.*, 1975
			balloon	Brewer and Thomson, 1972; Goldman *et al.*, 1973 c; Chaloner *et al.*, 1975
	>40 μm	<250	balloon	Hyson and Platt, 1974
	6.3 μM	1587	satellite	Wark *et al.*, 1974
	6.7–7.6 μm	1493–1316	rocket	Rogers *et al.*, 1977
	1.35 cm (22.2 GHz)	0.74	ground*	Deguchi and Muhleman, 1982; Gibbins *et al.*, 1982; Thacker *et al.*, 1981
H_2O_2	1.47 mm (204.57 GHz)	6.82	balloon	Waters *et al.*, 1981
OH	2.0 μm	5000	balloon	Moreels *et al.*, 1970
	1.062 μm	9416	ground*	Harrison *et al.*, 1970; Shefov, 1971
	727.45 nm		ground*	Takahashi *et al.*, 1974 and 1977
	1.4–2.12 μm	7143–4717	rocket	Rogers *et al.*, 1973
	622.5 nm		satellite (OGO-4)	Walker and Reed, 1976

Na	589.0–589.6 nm		ground*	SANDFORD and GIBSON, 1970; MEGIE and BLAMONT, 1977; KIRCHHOFF *et al.*, 1979
HNO_3	6.5–13.0 μm	1538–769	aircraft	MURCRAY *et al.*, 1975 b
	11.2 μm	893	aircraft	WILLIAMS *et al.*, 1972; MURCRAY *et al.*, 1973
	2.0–0.22 mm	5–45	balloon	HARRIES *et al.*, 1976
N_2O	264.7 μm	37.78	aircraft	HARRIES *et al.*, 1974 a
NO	5.33 μm	1876	balloon	CHALONER *et al.*, 1975
NO_2	6.18 μm	1618	balloon	CHALONER *et al.*, 1975
HCl	69.2–96.2 μm	145–104	balloon	TRAUB and CHANCE, 1981
HF	40.8–61.0 μm	245–164	balloon	TRAUB and CHANCE, 1981
ClO	1.079–1.245 μM (278–241 GHz)	9.27–8.03	aircraft	WATERS *et al.*, 1979
	1.47 mm (204.35 GHz)	6.82	balloon	WATERS *et al.*, 1981
CO	2.6, 1.3 mm (115, 230 GHz)	3.83, 7.69	ground*	KUNZI and CARLSON, 1982

*These ground-based (Infrared or microwave) observations measure the mesospheric abundance.

TABLE 8.6 Pairs of wavelengths used in Dobson spectrophotometer.

	λ_1	λ_2
A	325.4 nm	305.5 nm
B	329.1	308.8
C	332.4	311.45
D	339.8	317.6

at stations in the western world since IGY (International Geophysical Year 1957–1958).

The principle behind the Dobson instrument is to measure the ratio between the intensity of solar radiation at two wavelengths (λ_1 and λ_2). They are chosen close enough so that the intensity of incident radiation is about the same but one is subject to much greater absorption by ozone than the other; λ_1 is in the range where the radiation does not suffer from much ozone absorption, whereas λ_2 is in the range of stronger ozone absorption. There are several pairs of λ_1 and λ_2 used in the Dobson instrument as are shown in Table 8.6. Standard observations of total ozone are defined as those carried out by a Dobson spectrometer using the AD wavelength pairs with direct sun as a radiation source. The effects of aerosol attenuation at these pairs have been found to be very small (SHAH, 1976).

The measured intensity of radiation at a wavelength λ may be expressed by

$$I_\lambda = (I_\lambda)_\infty \cdot \exp\,(-\sigma_\lambda \Sigma\, O_3 \mu - \beta_\lambda m - \delta_\lambda m') \tag{8.1}$$

where $(I_\lambda)_\infty$ is the intensity outside the atmosphere, σ_λ the ozone absorption cross-section, and β_λ and δ_λ the extinction coefficients due to scattering by air molecules and aerosols, respectively. They represent the Rayleigh and Mie scattering, respectively. μ, m and m' are the air mass factors to convert the amounts of O_3, air molecules and aerosols in the slant path to those in the vertical path. These three values are not necessarily the same because the distributions of these three materials are different, but as a first approximation they are all close to $\sec\chi$ if χ is less than $\sim 80°$.

Measuring I_λ for the pair of wavelengths and assuming that δ_λ is the same for the two wavelengths, we have

$$\frac{I_1}{I_2} = \frac{(I_1)_\infty}{(I_2)_\infty} \exp\,\{-(\sigma_1 - \sigma_2)\mu \sum O_3 - (\beta_1 - \beta_2)m\} \tag{8.2}$$

where the suffices 1 and 2 indicate the values at the wavelengths λ_1 and λ_2, respectively. Using the measured values for the σ_λ's and the theoretical values for the β_λ's one can calculate ΣO_3 from (8.2). An error of up to 5% can occur for the total ozone measurement by the standard Dobson instrument because of the possible errors in the measurements of the σ_λ's, wavelength determination and the effects of absorption by other gases, aerosols and clouds (KOMHYR, 1980; DELUIS, 1980). The ratio of measured values of I_1/I_2 at wavelength A and D is usually used for the analysis, since the effect of a linear dependency of δ_λ on wavelength can then be eliminated.

8.2.1.2 Filter ozonometer

The filter technique to single out the desired wavelengths is simpler and less expensive than the Dobson spectophotometer which requires lenses and prisms made from crystal quartz. In general, however, the filter has a wide transmission band width and the ozone absorption occurs for a certain range of wavelength rather than for a discrete wavelength. In addition, the measurements are significantly affected by weather conditions and the geometry of the transmission, particularly at larger solar incident angles.

Although the measurements by filter ozonometers are less accurate, they are used extensively in ozone measurements on board aircraft, balloons and rockets (see Table 8.3) because of their simplicity and cost effectiveness. The instruments are also used widely at ozonometric stations in the Eastern (Soviet) bloc countries. Since these stations represent nearly a third of world-wide network stations of ozone observations, it is necessary to establish correction methods for data from filter ozonometers by comparing them with those of simultaneous observations by the standard Dobson spectrophotometer (BOJKOV, 1969). In this fashion we are able to make a world map of ozone distribution based on data from the world-wide ground based network stations.

8.2.2 Vertical distributions

8.2.2.1 Götz umkehr method

The vertical distribution of ozone may be measured by a balloon-borne (or rocket-borne) optical method by differentiating the measured profile (curve) of the total ozone with height. But this method is certainly not comparable in economy nor convenience with the Dobson or filter method that can carry out observations from the *earth's surface*. GÖTZ (1931) invented a clever method for measuring the vertical distribution of ozone using the Dobson spectrometer (or filter ozonometer) on the ground.

The Götz method uses measurements of scattered solar UV radiation from the sky by pointing the instrument at the sky instead of to the sun. The intensity of the forward scattered radiation entering the instrument is attenuated relative to the incoming solar radiation by (1) absorption by O_3 in the slant path from the sun to the scattering region (height) of the atmosphere, and also in the path between the scattering height and the observer, and by (2) scattering at the scattering height. The calculation for the attenuation of solar radiation by these processes is very complicated, particularly when the effects of multiple scattering are taken into account (see, for instance, CRAIG, 1965 and KHRGIAN, 1973 for detailed discussions).

The scattering height generally depends upon the incident angle of the light as well as the wavelength. GÖTZ *et al.* (1934) have shown that the ratio of the intensities of observed radiation at two wavelengths in the Huggins bands increases with the incident solar angle until the angle reaches some value between 80–90°, then decreases afterwards for some range, and finally turns back to increase again. Figure 8.1 illustrates the typical curves of intensity ratio against solar incident angle for the different pairs of wavelengths from Table 8.6. Because of the shape of these curves the Götz method is also called the *umkehr* method.

The *umkehr* observations are strongly affected by stratospheric and tropospheric aerosols; for instance, it is reported that an increased stratospheric aerosol concentration after the eruption of the Agung volcano caused an enormously low (and therefore false) ozone density in the upper stratosphere to be measured by the *umkehr* method (DELUISI, 1979).

8.2.2.2 Ozonesonde method

The direct *in situ* measurement of stratospheric ozone may be made by ozonesondes on board balloons or aircraft using the *electrochemical* or *chemiluminescent* method. The information obtained by these methods is radio transmitted to the ground. Neither method requires solar UV radiation and therefore measurements can be made even at night.

The electrochemical sonde is a wet chemical instrument developed by BREWER and MILFORD (1960) and KOMHYR (1969). It uses a very rapid reaction of potassium iodide decomposition in a water solution,

$$2KI + O_3 + H_2O \rightarrow 2KOH + I_2 + O_2 \tag{8.3}$$

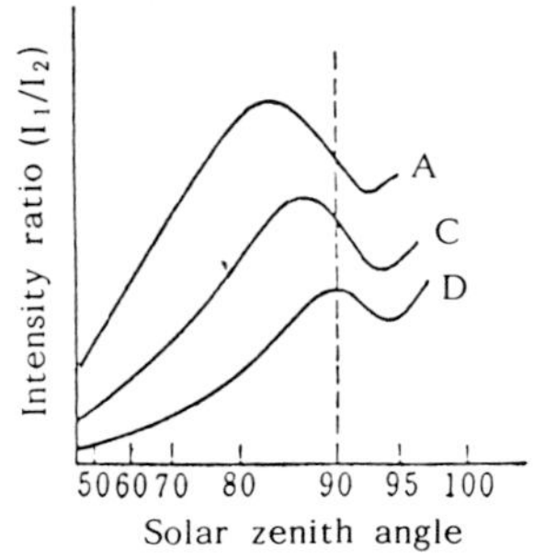

FIG. 8.1. Typical *Umkehr* curves for different pairs of wavelengths.

where about 5.3 μ gram of iodine is generated per 1 μ gram of ozone. The depolarization effect of I_2 at the cathode produces I^- ions, which are transported to the anode resulting in an electric current in the circuit. The intensity of the current is related to the ambient ozone density. The anode is made of silver to avoid regeneration of free iodine there.

Chemiluminescent-type ozonesondes have been developed by REGENER (1960, 1964), HILSENRATH and KIRSCHNER (1980), AIMEDIEU and BARAT (1981), AIMEDIEU (1983) and others. The observations can be made at all altitudes accessible to balloons and rockets. They are based on the chemiluminescent reactions of a dry substance (luminol, rhodamine, or oletin) with ozone. This method gives only a relative measure of the vertical ozone profile; absolute calibration must be achieved through reference to the ground-based spectrophotometer measurements. HERING (1964) and HERING and BORDEN (1964, 1965, 1967) have made extensive observations in North America by this method.

8.2.3 Satellite observations

Three methods have been developed for ozone measurements from orbiting satellites. They are the backscatter ultraviolet (BUV) method, the Infrared Interferometer Spectrometer (IRIS) method, and the stellar and solar occultation method.

The IRIS method observes the infrared emission at wavelengths of ~7–25 μm, which includes the 9.6 μm ozone absorption band. PRABHAKARA *et al.* (1970, 1976) have used this method to observe the total ozone, while GILLE *et al.* (1980b) have applied this method to the observations of limb radiance and to calculate the ozone density profile.

The solar occultation technique is useful for measuring mesospheric ozone (MILLER F. *et al.*;, 1979; GUENTHER *et al.*, 1977) and the stellar occulation technique for measurements of the nighttime ozone density (GILLE *et al.*, 1980b).

The BUV method is the most successful method for observing the global characteristics of the ozone density at high altitudes. The backscattered solar radiation that emerges from the atmosphere into the instrument on board the satellite contains information on the distribution of absorbing constituents (O_3) through which the radiation has passed. When scattering and absorption are wavelength dependent, measurements made at many wavelengths can be used to define the vertical profile of ozone density. In the Nimbus 4 BUV experiment the radiance has been measured at twelve discrete wavelengths between 255 nm and 340 nm (HEATH *et al.*, 1973).

The inversion of terrestrial atmospheric radiance to obtain vertical ozone profiles is possible since atmospheric radiance at a particular wavelength originates in a moderately well-defined and effective scattering layer. The

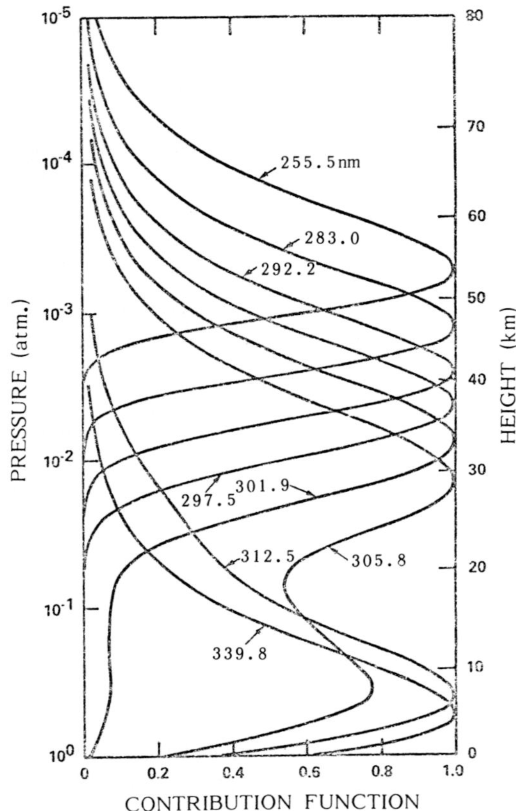

FIG. 8.2. Normalized intensity curves of the backscattered radiance in the nadir direction of the satellite from various heights of the atmosphere including multiple scattering (HEATH *et al.*, 1973).

height and width of a layer are functions of the ozone absorption coefficient, solar zenith angle (χ) and ozone distribution. Figure 8.2 illustrates an example of effective scattering layers for the case of $\chi = 60°$ and $\Sigma O_3 = 0.336$ cm NTP*.

Details of the mathematical and physical problems of inversion of UV atmospheric radiances to obtain a vertical ozone distribution are discussed by TWOMEY (1961), MATEER (1972) and others.

The total ozone can be measured by using a wavelength pair chosen in

*The total ozone given in units of cm NTP represents the thickness of the ozone layer that would be formed by the entire ozone in the column when reduced to the value at a standard pressure (760 mmHg) and a standard temperature (°C). The number is obtained by dividing the total ozone column content per cm^2 by the Loschmidt number 2.687×10^{19}.

the Huggins band; they have to be separated by about 20 nm to minimize scattering effects. The chosen pairs in the Nimbus satellite experiments are (A) 312.5, 331.2 nm and (B) 317.5, 339.8 nm (MATEER *et al.*, 1971; KRUEGER, 1983).

Measurements of ozone density by the BUV method are most accurate in the height range between ~30 km and ~55 km. Below these heights the effects of multiple scattering complicate the calculation of ozone density making the observations less accurate. Above these heights the intensity of backscattered radiance becomes too weak to make accurate observations.

The ozone density in the upper stratosphere as measured by the satellite BUV method do not necessarily agree with the umkehr data obtained by measurements on the surface. In fact, they show significant differences in some cases. However, they show a general agreement in seasonal variation (DELUISI and NIMIRA, 1978; DELUISI *et al.*, 1979). The total ozone can be inferred from BUV measurements at wavelengths near the long-wavelength end of the O_3 absorption band. The available evidence suggests that the standard error of the total ozone measurement may be less than 0.015 cm NTP for solar zenith angles less than 60° (MATEER *et al.*, 1971).

8.2.4 Additional methods of ozone observations

Ground-based lidar measurements of stratospheric ozone density by differential absorption technique have been developed by various groups. GIBSON and THOMAS (1975) and MEGIE *et al.* (1977) have used a transmitter of a flash-lamp pumped, frequency-doubled *dye laser* in the wavelength range 297–310 nm, whereas UCHINO *et al.* (1978) succeeded in observing the stratospheric ozone density by a discharge-pumped XeCl *excimer laser* with a 308 nm wavelength.

The receivers on the ground measure the intensity of the lidar beam backscattered from various heights of the atmosphere. Thus the method is similar to the BUV method of satellite observations, but the scattered lidar beams suffer from more complicated effects of aerosols (Rayleigh scattering, clouds, haze, fogs) and absorption of various gases in the lower atmosphere. In order to single out the effect of ozone absorption, the other effects must be eliminated by using the two wavelength method or by theoretical calculations; for instance, the attenuation of the beam due to Rayleigh scattering can be allowed for in the calculation of the expected signal.

The lidar techniques are very promising for measurements of many kinds of middle atmosphere constituents when high power lidar beams become available in both the UV and IR ranges. The measurements can be carried out even at night.

Observations of the dayglow emission at 1.27 μm originating from the

(0,0) band of molecular oxygen $O_2(a^1\Delta_g - X^3\Sigma_g^-)$ can be used to calculate the ozone density, provided that the quantum yield of $O_2(^1\Delta_g)$ production by ozone photolysis (J_{2a} and J_{2b}) and the loss coefficient by quenching are know. The other loss coefficient by spontaneous emission, known as the Einstein coefficient, is calculated theoretically to be $\sim 2.8 \times 10^{-4}$ sec^{-1}.

Measurements of the $O_2(^1\Delta_g)$ emission have been carried out by instruments on board balloons (EVANS *et al.*, 1968 and 1969) and rockets (WOOD *et al.*, 1970). The analysis of a rocket observed profile of the dayglow emission by EVANS *et al.* (1968) indicates that the emission is really due to ozone photodissociation below $\sim$80 km. Above this height, however, the observed excess emission requires either more ozone than predicted at that time or another source for $O_2(^1\Delta_g)$ emission. The measurement can also be made from satellites. It is particularly useful for obtaining the global distribution of mesospheric ozone. The Solar Mesosphere Explorer staellite launched in October, 1981, includes a 1.27 μm spectrometer to measure the global ozone distribution between 50 and 90 km (BARTH *et al.*, 1983; THOMAS *et al.*, 1983 a).

Another method to measure mesospheric ozone is by ground-based millimeter wave observations (SHIMABUKURO *et al.*, 1975; PENFIELD *et al.*, 1976; WILSON and SCHWARTZ, 1981). The millimeter wave method has been successful in radio astronomy in obtaining altitude distributions of molecules in the atmosphere (PENZIAS and BURRUS, 1973). The line shape integral of the emission intensity written as a Planck function with a brightness temperature yields the absorption coefficient when it is inverted mathematically. From the absorption coefficient one can calculate the mixing ratio distribution of the absorbers.

8.3 Result of Observations

The concentration profiles of individual constituents obtained by various observations do not necessarily agree with eath other. In some cases there is a systematic difference between observations obtained by different techniques. Although each observation includes experimental errors, there are some definite latitudinal and temporal (daily, seasonal and long-term) variations for some constituents. It is sometimes difficult to define typical profiles and their variations for each constituent. Here, we will summarize and discuss the main results of observations for representative constituents in the middle atmosphere.

8.3.1 Ozone and atomic oxygen

KRUEGER and MINZNER (1976) have constructed a model for an annual mean daytime ozone distribution at mid-latitude northern hemisphere based on data taken from rocket and balloon soundings at various latitudes and the latitude gradients determined from satellite observations. The model has been adopted for use in the U. S. Standard Atmosphere, 1976, and is shown in Fig. 5.2, (see also Appendix A).

The contour maps of ozone density distributions have been constructed on the meridional (latitude-height) plane for the height range 10 - 90 km, using observational data from various sources; the results are shown in Fig. 8.3 (a) and (b) for March and June, respectively. Below 30 km we use the monthly means of the vertical distribution of ozone density calculated by Wilcox *et al.*, (1977) from 3–8 years' observations by various ozonesondes over North America. In the upper stratosphere the satellite BUV data are probably the best for obtaining the latitudinal and seasonal variations in ozone density. Using the ozonesonde data at 30 km and the BUV data at 1 mb level (~45 - 50 km) deduced by McPeters (1980), we have calculated the ozone profiles between 30 km and 50 km by interpolation on the assumption that the ozone density decreases exponentially with height. In order to cover the entire latitude range the BUV data observed over the sunlit hemisphere have been extrapolated beyond the terminator into the dark side at high latitude in winter.

Above 50 km the observation of the 1.27 μm dayglow emission by an instrument on board the Solar Mesosphere Explorer (SME) satellite provides the global ozone data up to ~90 km (Barth *et al.*, 1983; Thomas *et al.*, 1983 a). Figure 8.3 (a) includes the ozone density distribution above 50 km observed by SME on March 14, 1982, whereas Fig. 8.3 (b) illustrates the same observed on December 22, 1981. In the latter case the December data in the northern and southern hemispheres are exchanged with each other for illustrative purposes in order to match the season with the data shown below 50 km (June).

It can be seen from Figs. 8.3 (a) and (b) that the latitudinal variation in ozone density is very smooth between ~25 km and ~75 km. This is due to the fact that the photochemical time constant of odd oxygen is very small in this height range (see Fig. 6.4). Outside this height range the dynamical effect influences the ozone distribution in a complicated fashion as the photochemical time constant increases. At lower heights (<25 km) transports from lower to higher latitudes as well as from upper to lower heights are essential for the ozone density distribution. Above ~75 km the ozone density distribution becomes irregular perhaps due to the effect of breakdown at these heights of gravity waves propagated from the troposphere.

The ozone density is largest in the lower stratosphere. The maximum ozone density in the vertical variation usually appears at ~20–25 km; this height depends on the latitude. Whereas the height of the maximum density

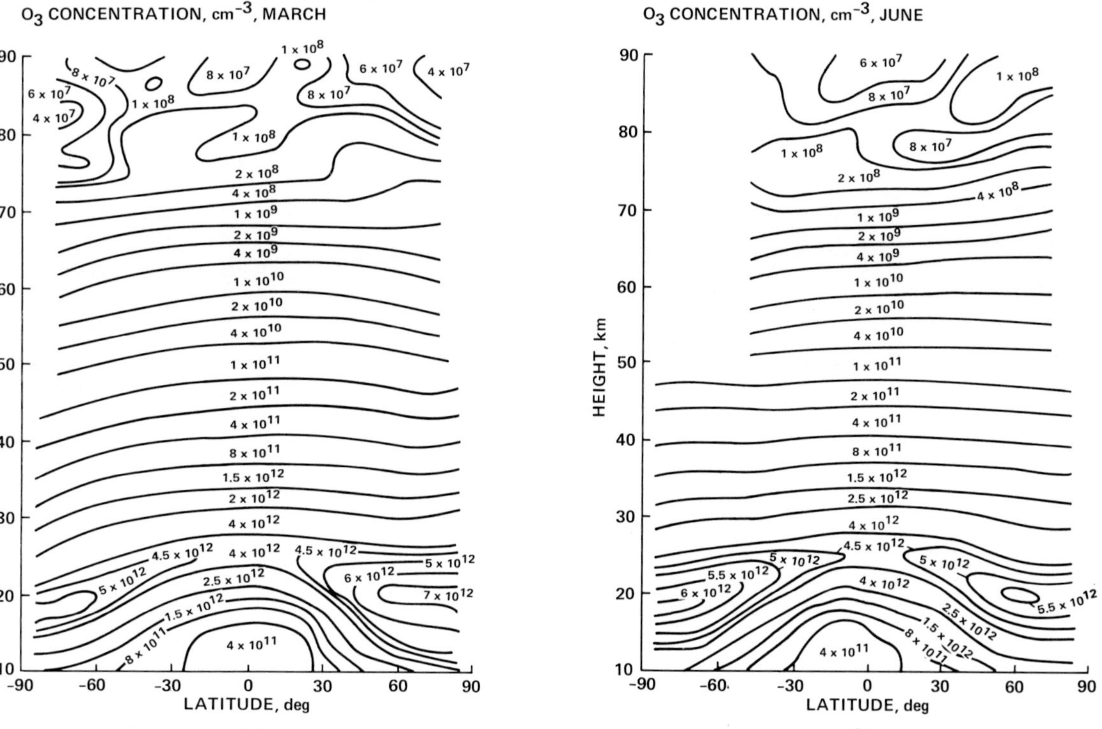

FIG. 8.3. (a) Meridional contour plots of the ozone density for March, illustrating the zonally averaged values based on ozone zonde and BUV data below 50 km and an individual data observed by SME on March 14, 1982 above 50 km (see text for the details). (b) The same as Fig. 8.3(a) but for June; above 50 km the SME data observed on December 22, 1981 is shown by reversing the latitude scale.

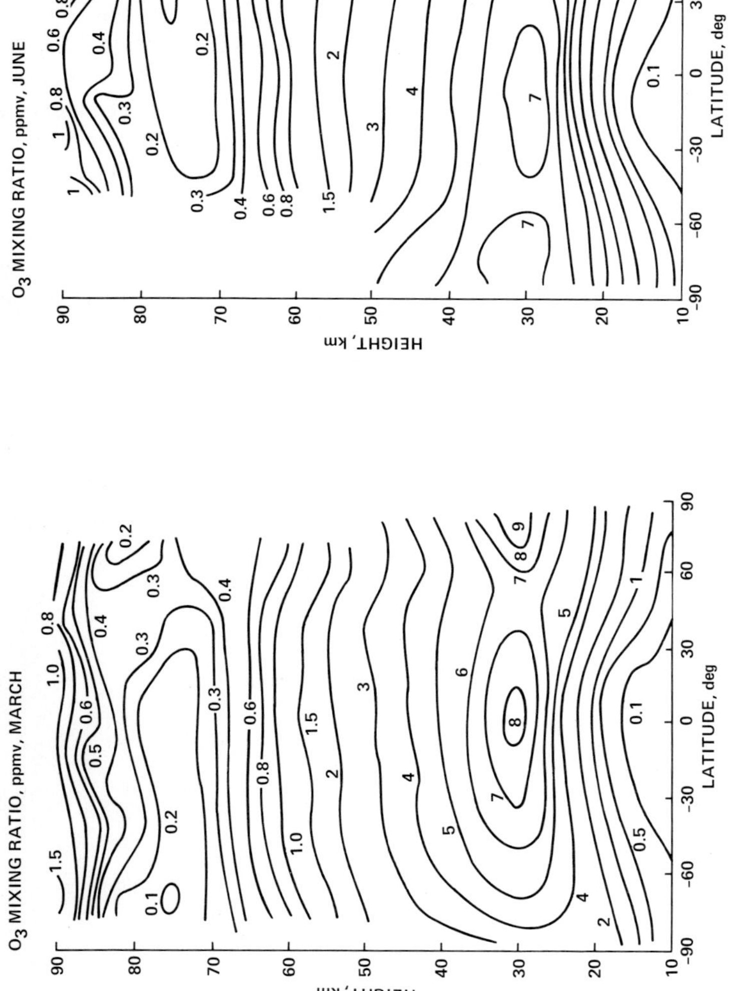

FIG. 8.3. (c) Meridional distribution of the ozone mixing ratio (in ppmv) for March. (d) The same as Fig. 8.3(c) but for June.

decreases from $\sim$25 km at low latitudes to below 20 km at high latitudes, the maximum ozone density increases as we go from lower to higher latitudes. Figures 8.3(c) and (d) illustrate the contour maps of the ozone mixing ratio for the two seasons. The maximum mixing ratio occurs at $\sim$30 km, which is appreciably higher than the height of the maximum ozone density.

The largest ozone density throughout all latitudes and seasons is found at high latitudes over the northern hemisphere in March. This tendency is clearly seen in Fig. 8.4 which illustrates the latitudinal variation of total ozone in the four seasons based on Fig. 8.3(a) and (b). Figure 8.5 shows the satellite data obtained by the Nimbus-4 BUV observations (KRUEGER *et al.*, 1980); the values are given in 10^{-3} cm NTP (Dobson units). It can be seen from Fig. 8.5 that the total ozone is greatest at high latitudes at all times but especially during late winter and early spring.

In order to illustrate more clearly the effect of latitudinal differences on the maximum ozone density and its height, we depict in Fig. 8.6 the altitude profiles of ozone density at several latitudes. It is now evident that the larger total ozone at high latitudes is caused by a large enhancement of ozone density in the lower stratosphere and upper troposphere. In the lower troposphere O_3 is lost by surface deposition (FABIAN and JUNGE, 1970) and chemical reactions with HO_x produced from a large amount of H_2O by reaction with $O(^1D)$.

The photochemical theory suggests that the ultimate source of middle atmosphere O_3 is the atomic oxygen produced in the upper and middle

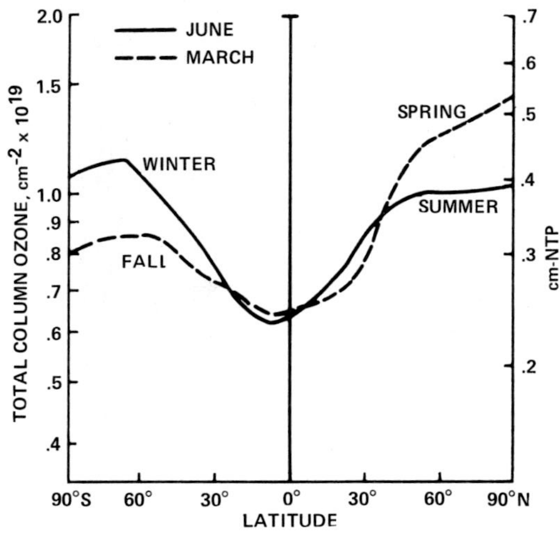

FIG. 8.4. Latitudinal distributions of the total ozone over four seasons.

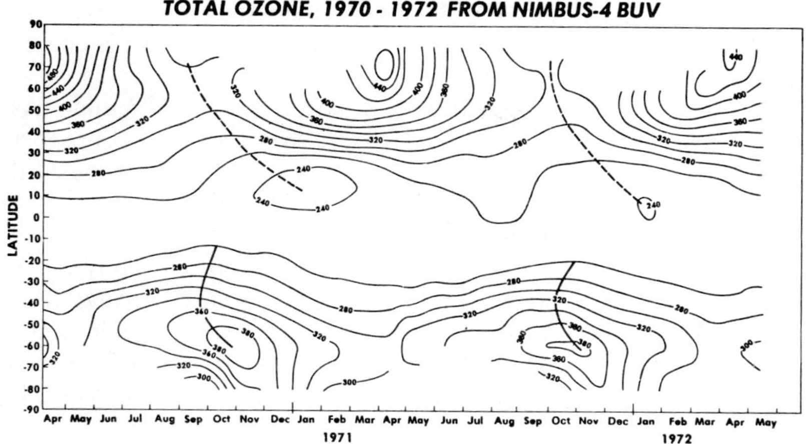

FIG. 8.5. Time-latitude cross section of zonally averaged total ozone obtained from the Nimbus 4 BUV from April 1970 to May 1972. Contours are drawn for every 20 cm NTP (KRUEGER *et al.*, 1980).

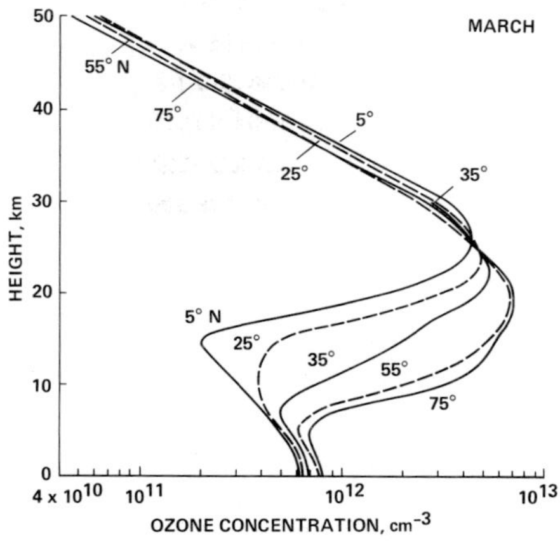

FIG. 8.6. Height variations of the ozone density at various latitudes in March.

stratosphere by O_2 photodissociation (see Chapter 5). Thus, the O_3 production rate should be larger where the intensity of the available solar radiation is stronger; i.e. more ozone is produced in the upper stratosphere than in the lower stratosphere, at lower latitudes than at higher latitudes, and in summer than in winter. All these theoretical expectations, however, are not realized in observed data. Actually, the observed latitudinal and seasonal variations are entirely contrary to the theoretical expectations. In particular, we find a very ozone rich region at high latitudes in winter, where no sunlight is available for the entire day. These discrepancies between the photochemical theory and observations may be explained by the effects of large scale ozone transport.

The total ozone also is not distributed uniformly with longitude. As is seen in Fig. 8.7 the value is particularly large over East Asia, Europe and eastern America. This may have been caused by the effects of the earth's topography on the propagation of planetary waves and the different cyclone activity at different longitudes (see Chapter 9).

The ozone distribution in the mesosphere has a marked diurnal variation, as predicted by SHIMAZAKI and LAIRD, (1970), and SHIMAZAKI and WUEBBLES (1973). The nighttime profiles have a large depression near the mesopause (~80 km) and an enhanced number density above that height. The nighttime profiles observed by various methods are compared in Fig. 8.8.

Atomic oxygen ($O(^3P)$) has been measured by radiance observations with a resonance lamp in both the stratosphere (ANDERSON, 1975) and in the lower thermosphere (DICKINSON *et al.*, 1974 and 1980; THOMAS and YOUNG, 1981). The results are shown in Fig. 5.2. The observations in the stratosphere (25–35 km) agree well with the photochemical model in magnitude, but the slope of the observed profile seems to be a little more moderate than the model prediction. The observations in the lower thermosphere (95–120 km) are consistent with the model result if the transport

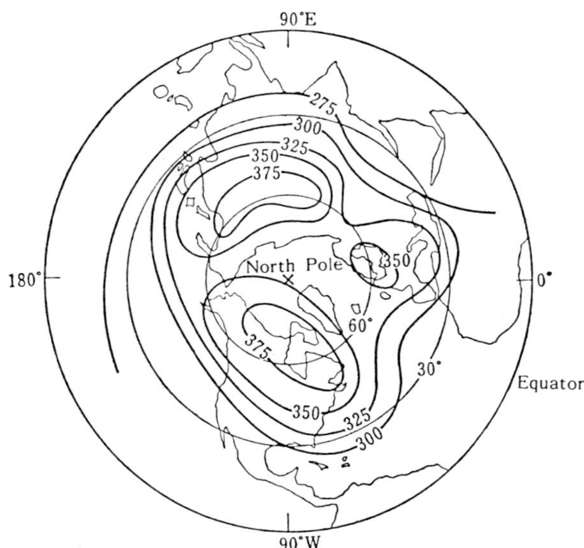

FIG. 8.7. Worldwide distribution of the total ozone, average for 1957–1972. The values are given in 10^{-3} cm NTP (WILCOX *et al.*, 1977).

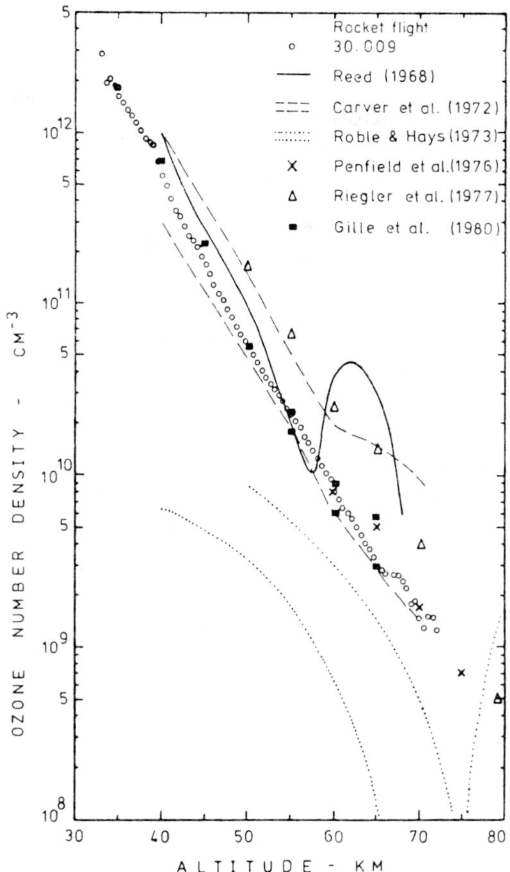

FIG. 8.8. Comparison of nighttime ozone densities measured by various experimental techniques (LEAN, 1983).

effects by eddy and molecular diffusion are taken into account. However, in order to explain the $O(^3P)$ density observed in the region 80–95 km we certainly need to consider the effect of chemical reactions, particularly the reactions of O with HO_x.

8.3.2 Nitrogen compounds

N_2O is pretty well mixed in the troposphere and the average mixing ratio is ~300 ppb (CICERONE *et al.*, 1978; GOLDAN *et al.*, 1978 and 1981; WEISS, 1981), although there are some seasonal and diurnal variabilities in the emission rate from soil (MATTHIAS *et al.*, 1979; BREMNER *et al.*;, 1980). The mixing ratio indicates a marked decrease in the stratosphere with substantial variabilities of the profile depending on the location and season as is shown in Fig. 8.9 (SCHMELTEKOPF *et al.*, 1977). These variabilities suggest a strong dependence of the N_2O profile on both stratospheric transport and photochemistry, and the observed N_2O profiles can be used to calculate the eddy diffusion coefficient profiles at various latitudes (see Section 7.1.1).

The NO density profiles in the stratosphere obtained by different observations are compared in Fig. 8.10. The large diversity in the measured profiles are partly attributed to experimental errors; *in situ* measurements by the chemiluminescent method may be in error because of heterogeneous recombination on the sampling tube (RIDLEY and SCHIEFF, 1981). Measurements by remote sensing should be considered to give the average over long slant paths during twilight when the NO density changes significantly.

The latitudinal variations in NO density have been observed at 18.3 km and 21.3 km by *in situ* measurements from a high-altitude research U2 aircraft (LOEWENSTEIN and SAVAGE, 1975). The result shown in Fig. 8.11 indicates that [NO] steadily increases towards the north at 21.3 km, whereas at 18.3 km it exhibits a broad minimum at $\sim 45°$ N.

In the mesosphere and lower thermosphere, the NO density has a marked depression near the mesopause as is seen from Fig. 8.12. The profile analyzed by BARTH (1966) did not show the minimum because of overestimation of the NO emission rate at these heights; part of the observed emission should have been attributed to the background Rayleigh scattering. The source of NO in the mesosphere lies in the ionic reactions in the ther-

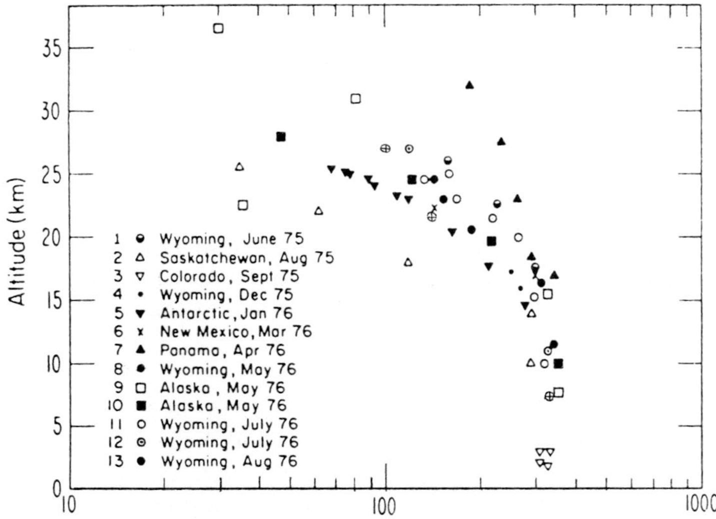

FIG. 8.9. Measured N_2O mixing ratios (ppbv) for various balloon flights at different locations and seasons (SCHMELTEKOPF *et al.*, 1977).

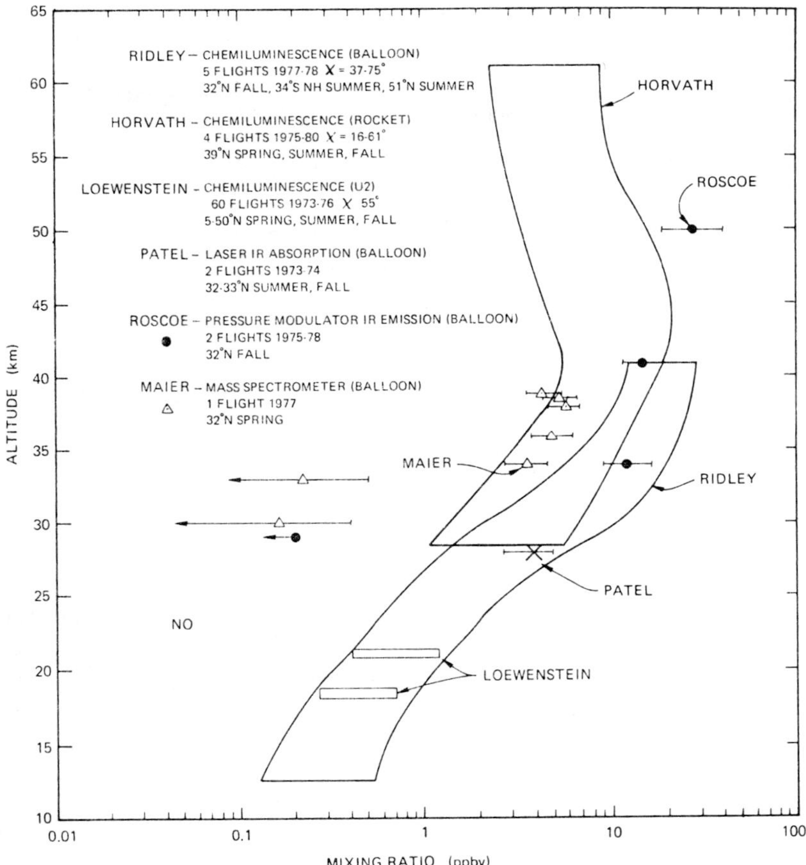

FIG. 8.10. Comparison of various NO measurements in the stratosphere (WMO, 1981).

mosphere (see Chapter 13); NO is thus transported downward into the mesosphere. The situation is similar to O_3 in the lower stratosphere whose photochemical source is in the upper and middle stratosphere.

NO_2 has been measured mainly by spectroscopic methods in the near UV and IR region. Figure 8.13 compares height profiles obtained by various observers. These height profiles suffer from inaccuracies caused by the long slant path length as was discussed for NO. Column amounts of NO_2 have been observed from the ground by measuring the NO_2 absorption at twilight (NOXON, 1979) and from aircraft platforms (COFFEY *et al.*, 1981 a). The result indicates that there is an increase by a factor of approximately two in the column density from equator to mid-latitudes. At high

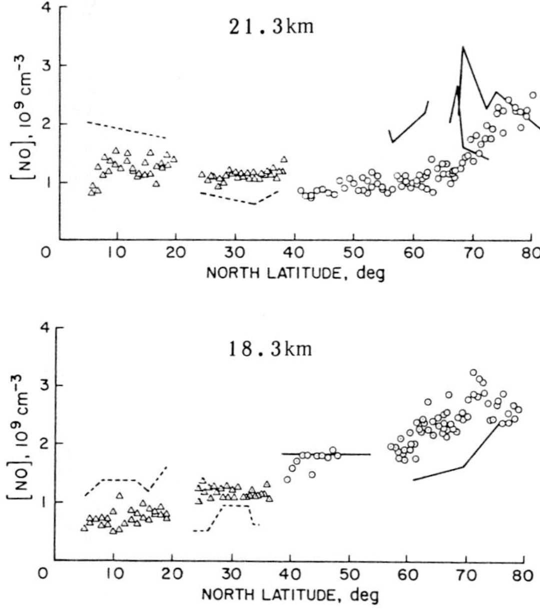

FIG. 8.11. Latitudinal variations of NO concentration at 18.3 km and 21.3 km at 122–158° west longitude (LOEWENSTEIN and SAVAGE, 1975).

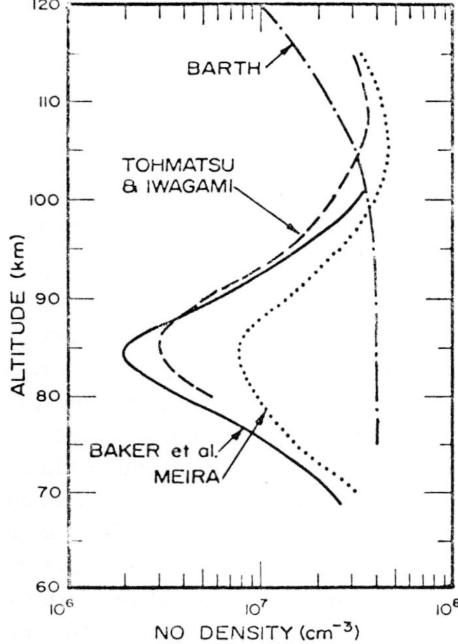

FIG. 8.12. Profiles of measured NO number density in the mesosphere and lower thermosphere (BARTH, 1966; MEIRA, 1971; TOHMATSU and IWAGAMI, 1976; BAKER *et al.*, 1977). Figure is taken from BAKER *et al.* (1977).

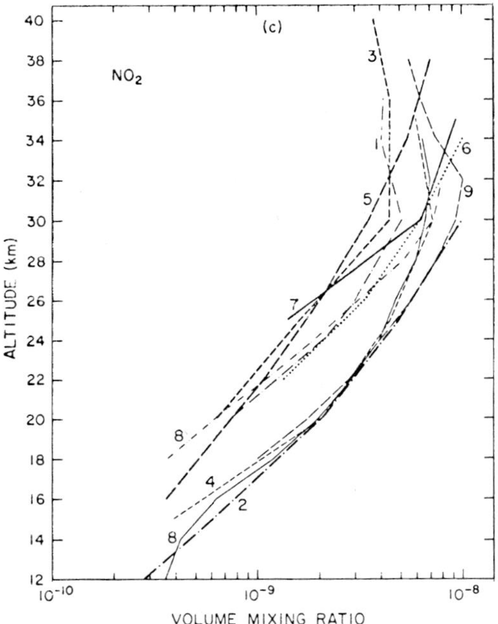

FIG. 8.13. Measured vertical profiles of NO_2 near 40°N by various investigators (1, ACKERMAN *et al.*, 1975; 2, BLOXAM *et al.*, 1975; 3, CHALONER *et al.*, 1975; 4, EVANS *et al.*, 1975; 5, FONTANELLA *et al.*, 1975; 6, GOLDMAN *et al.*, 1978; 7, HARRIES *et al.*, 1974b; 8, KERR and MCELROY, 1976; 9, LOUISNARD *et al.*, 1978). Figure is taken from COFFEY, MANKIN and GOLDMAN (1981a).

latitudes in winter, NO_2 abundance drops sharply near 45°, whereas it increases with latitude in summer; thus, there is a large seasonal variation in the total NO_2 at high latitudes (Fig. 8.14).

Figure 8.15 illustrates the measured mixing ratio of HNO_3 by various methods. *In situ* sampling methods by using filter-paper collection (LAZRUS and GANDRUD, 1974) seem to give smaller values than the IR measurements both in absorption and in emission (MURCRAY *et al.*, 1973), but both agree in giving a maximum mixing ratio at 20–25 km. Marked seasonal and latitudinal variations have been observed for the HNO_3 column density as is shown in Fig. 8.16 (MURCRAY *et al.*, 1975 b). The similar latitudinal variations have also been observed for HCl and HF (GIRARD *et al.*, 1982).

Since NO and NO_2 are strongly coupled by photochemical reactions and exchangeable with each other during the daytime (see Chapters 5 and 6), it is useful to measure their densities simultaneously in order to test the photochemical theory. Such simultaneous observations have been carried out, for instance, by ACKERMANN *et al.* (1975), DRUMMOND and JARNOT (1978) and BLATHERWICK *et al.* (1980). Simultaneous measurements of

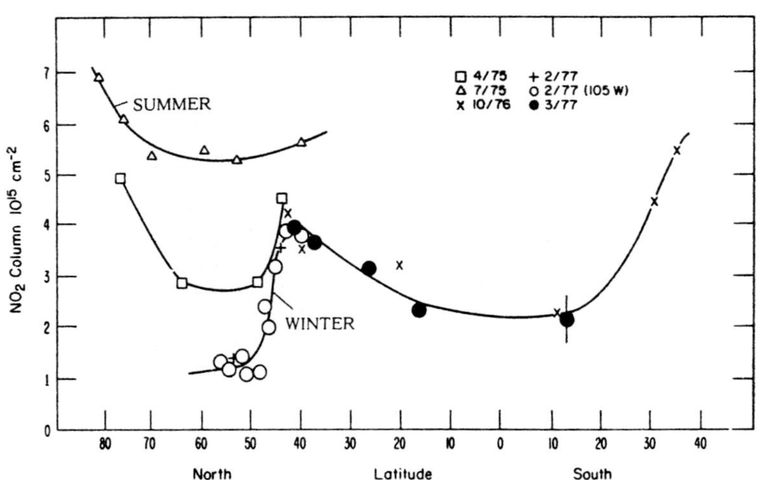

FIG. 8.14. Representative measurements of the total column abundance of stratospheric NO_2 at various latitudes and seasons. Note that ○ indicates the northern winter and △ the northern summer observations. Except as noted, measurements are all near 75°W (NOXON, 1979).

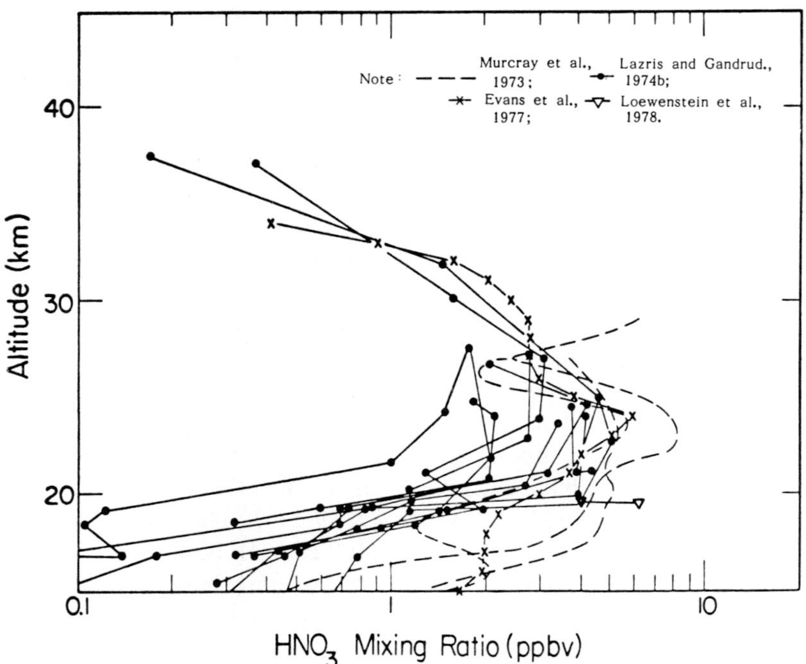

FIG. 8.15. Measurements of HNO_3 mixing ratio by various methods (NASA, 1979).

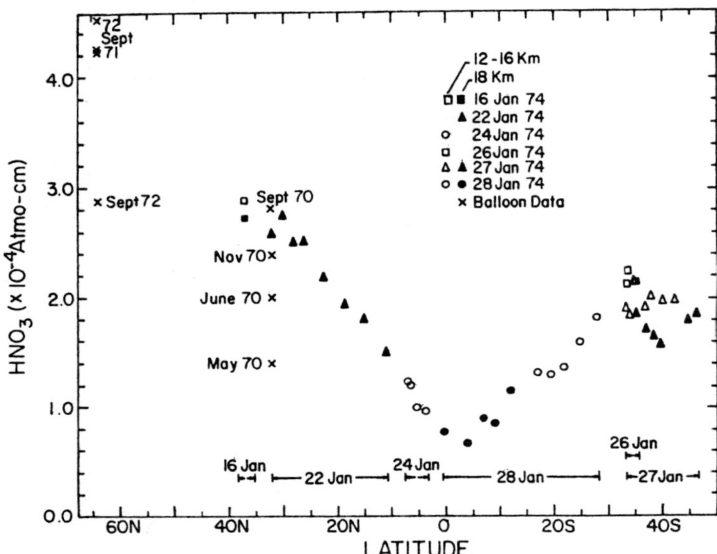

FIG. 8.16. Latitudinal variation in HNO_3 column density for January (MURCRAY *et al.*, 1975b).

HNO_3, O_3, and atmospheric temperature together with NO and NO_2 are even more valuable, since HNO_3 is a reservoir of NO and NO_2 and the ratio of [NO] to [NO_2] is controlled by [O_3]. Figure 8.17 illustrates an example of such a simultaneous observations by EVANS *et al.* (1981).

In an earlier study EVANS *et al.* (1977) concluded that the measured ratios of [NO_2]/[NO], [HNO_3]/[NO_2], and [HNO_3]/[NO] agreed with a theoretical model to better than a factor of two at most altitudes. Examining these ratios obtained from the observations at local noon by the sub-millimeter spectroscopic method (HARRIES *et al.*, 1976), HARRIES (1978) found that his values for [HNO_3]/[NO_2] were significantly larger than those obtained by EVANS *et al.* (1977) near sunset by NO_2 (visible) and HNO_3(IR) absorption measurements below ~25 km, although the two measurements agreed well with each other above this height. Harris also found that even his larger ratio [HNO_3]/[NO_2] is smaller than the result of theoretical calculations using the reaction rate constants available at that time. He suggested that the apparent agreement obtained by Evans *et al.* between their measurements and the model result was a result of a fortuitous calculated value of [OH]. In condition of photochemical equilibrium, the ratio is determined by the relationship

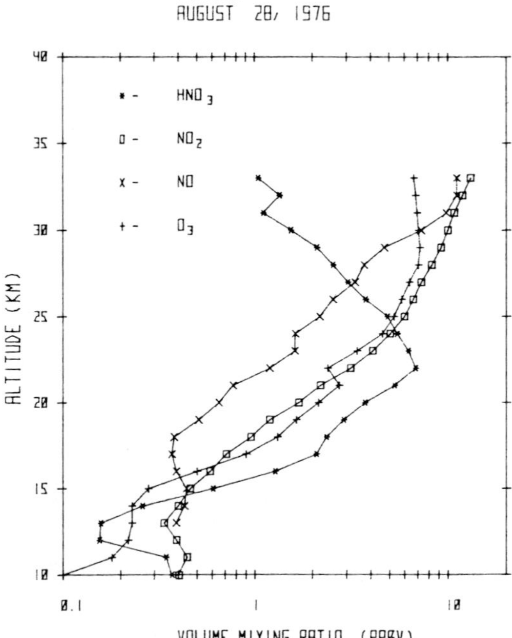

FIG. 8.17. Simultaneous measurements of O_3, NO, NO_2 and HNO_3 from Aug. 28, 1976 baloon flight at Yorkton (51°N). Note that the scale for O_3 is in ppmb (EVANS *et al.*, 1981).

$$\frac{[HNO_3]}{[NO_2]} = \frac{k_{37}[OH][M]}{J_{HNO_3} + k_{38}[OH]}. \tag{8.4}$$

Thus, the ratio depends on the concentration of OH present. The supposition of Harries was later confirmed by EVANS *et al.* (1981); these results seem to suggest that the OH density in the lower stratosphere should be less than those calculated with photochemical models at earlier times. The result is consistent with the result from recent laboratory measurements on reactions of OH with HO_2, HNO_3 and HNO_4; all these reactions seem to occur much faster than was previously thought, and therefore should reduce the model-predicted values for [OH]. The model developed in this book (Chapter 7) uses the new reaction rate constants, and it predicts the ratio $[HNO_3]/[NO_2]$ to be ~1.06 at 25 km at local noon (see Table 7.3). This values is very close to the value (~1.2) measured by HARRIES (1978) at local noon (see also Fig. 6 in EVANS *et al.*, 1981).

The nighttime column density of NO_3 has been observed by absorption measurement at 662 nm using the moon as a light source (NOXON *et al.*, 1978). The observations in March and April indicate a column abundance of $\sim 10^{14}$ cm^{-2}, but the July value seems to be less than the detection threshold of 3×10^{13} cm^{-2}.

The concentration of N_2O_5 has been measured by emission measurements in the 8.1 μm region (KING *et al.*, 1976); the result indicates that the upper limit of the lower stratospheric N_2O_5 concentration is $\sim 2 \times 10^8$ cm^{-3} ($\sim$0.1 ppbv). Murcray (unpublished data, 1978) noted that the column abundance of N_2O_5 above 18 km was less than 1.2×10^{15} cm^{-2}. The recent observation by ROSCOE (1982) tentatively suggests a volume mixing ratio of $1.1 \pm 0.5 \times 10^{-9}$ between 30 and 40 km, which is much larger than previous observations.

8.3.3 Hydrogen compounds

A large amount of observational data has been accumulated on stratospheric H_2O. Figure 8.18 summarizes recent observations of the vertical distribution of H_2O mass mixing ratio. The variability among data reflects observational difficulties; consirerable uncertainties are involved in the absolute values in each observation. However, most profiles agree in showing a minimum somewhere slightly above the tropopause. The increase of $[H_2O]$ above $\sim$20 km seems to indicate that there is a significant source for stratospheric H_2O in the stratosphere due to oxidation of CH_4 (see Section 5.4).

Since the atmospheric temperature reaches a minimum at the tropopause, it is expected that the water vapor density also has a minimum at around that height. The fact that the water vapor minimum appears appreciably above the tropopause has been a puzzle, but recently it has been observed that the temperature in the tropic region could have a deeper minimum above the regular (mean) tropopause when large cumulonimbus clouds penetrate through the tropopause into the lower stratosphere (DANIELSEN, 1982a). A large upward air motion accompanying the cumulonimbus causes a colder temperature by adiabatic cooling. DANIELSEN (1982b) has explained the observed two minima, one at the regular tropopause and the other (deeper minimum) at higher altitudes, by the mechanism of turbulent mixing during anvil formation of cumulonimbus clouds. Penetration of large cumulonimbus through the tropopause is a common phenomenon, particularly in the November to March period over western tropical Pacific and during the Monsoon season over India (NEWLL and GOULD-STEWART, 1981).

The observational data on latitudinal variation in stratospheric H_2O appear more ambiguous. Some data, particularly those obtained at lower heights, seem to indicate a larger amount of H_2O at lower latitudes than at

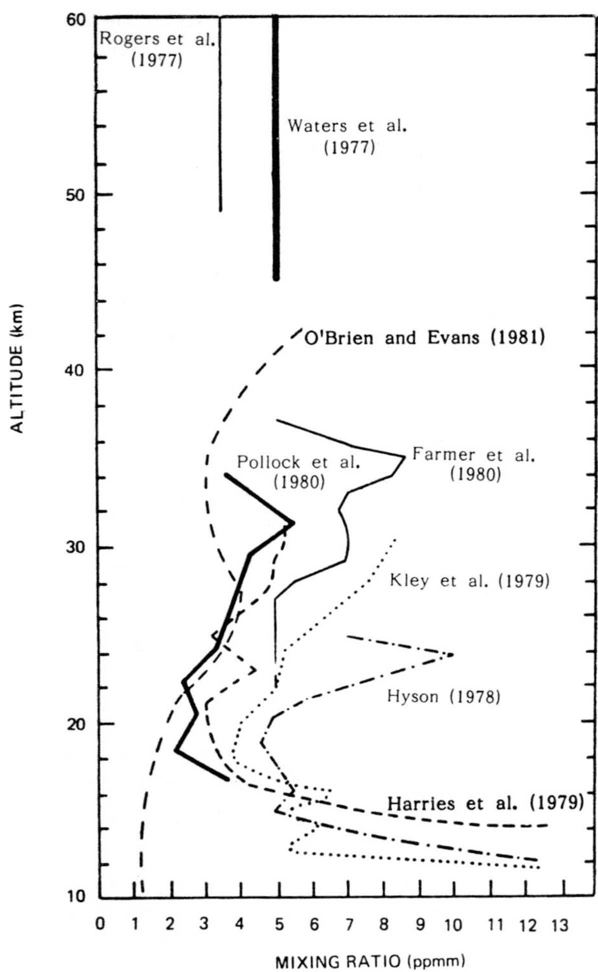

FIG. 8.18. Vertical distributions of stratospheric H_2O as measured by several authors since 1976 (NASA, 1979).

higher latitudes (FARMER, 1974; KUHN *et al.*, 1975; HARRIES, 1976), but measurements at higher altitudes (18 – 24 km) yield relatively flatter variations with latitudes (MCKINNON and MOREWOOD, 1970; MASTENBROOK, 1971). ELLSAESSER *et al.* (1980) argue that the observed gradient of H_2O mixing ratio decreasing with increasing latitude would contradict the mechanism of the stratospheric circulation of H_2O proposed by Brewer and Dobson (see Section 9.2.1). However, one must be cautious in interpreting the observational data made by different instruments at different times and locations.

The mixing ratio of stratospheric H_2 decreases with altitude at a slow rate of 4.8 ppb/km from an average of 0.54 ppm at the tropopause to ~0.45 ppm 35 km altitude (EHHALT *et al.*, 1977). Figure 8.19 illustrates simultaneously measured profiles of H_2, CH_4, CO, N_2O, $CFCl_3$, and CF_2Cl_2 from air samples taken during two balloon flights in June 1979 (FABIAN *et al.*, 1981). Earlier measurements over Texas by EHHALT and HEIDT (1973) have indicated a slight increase of H_2 mixing ratio with height in the stratosphere, reaching ~0.61 ppmv at 28 km, then decreasing again.

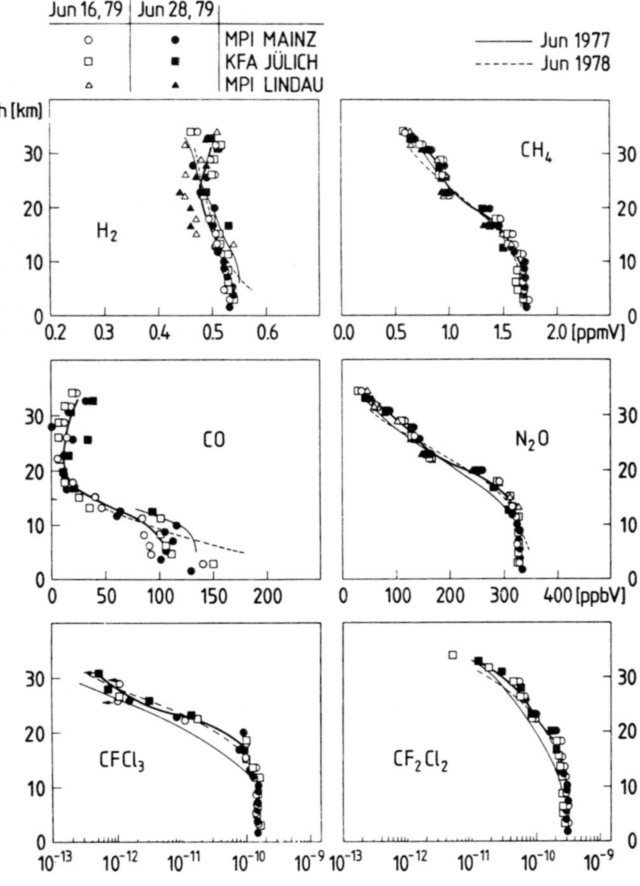

FIG. 8.19. Simultaneously measured vertical profiles of H_2, CH_4, CO, N_2O, $CFCl_3$ and CF_2Cl_2 taken from balloon data in June 1979. Supplementary tropospheric samples were taken from a commercial aircraft. Mean profiles are shown by heavy solid curves and are compared with mean curves from June 1977 and 1978 (FABIAN *et al.*, 1981).

Densities of stratospheric OH and HO_2 have been observed by measuring the molecular resonance fluorescence for the height range 28–42 km (ANDERSON, 1976; ANDERSON *et al.* 1981). The result is summarized in Fig. 8.20. The large difference among observations is partly due to the significant diurnal variations (see Fig. 7.5); the season and the solar zenith angle of each observation are indicated in Fig. 8.20. Both [OH] and [HO_2] have been observed to increase above 28 km (there are no observations below this height) and [HO_2] is generally larger than [OH] in the height range below 40 km; these characteristics are consistent with the model predictions (see Fig. 7.3(b)).

The H_2O_2 concentration was measured by WATERS *et al.* (1981) by the technique of microwave limb sounding using the emission at ~204.574 GHz; a tentative value for the stratospheric abundance has been determined to be ~1 ppbv at 30 km.

8.3.4 *Carbon compounds*

The tropospheric mixing ratio of CO_2 is ~332 ppmv for 1979 and is increasing by ~1 - 1.5 ppmv every year. The mixing ratio was observed to be height independent in the 21 - 39 km range by IR absorption measurements by FARMER *et al.* (1980), but measurements by mass spectrometer technique by BISCHOF *et al.* (1980) indicated a strongly decreasing mixing ratio with height; i.e. a decrease by ~7 ppmv between the tropopause and 33 km. The latter result is consistent with the *in situ* measurements in the up-

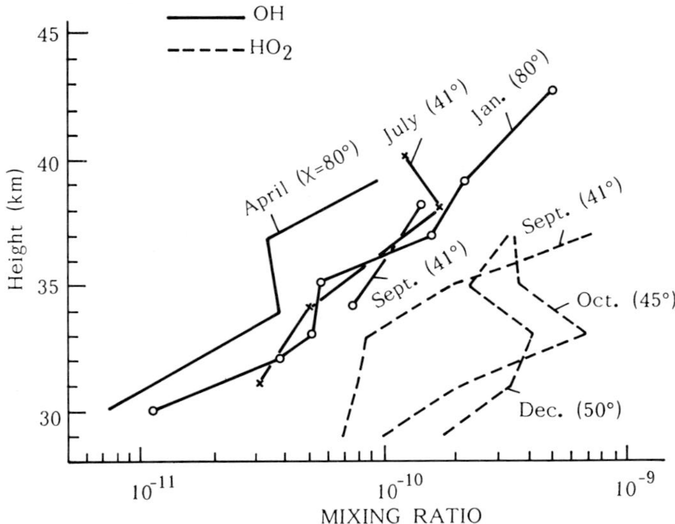

FIG. 8.20. Measured profiles of OH and HO_2 at different seasons and solar zenith angles (NASA, 1979).

per stratosphere by EHHALT *et al.* (1975 b), who observed a mixing ratio 316.2±2.8 ppmv at 40–50 km.

A summary of the measurements of the stratospheric mixing ratio of CO obtained during the last decade is shown in Fig. 8.21. Most observations seem to indicate an increase in the CO mixing ratio with height above 20 km. In the troposphere over the Pacific, HEIDT *et al.* (1980) have observed a large hemispheric difference in the CO mixing ratio. The mixing ratio increases with latitude in the Northern hemisphere and the average value of 175 ppb poleward of 30° N is about 3 times larger than the almost flat value (~57 ppv) south of 10° S.

The mixing ratio of stratospheric CH_4 is observed to decrease from ~ 1.65 ppm at the tropopause to ~1.3 ppm at 21 km and to ~0.45 ppm at 37

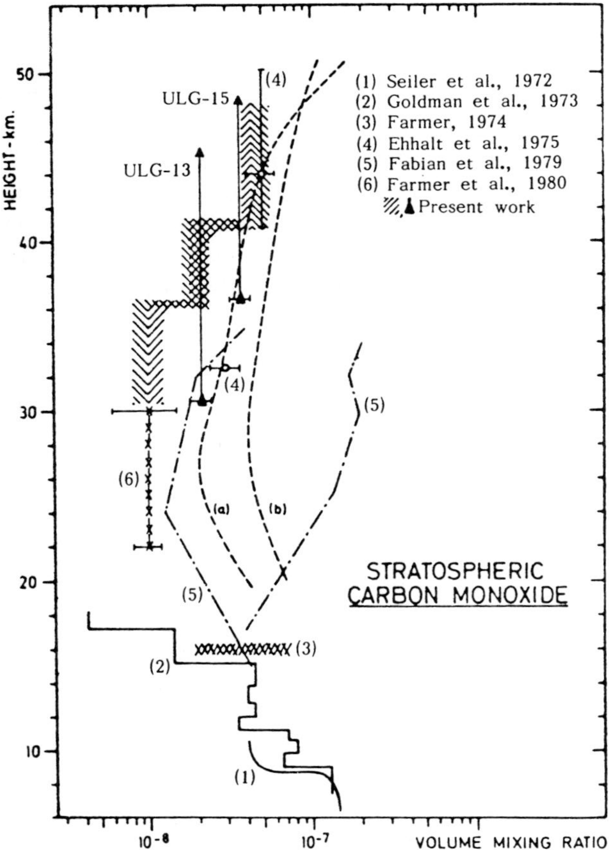

FIG. 8.21. Summary of measurements of stratospheric CO mixing ratios obtained during the last decade (ZANDER *et al.*, 1981).

km (FABIAN *et al.*, 1979a; FARMER *et al.*, 1980). CH_4 is destroyed in the stratosphere mainly by reactions with OH and $O(^1D)$. These CH_4 oxidation may be a source for stratospheric H_2O, CO and H_2 (see Section 5.4).

8.3.5 *Chlorine compounds*

A comparison of the HCl mixing ratio observations in the stratosphere is shown in Fig. 8.22. There is some general agreement among the spectroscopic results of ACKERMAN *et al.* (1976), WILLIAMS *et al.* (1976) and FARMER *et al.*, (1976, 1980). However, the measurements by *in situ* sampling techniques (LAZRUS *et al.*, 1977) and by a pressure modulated radiometer (EYRE and ROSCOE, 1977) indicate poor agreement with the spectroscopic results above ~25 km. In fact, these two results also conflict with each other; the data by the sampling method indicate a sharp minimum near 32 km, whereas the data by the radiometer show a clear maximum at around the same height. The data by the spectroscopic method indicates a steady increase in the HCl mixing ratio in the height range near 32 km. Thus, it is strongly suggested that there is a significant systematic error in at least one of the techniques.

The results of 10 balloon-borne *in situ* measurements of ClO in the stratosphere obtained by the resonance fluorescence method are summarized in Fig. 8.23 (WEINSTOCK *et al.*, 1981). The ClO mixing ratio generally increases from 25 km to 40 km, but the measured values change widely from experiment to experiment; there is more than an order of magnitude

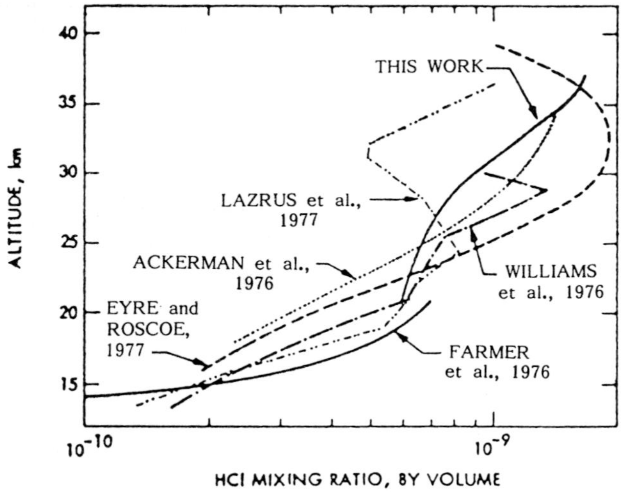

FIG. 8.22. Comparison of stratospheric HCl measurements obtained by various techniques (FARMER *et al.*, 1980).

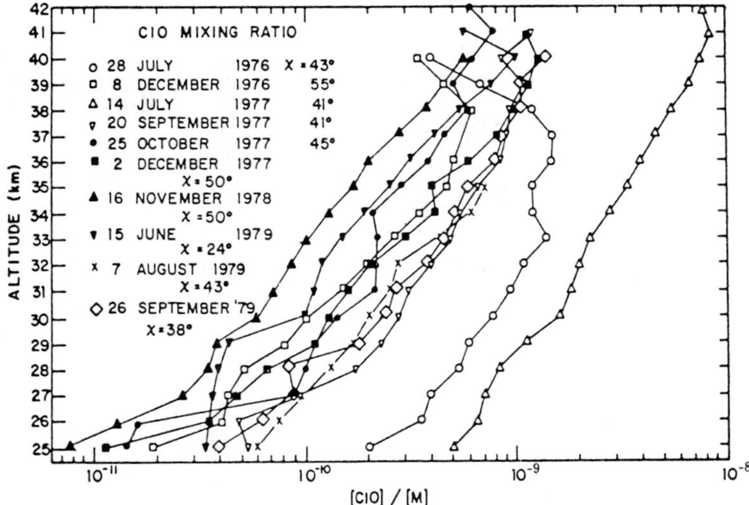

FIG. 8.23. Vertical distribution of ClO obtained by 10 balloon experiments between July 28, 1976 and Sept. 26, 1979 using *in situ* resonance fluorescence methods (WEINSTOCK *et al.*, 1981).

difference between the maximum and minimum ovserved values at a given altitude.

Using a balloon-borne laser heterodyne radiometer, MENZIES (1979) and MENZIES *et al.* (1981) observed a peak mixing ratio of ClO in excess of 1 ppbv above 34 km and a rapid decrease below that height. Re-evaluation of the data based on the more recent spectroscopic observastions (MENZIES, 1983) indicates a measured ClO mixing ratio falling off more rapidly with decreasing height than the previous results indicated. WATERS *et al.* (1979; 1981) detected an emission of milimeter wavelength from ClO, which indicated that the profile peak did not exceed 1 ppbv. The extremely large values ($\geqslant 1$ ppbv) in some curves in Fig. 8.23, particularly those observed in July, 1977, are difficult to comprehend in light of the photochemical role of ClO in the stratosphere currently understood; model calculations indicate that if the ClO mixing ratio were as large as 13.5 ppbv at 40 km, the odd oxygen loss rate due to the catalytic cycle of ClO_x chemistry should have constituted ~90 % of the total O_x loss at that height, and it should have resulted in O_3 concentration of $\sim 4.4 \times 10^{10}$ cm^{-3}, which is certainly much smaller than the observed concentration (5.2 -6.8$\times 10^{11}$ cm^{-3}, see Fig. 5.2). Thus, some problems remain in the measurements of stratospheric ClO, although the average of the profiles sown in Fig. 8.23 is acceptable. WATERS *et al.* (1981) also observed a sudden large decrease of the ClO concentration at sunset, which supports a theory of formation of $ClONO_2$

from ClO at night.

For $ClONO_2$, MURCRAY *et al.* (1979) have observed the concentration distribution in the stratosphere, which shows a maximum mixing ratio of 0.8 ppbv at around 28 km.

The mixing ratios of $CFCl_3$ and CF_2Cl_2 in the troposphere are observed to be $\sim 1.5 \times 10^{-10}$ and $\sim 2.5 \times 10^{-10}$, respectively. Above the tropopause they decrease with height with a much more rapid slope for $CFCl_3$ than for CF_2Cl_2. Fig. 8.24 illustrates the vertical profiles obtained at different latitudes during the period from 1976 to 1979. The ordinate indicates the height relative to the tropopause. There is a tendency for the mixing ratio to

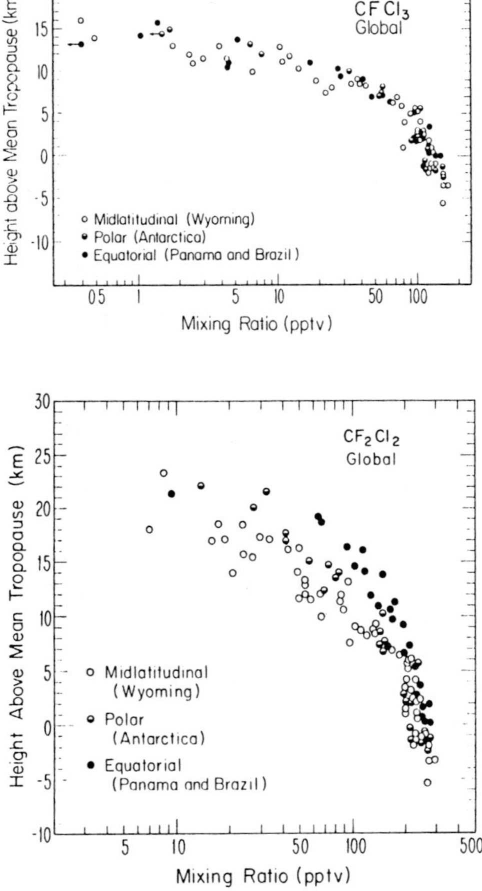

FIG. 8.24. Mixing ratios of $CFCl_3$ and CF_2Cl_2 versus height above the mean tropopause at three different latitudes (GOLDAN *et al.*, 1980).

decrease with height more slowly in equatorial regions than at middle and high latitudes, particularly for CF_2Cl_2.

It has been pointed out by MILLER *et al.* (1981) and WOFSY and LOGAN (1982) that the observational data of $CFCl_3$ show much more rapid decrease with altitude than standard models predict. Reasons for the discrepancy may be found in (i) increased penetration of UV radiation caused by smaller O_2 absorption cross section in the Herzberg continuum (FREDERICK and MENTALL, 1982; FROIDEVAUX and YUNG, 1982), (ii) reduced eddy diffusion coefficient (FROIDEVAUX and YUNG, 1982), (iii) effects of random walk of air parcels (HUNTEN, 1983), or combination of the above three effects.

Figure 8.25 illustrates a contour map of the mixing ratio of $CFCl_3$ observ-

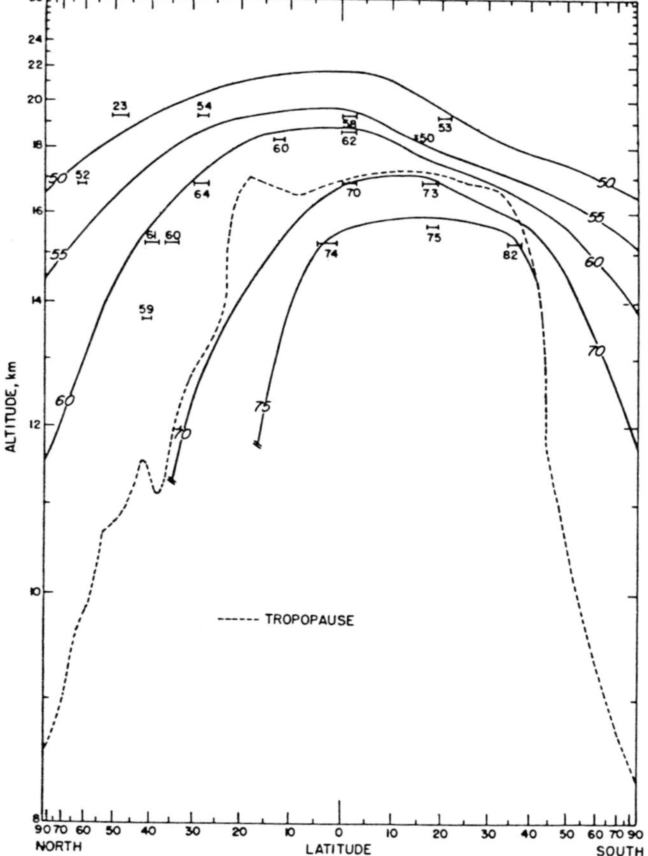

FIG. 8.25. Contour map of the $CFCl_3$ mixing ratio (10^{-12} parts by volume) in April, 1974. Note that the altitude scale is logarithmic (KREY *et al.*, 1976).

ed in April, 1974. There is a marked latitudinal variation; for the same altitude the mixing ratio is larger at lower latitudes than at higher latitudes. The main reason for this is the change in the tropopause height with latitude. A marked asymmetry can also be seen between the northern and southern hemisphere. TYSON *et al.* (1978 b) have found from observations made at latitudes from 74°N to 62°S that the average mixing ratios in the altitude range 9.14 to 11.89 km over the northern hemisphere are about 6% larger than the corresponding values over the southern hemisphere.

Chapter 9

DYNAMICS IN THE MIDDLE ATMOSPHERE

Dynamics play an important role in determining the distributions of minor constituents when their photochemical time constants are greater than the characteristic times for atmospheric transport. This happens, for instance, to odd oxygen (mainly O_3) below ~25 km (see Fig. 6.4). In this chapter we will discuss mainly *large scale meridional transport*, which controls the *global distribution* of minor constituents in the middle atmosphere. We first review the observed motions, and then discuss some theoretical aspects of the large scale (global) circulation of the middle atmosphere relevant to the problems of meridional transport of minor constituents and also of heat. These are not only of considerable interest in getting a general knowledge of atmospheric motion, but also useful for practical purposes to construct a better representation for transport in 2- and 3-dimensional models.

9.1 Observed Motions of Atmospheric Tracers

Any substance having a long life time in the atmosphere can serve as a tracer to detect atmospheric motions. These include chemical tracers (O_3, H_2O, aerosols etc.), radioactive tracers (^{185}W, ^{90}Sr, ^{14}C etc.), and hydrodynamic tracers (potential temperature, potential vorticity etc.). The transport processes of these tracers have been reviewed, amongst others, by REITER (1971, 1972).

9.1.1 Chemical tracers

Ozone (odd oxygen in a strict sense) has long been used as a tracer by meteorologists to observe atmospheric motion. The observed ozone density is generally greater in the lower stratosphere than in the upper stratosphere, at higher latitudes than at lower latitudes, and in winter-spring than in summer-fall (see Section 8.3.1). All these observed features contradict predictions from photochemical theory. In particular, there is no solar UV radiation available for O_x production at high latitudes (above ~66.5°) in winter, but this is the region where the greatest O_3 concentration is observed. Clearly, one cannot understand the existence of such a large O_3 concentration there without invoking the effects of transport from the source region in the lower latitudes where O_x is produced by solar UV radiation.

Figure 9.1 illustrates height profiles of O_3 concentration and the odd oxygen production rate, $Q(O_x)$, predicted for mid-latitudes. While $Q(O_x)$ had a peak at ~40–42 km, the O_3 concentration has a maximum at ~20–22 km. It must be noted, however, that O_x produced above 40 km may be hard to transport all the way down to the heights near 20 km, since the photochemical time constant of O_x above ~35 km is much shorter than a day (see Fig. 6.4). It must also be noted that it is the gradient of the mixing ratio, not of the concentration, that determines the eddy diffusion

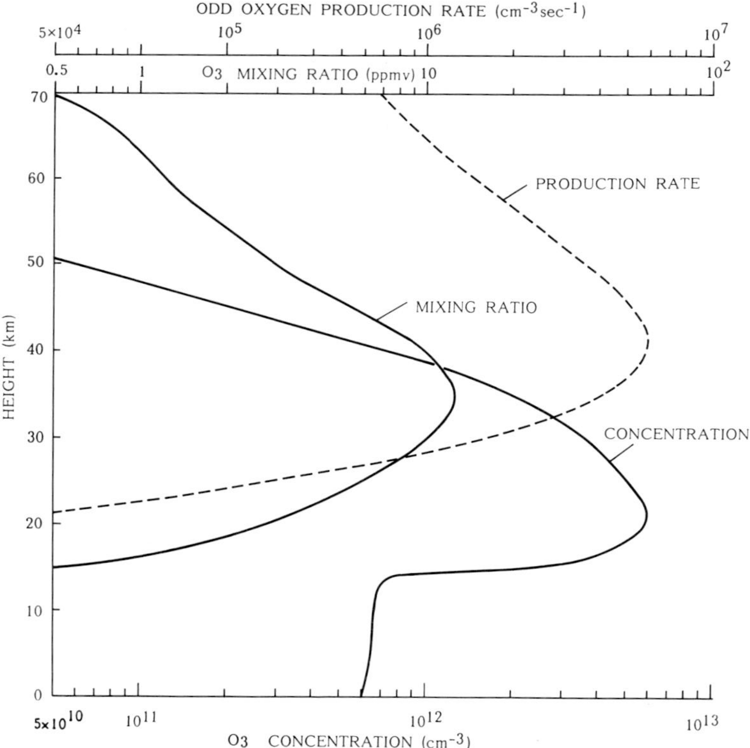

FIG. 9.1. Height variations in O_3 concentration, mixing ratio and O_x production rate predicted in one-dimensional diurnally averaged model discussed in Chapter 7.

transports. The O_3 mixing ratio has a peak at ~ 35 km as is seen in Fig. 9.1; thus, O_3 produced up to heights ~ 35 km can be transported downward by eddy diffusion.

A meridional distribution of average ozone mixing ratio observed for spring 1962–1964 is shown in Fig. 9.2 (HERING, 1966). The global maximum appears near the equator, and at ~30 km the average O_3 mixing ratio at the equator is almost twice as large as the polar region values. Thus, it is evident that O_3 can be transported from lower to higher latitudes by horizontal eddy diffusion.

The meridional velocity at a given latitude is generally a function of time and longitude, and it can be expressed as the sum of the mean velocity and the departure from the mean so that

$$v=\bar{v}+v', \quad \overline{v'}=0. \tag{9.1}$$

The deviation of concentration from the mean also occurs due to the fluctuating velocity such that

$$n=\bar{n}+n', \quad \overline{n'}=0. \tag{9.2}$$

The eddy transport flux of n is generally expressed as

$$\overline{n'v'}=\overline{nv}-\bar{n}\cdot\bar{v} \tag{9.3}$$

since averages of n' and v' should vanish.

Using data on temporal variations in n (ozone density) and v observed at a particular point and tacking the means with respect to time, we can calculate the *transient eddy flux*. From the variations around latitude circles and the means with respect to longitude, one can calculate the *standing eddy flux.*. NEWELL (1963) has shown that both transient and standing eddy fluxes calculated for the January-March period transport ozone northwards in the northern hemisphere, and that northward-movong parcels ($v>0$) are sinking, whereas southward-moving parcels ($v<0$) are rising.

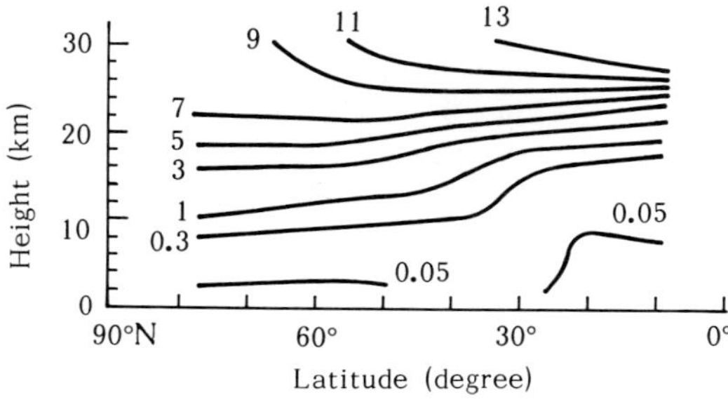

FIG. 9.2. Average ozone mass mixing ratio distribution observed for spring 1963–1964. The unit is μg/g or ppm (HERING, 1966).

BREWER(1949) observed the very dry air just above the troposphere over England with a frost-point of about −80°C, which corresponds to a humidity mixing ratio of about 0.002 g/kg. This temperature is close to that observed at the tropopause near the equator (~195°K, see Fig. 1.2). Brewer postulated that the air had been dried when it rose through the equatorial tropopause region which acts as a *cold trap* and that the air then moved northwards in a mean meridional circulation and subsided over southern England. DOBSON (1956) extended this circulation model to account for seasonal variation in the total ozone over temperate and polar regions. He suggested that the cold pool over the winter pole stores ozone-enriched air which arrived from the ozone-producing region via the meridional circulation. This air subsides into the lower stratosphere of temperate latitudes in the late winter or spring and so accounts for the spring peak in the total ozone observed there.

The mechanism of meridional transport postulated by Brewer and Dobson to explain the global distributions of H_2O and O_3 is called the ''Brewer-Dobson'' circulation. Although long distance O_3 transport in the *upper* stratosphere (>35 km) is unlikely because of a short photochemical time constant, the *middle* stratosphere O_3 can be transported from the equator to high latitudes. It is generally established by recent observational and theoretical studies that tropospheric materials enter the stratosphere mainly in equatorial regions (the Intertropical Convergence Zone, ITCZ) and that they are transported to middle and high latitudes within the stratosphere accompanying the downward motion.

Additional observational evidence for the ''Brewer-Dobson'' circulation may be seen in the global distribution of stratospheric aerosols. As is discussed in Appendix J a stratospheric aerosol is usually a *large* particle. Since only smaller size *Aikin* particles can rise into the stratosphere from the troposphere, stratospheric aerosols must necessarily have been produced within the stratosphere. However, Aikin particles transported from the troposphere into the stratosphere could be important as condensation nuclei for producing stratospheric aerosols.

The size distributions of stratospheric aerosols observed at different latitudes are shown in Fig. 9.3. The particle concentration decreases uniformly with the particle radius r in the tropical region, whereas a peak concentration is observed at $r \sim 0.3\ \mu m$ at middle and high latitudes. This may imply that the stratospheric aerosols are produced as the air containing Aikin particles as condensation nuclei enters the stratosphere in equatorial regions, and moves from lower to middle and high latitudes within the stratosphere.

The vertical distribution of the stratospheric aerosol has a peak around 20 km, called the ''Junge layer'' (JUNGE *et al.*, 1961; JUNGE, 1963). The

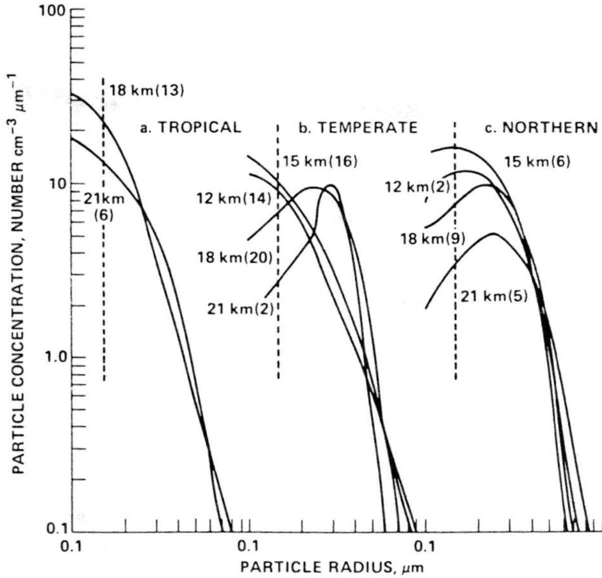

FIG. 9.3. Average size distributions of aerosol particles at various altitudes at three latitudes observed during 1976–1977. The numbers in parentheses are the numbers of observations used for calculating the averages (FARLOW *et al.*, 1979).

height of the Junge layer changes with latitude as is seen in Fig. 9.4; the height generally decreases from lower to higher latitudes. Latitudinal distributions of both the size distribution and the vertical profile of concentration may well be explained by the mechanism of the "Brewer-Dobson" circulation, which transports the condensation nuclei (Aikin particles) from the troposphere into the stratosphere at equatorial regions, and then to higher latitudes within the stratosphere with a coincident sinking motion.

9.1.2 Radioactive tracers

Many kinds of radioactive substances can be produced in the upper atmosphere by nuclear weapon tests. Strontium-90 and Carbon-14 have half-lives of about 28 years and 5,760 years, respectively, and are particularly useful as tracers. Much strontium has been injected by high yield tests whose clouds penetrated higher into the stratosphere. Many experiments have been carried out at both high and low latitudes, but monthly mean meridional profiles of the intensity of the radioactivity of the air measured near the ground indicates the maximum is in the vicinity of the middle latitude tropopause gap. Also, the intensity is highest in the spring. The

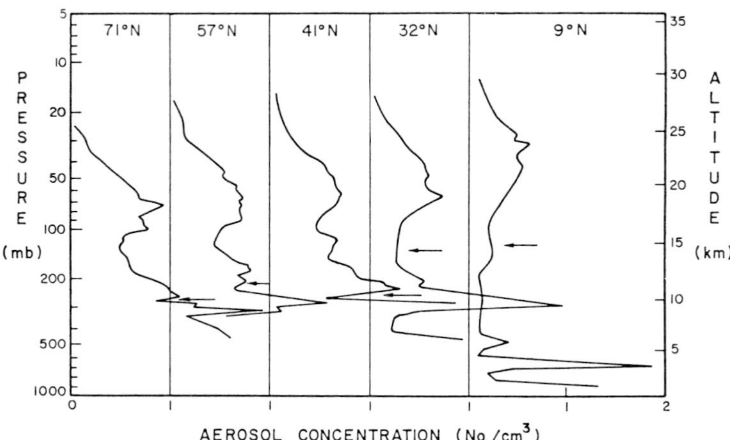

FIG. 9.4. A comparison of five soundings of aerosol distributions made at different latitudes in June of 1973. The arrows mark the observed position of the tropopause (ROSEN *et al.*, 1975).

^{90}SR content of soil samples taken from the entire globe shows a similar mid-latitude maximum. These suggest that much of the stratospheric air has fallen into the troposphere in the middle latitude. There is considerable observational evidence indicating that much of the mass exchange between the stratosphere and the troposphere takes place in the region of the jet stream and tropopause gap at middle latitudes.

Tungsten-185 was injected into the lower stratosphere and troposphere in the equatorial pacific region (~11°N) in summer, 1958. ^{185}W is not a fission product but is produced in the bombardment of ^{184}W by neutrons. Its half-life is only 74 days, but it is particularly useful for investigating transport in the lower stratosphere, since the source regions of ^{185}W can be pinpointed and the time of its injection is well known; no nuclear tests other than that of summer, 1958 have produced ^{185}W (FEELY and SPAR, 1960). Figure 9.5 illustrates the distribution of ^{185}W measured in November-December, 1958, about half a year after the injection. The maximum still remains over low latitudes, shifted slightly southwards from the injection area. NEWELL (1963) considered that this was eivdence to indicate that the transport of ^{185}W occurred mainly by a large scale eddy motion rather than by an organized mean meridional circulation. Two secondary maxima in middle latitudes in both hemispheres indicate the regions of strong stratospheric-tropospheric exchange.

The excess of ^{14}C over that produced naturally by cosmic rays was injected into the high latitude stratosphere by Russian neclear bomb tests (1961–1962 series). This is useful for investigating the stratospheric

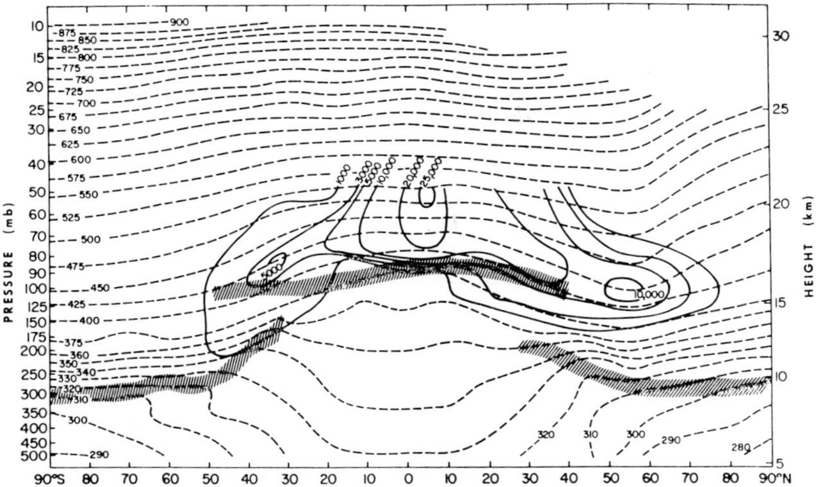

FIG. 9.5. Distribution of tungsten-185 (solid curves) and potential temperature (dashed curves) in the stratosphere observed for Nov.–Dec., 1958. Units are disintegrations per min. per 1,000 standard cubic feet of air. Shaded area represents the region of tropopause (NEWELL, 1963).

transport near the polar region. JOHNSTON *et al.* (1976) have noted the advantage of using ^{14}C data over ^{90}Sr data because of the apparent larger residence time. ^{14}C is formed in the gas phase (^{14}CO or $^{14}CO_2$) and is therefore more adequate for tracers of atmospheric gas than ^{90}Sr, which is lodged on solid particles. Figure 9.6 shows the time series of the excess ^{14}C distribution. During the one year period from January 1963 to January 1964, the maximum intensity over the polar region was reduced by more than half, but the maximum still remains at about the same location. The excess ^{14}C has spread over an extended area and there is an indication of across-the-equator transport.

Rhodium 102 was produced in the upper part of the middle atmosphere by two high-yield weapons fired from a rocket at ~43 km, 16°N in August, 1958. ^{102}Rh is not a fission product, but is produced probably by a reaction of ^{103}Rh with neutrons (KALKSTEIN, 1962); its half-life is ~210 days. The clouds apparently reached above 100 km and produced ^{102}Rh at all heights in the mesosphere and lower thermosphere. Measurements at ~20 km indicate that ^{102}Rh started to increase in late October, 1959 reaching its maximum around January, 1960. The intensity seemed to be strongest north of ~60°N. By mid-1960 the distribution pattern became relatively uniform, but equal, high concentrations at high latitudes in both hemispheres appeared with a rather sharp drop to low concentration at low latitudes.

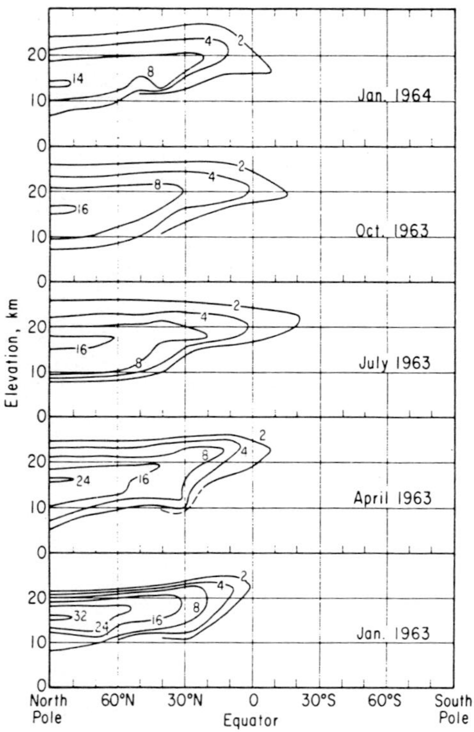

FIG. 9.6. Meridional distribution of the zonal average of excess carbon 14 (units of 10^3 molecules per cm^3) for the indicated times (JOHNSTON *et al.*, 1976).

Thus, the distribution is similar to that of ozone, which also has its source region above the middle stratosphere.

9.1.3 Hydrodynamic tracers

From the primitive equations of hydrodynamics and thermodynamics, we can derive air-mass properties that are conserved under certain conditions (assumptions); such properties can be used effectively as air-mass tracers. The best known conservative quantity is the *potential temperature*, which is the temperature of an air parcel when compressed to a pressure of 1,000 mb and is expressed as

$$\theta = T\left(\frac{1000}{P}\right)^{R/c_p} \tag{9.4}$$

where R is the specific gas constant and c_p the specific heat at constant pressure. The value R/c_p is ~0.286 (see Chapter 2). If there is no heat transfer between the air parcel and its environment, i.e. if changes occur adiabatically, the θ of that air parcel remains constant. The dashed curves in Fig. 9.5 represent surfaces of constant θ (*isentropic surfaces*) for December, 1957. It seems that ^{185}W was transported polewards more or less along the isentropic surfaces with the greatest transport occurring in the winter hemisphere. The isentropic surface runs almost horizontally. A surface of, for instance, θ=360°K is above the tropopause at higher latitudes but below the tropopause at lower latitudes; therefore, it penetrates the tropopause gap in middle latitudes. Thus, it is much easier for air parcels to be exchanged between the stratosphere and the troposphere in the vicinity of the tropopause gap associated with mid-latitude cyclones.

Another conservative quantity to be used as a hydrodynamic tracer is the *potential vorticity* given by

$$p_\theta = -\frac{\partial\theta}{\partial p}(\zeta_\theta + f) \tag{9.5}$$

where ζ_θ is the vertical relative vorticity computed from a stream function for the actual wind, and f is the vertical component of the solid earth's vorticity, namely 2Ω sin ϕ (Ω is the angular velocity of the earth and ϕ is the latitude). $-\partial\theta/\partial_p$ is a measure of the mean stability of a parcel against vertical displacement.

The conservation of potential vorticity is a powerful constraint on the large scale motions of the atmosphere (HOLTON, 1979); it is ideal for identifying air masses originating either in the stratosphere or in the troposphere. Typical values of p_θ for a stratospheric air mass are greater than 20×10^{-9} cm sec deg/g, whereas tropospheric values are usually smaller by more than one order of magnitude (REED and DANIELSEN, 1959; REITER, 1961). Figure 9.7 illustrates isolines of P_θ and the ozone mixing ratio, and the Zirconium 95 activities at the individual sampling points (DANIELSEN *et al.*, 1970). It is seen that the stratospheric air is falling down into the troposphere through the region of folded tropopause developing near the upper cold front, and the larger values of ozone mixing ratio and zirconium activity in that vicinity indicate that O_3 and ^{95}Zr are being brought down into the troposphere by that downward motion.

It has long been observed that the air at the rear of a cyclone is generally ozone rich, whereas there is a region of smaller ozone density at the head of a warm front. These are associated with the downward motion developed after the cold front in the cyclone system; strong flows in cyclones could extend to high altitudes and the downward motion carries along air masses with ozone from the stratosphere to the troposphere. Because of the effect of topography and the difference of activity of cyclones with longitude, the

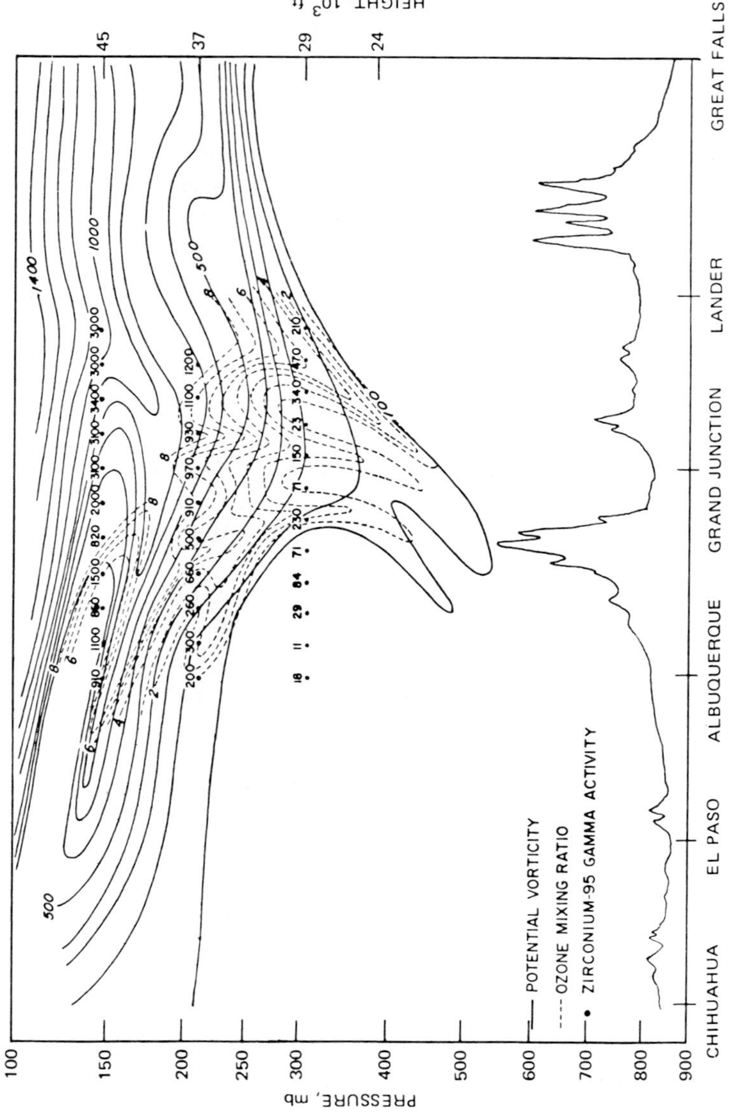

FIG. 9.7. Meridional cross section of potential vorticity (10^{-10} cm sec deg/g), $^{95}Z_r$ activities (disintegrations per min. per 1,000 cubic feet of air) and ozone mixing ratio (10^{-7} g/g) (DANIELSEN *et al.*, 1970).

ozone rich area will not be produced uniformly. Thus, regions of relatively larger ozone concentration have resulted semi-permanently over the Eastern United States, Europe and Eastern Asia where the cyclone activity is high (see Fig. 8.7).

The position of these ozone maxima has a rather close correspondence to the position of the major standing trough in the mean flow at 500 mb level. It is generally believed that the locations of these maxima are a consequence of the dynamics of the flow in the troposphere and lower stratosphere. This hypotheses is reinforced by observations that both tropospheric flow and ozone distributions are more zonally symmetric in the southern hemisphere than in the northern hemisphere. The appearance of the trouths at 500 mb level must prevail particularly in winter in the northern hemisphere when the stronger cyclone system develops. MANTIS *et al.* (1980/81) found a similar but weaker ozone maximum in the lower stratosphere over Northwest Europe in summer; they attributed this to a strong horizontal advection of ozone by the mean flow.

9.2 General Circulation of the Middle Atmosphere

The mean (or time averaged) hemispheric circulation of the atmosphere is called the "general circulation." We are particularly interested in meridional circulation which could transport minor constituents from lower to higher latitudes. The complex mathematics and detailed discussion involved are beyond the scope of this book (see e.g. HOLTON, 1975), but an attempt is made in this section to describe the essential features of this very complicated subject using as little mathematics as possible.

9.2.1 Eulerian mean circulation

Under a Eulerian coordinate system, air parcels are assigned to points in space at each given time; therefore, the Eulerian mean is the time average of properties of many *different* parcels which occupy the same point at different times. Since most meteorological observations (wind speed, temperature, etc.) are made at fixed points at specified time intervals, a Eulerian system is usually more convenient for analyses of observational data, such as the calculation of mean velocity and mean temperature. Furthermore, general circulation models usually solve primitive equations, which are a set of equations (equations of motion, the thermodynamic energy equation, the continuity equation, the hydrodynamic approximation and the equation of state) written in Eulerian coordinates; therefore, predictions of various properties of fluids in Eulerian coordinates are the direct, initial result from model calculations.

The general circulation of the middle atmosphere is much simpler than

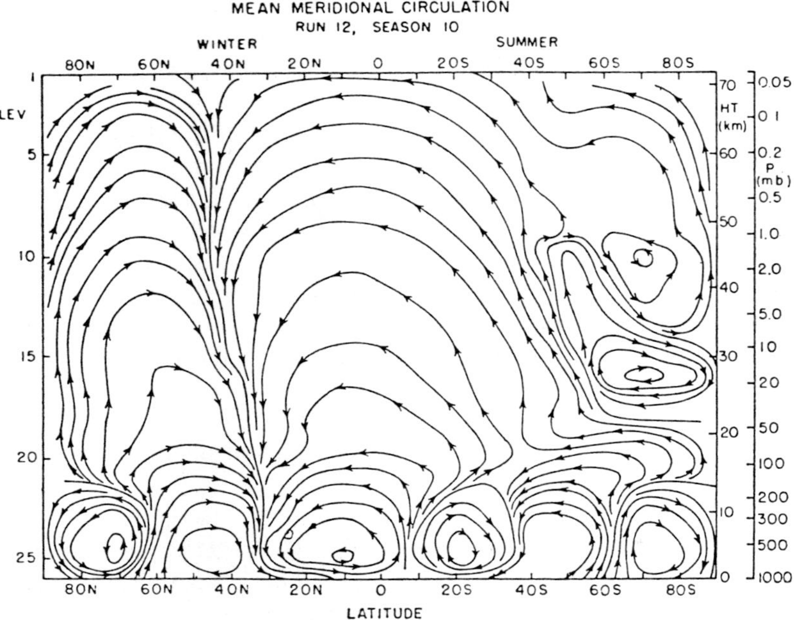

FIG. 9.8. Height-latitude section of the mean meridional streamlines computed by a 3-dimensional model for the solstice seasons (CUNNOLD *et al.*, 1975).

that of the troposphere. The main reasons for this are fourfold; first, there is little water vapor in the stratosphere; secondly, no large *baroclinic instability** develops in the stratosphere; thirdly, there is no heat exchange between the stratospheric air and the ocean; and fourthly, there is no direct effect of topography in the motion of stratospheric air.

The mean meridional circulation of the atmosphere up to the height of 70 km as computed by a 3-dimensional model is illustrated in Fig. 9.8 (CUNNOLD *et al.*, 1975). The general circulation of the middle atmosphere is composed mainly of a large scale global cell caused by the excess heating over the summer hemisphere and the net cooling over the winter hemisphere, superimposed on an indirect cell (counter circulation) which develops at

* The air motion is deflected normal to the velocity by the Coriolis force. Because of this deflecting effect the Hadley cell motion in the troposphere cannot reach the high latitudes. Thus, the meridional circulation cannot transport heat energy from the lower to higher latitudes sufficiently enough to remove the heat inbalance between equatorial and polar regions, and a large north-south temperature gradient would result in middle latitudes. If this temperature difference exceeds a certain limit, the atmosphere becomes unstable against large scale wave motions. This instability is called the *baroclinic instability* and can cause polar jet streams in winter.

high latitudes, particularly in winter.

There is a similar but very much weaker counter circulation (indirect cell) in the summer high latitudes. The two-(or three-) cell structure over the entire globe in and above the stratosphere is in sharp contrast to the general circulation of the troposphere, which is composed of three-cell motions in each hemisphere.

Air parcels can be transported efficiently from lower to middle latitudes by the mean meridional circulation developed in these latitudes, but the counter circulation at high latitudes seems to transport air parcels from polar regions to middle latitudes. In this region, however, a large eddy flux represented by (9.3) tends to transport the air parcels from middle to high latitudes. In fact, these two models of transport almost cancel each other out. This *cancellation effect* was first noted in a stratospheric general circulation model by MANABE and HUNT (1968). They found that the divergence of large scale eddies was almost compensated by the advection of absolute angular momentum by the meridional circulation (see Fig. 9.9). Similar cancellation effect between the mean meridional circulation and the large scale eddies were also found in the heating rate (MANABE and HUNT, 1968) and in the rate of change of ozone concentration (HUNT and MANABE, 1968). The latter can be seen even in a simple 2-dimensional model by SHIMAZAKI *et al.* (1973).

The counter circulation at high latitude in the winter stratosphere is caused by the upward propagation of planetary waves originating in the troposphere. The northward heat transport by planetary waves in the middle and high latitudes of the northern winter hemisphere is indeed much larger than that required by the differential diabatic heating; therefore, an upward mass motion is produced over the polar region where convergence of eddy transport occurs, whereas a downward mass motion is produced over the middle latitudes where divergence of eddy transport occurs. Thus, the counter circulation develops from higher to middle latitudes in the middle atmosphere over the winter hemisphere. A counter circulation is also developed in the troposphere but the mechanism of its production is different, i.e. the tropospheric counter circulation is produced by large scale eddy motions due to baroclinic instability.

CHARNEY and DRAZIN (1961) have shown theoretically that baroclinically unstable wave disturbances originating in the troposphere do not propagate energy vertically at all. They also have shown that wave disturbances due to planetary wave propagation do not affect the zonal flow when the wave disturbance is a small stationary perturbation on a zonal flow that varies vertically but not horizontally. The result of Charney and Drazin implies that the effect of planetary waves on the motion of air parcels completely cancels out the effect of the mean meridional circulation, if the wave

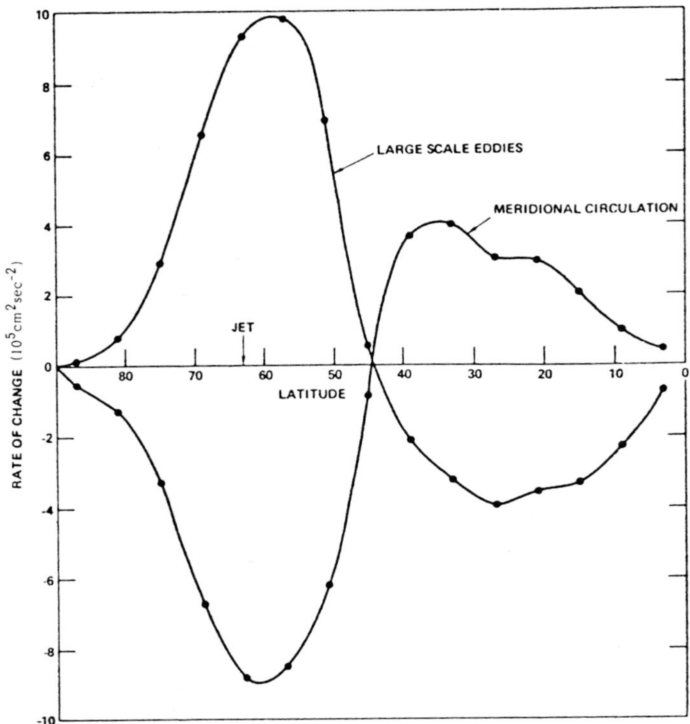

FIG. 9.9. The variations with latitude of the rate of change with time of the divergence of large scale eddies and of the absolute angular momentum due to the mean meridional circulation in the top level (37.5 Km) of the model (MANABE and HUNT, 1968). Two effects are cancelled out with each other.

disturbance is small enough and if there is no net heating or cooling in the atmosphere. ANDREWS and MCINTYRE (1978) later showed that the cancellation holds exactly for finite amplitude waves. The air parcels are simply moved up and down along a circular orbit by the waves and cannot be transported for long distances. Thus, no large scale mass transport can occur by planetary waves (or eddy motions) in the stratosphere; it can occur by the meridional circulation only when there is an imbalance between heating and cooling in the atmosphere.

9.2.2 Lagrangian mean circulation

For investigating the long distance transport of air parcels it is physically more tenable to follow the trajectories of parcels using Lagrangian coordinates, in which air parcels are identified at all times during their motions.

Consider the air parcels P located on a rod along a latitude circle (Fig.

9.10). After some time these parcels will be displaced from their initial positions and will be located at the positions Q along a curved rod. The average properties of air parcels at Q calculated by applying a weight to each point proportional to the longitudinal displacement is called the "generalized Lagrangian mean", and the motion of the center of mass of these parcels at Q is defined to be the "Lagrangian mean motion" (ANDREWS and MCINTYRE, 1978; DUNKERTON, 1978; MCINTYRE, 1980). Since the equation for a Lagrangian-mean quantity does not include eddy flux terms, its change can be calculated without suffering inconvenience from the troublesome cancellation effect between eddy and mean motions; it is determined simply by the source and sink terms so that

$$\left(\frac{\partial}{\partial t}+\bar{v}^L\frac{\partial}{\partial y}+\bar{w}^L\frac{\partial}{\partial z}\right)\bar{\chi}^L=\bar{Q}^L \tag{9.6}$$

where $\bar{A}^L$ denotes the generalized Lagrangian mean of a quantity A, and Q represents a source and sink of A.

The ultimate origin of the large scale motion of air parcels is diabatic heating or an imbalance between heating and cooling in the air. Additional motion could occur in the stratosphere by eddy motions which consist primarily of ultralong quasistational planetary waves. These waves are generated in the troposphere by topography and land-sea heating contrasts and are forced to propagate vertically through the stratosphere (CHARNEY and DRAZIN, 1961; MATSUNO, 1970). It is shown theoretically (see, for instance, HOLTON, 1979) that vertical propagation of stationary waves can occur only in the presence of westerly wind weaker than a critical value which depends on the horizontal scale of the wave. In the summer hemisphere the stratospheric mean zonal winds are easterly so that stationary planetary waves are all trapped and cannot propagate into the stratosphere. Thus, there are small eddy motions in summer high latitude as is seen in Fig. 9.8. In the winter hemisphere the mean zonal winds are westerly but generally are so strong that only the largest scale waves (wavelength longer than 10,000 km) can propagate through the stratosphere. Thus, the eddy motions of long wavelength (wave number 1 or 2) can develop in the stratosphere only in the winter hemisphere.

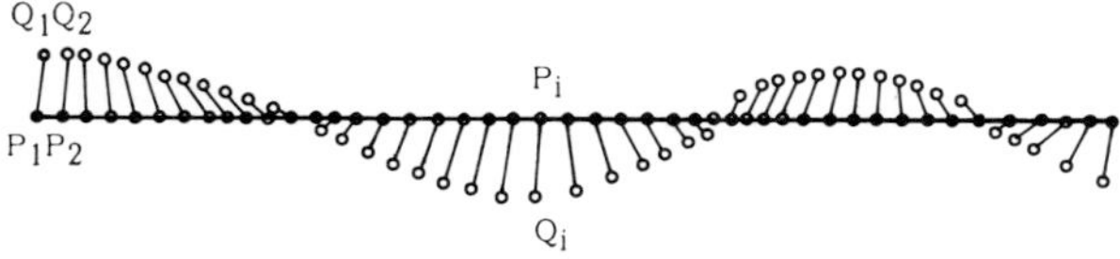

FIG. 9.10. Illustrating displacements of air parcels from the points along a latitude circle, to help in understanding the calculation of Lagrangian mean.

The activity of planetary waves has a maximum around 60° in latitude (MATSUNO, 1980). Projections of parcel trajectories (circular motion) caused by planetary waves on the meridional plane are shown schematically by ellipses in the upper panel of Fig. 9.11. Since the counter-clockwise circular velocity is largest for the air parcel near 60°(located at the origin of the y-axis in the figure), the upward motion dominates the downward motion at any point in the region poleward of 60° (the region of $y>0$) and, therefore, a net *upward* motion occurs there. Similarly, a net *downward* motion is produced in the region equatorward of 60° ($y<0$).

If the second order terms of wave amplitude are included in the analyses, the trajectory usually contains non-periodic motion, and the projection on the meridional plane is shown in the bottom panel of Fig. 9.11. The spiral motion takes place by the change in the curvature of elliptic motion with latitude. Since the curvature has a minimum at ~60°, the net drift by the spiral motion is in the downward direction at higher latitudes and in the upward direction at lower latitudes. These drift motions are called the "Strokes drift". It is in the opposite direction to the net velocity due to elliptic motions discussed in the previous paragraph, and these two velocities cancel each other out at each latitude. The same result is obtained for the horizontal components of the net velocity of elliptic motions and the Stokes drift (MATSUNO, 1980). Thus, no large scale net motion can occur by planetary waves. This is the Lagrangian point of view for the cancellation effect discussed in the previous section.

MURGATROYD and SINGLETON (1961) calculated the meridional circulation by assuming that the differential diabatic heating pattern was balanced

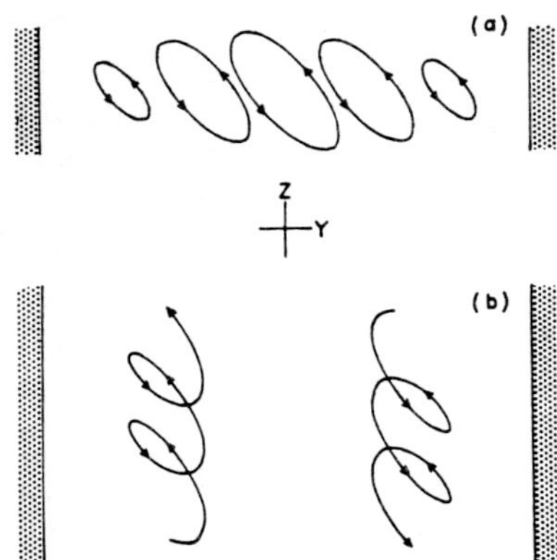

FIG. 9.11. Schematic illustration of projection on meridional plane of trajections of air parcels caused by planetary waves in the stratosphere; (a) is correct to the first order of wave amplitude, (b) includes the second order quantity (MATSUNO, 1980).

by adiabatic (dynamic) cooling and heating in the respective upward and downward motions. They have constructed a pattern of radiative heating and cooling in the meridional plane at solstice based on radiative energy calculation by MURGATROYD and GOODY (1958) and OHRING (1958) as is shown in Fig. 9.12. The absolute values in Fig. 9.12 are probably too large particularly near the regions of largest heating and cooling, but the general feature of global distribution of the net radiative heating may properly be represented by this figure (see, for instance, Fig. 9.8 of APRUZESE *et al.*, 1982). Mass transport should occur from the region of heating (shaded area in Fig. 9.12) to the region of cooling (blank area). Thus, it is easily inferred from the figure that the air moves from the summer to the winter hemisphere above ~25 km and from equator to pole below that height.

The model of MURGATROYD and SINGLETON (1961) has long been criticized because of the neglect of transport by eddy motions; that neglect was noted as a deficiency of the model even by the authors themselves. We now know, however, that eddy motion in the stratosphere is caused by planetary waves, which do not transport heat and momentum for long distances. What Murgatroyd and Singleton calculated actually is the Lagrangian mean circulation produced by radiative sources and sinks, and the result of the calculation shows a similar circulation to the Brewer-Dobson circulation.

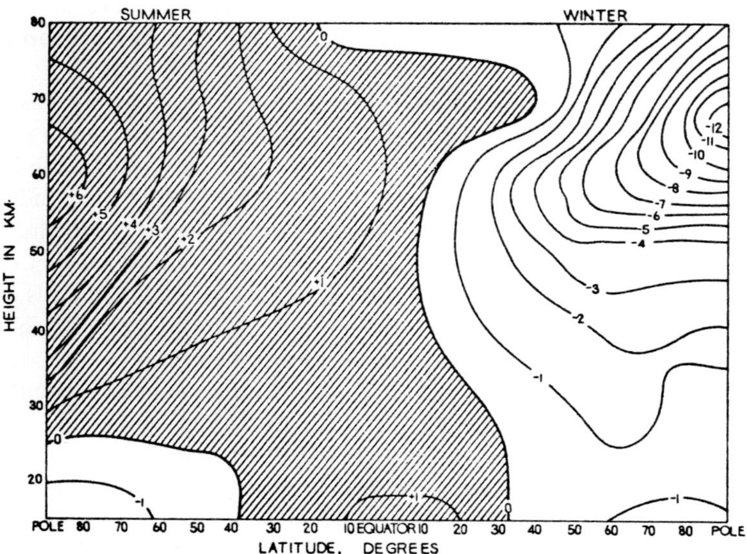

FIG. 9.12. Radiative heating and cooling in °K day^{-1}. Regions of heating are shaded (MURGATROYD and SINGLETON, 1961).

The vertical velocity of the Lagrangian mean motion in the above sense is calculated by (DUNKERTON, 1978)

$$\overline{w^L}=\overline{Q}/(\Gamma_d-\Gamma) \tag{9.7}$$

where $\bar{Q}$ is the Eulerian mean diabatic heating, and Γ_d and Γ are the temperature gradient given by (2.1) and (2.4), respectively. This velocity is actually the velocity of adiabatic mean circulation, but judging from the characteristics of the stratospheric dynamics it is anticipated that the velocity coincides with the advective part of the Lagrangian mean motion (DUNKERTON, 1978; MATSUNO, 1980; KIDA, 1983a). The meridional component of the Lagrangian mean velocity is then calculated from the continuity equation as follows:

$$\overline{v^L}=-\frac{1}{\cos\phi}\int_{-\frac{\pi}{2}}^{\phi}\frac{a\cos\phi'}{\rho}\frac{\partial}{\partial z}(\rho\overline{w^L})\,\mathrm{d}\phi'. \tag{9.8}$$

$\bar{v}^L$ is assumed to be zero at $\phi=-1/2$ (south pole), but (9.8) does not necessarily guarantee that $\bar{v}^L$ becomes zero at $\phi=1/2$ (north pole). In order to satisfy the zero velocity condition at both poles, small corrections are necessary to the radiative heating rates (MURGATROYD and SINGLETON, 1961). Fig. 9.12 includes such corrections. The Lagrangian mean circulation calculated from Fig. 9.12 is shown in Fig. 9.13 (MATSUNO and SHIMAZAKI, 1981). The circulation pattern calculated here is correct, although velocities are quantitatively inaccurate (they are probably too large, see p. 231).

A Lagrangian mean meridional circulation of air parcels has been calculated by KIDA (1977; 1983 a) with a simplified 3-dimensional general circulation model (GCM) representing the annual mean condition. Recognizing that the motion of air parcels is affected by advection and diffusion and that the diffusion process occurs symmetrically about the time, KIDA (1983 a) has developed a method to deduce the advective part of the parcels' motion by eliminating the effect of diffusion. The method is essentially to take the average of the two velocity vectors of the Lagrangian mean motion calculated in the forward and backward directions in time, respectively. By this approach he was able to calculate a systematic circulatory advective mean motion of air parcels in the lower stratosphere.

In a long-term trajectory analysis (KIDA, 1983 b) variations of 30 day periods have been chosen from a 100 days' integration of GCM so that they may represent the typical quasi-equilibrium condition. Using the GCM data of the same 30 day variation repeatedly, he has calculated the trajectories of the same air parcels for a long period. Actually, trajectories of the same parcels are followed only for each 30 day period. For instance, for the

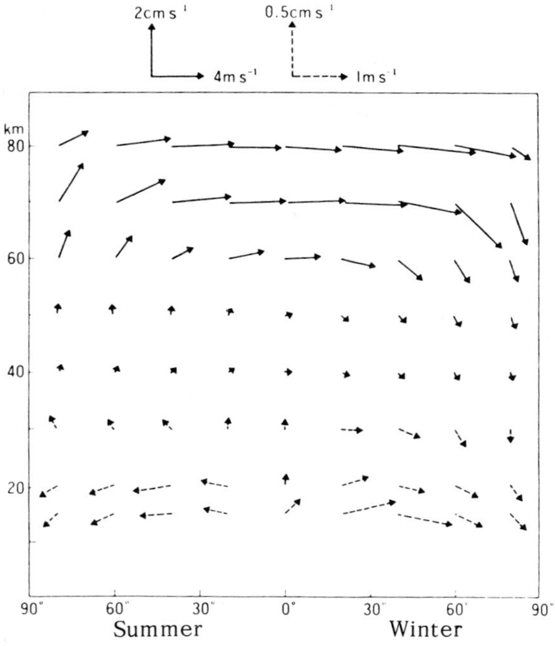

FIG. 9.13. Lagrangian mean circulation calculated from the differential adiabatic heating pattern of Fig. 9.12 (MATSUNO and SHIMAZAKI, 1981).

second 30 days, each parcel starts from the location arrived at the end of the first 30 days, but the GCM data (velocity profiles) change suddenly to the same data as was used at the beginning of the first 30 days. Although the identity of the air parcel is lost between the two 30 days periods, the result of long term integration by this method should well represent the statistical distribution of air parcels originating from the same location. Since it is still formidable to perform a very long term integration for trajectories of specific air parcels, Kida's method is useful for investigating the statistical properties of the movements of air parcels.

Figure 9.14(a) illustrates the meridional distribution of ~2,000 air parcels started from initial locations at 900 mb (~1 km in height) in middle latitude (35 - 40°), uniformly distributed on grid points separated by 0.5° in longitude. Air parcels move to both higher and lower latitudes. Upward motion also occurs, but it generally stops at the tropopause, although some parcels can penetrate through. It is interesting to see that air parcels reaching the equatorial region can move up to the highest level.

Figures 9.14(b) and (c) illustrate the result from air parcels starting from the 100 mb (~16 km) and the 3 mb (~24 km) levels, respectively, both near

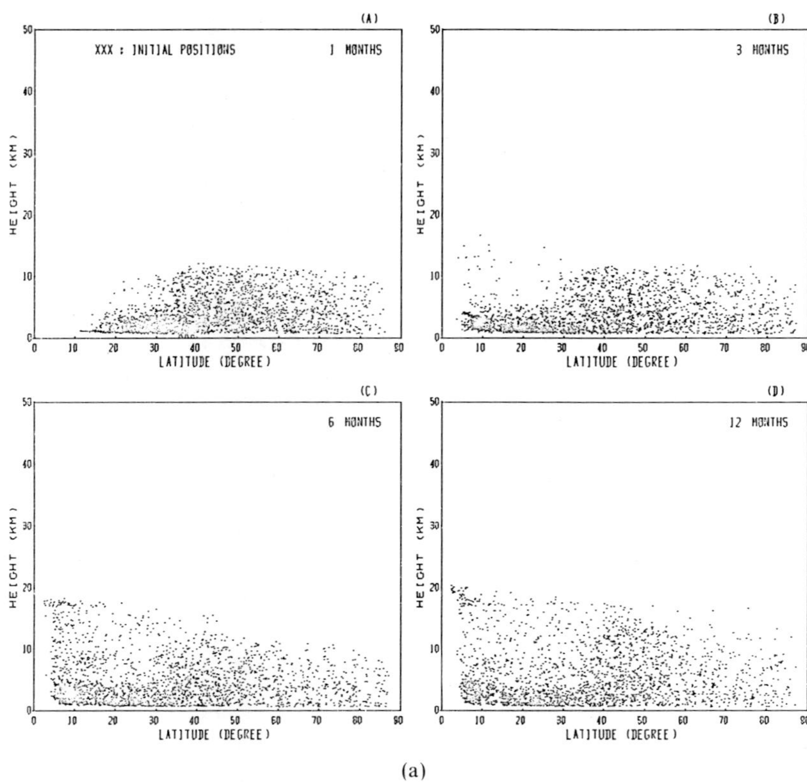

(a)

FIG. 9.14. Distribution of air parcels followed by a numerical method of Lagrangian trajectory analysis: (a) initial locations are at 900 mb(~2 km) in latitudes 35–40°, (b) initial locations are at 100 mb(~16 km) in latitudes 2.5–7.5° and (c) initial locations are at 30 mb(~24 km) in latitudes 2.5–7.5° (KIDA, 1983b).

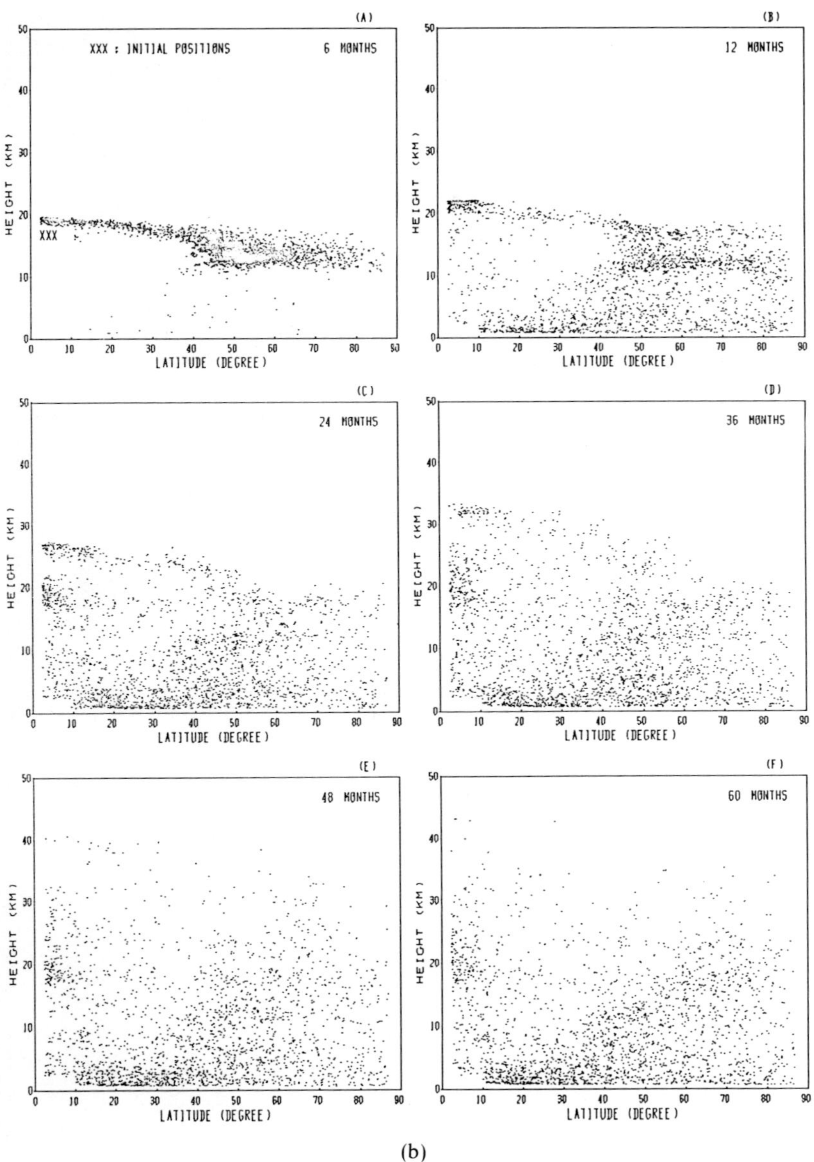

(A)
XXX : INITIAL POSITIONS
6 MONTHS
XXX
HEIGHT (KM)
LATITUDE (DEGREE)
(B)
12 MONTHS
HEIGHT (KM)
LATITUDE (DEGREE)
(C)
24 MONTHS
HEIGHT (KM)
LATITUDE (DEGREE)
(D)
36 MONTHS
HEIGHT (KM)
LATITUDE (DEGREE)
(E)
48 MONTHS
HEIGHT (KM)
LATITUDE (DEGREE)
(F)
60 MONTHS
HEIGHT (KM)
LATITUDE (DEGREE)

(b)

FIG. 9.14.

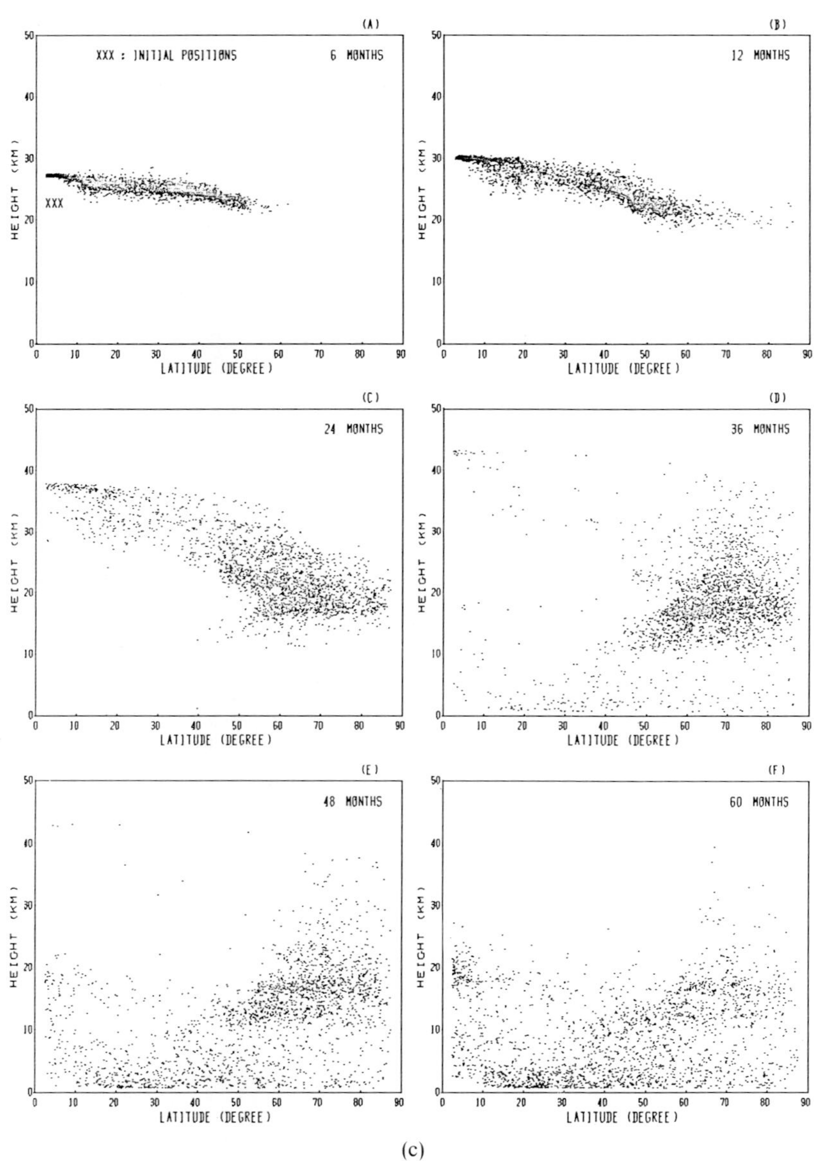

(A)
XXX : INITIAL POSITIONS
6 MONTHS
XXX
HEIGHT (KM)
LATITUDE (DEGREE)
(B)
12 MONTHS
HEIGHT (KM)
LATITUDE (DEGREE)
(C)
24 MONTHS
HEIGHT (KM)
LATITUDE (DEGREE)
(D)
36 MONTHS
HEIGHT (KM)
LATITUDE (DEGREE)
(E)
48 MONTHS
HEIGHT (KM)
LATITUDE (DEGREE)
(F)
60 MONTHS
HEIGHT (KM)
LATITUDE (DEGREE)

(c)

FIG. 9.14.

the equator (2.5 – 7.5°). It is evident that for the first year or so air parcels move mainly to higher latitudes, although some are lifted upwards slightly near the equator. When reaching latitudes greater than ~50°, they are largely transported downward and fall into the troposphere. Thus, after more than two years the largest proportion of air parcels exists in the lower stratosphere at high latitudes; this is particularly the casse for the air parcels started from the middle stratosphere near the equator (see Fig. 9.14(c)).

The result of the analysis shown here clearly suggests the mechanism for global transportation of O_3 from the source region (middle and upper stratosphere in the lower latitudes) to the region where it is most abundant (lower stratosphere at higher latitudes). The result also confirms, for the first time by numerical analysis, the existence of the Brewer-Dobson type circulation, although the predicted mean velocity is somewhat smaller than the motions observed during special events by DANIELSEN (1981). The difference is attributed mainly to the fact that the Kida's calculation simulated the annual mean condition of a hemisphere, which obscured larger motions that could occur in the winter season and between summer and winter hemispheres.

9.3 Eddy Transport Flux Due to Planetary Waves

In two-dimensional models the eddy diffusion flux plays an important role for poleward transport. The eddy flux expressed by (9.3) is usually connected to the mean gradients of $\bar{n}$ by

$$\overline{n'v'_y} = -K_{yy}\frac{\partial \bar{n}}{\partial y} - K_{zy}\frac{\partial \bar{n}}{\partial z} \tag{9.9}$$

$$\overline{n'v'_z} = -K_{yz}\frac{\partial \bar{n}}{\partial y} - K_{zz}\frac{\partial \bar{n}}{\partial z} \tag{9.10}$$

where y and z indicate the meridional (horizontal) and vertical components, respectively. In the mixing-length theory, the components of the K-tensor can be defined by (cf. (7.3))

$$\begin{aligned} K_{yy} &= \overline{l_y v'_y}, \quad K_{zy} = \overline{l_z v'_y} \\ K_{yz} &= \overline{l_y v'_z}, \quad K_{zz} = \overline{l_z v'_z} \end{aligned} \tag{9.11}$$

where $\vec{l}(l_y, l_z)$ represents the displacement vector.

REED and GERMAN (1965) assume that the mixing lengths l_y and l_z are smaller than the sizes of eddies involved in the mixing process and that the velocity vector $\vec{v}$ and $\vec{l}$ have the same direction so that

$$\frac{v'_y}{v'_z} = \frac{l_y}{l_z} \tag{9.12}$$

from which we can derive the relationship of a symmetric tensor

$$K_{yz} = K_{zy}. \tag{9.13}$$

The right hand side of (9.9) is a scalar product of two vectors in the meridional plane; i.e. a concentration gradient $\nabla\bar{n}(\partial\bar{n}/\partial y,\ \partial\bar{n}/\partial z)$ and a vector $\vec{K}_y(K_{yz},\ K_{zy})$. In order to make the meridional flux occur to the poleward direction ($\overline{n'\,v_y'} > 0$ in the northern hemisphere), the two vectors $\nabla\bar{n}$ *and* $\vec{K}_y$ must be at an obtuse angle ($>90°$). Thus, since K_{yy} is positive, K_{zy} must be negative.

Recently, it was shown by MATSUNO (1980) and DANIELSEN (1981), among others, that the eddy fluxes caused by wave motions should have an anti-symmetric K-tensor ($K_{yz} = -K_{zy}$) and that the eddy transports by this tensor tends to cancel out the counter transports by mean motion originated from the planetary wave.

Trajectories of air parcels generated by a sinusoidal wave (e.g. planetary wave) may be expressed as

$$y(t, \varphi) = A_y[\sin(\omega t + \varphi) - \sin\varphi] \tag{9.14}$$

$$z(t, \varphi) = A_z[\sin(\omega t + \varphi') - \sin\varphi'] \tag{9.15}$$

where

$$\varphi' = \varphi + \varphi_0 \tag{9.16}$$

We now define a mixing time τ as the time needed for a parcel to merge with the surrounding air by

$$\frac{d\chi}{dt} = \frac{1}{\tau}(\bar{\chi} - \chi) \tag{9.17}$$

where χ is any physical quantity and $\bar{\chi}$ is an environmental value of χ. In general, τ may be expressed as a combination of a time constant due to dynamical mixing τ_m and a photochemical time constant τ_c by

$$\frac{1}{\tau} = \frac{1}{\tau_m} + \frac{1}{\tau_c}. \tag{9.18}$$

For chemically active species, τ may represent a photochemical relaxation time and $\bar{\chi}$ may be a photochemical equilibrium value. We now approximate $\bar{\chi}$ by a linear function in the neighborhood of the origin, i.e.,

$$\bar{\chi}(y,z)=\bar{\chi}_y \cdot y+\bar{\chi}_z \cdot z \tag{9.19}$$

where $\bar{\chi}_y$ and $\bar{\chi}_z$ are constants.

A general solution of (9.17) is given by

$$\chi(t)=\frac{1}{\tau}\int_{-\infty}^{t} e^{(t'-t)/\tau}\bar{\chi}(t')\, dt' \tag{9.20}$$

and performing integration on (9.20) for $t=0$ by inserting (9.19) along with (9.14) and (9.15) we obtain

$$\begin{aligned}\chi(0,\varphi)=&(-\bar{\chi}_y A_y \cos\varphi-\bar{\chi}_z A_z \cos\varphi')\Phi_1\\&-(-\bar{\chi}_y A_y \sin\varphi+\bar{\chi}_z A_z \sin\varphi')\Phi_2\end{aligned} \tag{9.21}$$

where

$$\Phi_1=\frac{\omega\tau}{1+\omega^2\tau^2},\quad \Phi_2=\frac{\omega^2\tau^2}{1+\omega^2\tau^2}. \tag{9.22}$$

Taking the average with respect to φ of the product of $v(0,\varphi)$ and $\chi(0,\varphi)$ we obtain the horizontal flux of eddy motions as follows:

$$\begin{aligned}\overline{v\chi}&=\overline{\frac{\partial y}{\partial t}\chi}=\omega A_y\overline{(\cos\varphi)\cdot\chi}\\&=\frac{\omega A_y}{2}[(-\bar{\chi}_y A_y-\bar{\chi}_z A_z\cos\varphi_0)\Phi_1-\bar{\chi}_z A_z\sin\varphi_0\Phi_z].\end{aligned} \tag{9.23}$$

Similarly, the vertical flux can be calculated as

$$\overline{w\chi}=\frac{\omega A_z}{2}[(-\bar{\chi}_y A_y\cos\varphi_0-\bar{\chi}_z A_z)\Phi_1+\bar{\chi}_y A_y\sin\varphi_0\Phi_z]. \tag{9.24}$$

The K-tensor defined in $\overline{V\chi}=-K\nabla\chi$ can now be written as

$$K=\begin{pmatrix}\frac{\omega}{2}A_y^2 & \frac{\omega}{2}A_yA_z\cos\varphi_0\\ \frac{\omega}{2}A_yA_z\cos\varphi_0 & \frac{\omega}{2}A_z^2\end{pmatrix}\Phi_1+\begin{pmatrix}0 & \frac{\omega}{2}A_yA_z\sin\varphi_0\\ -\frac{\omega}{2}A_yA_z\sin\varphi_0 & 0\end{pmatrix}\Phi_2 \tag{9.25}$$

which indicates that the K-tensor is generally composed of two terms; a symmetric tensor and an anti-symmetric tensor. The symmetric tensor (the first term of (9.25)) dominates if $\omega\tau \ll 1$, whereas the anti-symmetric term dominates if $\omega\tau \gg 1$. A planetary wave of zonal wave number one imbedded in a westerly flow of 30 m/sec would give $\omega \simeq 10^{-5}\text{sec}^{-1}$. Thus, for constituents having a life time much longer than $\sim 10^5$sec, the K-tensor is represented mainly by the anti-symmetric diffusion, which produces a transverse-gradient flux and does not contribute to the actual diffusion (MATSUNO, 1980).

The anti-symmetric tensor is independent of the species being transported, depending only on the wave structure. On the other hand, the symmetric tensor would depend on the photochemical lifetime as well as the wave structure, and is, therefore, different for different species. The new approach for expressing the K-tensor based on eddy motions caused by planetary waves is physically more tenable than the classical expression by REED and GERMAN (1965), and attempts are being made to apply them in two-dimensional models (e.g. PYLE and ROGERS, 1980 a and b).

9.4 Internal Gravity Waves and Eddy Diffusion in the Mesosphere and Lower Thermosphere

Gravity waves arise when a parcel is displaced vertically in a statically stable atmosphere. The upward moving air parcel will be cooled by adiabatic expansion, becoming heavier than the surrounding air. Similarly, the downward moving air parcel will be warmed up by adiabatic compression, resulting in lighter gas than the surrounding air. As a result, a gravitational restoring force will be exerted on the air parcel, producing an oscillation.

Gravity waves that can propagate vertically are called *internal gravity waves*. In absence of damping the energy of such waves ($1/2\ \rho v^2$) remains constant with height. Since the air density ρ decreases with altitude, the amplitude of internal gravity waves should grow very rapidly with height in inversely proportion of $\sqrt{\rho}$. Thus, internal gravity waves propagating from the lower atmosphere are greatly amplified in the middle atmosphere, particularly in the mesosphere and lower thermosphere. These waves could at some height reach the amplitude for which the wave fields themselves would become strongly unstable, and the breakdown of waves would occur so that further wave growth may be prohibited above this height (LINDZEN, 1981; HOLTON, 1982).

HODGES (1969) and HINES (1970) have argued that the normal growth of amplitudes of internal gravity waves with height will be offset by energy lost to turbulence, so that wave amplitude may be kept unchanged with altitude. The amplitudes of internal gravity waves measured near the mesopause are often sufficiently large to have caused instabilities to occur, leading to the generation of turbulence (KOCHANSKI, 1964).

The attenuation of the wave energy by turbulence can be approximately attributed to an average eddy diffusion coefficient. Such a coefficient is limited to that which would cause the wave to maintain constant amplitude with altitude (HODGES, 1969). HINES (1970) has formulated the eddy diffusion coefficient obtained by Hodges in the form

$$K_D = 1.4 \times 10^{-2} \tau_g^{-1} H^{-1} \lambda_x^4 \lambda_z^4 (\lambda_x^2 + \lambda_z^2)^{-\frac{5}{2}} \tag{9.26}$$

where τ_g is the isothermal Brunt-Väisälä period (the upper limit to the internal gravity wave, ~280 second), H is the scale height for atmospheric density, and λ_x and λ_z are the horizontal and vertical wavelengths of upward propagating gravity waves, respectively. A K_D of the order of $10^7 \mathrm{cm}^2 \mathrm{sec}^{-1}$ is readily calculated by (9.26) near the mesopause. HODGES (1969) attributes this excessively large K_D to the sporadic occurrence of turbulence caused by instabilities in gravity waves.

Observations of periodic wind components at meteoric heights, however, often indicate variable wave amplitudes between altitude increments (ROPER and ELFORD, 1965). Recognizing that no amplitude growth is a poor approximation in the middle atmosphere, and considering that the vertical gradient of the specific wave energy is balanced by the effective turbulent viscosity, ZIMMERMAN (1974) has derived the following expression for the vertical eddy diffusion coefficient:

$$K_z = \frac{\lambda_z^3}{4\pi^2 \tau} \left(\frac{1}{H} - \frac{1}{z} \log_e \frac{V^2}{V_0^2} \right) \tag{9.27}$$

where τ is the period of the gravity wave, V is the perturbation velocity at the level z, and V_0 is the value of V at $z=0$. If amplitude is independent of height ($V = V_0$), the second term on the right-hand side of (9.27) vanishes and K_z should be equal to K_D. In general, however, there is some amplitude growth with height ($V > V_0$) and K_z is smaller than K_D. Thus, the eddy diffusion coefficient K_z can be in the order of $10^6 \mathrm{cm}^2 \mathrm{sec}^{-1}$ near the mesopause. This is consistent with the result of a theoretical study on gravity wave breakdown in the mesosphere by LINDZEN (1981).

The downward heat flux due to turbulent mixing may be expressed as

$$F = - c_P \rho K \left(\frac{\mathrm{d}T}{\mathrm{d}z} + \frac{g}{c_P} \right). \tag{9.28}$$

JOHNSON and WILKINS (1965) assume that the net heat input due to absorption of solar energy by O_2 and O_3 is transported downward by eddy conduction in the mesosphere and lower thermosphere. On this assumption they have evaluated the maximum acceptable value of the eddy diffusion coefficient, which increases from $\sim 2 \times 10^5 cm^2 sec^{-1}$ at 50 km to $\sim 10^7 cm^2 sec^{-1}$ at 100 km. The estimation of the flux F is a difficult task, however, since it represents the difference between the two competing terms, i.e. radiative and conductive heating and cooling. The resultant eddy diffusion coefficient may reflect the large uncertainty in this estimate (HUNTEN, 1974).

HUNTEN (1974) also argues that heat generated by the dissipation of eddy motion seems to exceed the heat transmitted by eddy conduction, i.e. turbulence is a net heat source in the lower thermosphere and not a net cooling mechanism as was considered by JOHNSON and WILKINS (1965).

The criterion for growth or maintenance of turbulence is given by the Richardson number, which is related to the rate of work done against buoyancy forces (or gravitational stability). Theoretical studies have placed the critical Richardson number at ~ 0.25 with instability (or growth of infinitesmal disturbances) for smaller values. JOHNSON (1975) has suggested that the conclusion of HUNTEN (1974) depends largely upon the Richardson number used (0.2). Johnson argues that in a situation where turbulence is continuously present in a steady state, the Richardson number is likely to be near to, but less than, 1. The calculation by CHANDRA (1980) indicates that both the heating rate by eddy dissipation and the cooling rate by eddy conduction are large in the lower thermosphere and that their magnitudes are about equal to each other.

Chapter 10

TEMPORAL VARIATIONS IN STRATOSPHERIC OZONE

Changes in stratospheric ozone density are due to various natural and man-made causes. It is important to examine the extent of long term natural variations (ozone trend) before any attempt could be made for determining whether or not anthropogenic perturbations have actually occurred or are occurring. In this chapter we first discuss the problems associated with identifying the ozone trend; the observed increase in total ozone over the period 1963 – 1970 is discussed particularly on the possible impact of nuclear explosions.

We then discuss short term variations; detections of short term excursions in stratospheric ozone density attributable to some known causes are useful to prove (or disprove) that the photochemical mechanisms of ozone changes theoretically postulated can actually occur in the middle atmosphere. The solar proton events are probably the only known causes so far confirmed to have produced the effects on stratospheric ozone predicted by the photochemical theory. Finally, the important but controversial relationships between the stratospheric ozone and solar activity (solar cycle and monthly variations in solar fluxes) are discussed concerning both observational and theoretical works.

10.1 Long Term Variations

Long term variation refers to the residual after corrections have been made for all known cyclic variations, systematic variations due to drifts of instrument calibration, and random measurement noise.

10.1.1 Ozone trend

Analysing the time series of mean monthly total ozone observed at *several* stations, KOMHYR *et al.* (1971) revealed that total ozone increased significantly during the 1960s; the rate of increase varies from station to station but is in the range of 2.4 - 10% per decade. Although a possibility remains that a part of the calculated rate of ozone increase is due to long term drift in spectrophotometer calibrations, it seems unlikely that all of the stations analysed had the drifts in the same direction. Thus, the trend of ozone increase at these stations may be true in large measure. However, the ozone increases found in a selected area may have been compensated by decreases in another area. In order to confirm the global trend, a more comprehensive analysis using data from a worldwide network of total ozone observations is desirable.

Large amounts of data are available from world network stations on ground based total ozone observations since 1958 (International Geophysical Year). ANGELL and KORSHOVER (1973, 1976, 1978a) have used monthly mean total ozone data up to 1976 obtained at worldwide network stations. Classifying these data into five different latitude groups (north and south polar, north and south temperate, and tropical) they determined the ozone trend for each group by eliminating the annual and quasibiennial oscillations using the method of running averages. The result is shown in Fig. 10.1. In summary the total ozone in the Northern hemisphere has increased by ~5% from early 1960 to 1970, decreased by 1 - 2% for 1970–1972, and remained almost unchanged after 1972. In the Southern hemisphere the total ozone has increased by ~2% from 1960 to 1968, and decreased by ~1% thereafter.

HASEBE (1980 a and b) worked out a method to incorporate the data from M-83 filter ozonometers into analyses together with those of Dobson spectrophotometers in order to improve the global data coverage. Applying the method of *numerical filtering* to the time series of monthly data from 1962 to 1976, he has calculated the dominant frequency components of variations for the hemispheric and global means of total ozone. Figure 10.2 illustrates the result for Northern and Southern Hemispheric means, on the left and right respectively. In each illustration the panel (a) indicates the original variation, (b) the non-stationary annual oscillation, (c) the quasi-biennial oscillation and (d) the long term variation. The result in panel (d) indicates that the total ozone increased markedly between 1962 and 1970, particularly in the Northern Hemisphere, but it tends to decrease slightly after 1970. The characteristics of these long term variations basically agree with those obtained by ANGELL and KORSCHOVER (1978a).

Satellite data are certainly better suited for analyses of variations in the global ozone trend, since data are more uniformly and densely distributed, and are more continuous in time, and also because observations are made by a *single* instrument on board the satellite. TOLSON (1981), HILSENRATH and SCHLESINGER (1981) and KEATING *et al.* (1981) have analysed the mon-

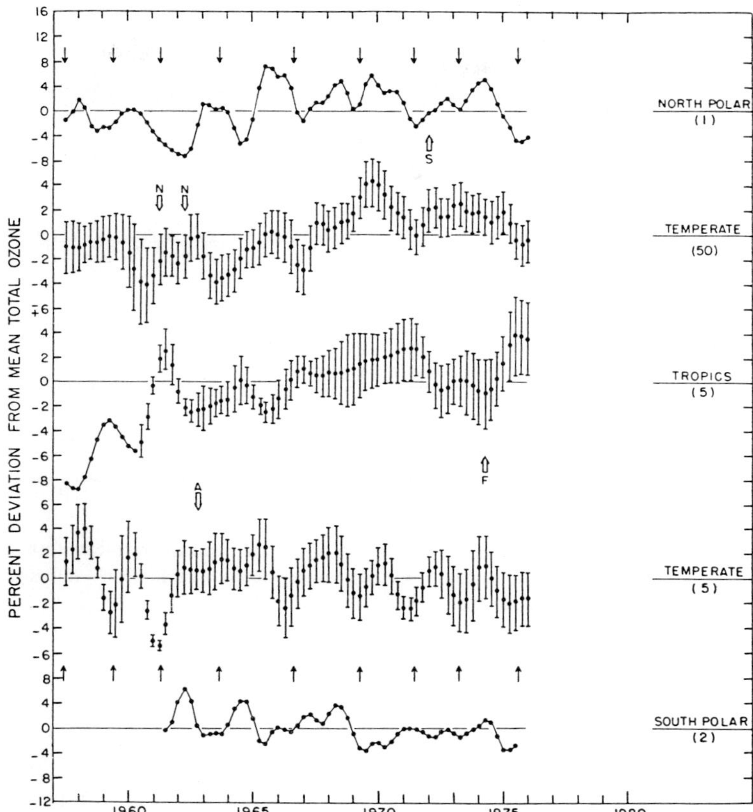

FIG. 10.1. Variations in seasonal-mean total ozone by climatic zone. Double-shafted arrows indicate the dates of large nuclear tests (N), volcanic eruptions (A and F) and an intense solar proton event (S). Single arrows indicate the time when the westerly becomes strongest at the 50 mbar level (ANGELL and KORSHOVER, 1978a).

thly mean total ozone data derived from seven years (1970–1977) of Nimbus 4 BUV measurements using ninth degree spherical harmonic models and determined spatial and temporal variations. A summary of the result is shown in Fig. 10.3. Filtering out the semiannual, annual and quasibiennial oscillations from variations in the monthly global mean (bottom panel of Fig. 10.3) by the method of 6-month running averages, KEATING *et al.* (1981) found a clear trend of ozone decrease from 1970 to 1975, which is consistent with the ground based observations.

It seems certain from both ground-based and satellite observations that

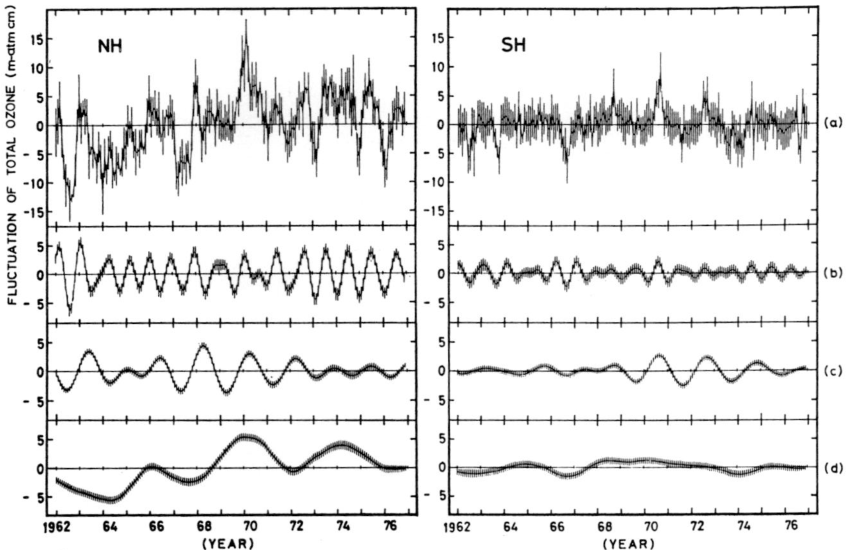

FIG. 10.2. Variations in northern and southern hemisphere mean total ozone (a), stationary annual oscillation (b), quasi-biennial oscillation (c), and long term variation (d) (HASEBE, 1980b).

the global total ozone has *increased markedly* during the 1960s and then *decreased slightly* after 1970. Thus, there may be a tendency for a slight net increase throughout the 1960s and 1970s. PARRY (1977) found a long term trend of a net ozone increase of ~4% during the period 1958 to 1975. JOHN *et al.* (1981) found from 12-month running averages at 37 Dobson stations that the ozone trend showed an average increase of 1.5% from 1958–1979.

An upward ozone trend in the last two decades obtained by various workers appears to contradict the ozone depletion theory which predicts a downward trend due to the active release of chlorine-bearing materials. However, there is a possibility that the observed trend is affected by a solar cycle (11 years) variation (see Section 10.3). Furthermore, using Nimbus 4 and 7 BUV data, HEATH (1981) observed that the middle stratospheric ozone decreased, whereas the upper stratospheric ozone increased, during the period from 1970 to 1979. Thus, chemicals of human origin may have decreased the middle stratospheric ozone density after 1970. However, a more careful examination for instrumental drift and observations of further extended periods are necessary before we can reach a final conclusion (NORMAN, 1981; DICKSON, 1981).

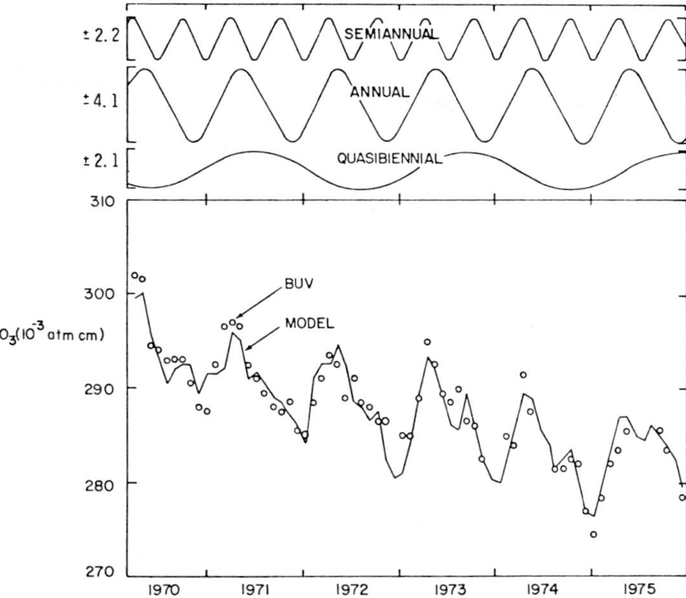

FIG. 10.3. Monthly global mean values of total ozone as calculated from BUV data (bottom). Model is a least mean square fit. The top panel shows the semiannual, annual and quasi-biennial components of variations (KEATING *et al.*, 1981).

10.1.2 *Effects of nuclear explosions*

The ozone increase over the period 1963–1970 has been interpreted by JOHNSTON *et al.* (1973) to be due to the stratosphere returning to normal after NO injections by the nuclear bomb explosions of 1961–1962. These explosions yielded the equivalent of ~334 M ton of TNT, mostly at high latitudes, due to U.S.S.R. tests. FOLEY and RUDERMAN (1973) estimated that high-yield nuclear explosions have injected into the stratosphere about 10^{34} molecules of NO, comparable to that generated by 500 SSTs. GILMORE (1975) calculated the upper limit production rate be $\sim 1.5 \times 10^{32}$ NO/Mton, and the lower limit to be $\sim 0.4 \times 10^{32}$ NO/Mton. Large O_3 reductions could have occurred by injection of such large amounts of NO into the stratosphere through the NO_x catalytic cycle.

Recent studies using BUV observations from the Nimbus 4 satellite, however, have shown that little change in total ozone was observed at the time of the nuclear explositions, less than might have been expected from current models of NO_x catalytic depletion (CHRISTIE, 1976). Measurements of NO and O_3 were made by special instruments on board a RB-57 F aircraft on November 27, 1976, 7 days after a 4 Mton nuclear test explosion in

China. No measurable changes from the ambient density of either O_3 or NO were found (McGHAN *et al.*, 1981).

Various reasons have been presented for the apparent non-existence of the effect of nuclear explosions. NO_x yields may have been overestimated or there may be an efficient unknown NO_x sink in the stratosphere (CHRISTIE, 1976). JOHNSTON (1977) suggested that the bomb-produced NO_x may have consumed virtually all HO_x radicals to form HNO_3, and thus HO_x catalyzed reduction of O_3 may have been temporarilly terminated. CHANG *et al.* (1979a) have shown that model calculations carried out with recent chemistries result in ozone reduction that are more easily consistent with the observations.

10.2 Short Term Variations

Short term (less than approximately one month) natural events may sometimes produce the corresponding changes in densities of ozone and other minor constituents in the middle atmosphere. Investigations of cause-and-effect relationships and searches of such effects in observed data are useful for verifying the photochemical theory, particularly the mechanism of catalytic destruction of stratospheric ozone.

10.2.1 Solar proton events (SPE)

CRUTZEN *et al.* (1975) had predicted that the formation of NO in the upper stratosphere by ion chemistry due to the increased amount of protons during large proton events could reduce the stratospheric ozone at high latitudes. During the intense SPEs, substantial fluxes of secondary electrons with energies in the range of tens to hundreds electron volts are produced. The rate of energy loss of these fast protons in producing an ion pair in air is very nearly independent of proton energy and the value is $\sim$36 eV per ion (BAILEY, 1959; see Section 13.2 for ion chemistry during the SPEs).

Evidence of an ozone decrease associated with the major solar proton event of August 4, 1972 was found in Nimbus 4 satellite data (HEATH *et al.*, 1977). The highest energy measured during that event was $\sim$60 MeV and these protons could reach the upper stratosphere (see the right-hand scale of Fig. 13.1). Figure 10.4 illustrates the Nimbus 4 data which clearly indicate a sudden decrease of total ozone at high latitudes on day 219 (August 6). No data is available for Aug. 5 when high radiation-induced dark currents occurred within the photomultiplier. Such direct effects of high energy protons on the instrument may have caused serious errors in measurements of radiances below 300 nm, and therefore in ozone measurements, on all days after Aug. 5, when the calibration may have been affected.

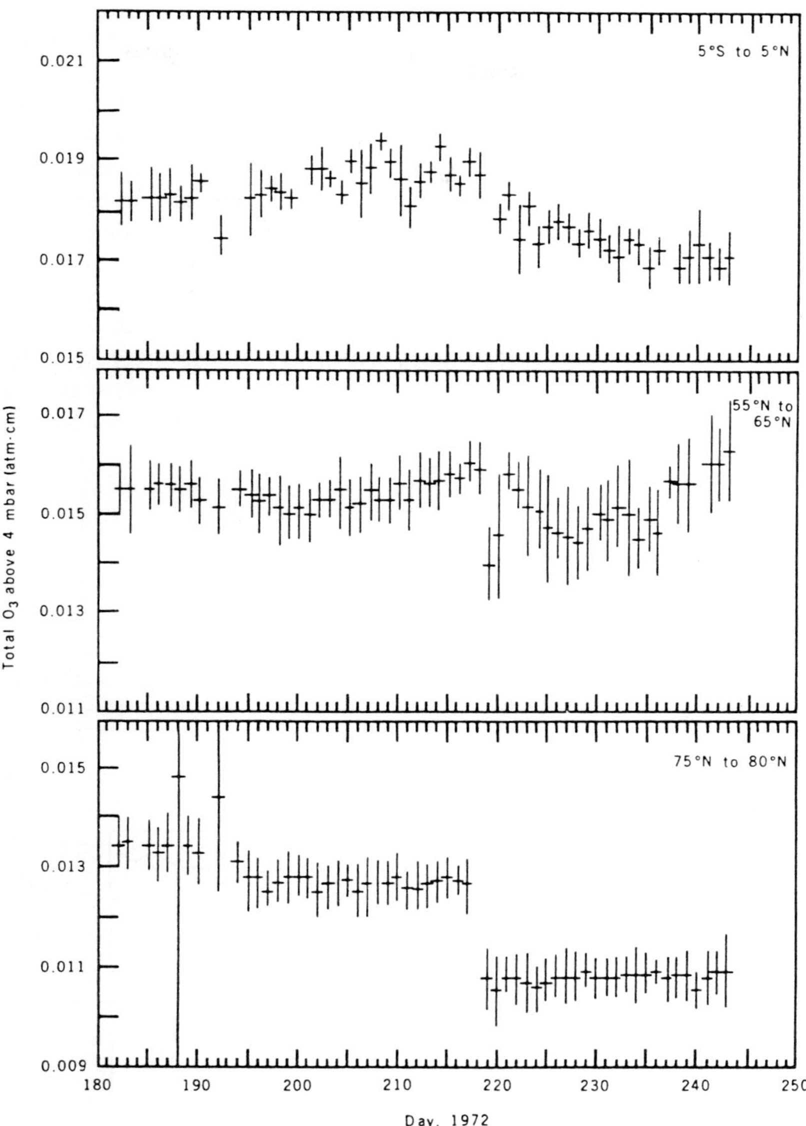

FIG. 10.4. Zonally averaged total ozone above the 4-mbar level during July and August 1972 at northern high latitudes. The solar proton event occurred on August 4, day 217 (HEATH *et al.*, 1977).

McPeters *et al.* (1981) revised the analyses of the Nimbus 4 data, determining the direct effect of high energy protons by comparing nighttime BUV counting rates with particle data from IMP 6 (Explorer 43 satellite), which measures integral fluxes of protons for energies >60, >30, and >10 MeV. The IMP derived counting rates of solar protons and the ozone mass mixing ratio at four height levels during August 1972 are shown in Fig. 10.5. The ozone data are one day zonal averages for a latitude band at 69 to 77°N (geomagnetic latitudes greater than 60°). The dotted curves represent the "normal" ozone mixing ratio of August 1970. It is evident that the ozone depletion occurred after the SPE at altitudes below 0.5 mb ($\sim$55 km) and that it lasted at least a month.

Model calculation of the August 1972 SPEs have been made by various workers (Crutzen *et al.* 1975; Borucki *et al.*, 1978; Fabian *et al.*, 1979 b; Solomon and Crutzen, 1981; McPeters *et al.*, 1981; Rusch *et al.*, 1981). In Fig. 10.6 the result of Rusch *et al.*, (1981) is shown for com-

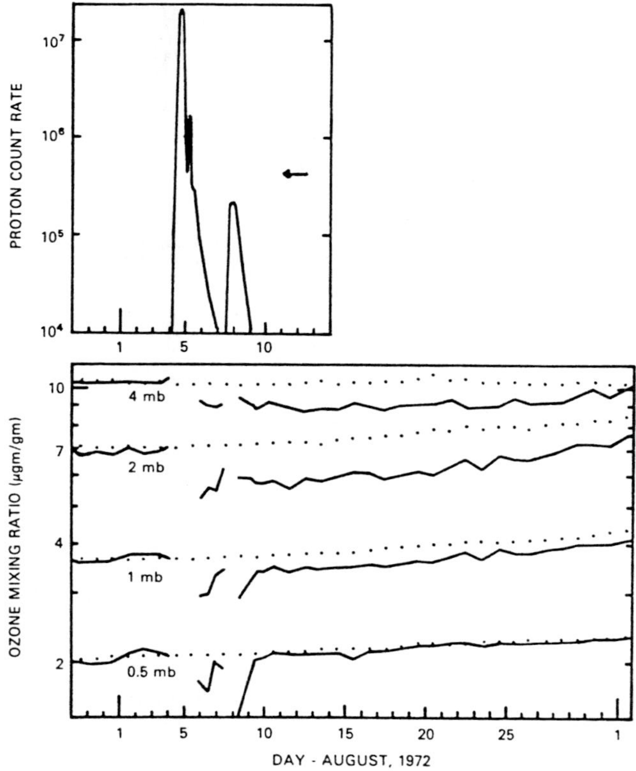

Fig. 10.5. The changes in O_3 mixing ratio at 4 levels during August 1972 plotted below the IMP derived proton counting rates (McPeters *et al.*, 1981).

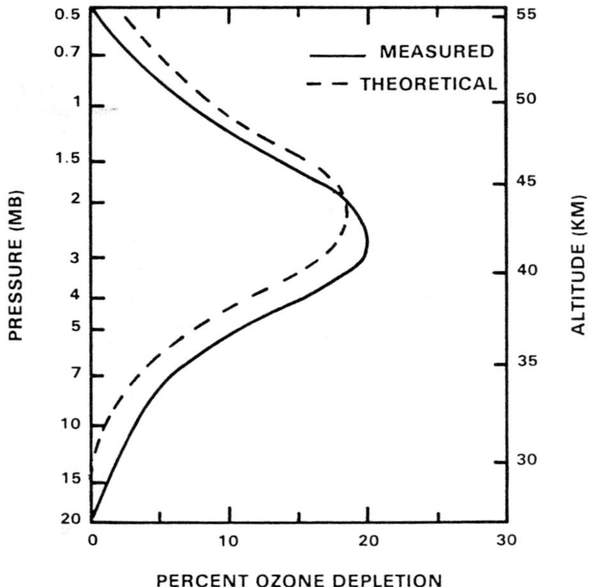

FIG. 10.6. The height profiles of percentage ozone decrease 2 weeks after the August 1972 SPEs (MCPETERS *et al.*, 1981).

parison with observations. A large depletion rate (~10%) at ~30 km obtained in an early analysis by HEATH *et al.* (1977) is now completely removed. The ozone depletion observed near 3 mb is greater than predicted. Similar results were inferred from the rocket data by WEEKS *et al.* (1972). SOLOMON *et al.* (1981) pointed out that fluctuating temperatures could be responsible for the apparent differences.

SOLOMON and CRUTZEN (1981) examined in detail the effects of the August 1972 SPEs on stratospheric ozone using more recent chemical data than earlier models and including the effect of temperature feedback. The calculated ozone changes are somewhat sensitive to background ClO_x concentrations. The faster reaction constant for $HO_2+NO \rightarrow NO_2+OH$ resulted in more ozone depletion above ~35 Km and made the agreement between the model and the observations better. The calculated ozone decreases are expected to produce temperature decreases, and consequently affect the reaction rate constants. The inclusion of temperature feedback further improved the agreement between the calculated and the observed ozone depletion above ~40 km.

MCPETERS *et al.* (1981) noted that a long term ozone depletion following the August 1972 SPEs was very clear and lasted at least a month in both northern and southern polar regions. Such an ozone depletion having the time constant of the order of a month is explained as catalytic destruction of O_3 by NO_x produced during the SPE. Thus, they conclude that ozone catalysis by NO_x injection has been demonstrated by the ozone changes taking place during and after the SPEs.

During the SPE of November 1969 WEEKS *et al.* (1972) observed by rocket experiments a decreased ozone concentration in the 50 - 70 km region at Fort Churchill, Canada. MCPETERS *et al.* (1981) and REAGAN *et al.* (1981) noted relatively sharp ozone decrease in the mesosphere during initial hours of the August 1972 SPEs and a short-term ozone decrease during the small SPE in January 1971. From the continuous observation of the mesospheric ozone density from the SME satellite, THOMAS *et al.* (1983b) have detected a clear decrease of ozone density during the SPE event of July 13, 1982; a maximum depletion of ~70 % was observed in the northern high latitudes for the period of less than a day.

SWIDER and KENESHEA (1973) first attributed these changes in mesospheric ozone density during the SPE events to HO_x chemistry; smaller SPEs usually affect higher regions, where HO_x catalysis is more important than NO_x catalysis and the lifetime of HO_x is generally much shorter than NO_x. SOLOMON *et al.* (1983) have shown that the expected ozone density change should be roughly related to the relative magnitude of odd hydrogen production rate by the ions versus the normal background production rate due to photolysis of water vapor.

10.2.2 Solar eclipses

Solar eclipses provide a unique opportunity to test chemical reactions in the middle atmosphere and their reaction rate constants. Since the solar radiation flux in the atmosphere diminishes in the middle of the day, large concentration changes may occur for some constituents during the short period of eclipse. RANDHAWA (1973) observed a partial solar eclipse on July 10, 1972 over Alaska, but recorded no significant effect on ozone concentration in the stratosphere. This is further confirmed by measurements during the total eclipse of February 26, 1979 by STARR *et al.* (1980). The result is reasonable since the photochemical time constant of O_3 increases rapidly towards the large nighttime values as the eclipse proceeds. Even under sunlit conditions the time constant of O_x, by which the change in O_3 concentration is controlled, is much larger than the duration of the eclipse (see Fig. 6.4).

STARR *et al.* (1980) have observed a reduction of at least a factor of 25 in the NO mixing ratio at an altitude of 19.8 km during the maximum of a solar eclipse. This is consistent with the results of model calculations (WUEBBLES and CHANG, 1979; HERMAN, 1979). The model also predicts increases of NO_2 and decreases of OH, HO_2, Cl, and ClO, all of which have time constants less than or of the order of a minute, which is smaller than the duration of an eclipse. If changes in NO_2 and HNO_3 were measured simultaneously with that in NO, it would be useful to examine the validity of the theory of family conservation in NO_x.

10.3 Stratospheric Ozone and Solar Activity

10.3.1 Solar cycle variations

It is logical to expect a correlation between stratospheric ozone density and solar activity, since the primary mechanism for the production of stratospheric ozone is through solar UV radiation. The relationship of total ozone to the solar 11-year cycle was studied by WILLETT (1962) by using monthly mean ozone data from a large number of stations throughout the world, covering various time intervals during the 26-year period 1933–1959. The period was later extended to 1971 by CHRISTIE (1973), whose result is shown in Fig. 10.7. A solar cycle variation can be seen clearly in the residual (bottom panel) up to ~1965, but there is a marked difference between the time of sunspot maximum and ozone maximum. Cross correlation analysis indicates that the highest positive correlation (with a coefficient of ~0.45) occurs with a 75 month lag of sunspot number to ozone. Similar results were obtained by LONDON and OLTMANS (1973) from the analysis of continuous ozone data at Arosa, Switzerland since 1932 and Oxford, England since 1951. The result seems to be difficult to understand, however, since it implies that sunspot variations follow ozone variations.

RUDERMAN and CHAMBERLAIN (1975) proposed that the modulation of cosmic rays deposited in the stratosphere over a sunspot cycle by the solar magnetic field would produce an oscillating source of stratospheric NO with an 11-year period and that the resulting modulation of O_3 over this period may explain the time lag between sunspot maximum and ozone maximum.

Studies of the correlation between O_3 variations and solar activity have been controversial. Particularly, variations in O_3 in the lower stratosphere, and therefore those of total ozone, reflect the important intervention of related stratospheric and tropospheric circulation processes, thus obscuring any direct link between the total ozone and solar variability (LONDON and OLTMANS, 1973). Variations in O_3 in the upper and middle atmosphere (above about 30 km) may respond more directly to solar activity, since the photochemical time constant of O_x is much less than the transport time constant (see Fig. 6.4).

Analyses of Umkehr data above 36 km during 1958–1962 (RANGARAJAN, 1965) and those above 30 km during 1956–1964 (DÜTSCH, 1969) have shown pronounced variations in O_3 with the solar cycle. CHAKRABARTY

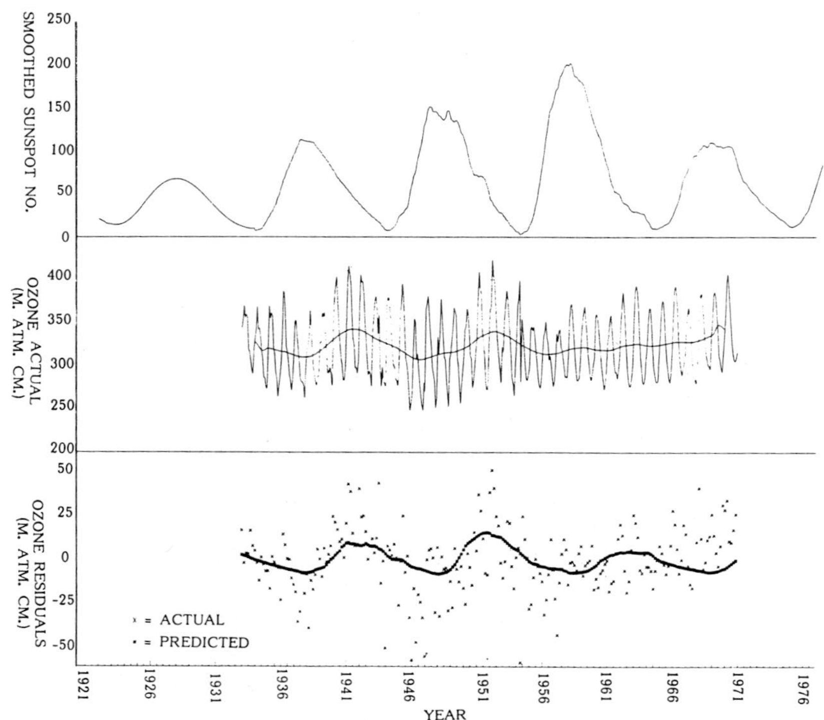

FIG. 10.7. Time series of smoothed sunspot number (top), unweighted average total ozone with the curve representing the smoothed annual cycle (middle) and residuals representing the departure of monthly ozone values from the seasonal mean values (CHRISTIE, 1973).

and CHAKRABARTY (1982) have used Umkehr ozone data obtained at Mt. Abu in Ahmedabad, India, from 1953 onwards to study the effect of solar activity on ozone amount at different altitudes. The result indicates that in the upper stratosphere there is a trend of increasing ozone (daily value) with an increased level of solar activity (10.7 cm solar flux). In the lower stratosphere, however, this trend is seen only when the monthly averaged values are considered.

Using balloon sounding data over Central Europe, PAETZOLD *et al.* (1972) calculated a positive correlation at 20–30 km altitude for the years from 1951 to 1972; the solar cycle ozone variation was ~22 % at 25–30 km and ~6.6 % at 20–25 km. However, they have obtained a negative variation (i.e. ozone density is smaller at sunspot maximum) of ~6.3% above 30 km. Thus, the variation over a solar cycle for total ozone is calculated to be small (less than 3 %).

From the search for solar cycle-ozone relationships DÜTCH (1979) concluded that the strongest indication for a direct correlation is found in the upper stratosphere but even that connection has not been proven beyond doubt. He also concluded that most other correlations with the total ozone (found from time series that were too short) were produced indirectly by variations in the stratospheric circulation within the solar cycle.

A strong connection between the sunspot number and stratospheric temperature over Berlin in winter was reported by SCHWENTEK (1971). From analyses of rocketsonde data, ANGELL and KOVSHOVER (1978b) found that there was a general warming trend in the middle and upper stratosphere prior to 1970 and a somewhat larger cooling trend thereafter, yielding a temperature variation basically in phase with sunspot number. The radiosonde data for the lower stratosphere, however, do not indicate such a variation. NAUJOKAT (1981) reported from an analysis of stratospheric geopotential height and temperature during the last two sunspot cycles that no evidence exists to support a relationship between the stratospheric temperature below the 10-mb level and solar activity.

10.3.2 Monthly variations

KEATING (1978) and BLACKSHEAR and TOLSON (1978) have shown that there is a good correlation between the monthly variations of satellite ozone data and the solar activity index represented by eighter the 10.7 cm solar flux ($F_{10.7}$) or the Lyman-α intensity. Data from the IRIS (infrared interferometer spectrometer) instrument aboard the Nimbus 4 satellite were averaged into 1,296 cells of 5° latitude by 10° longitude and area weighted by each month. The resulting data on the global distribution of ozone were fitted by weighted least-mean squares to a 9th order, 9th degree spherical harmonic expansion. The first term (zero order, zero degree) of the 100-term spherical harmonic expansion represents the global mean ozone, $\bar{O}_3$. Variations in $\bar{O}_3$ (monthly mean) are found to be remarkably consistent with the variations in $F_{10.7}$ or the Lyman-α index for the period from April, 1979 to January, 1971. KEATING (1978) has calculated a correlation coefficient of $\sim$0.75 between $\bar{O}_3$ and $F_{10.7}$, whereas a value 0.94 has been obtained between $\bar{O}_3$ and Lyman-α by BLACKSHEAR and TOLSON (1978).

The apparently high correlation between solar activity and total ozone on mean monthly variations seems to be particularly surprising in view of the fact that, on the average, about 80 % of the total ozone lies below 30 km and that the photochemical time constant increases from about one month at 30 km to about one year at 20 km (see Fig. 6.4). LONDON and REBER (1979) have argued that the results by KEATING (1978) and BLACKSHEAR and TOLSON (1978) should be valid only for a limited time period. Extending the period of analysis up to April 1972 using the BUV satellite data (instead of IRIS data), LONDON and REBER (1979) have found the apparent high correlation falls to insignificance. The correlation coefficient between $F_{10.7}$ and BUV data is 0.47 for the period April 1970 - Dec. 1970, but falls to 0.11 for April 1970 - March 1971, and to −0.01 for April 1971 - April 1972. The correlation coefficients between Lyman-α intensity and BUV data for the same three period are 0.71, 0.30 and −0.04, respectively.

KEATING *et al.* (1981) reported, however, that a high correlation still remained if the shorter period (6-month) variations were filtered out; they actually calculated an even higher correlation coefficient (~0.97) for the period from 1970 to 1975. Analysing nearly six year (1970–1976) data from the Nimbus 4 BUV observations, REBER and HUANG (1982) have obtained relatively high correlation between total ozone and solar activity, ranging from 0.68 for the monthly means to 0.94 for the 6-month running averages. They found, however, that these high correlations were due almost entirely to the long-term decreasing trends in both data sets. They conclude that it is unrealistic to draw a firm conclusion about solar activity influence on total ozone from the Nimbus 4 data set, since there is a great uncertainty in the long-term stability of the Nimbus 4 instruments.

10.3.3 Model studies

Theoretical studies on the possible effects of solar UV flux on ozone concentrations in the middle and upper stratosphere have been carried out by CALLIS and NEALY (1978), PENNER and CHANG (1978) and CALLIS *et al.* (1979). Based on the variability in solar UV flux below 300 nm between 1964 and 1972 reported by HEATH and THEKAEKARA (1976) they calculated the corresponding changes in temperature and concentrations of ozone and other minor constituents including N_2O, by a steady-state 1-D radiative-convective-photochemical model. Temperature feedback to the ozone concentration through changes in reaction rate constants, and the effects of changes in N_2O on ozone concentration are included in the model calculations. Although there is no conclusive evidence for a solar cycle variation in UV radiation of the magnitude reported by HEATH and THEKAEKARA (SIMON, 1978), model calculations are very useful and important; they allow us to look into physical and chemical mechanisms for the relationship between solar activity and climate-related phenomena (CALLIS *et al.*, 1979).

The ozone concentration in the middle and upper stratosphere changes in proportion to the square root of the ratios J_{O_2}/J_{O_3} and k_2/k_3 as is seen from (5.9). J_{O_2} changes as the solar UV flux near 200 nm changes, whereas J_{O_3} depends mainly on the flux near 300 nm. Thus, the relative change between fluxes near 200 nm and 300 nm is an important factor controlling ozone concentration. Since the solar flux variability is larger for shorter wavelengths, the main controlling factor is the change in the solar flux near 200 nm. HEATH and THEKAEKARA (1976) reported a change in the flux near 200 nm by a factor of ~2.0 between sunspot minimum (1964–1966) and sunspot maximum (1969–1970).

The rate constants k_2 and k_3 have opposite temperature dependence; while k_2 decreases with temperature, k_3 increases. Thus, for instance, the ratio k_2/k_3 decreases by a factor of ~0.78 if temperature increases from 260K to 270K. The predicted temperature change from solar minimum to solar maximum is shown in the right panel of Fig. 10.8. and the changes in O_3 concentration are illustrated in the left panel. The percentage change in O_3 decreases if the effect of temperature feedback is taken into consideration; i.e. ozone concentration then decreases by the temperature increases in the upper stratosphere. However, the O_3 increase by the increase of solar flux near 200 nm generally dominates the O_3 decrease by the temperature feedback effect.

Calculated O_3 variations at 32–46 km are compared with observations in Fig. 10.9 (a) for the period 1962 to 1973. The curve HT indicates that the solar flux variability proposed by HEATH and THEKAEKARA (1976) has been used, whereas the curve MOD indicates a model in which the change in solar flux near 200 nm is reduced to a factor of ~1.55 from 2.0. The predicted O_3 variations with MOD agree better with observations.

Figure 10.9 (b) illustrates a comparison between the predicted and observed variations in temperature; the agreement is again very good. Thus, there is general agreement between model predictions and observations for both ozone and temperature in the middle and upper stratosphere for the period 1962–1972. However, CALLIS *et al.* (1979) noted that there are significant differences between model results and observations, particularly with

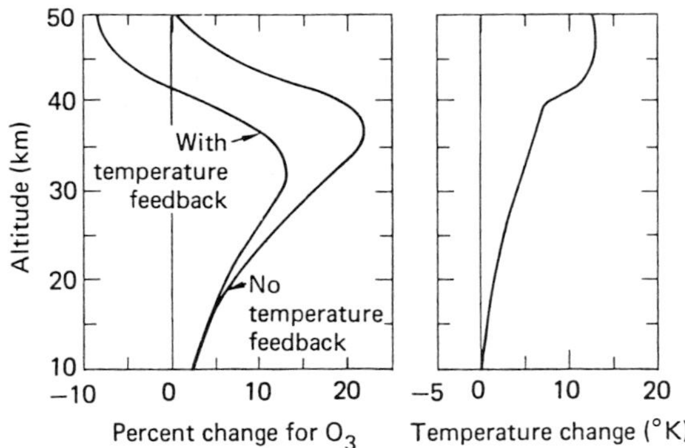

FIG. 10.8. Predicted changes on O_3 (in %) and temperature from sunspot minimum to maximum (PENNER and CHANG, 1978).

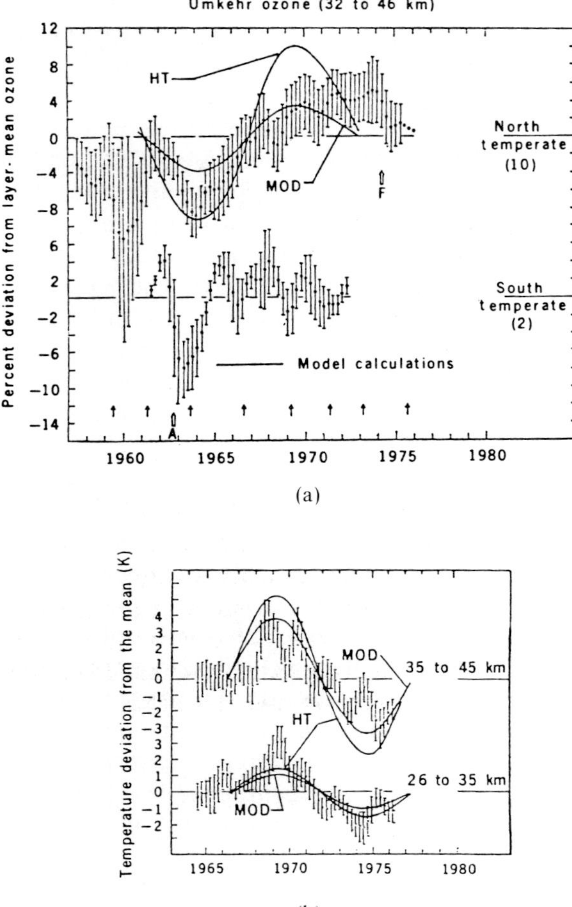

FIG. 10.9. Variations in (a) ozone and (b) temperature predicted for the change in solar flux (solid curves) are compared with observations (vertical lines). Two models of the solar flux variability have been used (CALLIS *et al.*, 1979).

regards to O_3 variations prior to 1962 and after 1972. Thus, there are still unsolved problems for the relationships between solar activity and stratospheric ozone.

Chapter 11

ENVIRONMENTAL IMPACTS OF CHANGES IN STRATOSPHERIC OZONE

Stratospheric ozone is subject to considerable variations both by natural causes and anthropogenic perturbations. In the past decade or so it has been suggested that various human activities may be destroying the stratospheric ozone. In this chapter we will discuss model predictions of the effects of various anthropogenic perturbations on stratospheric ozone and the characteristics of the potential impacts of these ozone changes on UV radiation and life on the earth's surface. Then, model studies of the evolution of the ozone layer during geological time will be discussed. Finally, ozone in the atmosphere of other planets, particularly on Mars, will be discussed briefly.

11.1 Model Predictions of Anthropogenic Perturbations of Stratospheric Ozone

Model predictions of changes in stratospheric ozone by various human activities have been carried out mainly using one-dimensional models. Although the chemistry can be treated most precisely in 1-D models, the representation of dynamics is so poor that the results may be seriously in error. In making a reliable assessment, it is essential for the model to be able to predict realistically the transport of the injected materials to the region where they actually undergo photochemical reactions and thereby affect the stratospheric ozone. The simulation of such transport is possible only through 3-D general circulation models. With GCMs, however, the treatment of chemistry is very limited, and it is costly to run the GCM model for various conditions of injected material. Actually, no GCMs are available for such extensive studies at the present time.

Studies by 1-D models are simpler and more economical; they are still worthwhile for looking into some detailed photochemical mechanisms in the middle atmosphere. We will discuss here perturbations on stratospheric ozone by NO_x emissions from supersonic transports (SST), Cl_x emissions from chlorofluoromethanes (CFM), and increased human usage of nitrogen fertilizers. Although it is not the intention of this chapter to present any conclusions and the *numerical* results may be far from accurate, the studies should be useful in understanding the nature of the problem and clarifying various photochemical links among minor constituents in the middle atmosphere.

11.1.1 Supersonic transports (SST)

It was first suggested by CRUTZEN (1970) and JOHNSTON (1971) that ozone in the stratosphere may be reduced to a dangerous degree by catalytic reactions involving NO_x emitted from SST vehicles. The problem has been studied intensively and extensively by the Climate Impact Assessment Program (CIAP) of the U. S. Department of Transportation during the period 1972–1974. The related problems have been reviewed in CIAP (1974) and NAS (1975).

Aircraft operations projected for 1990 would inject NO_x into the atmosphere at the emission rates shown in Table 11.1 (OLIVER *et al.*, 1977; high estimate). The emission occurs mainly below ~20 km, where the emission rate greatly exceeds the production rate of NO by the reaction $N_2O + O(^1D)$, which is the main natural source for the stratospheric NO_x (MCELROY and MCCONNEL, 1971). The NO molecules emitted from aircraft engines quickly attack O_3, producing NO_2 while destroying O_3. The product NO_2 further reacts with O, reproducing NO while destroying O. The net effect is NO_x catalytic cycle V destroying odd oxygens (see Chapter 5).

In the troposphere and lower stratosphere where O is scarce, however, the photolysis of NO_2 should dominate over the reaction of NO_2 with O. Since the NO_2 photolysis produces O as well as NO, there is no net effect on O_x of the SST-emitted NO. Of course, the NO_x catalytic reaction should be important if these NO molecules are transported up into the middle and upper stratosphere.

An important modification has been developed after CIAP to the theory of the SST impact on the stratospheric ozone. As is seen in Fig. 5.6, the main catalytic cycle destroying O_x below ~20 km occurs through the HO_x cycle IV, i.e.

$$\begin{array}{rlr} & OH + O_3 \rightarrow HO_2 + O_2 & (R_9) \\ + & HO_2 + O_3 \rightarrow OH + 2O_2 & (R_{10}) \\ \hline \text{net:} & 2O_3 \rightarrow 3O_2. & \end{array}$$

TABLE 11.1 Projected 1990 aircraft emissions of NO_x (OLIVER *et al.*, 1977) compared with the natural NO production rate by the reaction $N_2O + O(^1D) \rightarrow 2\,NO$.

height (km)	NO_x emission rate from SST ($cm^{-3}sec^{-1}$)	Natural NO production rate ($cm^{-3}sec^{-1}$)
50		13.9
45		40.6
40		67.2
35		75.6
30		48.8
25		28.5
20		14.5
19	8	
18	29	9.98
17	43	
16	33	5.49
15	18	3.37
14	18	1.16
13	75	1.09
12	520	1.15
11	1161	
10	1167	1.43
9	665	
8	265	1.73
7	179	
6	90	1.93
5		2.00

In this cycle, the HO_2 molecule produced by (R_9) reproduce OH by (R_{10}), and OH and HO_2 effectively change two ozone molecules to three molecular oxygens.

Soon after the completion of the CIAP studies, the reaction rate constant for (R_{43}) $HO_2 + NO \rightarrow OH + NO_2$ was found to be almost 40 times larger than the previously measured values (HOWARD and EVENSON, 1977; see also Section 5.5.3). Thus, HO_2 produced by (R_9) would react with NO faster than with O_3; this is particularly true if the NO concentration is increased by emissions from aircraft engines. Then, the following series of reactions would occur instead of (R_9) and (R_{10}):

$$OH + O_3 \rightarrow HO_2 + O_2 \qquad (R_9)$$
$$HO_2 + NO \rightarrow OH + NO_2 \qquad (R_{43})$$
$$NO_2 + h\nu \rightarrow NO + O \qquad (J_5)$$
$$+ \quad O + O_2 + M \rightarrow O_3 + M \qquad (R_2)$$

net: no change.

Thus, no destruction of O_x occurs by the engine exhaust gases if reaction (R_{43}) is much faster than reaction (R_{10}). The SST-emitted NO now tends to lessen the effect of HO_x catalytic cycle, and therefore, to increase the ozone density in the lower stratosphere. Furthermore, ozone may be produced by smog reactions (see Section 5.4).

The CIAP report (1974) concluded that 500 aircraft flying 8 hours per day at 16 km would result in a global scale reduction of atmospheric O_3 by ~4%. The reduction rate predicted by model calculations after 1975 was much lower because they used the faster reaction rate constant for reaction (R_{43}). We may call this new calculation a *second stage prediction*. In some cases, particularly in the troposphere, an O_3 increase has been predicted in the second stage prediction.

More recently, larger rate constants have been measured for reactions $OH+HO_2$, $OH+HNO_3$ and $OH+HNO_4$ (WMO, 1981; see Table 5.2). Incorporating these newly measured rate constants in the model results in smaller OH and HO_2 concentrations; therefore, the O_3 reduction by catalytic cycles involving HO_x becomes even more ineffective. The net effect of this *third stage prediction* of SST impact is to increase slightly the ozone reduction rate compared to the second stage prediction, but the reduction rate is still smaller than the first stage (CIAP) prediction (TURCO *et al.*, 1981).

Models of SST impact have been carried out by various workers including MCELROY *et al.* (1974), SHIMAZAKI and OGAWA (1974 c), WHITTEN and TURCO (1974 c), HESSTVEDT (1974), BORUCKI *et al.* (1976). OLIVER *et al.* (1977), TURCO *et al.* (1978) and LUTHER *et al.* (1979). We will present and discuss here the result obtained by the time-dependent diurnally-averaged model calculated by the method discussed in Section 7.1 with following modifications: the lower boundary condition for O_x is now changed from a fixed concentration to the flux condition. A downward flux of 2×10^{11} cm^{-2}sec^{-1} is assumed based on Table 7.2; this is equivalent to assuming a global O_3 destruction rate of $\sim10^{30}$ molecules sec^{-1} (see Table 7.2). The model now allows for the surface O_x concentration to change following the artificial addition to the atmosphere of NO_x and other substances that are reactive with O_3.

In the current model calculation, we essentially have made a third stage prediction using NO_x emission rate given in Table 11.1. A steady state is reached after 3–4 years, and the predicted changes in ozone density at various heights after 5 years are approximately +14% (the maximum increase) at the surface, +6% at 10 km, −0.65% at 20 km, −0.68% (the maximum reduction) at 37 km, and −0.56% at 40 km. The increase in the entire troposphere and the decrease in the entire stratosphere nearly cancel each other out. Thus, there is almost no effect on total ozone due to the

NO_x emission form SST projected for 1990; the net effect is to increase total ozone only slightly (+0.06%).

TURCO *et al.* (1978) and LUTHER and DUEWER (1978) have reported that the injection of water vapor from SST's may be an important factor, since the injection occurs in the height region where the HO_x catalytic cycle is most important for odd oxygen destruction. They also report that calculated ozone changes due to added NO_x and H_2O are quite sensitive to the adopted background chlorine level. In general, as the chlorine level rises, the ozone enhancement following NO_x injection, and the ozone decrease following H_2O injection, both become larger (POPPOFF *et al.*, 1978). With a background volume mixing ratio of 2×10^{-9} for chlorine, they calculated a 4% increase of total ozone if only NO_x injection (at the global rate of $\sim 1.3\times 10^9$kg yr^{-1}) is considered at 20 km. If NO_x and H_2O are simultaneously injected at the rate of ~550 H_2O molecules released for each NO molecule, the net effect on total ozone becomes much less than 1%. The result of our model calculation discussed in the previous paragraph used a much smaller mixing ratio for background ClO ($\sim 1.2\times 10^{-10}$ at 40 km).

In order to demonstrate the effect of an aircraft's flight height, case studies have been made assuming NO emission at four different height ranges; i.e. two km ranges centered at 20.5 km, 18.5 km and 16.5 km in models A, B and C respectively, and a six km range centered at 12.5 km in model D. In each case, an emission rate of 10^3cm^{-3}sec^{-1} was assumed; this rate is certainly too large compared with the estimated emission rate particularly for the stratosphere (models A, B and C). Thus, the result of model calculations shown in Fig. 11.1 indicates unrealistically much larger ozone changes. Calculations of amplified effects, however, are useful as a sensitivity study in order to see the characteristics of ozone changes due to additional NO_x at different heights.

The NO_x emission in the troposphere (model D) increases [O_3] in the troposphere but decreases stratospheric [O_3]; the net effect on total ozone is minimal (~−0.5%). Injection of NO_x into the lower stratosphere (models A, B and C) generally decreases [O_3] in the stratosphere, whereas the effect on the troposphere is minimal. The percentage decreases of total ozone calculated for models A, B and C are ~15.4%, ~10.9% and ~5.5% respectively. In general, the decrease in total ozone becomes larger as the injection height rises. Thus, SST flights at 20 km (model A) should reduce the total ozone much more than SST flights at 16 km (model C).

11.1.2 Chlorofluoromethanes (CFM)

At about the same time as the CIAP began its research into the effects of NO_x emission from SST's the attention of scientists was also directed to the

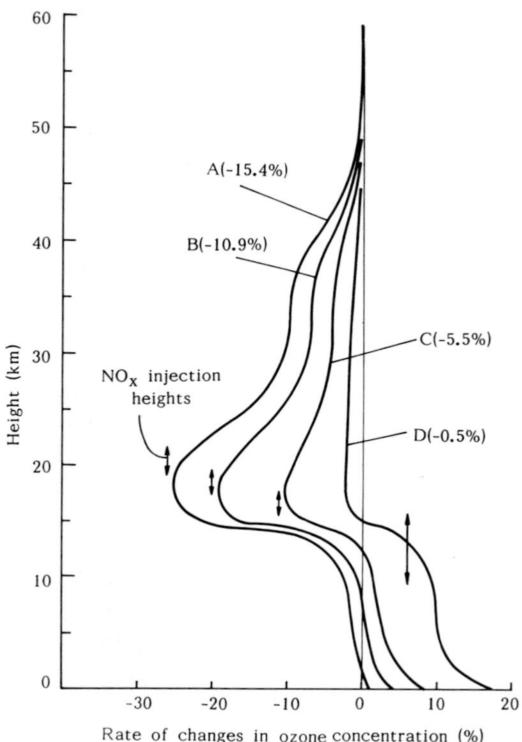

FIG. 11.1. Predicted percentage changes in ozone concentration due to NO_x emission from aircraft operation at the rate of 10^3 cm^{-1} sec^{-1} for the height range indicated. Numerals in parentheses are the reduction percentage of total ozone in each oase.

possible environmental effects of the National Aeronautics and Space Administration's (NASA) Space Shuttle. Although the shuttles' solid fuel does not emit NO_x, concern has centered around emissions of HCl from which Cl atoms are produced (STOLARSKI and CICERONE, 1974; WOFSY and MCELROY, 1974). Cl atoms can initiate a catalytic cycle (VI) by Cl and ClO destroying odd oxygen. The reaction rate constant of (R_{36}) ClO+O has been found to be about 2.5 times larger than that of (R_{25}) NO_2+O (see Table 5.2). However, because of the infrequent number of shuttle operations, they may result in little chance of permanent change to the stratospheric ozone (WHITTEN *et al.*, 1975).

The ClO_x catalytic cycle again became a major concern to scientists when MOLINA and ROWLAND (1974) and CICERONE *et al.*, (1974) warned that freons released into the atmosphere, following their use as propellants for aerosol spray cans and in refrigerators, can break down to form Cl due to photolytic dissociation by UV radiation around 200 nm in the upper stratosphere.

When measuring the concentrations of atmospheric fluoride compounds around metropolitan areas during his research into air movement related to air pollution problems, LOVELOCK (1971) found, against his expectation, that their densities did not decline in rural areas. Surprisingly, later observations indicated that the concentration of atmospheric halocarbons over the ocean were not much different from those on land (LOVELOCK *et al.*, 1973). LOVELOCK (1974) also found a fair abundance of halocarbons in the stratosphere and discussed their relationship to stratospheric ozone.

Chlorofluoromethanes (CF_2Cl_2, $CFCl_3$ etc., usually called "freons", the brand name of products manufactured by Du Pont, Inc.) have very long atmospheric residence times. Although they may be lost in the troposphere to some extent by photodecomposition after being absorbed on silicate surfaces, particularly in the sands of desert areas (AUSLOOPS *et al.*, 1977; PIEROTTI *et al.*, 1978), most freons eventually rise into the stratosphere. When they reach the upper stratosphere, they are decomposed by solar UV radiation around 200 nm and chlorine atoms are released, which initiate an extensive catalytic chain cycle (VI) leading to the net destruction of stratospheric O_3. The significant fact is that these reactions occur mainly in the upper stratosphere, where O_3 is produced most effectively; thus, the impact of freons is far more important than the impact of SST, whose NO_x emission occurs mainly in the lower stratosphere, where O_3 is controlled mainly by dynamics. Any loss of O_3 in that lower region may easily be replenished by transport from the source region in the upper stratosphere. Problems of CFM's impacts on stratospheric ozone have been reviewed in the reports of NAS (1976), NASA (1977) and NAS (1979).

Figure 11.2 shows the global annual production rates of F_{12} (CF_2Cl_2) and F_{11} ($CFCl_3$) for each year since 1935, estimated by the Chemical Manufacturers Association (CMA). The rates of actual release of freons into the atmosphere are slightly less than the production rates as shown by the dashed curves, but both rates have been steadily increasing at the rate of $\sim$10% per year until 1974. The rates have decreased after 1974 because major U. S. industries have reduced or stopped production of freons due to warnings from scientists and public pressure (the U. S. federal government, in 1979, banned the production and sale of freons as propellantion spray cans). However, the production of CFM in other countries continues at increasing rates, and the amount of freons in remote areas such as Antarctica has been increasing even after 1974 as is seen in Fig. 11.3 (ROBINSON *et al.*, 1979). In fact, there are some recent evidences to indicate that the emission rate of F_{12} into the atmosphere has reached the maximum of more than

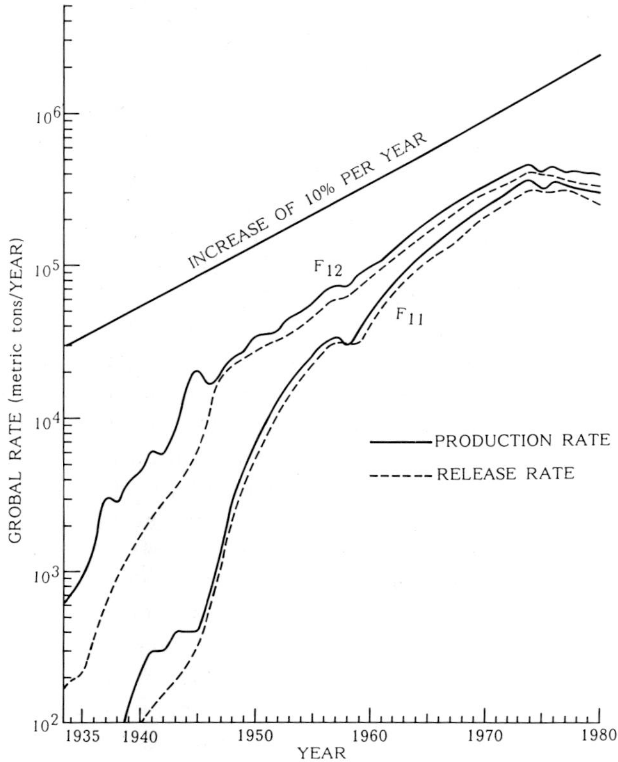

FIG. 11.2. Global production and release rates of F_{11} ($CFCl_3$) and F_{12} (CF_2Cl_2) for each year between 1935 and 1980 (CHEMICAL MANUFACTURERS ASSOCIATION, 1981). The straight line indicates a rate of increase of 10% per year.

5×10^5 tons/year, in 1978 or later, in contrast to the CMA maximum ~ 4.2×10^5 tons/year during 1974 (ROWLAND *et al.*, 1982).

Model studies of the impact of CFM's on stratospheric ozone have been made by ROWLAND and MOLINA (1975), STOLARSKI and RUNDEL (1975), TURCO and WHITTEN (1975), SZE and WU (1976), LIU *et al.* (1976), CRUTZEN *et al.* (1978), LOGAN *et al.* (1978), RAO-VUPPUTURI (1978/1979), MILLER C. *et al.* (1979, 1981), BORUCKI *et al.* (1980), CALLIS and NATARAJAN (1981), GIDEL *et al.* (1983) and others. In order to study the impact of CFM's one has first to calculate a model which is not influenced by ClO_x produced from CFM. For this purpose, we adopt the *diurnally averaged steady state* model that was calculated in Section 7.1.3. The upward fluxes of F_{11} and F_{12} at the surface calculated in this model are given in Table 7.2. They are about the same as the injection rates of these molecules in 1955.

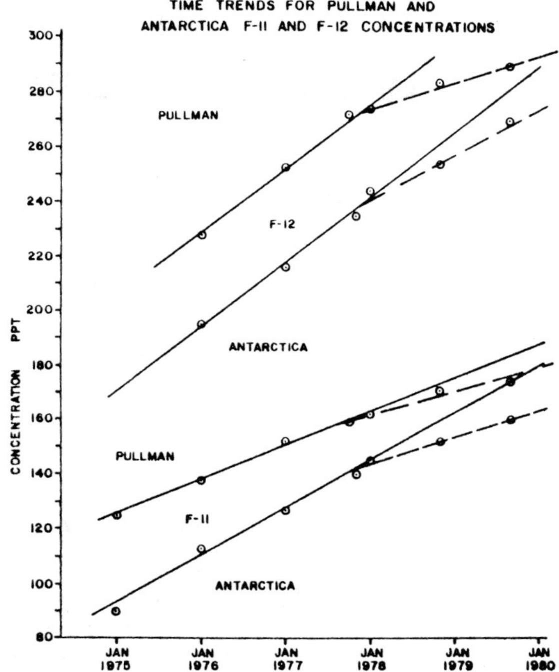

FIG. 11.3. Increase in the mixing ratios of F_{11} and F_{12} from 1975 to 1980 observed at Pullman, Washington, U.S.A. and in Antarctica (ROBINSON *et al.*, 1979).

Thus, the model may represent the stratospheric ozone unperturbed by CFM's. We label this condition "state *A*" of the model.

We now use the *time-dependent diurnally averaged* model by specifying the upward (injection) fluxes on the ground as the lower boundary conditions for F_{11} and F_{12}, and calculate the time series of the model by increasing the injection rate by 10% every year. After about 20 years the fluxes reach the values close to those observed in 1974; we call this status of the model "state *B*." The total ozone calculated at state *B* is changed little (~ −0.2%) from that at state *A*, as is seen from Fig. 11.4.

If we continue increasing freon injection at the same rate (i.e. 10% every year), the total ozone starts to decrease dramatically ~10 years after state *B* (see Fig. 11.4). After ~50 years, the model reaches state *C*, where an approximately 50% decrease is predicted for the total ozone. Realizing that the ozone reduction is dangerously large, we stop freon injection at state C_0, which occurs at ~45 years after state *B* and where the reduction rate is ~ 25%. However, there are further decreases in the total ozone for another few years. Reaching the lowest reduction rate (~44%) at ~54 years (state

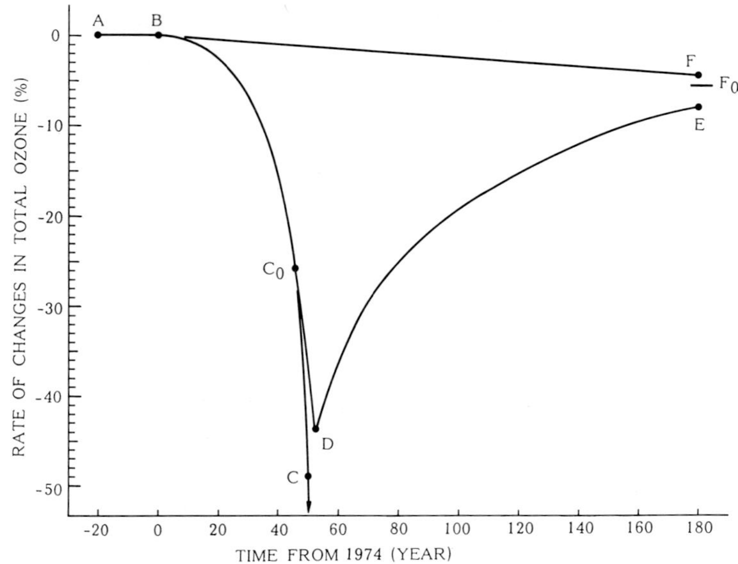

FIG. 11.4. Time evolution of the predicted percentage changes in total ozone due to freon injection. Curve ABC indicates the change when the injection rate is increased by 10% each year. Curve C_0DE indicates the change when the injection is terminated after C_0. Curve BF indicates the change when the injection rate remains constant after B. F_0 represents a steady-state model with the injection rate of status B.

D), the reduction rate recovers slowly and arrives at ~8% at 180 years (state *E*).

If freon injection continues after state *B* at a constant rate (instead of a 10% increase per year after 1974), the total ozone decreases very slowly along curve BF in Fig. 11.4; it approaches a status F_0 that is the result of a *steady-state* model calculation with the constant freon fluxds of 1974 at the lower boundary. The reduction in total ozone for state F_0 is ~−5.2%. Thus, whether the change occurs along DE or BF, the ultimate effect is a ~5–6% decrease in total ozone after more than 200 years. This is consistent with the result obtained by various other modelers as summarized in Fig. 11.5 (WMO, 1981). Each bar in the figure represents the range of model results using different chemical reaction rate constants. The top one uses the NASA (1979) chemical kinetics which correspond to the chemistry used in the second stage prediction of the SST's impact in the previous section; it shows a large total ozone reduction (15–19%). The other three cases use the chemistry that was employed for the third stage prediction of SST's; the second from the top takes into account the faster rate constant for

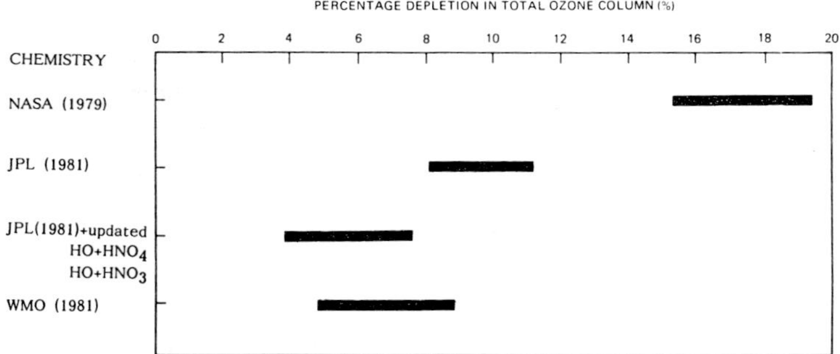

FIG. 11.5. The variation in steady-state total ozone depletion estimates due to freon release calculated using different chemical rate coefficients. Each bar indicates the range of evaluations by different modelers (WMO, 1981).

$OH+HO_2$, the third one the faster rate constants for $OH+HNO_3$ and $OH+HNO_4$, and the bottom one the slower rate constant for $ClO+NO_2$. The main reason why the ozone loss is much smaller in these three cases than the top one is the reduced OH abundance below ~30 km, which slows reaction R_{47} ($OH+HCl$); this reaction releases the ozone-active chlorine (WHITTEN *et al.*, 1981). The result of our model calculation discussed earlier should be compared with the bottom case in Fig. 11.5, which indicates the depletion of total ozone to be in the range ~5 - 9 %.

The percentage change in $[O_3]$ at each altitude is compared among different states of the model in the left frame of Fig. 11.6. The right frame shows the profiles of [ClO] predicted for three states, *B*, *D* and F_0. The volume mixing ratio at 30 and 40 km are also shown. The largest percentage change in $[O_3]$ always occurs at ~40 km. Furthermore, the percentage change in total ozone is proportional to the change in $[O_3]$ at this height. The latter change is inversely related to the magnitude of [ClO] at 40 km predicted for each state. Large changes with irregular variations are observed below ~25 km in states *C* and *D*. In these two cases a greatly reduced ozone density in the upper stratosphere allows deeper penetration of solar radiation, increasing the O_x production rate in the lower regions. This causes a major modification to the balance among three (production, loss and transport) terms in the continuity equation. Inspection of the model results reveals that large, irregular changes below ~25 km are caused mainly by changes in the transport term. Since the transport in 1-D models is not realistic, the predicted changes in $[O_3]$ below ~25 km in states *C* and *D* may involve some uncertainties. Thus, the prediction of a total ozone depletion of more than ~40% shown in Fig. 11.4 may be seriously affected by transport and must be considered with caution. However, the change in

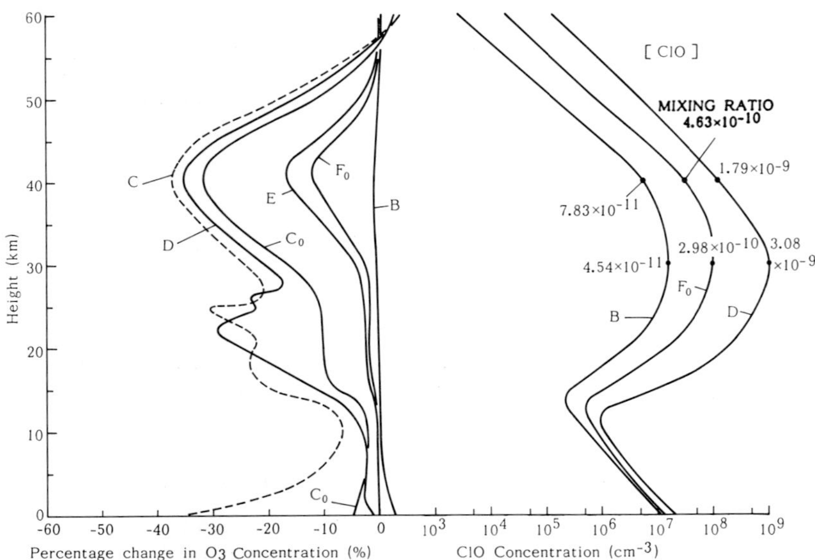

FIG. 11.6. Percentage ozone reduction rate (left frame) and ClO concentration (right frame) as a function of altitude predicted at different statuses of the model in the time series shown in Fig. 11.4.

ozone density at ~40 km is always chemically determined; therefore, its observation is certainly useful to detect the ozone change due to industrial release of CFM's into the atmosphere.

Other fluorocarbons including methyl chloroform (CH_3CCl_3) could also affect stratospheric ozone. Bromine compounds also presumably have a destructive catalytic effect on stratospheric ozone; in fact the measured reaction rate constant for BrO+O is $3\times10^{-11}\mathrm{cm}^3\mathrm{sec}^{-1}$ (CLYNE *et al.*, 1976), which is about 3.2 times larger than that for NO_2+O, and is even larger than that for ClO+O. Concentrations currently observed for these halocarbons are not large enough to have significant impact on stratospheric ozone. However, a careful surveillance on the amounts of these substances in the atmosphere is very important.

11.1.3 Nitrogen fertilizers

It has been known since the late 1960's or early 1970's that NO_x plays an important role in determining the stratospheric ozone density and that the main natural source for NO_x in the atmosphere is nitrous oxide (N_2O) released at the earth's surface, primarily by biological processes in the soil and in the oceans.

Nitrogen makes up a large part of our atmosphere but mostly in the form of an inert gas, N_2, that cannot be directly utilized by living things. This nitrogen must be converted into useful forms such as ammonia (NH_3) or nitrates (NO_3^-) before it can be incorporated into living tissue; this process is called *nitrogen fixation*. In nature, nitrogen fixation is accomplished mainly by soil bacteria that lives in the roots of leguminous plants such as beans and peas. Blue-green algae also fix nitrogen to a lesser extent.

The nitrogen cycle in nature is illustrated in Fig. 11.7. The fixed nitrogen is taken up from the soil by plants through *assimilation* processes; these plants provide protein for animals. It is returned to the soil through the excrement and excreta of animals, producing microorganisms. Microorganisms are also produced by the decay of plants in the soil. Ammonia is produced from this organically bound nitrogen by a process called *ammonification*. The oxidation of ammonia to form nitrate is termed *nitrification*. The utilization of ammonia and nitrate by plants to reproduce organic substances constitutes *ammonia assimilation* and *nitrate assimilation*, respectively. A part of the nitrate is returned to the atmosphere in the form of either N_2 or N_2O through *denitrification* under anaerobic conditions.

Total nitrogen fixation for the year 1974 was estimated to be ~237 MT/yr (millions of metric tons per year), of which about 63% comes from natural processes on agricultural, forested and unused land and about 33% from industrial processes including fertilizers and combustion (PRATT *et al.*, 1977).

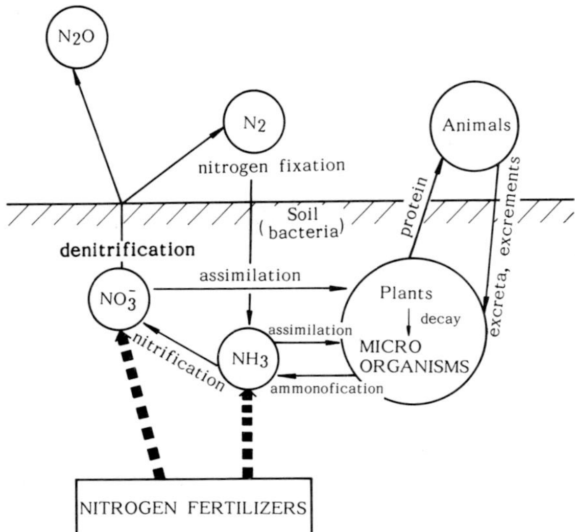

FIG. 11.7. Nitrogen cycle in the atmosphere and soil.

The proportion of total nitrogenous gases that is liberated as N_2O during denitrification in soils varies between 0 and 100% depending upon the conditions in the soil. However, it is suggested that a nitrogen to nitrous oxide ratio of 16 (or about 6% N_2O) would be typical for conditions normally found during denitrification episodes in agricultural soils (PRATT *et al.*, 1977). If the same ratio is applied to all conditions, the global N_2O production rate would be ~14 MT/yr, which corresponds to a molecular flux of $\sim 1.2 \times 10^9 cm^{-2}sec^{-1}$. This flux is close to the value calculated in our diurnally averaged steady state model, $1.09 \times 10^9 cm^{-2}sec^{-1}$ (see Table 7.2).

Use of nitrogen fertilizers add NH_3 and NO_3^- directly to the soil, and therefore, could enhance the production of N_2O from the soil. The recent measurements of the tropospheric N_2O indicate a 0.2 – 0.4 % increase per year during the period 1975 – 1982 (WEISS, 1981; KHALIL and RASMUSSEN, 1983). The worldwide consumption of nitrogen in fertilizers has increased from 11.5 to 38.7 MT/yr over the period 1962 through 1974, a rate of increase of ~10.7% per year. It is difficult to make accurate long-range predictions on the amount of nitrogen that will be used as fertilizers in the chemical industry and how much N_2O will be produced from them. To obtain a general picture of the effects of increasing N_2O, however, we have calculated two steady state models, A and B, by increasing the influx at the ground to 2.5×10^9 and 5.0×10^9 $cm^{-2}sec^{-1}$, which are about 2.3 and 4.6 times greater than the value in the standard model simulated for 1955 (see Section 7.1.3).

It is found that these two models decrease the total ozone by ~15.5% and ~29.7%, respectively. Height profiles of the percentage change are illustrated in Fig. 11.8. The ozone reduction occurs mainly in the region between 15 and 45 km, but the *percentage reduction* is slightly less around 30 km. This is because the NO_x catalytic cycle has its largest effect (~69% of the total O_x loss) at ~30 km in the standard model as is seen in Fig. 5.7. Therefore, the effect of *additional* N_2O is relatively small around this height compared to surrounding height regions. The *absolute changes* in the stratospheric ozone is largest at ~20 km. If the observed yearly increase (0.2 – 0.4 %) of atmospheric N_2O continues, a 10 % increase of N_2O is predicted to occur within 20 – 40 years. The result of model claculations suggests that this N_2O increase may cause an ozone column density to decrease by ~1.5 %, if a roughly linear relationship is assumed between the fractional increase of N_2O and the ozone loss.

The dashed curve C in Fig. 11.8 indicates the effect of simultaneous increases in N_2O and freons. The model assumes the same N_2O influx as model A and constant freon injection rates at ~1974 levels; the latter is about 7 times larger than in models A and B which assume the injection rate at ~1955 levels. The single effect of the freon injection rates at 1974

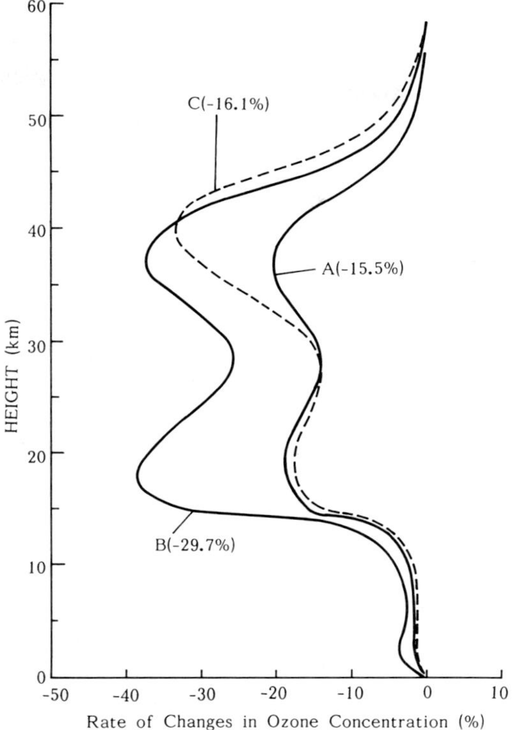

FIG. 11.8. Predicted percentage changes in ozone concentration due to the increased N_2O flux from the soil to the atmosphere; model A and B have the N_2O flux 2.3 and 4.6 times larger than the standard model. Model C has an increased freon flux as well as the increased N_2O flux. Numerals in parenthesis are the reduction percentage for total ozone.

levels was calculated to be $\sim -5.2\%$ for total ozone in the previous section (status F_0). Thus, the predicted percentage decrease -16.1% in model C is much smaller than the simple addition of percentage changes due to the effect of N_2O alone (-15.5%) and freons alone (-5.2%).

In general, if N_2O, freons, and H_2O all increase, the combined effect is smaller than the simple addition of the effects calculated individually. This is partly because of rapid exchange reactions among family components of NO_x, HO_x and ClO_x such as

$$NO + HO_2 \rightarrow NO_2 + OH$$

and

$$NO + ClO \rightarrow NO_2 + Cl.$$

Another reason is that some reactions among family components result in the formation of constituents that are non-reactive with O_3 (WHITTEN *et al.*, 1980); these include

$$OH + NO_2 + M \rightarrow NO_3 + M$$
$$HO_2 + NO_2 + M \rightarrow HNO_4 + M$$

and

$$NO_2 + ClO + M \rightarrow ClONO_2 + M.$$

11.1.4 Increase in CO_2

The mixing ratio for CO_2 in the current atmosphere is ~330 ppm, but it is believed that the mixing ratio was 260–270 ppmv prior to ~1925 when worldwide industrialization started. As of 1975 the CO_2 level is increasing at the rate of ~0.6 ppm/yr (MCRAE and GRAEDEL, 1979), and it is expected to reach ~600 ppm by 2030, if the fuel-consumption rate increases by 4.3 percent every year (WOODWELL, 1978). This is more than double the pre-industrialization value.

CO_2 plays an important role in increasing the global surface temperature through the greenhouse effect. In and above the middle stratosphere, however, CO_2 is the principal constituent that contributes to stratospheric cooling via escape of IR radiation into space (see Chapter 12). Thus, an increase in stratospheric CO_2 will lead to a cooling of the stratosphere, which will cause an ozone increase through changes in the chemical reaction rate constants (k_2 in (5.9) increases, while k_3 decreases). This effect is particularly large in the upper stratosphere.

CALLIS and NATARAJAN (1981) have calculated the changes in temperature and ozone concentration due to the increase in CO_2 after 1954 given by

$$[CO_2]_{ppm} = 321 \cdot e^{4.69 \times 10^{-6}(y-1954)^{2.66}} \tag{11.1}$$

where y represents the calendar year. The result is shown in Fig. 11.9. Whereas the temperature at 41.5 km decreases monotonically, both total ozone and $[O_3]$ at 41.5 km increase due to the continuous increase of $[CO_2]$. GROVES and TUCK (1980 a) have also calculated the increase of annually averaged ozone column density by ~10 % as the CO_2 volume mixing ratio is increased from 250 to 700 ppmv. This is thought to be resulted from cooling of the upper stratosphere by up to 20 K over this range of the CO_2 change.

The effect of temperature feedback on stratospheric ozone is very important. Such an effect can also occur by a change in the ozone concentration itself, since the absorption of solar UV radiation by O_3 is an important heat

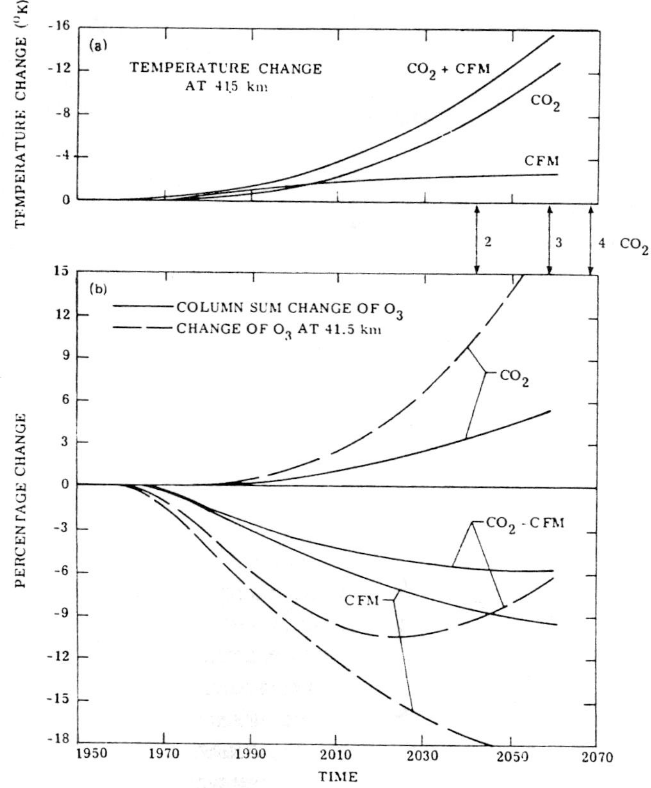

FIG. 11.9. Time variations in temperature and ozone for CO_2, CFM, and CO_2+CFM scenarios (CALLIS and NATARAJAN, 1981).

source in the stratosphere. The upper stratospheric ozone has a self-healing effect through temperature feedback; the decrease in $[O_3]$ decreases the temperature, which in turn increases $[O_3]$ (see Section 12.3.1 for more discussion on this subject).

Figure 11.9 also illustrates the effect of CFM release with the historical release rates from 1954–1977 (see Fig. 11.2), after which the release rate is fixed at the 1977 level. Note that these calculations primarily use the rate constants of NASA (1979); therefore, the effect of CFM on stratospheric ozone is overestimated (see Fig. 11.5). CALLIS and NATARAJAN (1981) found that when CO_2 and CFM levels increase simultaneously, the ozone behavior is quite different. As is seen in Fig. 11.9, the total ozone depletion tends to level off at a reduction ~5.5%. The ozone depletion at 41.5% km reaches a maximum around the year 2020 and then the concentration tends to increase.

Including the chlorine chemistry in a photochemical radiative column model for the study of stratospheric O_3 - CO_2 coupling, GROVES and TUCK (1980 b) have found that there is an alleviation of the CFM-induced ozone reduction attributable to the increased CO_2. Two perturbations upon ozone column density are not quite linearly additive; the net effect on ozone column density consists of larger percentage decrease in the upper stratosphere almost compensated by smaller percentage increase in the lower stratosphere where much more ozone is resident. More model calculations and discussions on the nonlinear response of stratospheric ozone column to coupled anthropogenic perturbations of CFM, N_2O and CO_2 are given by NICOLI and VISCONTI (1982), WUEBBLES *et al.* (1983) and CICERONE *et al.* (1983).

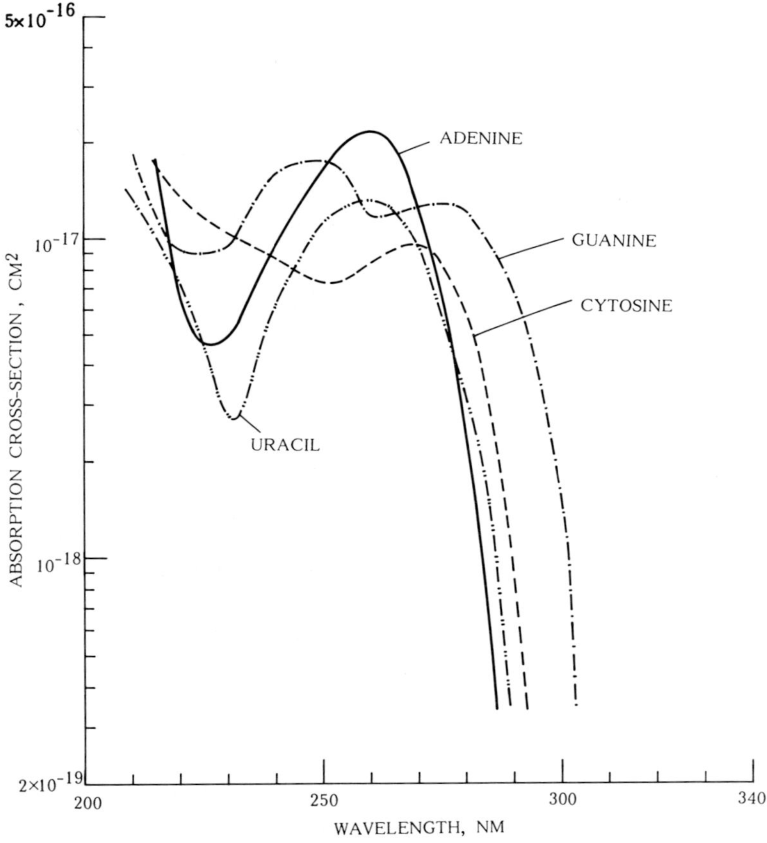

FIG. 11.10. Spectra of the absorption cross-sections of DNA molecules including adenine ($C_5H_5N_5$), cytosine ($C_4H_5N_3O$), guanine ($C_5H_5N_5O$) and uracil ($C_4H_4N_2O_2$).

11.2 Ozone Screening and Life on the Surface

Ozone absorbs almost all solar UV radiation below ~320 nm, thus preventing that radiation from arriving at the earth's surface. As a result, deoxyribonucleic acid (DNA) can survive on the ground surface without being destroyed by solar UV radiation, although it has a strong absorption band near 260 nm as is seen in Fig. 11.10, which illustrates the absorption cross-sections of various DNA molecules including adenine ($C_5H_5N_5$), cytosine ($C_4H_5N_3O$), guanine ($C_5H_5N_5O$) and uracil ($C_4H_4N_2O_5$). DNA is an essential component of all living matter and a basic material in the chromosomes of the cell nucleus; it contains the genetic code and transmits the hereditary pattern. It is interesting to note from the comparison with Fig. 4.1 that the peaks for the absorption bands for O_3 and DNA occur in about the same spectrum range.

The intensity of solar UV radiation arriving at the surface has been calculated for various levels of total ozone as is shown in Fig. 11.11 for the equinoxes. The calculation includes the effect of diurnal variations. Curve 1 in the figure represents approximately the current level of total ozone (~ 0.372 cm NTP), whereas curve 0 indicates the case of no atmospheric ozone. Various levels of total ozone between these two cases are shown in the figure.

When we discuss the biological effects of UV radiation, it is convenient to subdivide the spectral range into three ranges; i.e. UV-A (320–400 nm), UV-B (280–320 nm) and UV-C (<280 nm). It can be seen from Fig. 11.11 that the entire range of UV-B radiation would arrive at the earth's surface if total ozone is reduced to 10% of the currently observed value. The reduction of total ozone to 1% would eliminate the screening effect of O_3 for the entire range of UV-A to UV-C radiation.

The UV-A radiation is not absorbed by O_3; therefore, the intensity of radiation arriving at the surface will not be greatly affected by a change in stratospheric ozone. Since the UV-C radiation is strongly absorbed by O_3, it is almost completely absorbed by stratospheric ozone at the currently observed level. It cannot reach the surface unless the atmospheric ozone is depleted almost completely (by more than 90%) or its density is very low, as might have been the case in the very early stages of atmospheric evolution (see next section).

The UV-B radiation is most important since the amount of absorption may vary over a wide range due to the potential changes in stratospheric ozone. Figure 11.12 indicates the percentage increase in atmospheric transmission in the UV-B range due to decreases in total ozone by various factors (1/2, 1/4, ..., and 1/64). Although it is unlikely that the UV-B

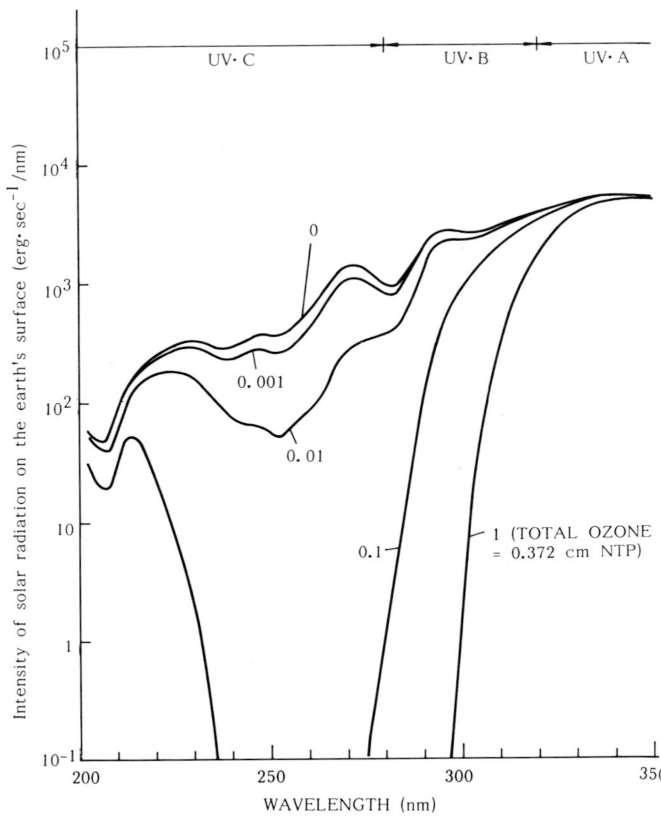

FIG. 11.11. Intensity of solar radiation in the spectrum range 200–350 nm predicted on the earth's surface for equinoxes and for various levels of total ozone. Effects of diurnal variation is included.

range below 300 nm changes substantially due to human or naturally caused changes in stratospheric ozone, the same may not be true for UV-B radiation in the 300–320 nm range. A change in this radiation near the surface would affect life on the earth's surface.

It is certainly beyond the scope of this book to discuss in detail the biological impacts that may be produced by changes in stratospheric ozone. In fact, information on biological impacts is still limited, although some progress has been made recently, particularly in the relationship between skin cancer and UV-B radiation (NAS, 1982).

The inflammation of human skin (sunburn) tends to increase as the wavelength of UV-B radiation to which the skin is exposed decreases. On the other hand, the intensity of the incoming UV-B radiation is stronger as

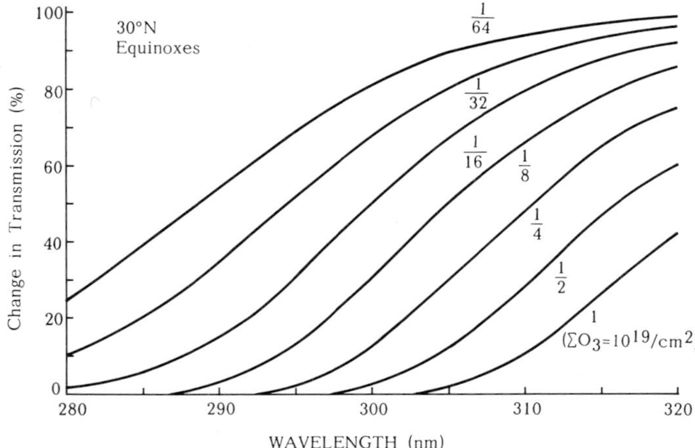

FIG. 11.12. Percentage increase in atmospheric transmission in the spectrum range 280–320 nm due to the decrease of total ozone by various factors.

the wavelength increases. Thus, the maximum effect should occur somewhere in the middle of the UV-B range, i.e. at around 305–310 nm. We now claculate the change in radiation intensity of this spectral range at the surface due to a change in total ozone.

The intensity of solar radiation on the surface can be calculated from (3.4) by

$$I_0 = I_\infty e^{-\tau_{O_3}(0)\cdot \sec\chi} \tag{11.2}$$

where $\tau_{O_3}(0)$ is the optical depth of ozone absorption at the surface and is given by

$$\tau_{O_3}(0) = \sigma_{O_3}\cdot N(O_3) \tag{11.3}$$

where $N(O_3)$ is the total ozone. From these equations one can calculate the rate of change in I_0 due to the change in $N(O_3)$ as follows:

$$\frac{\Delta I_0}{I_0} = -\tau_{O_3}\cdot \sec\chi \frac{\Delta N(O_3)}{N(O_3)}. \tag{11.4}$$

This indicates that the rate of increase in the intensity of radiation on the surface is τ_{O_3}, secχ times the rate of decrease in total ozone.

Table 11.2 shows the calculated values of τ_{O_3}, secχ and their product, using $\sigma_{O_3}=2\times10^{-19}$cm^2 for the spectral range 305–310 nm, and observed values of $N(O_3)$ representative of different χ at the equinoxes. The product changes from 1.26 at $\chi=0°$ to $\sim$3.25 at $\chi=60°$; thus, the rate of change in I_0 is about 2.5 times larger at high latitudes than at lower latitudes for the

TABLE 11.2 Total ozone and optical depth on the surface at different latitudes (or zenith angles in equinoxes) and the factor to convert the percentage change in total ozone to the percentage change in solar UV-B radiation at 305–310 nm on the ground surface (rightmost column).

solar zenith angle χ	sec χ	total ozone $N(O_3)$	optical depth τ_{O_3}	$\tau_{O_3} \cdot \sec\chi$
0°	1.000	6.3×10^{18} cm^{-2}	1.260	1.260
10°	1.0154	6.3×10^{18}	1.260	1.279
20°	1.0642	6.68×10^{18}	1.3366	1.422
30°	1.1547	7.24×10^{18}	1.448	1.672
45°	1.4142	8.51×10^{18}	1.702	2.407
60°	2.000	8.128×10^{18}	1.6256	3.251

same rate of change in total ozone. Adopting the value of 2 as a representative global average, we may conclude that for every 1 percent depletion in the total ozone there will be an approximately 2 percent increase in the amount of UV-B radiation in the 305–310 nm range reaching the earth's surface.

The increase in cancer incidence as a result of this increase in UV-B is not yet known, but the report of NAS (1982) estimates that there will be a ~2–5 percent increase in the incidence of basal cell skin cancer and ~4–10 percent increase in the more serious squamous cell skin cancer (malignant melanoma) for every 1 percent decrease in total ozone.

11.3 Evolution of the Ozone Layer

The production of O due to photolysis of O_2 by solar UV radiation is the main natural source of stratospheric ozone (or ozone layer). Therefore, the evolution of the ozone layer is directly related to the evolution of atmospheric O_2. It is now generally accepted that the earth's primordial atmosphere was quickly lost to space by escape due to high temperature and/or by the effect of the solar wind, and that the primitive atmosphere was derived from internal secondary sources, primarily through volcanic effluents. Volcanic gases include H_2O, SO_2, H_2S, HCl, CO_2, CO, NO, N_2, CH_4, Cl, CS_2 and COS, but very little oxygen. Thus, it is evident that the earth's primitive atmosphere had no ozone layer to protect life on the surface.

Photolysis of H_2O can produce O_2, but since H_2O and O_2 absorb UV radiation at almost identical wavelengths, the concentrations of O_2 produced by H_2O photolysis in the primitive atmosphere must be self-regulating; the O_2 that is produced tends to prevent further photodissociation of H_2O (the Urey effect). BERKNER and MARSHALL (1964) have concluded from a calculation of Urey effects together with other geochemical and biological evidence that the oxygen level in the primitive atmosphere of the earth was less than 10^{-2} PAL (present atmospheric level). Considering the balance between the rate of O_2 production from H_2O photolysis followed by hydrogen escape into space and the loss due to oxidation of reduced volcanic gases (primarily H_2 and CO), KASTING *et al.* (1979) calculated the ground level O_2 mixing ratios to be of the order of 10^{-12} PAL or less. Taking into account the OH radical reactivity with trace species (CO, HCl, SO_2, H_2S and CH_4), WOOD and THIEMENS (1980) also estimated that the formation of O_2 may have been prevented at a level as low as 10^{-10} PAL.

The subsequent large increase in the O_2 level can be attributed only to green-plant photosynthesis, which produces O_2 from H_2O and CO_2 essentially by the reaction written as

$$HO_2 + CO_2 \rightarrow CH_2O + O_2.$$

The oxygen partial pressure would have risen slowly at first, as the extra oxygen reacted with atmospheric hydrogen resulting in the loss of O by reaction R_{18} ($H_2 + O \rightarrow H + OH$). When the oxygen source became comparable to the hydrogen source, however, the rise in atmospheric oxygen would have become very rapid.

In the present atmosphere the global O_2 production rate by photosynthesis is estimated to be $\sim 2.66 \times 10^5$ Mton yr^{-1}, which is large enough to produce the total amount of O_2 in the current atmosphere ($\sim 1.18 \times 10^9$ Mton) within 4,000 years. In actuality, this period would have to be much longer, if we consider the loss of oxygen by various oxidation processes and respiration. However, it seems certain that the build up of oxygen to its current level occurred in much a short period when compared to geological time.

BERKNER and MARSHALL (1965, 1966) have argued that a biologically effective UV shield in the ancient atmosphere would have been provided by a total ozone level of ~ 0.2 atm. cm, which is about 54% of the currently observed value (~ 0.372 atm. cm). So, our problems are two-fold, i.e. (1) at what O_2 level could an ozone layer of 0.2 atm. cm have been produced and (2) when, in the earth's history did the atmosphere attain such an O_2 level (or total ozone), to allow life on earth to flourish? The rest of this section will be devoted primarily to the first question; the answer to the second question can be obtained from investigations of various fossils embedded in geological strata developed at different periods of geological time.

Model calculations of various degrees of sophistication have been published for the evolution of atmospheric ozone. Using a semi-empirical method in order to relate total ozone to atmospheric oxygen levels based on the calssical Chapman reaction scheme, BERKNER and MARSHALL (1965) found that the critical level of total ozone (0.2 atm.cm) should have been achieved at ~0.1 PAL of atmospheric oxygen. By a more realistic model calculation, but still using the Chapman reaction scheme and neglecting the effects of transport, RATNER and WALKER (1972) predicted that their critical level of O_3 column density 7×10^{18} cm^{-2} (or 0.26 atm.cm) should have been achieved at an oxygen content of only 10^{-3} PAL. HESSTVEDT *et al.* (1974) included the HO_x chemistry and the effects of vertical eddy transport and found that the full ozone shield would not have been developed until the oxygen level exceeded 2×10^{-2} PAL.

BLAKE and CARVER (1977) developed a model including the photochemistry of HO_x, NO_x, CO_2 and CH_4, in addition to the Chapman ozone chemistry but ignoring the effects of vertical transport in most cases. They concluded that biological evolution took place under a protective ozone shield even when the total oxygen content in the atmosphere was as low as 10^{-3} PAL.

The change in the temperature structure of the evolving atmosphere has been investigated by MORSS and KUHN (1978) and KATAMORI (1979). They found that temperature decreases monotonically with altitude and that neither the tropopause nor stratopause is produced, if the oxygen level is less than 10^{-3} PAL. However, their calculations neglected the effects of transport on both composition and temperature.

LEVINE *et al.* (1979) and KASTING and DONAHUE (1980) both developed more sophisticated models including the effects of vertical eddy transport and the complex chemical reaction scheme established by recent stratospheric studies. For oxygen levels less than 10^{-1} PAL, both models assume a temperature profile in which the temperature decreases linearly from the tropopause to the mesopause. However, there are some differences in detail between the two models. The model of Kasting and Donahue (hereinafter referred to as the K-D model) considers the possibility of increased eddy diffusivity due to the absence of the temperature inversion. Both models modify the amounts of H_2O, N_2O and NO_x (also of CH_4 in the K-D model) as a function of oxygen level but in somewhat different fashions (note that how they actually changed is not known). The K-D model takes into account the production of NO_x by lightning, loss of water-soluble species by rainout, and the surface loss of O_3, whereas the zero flux condition is assumed at the surface for O_3 in the model of Levine *et al.*

The semi-empirical model by BERKNER and MARSHALL (1965) indicates that the total ozone decreases monotonically as the level of O_2 decreases, and that the critical value for the protective ozone shield occurs around ~ 0.1 PAL of atmospheric oxygen. Most of the later model calculations have shown, however, that total ozone reaches a maximum at an oxygen level of ~0.1 PAL; for instance, the result of LEVINE *et al.* (1979) indicates the total ozone for 0.1 PAL exceeded that in the present atmosphere by ~ 40%, as is seen by the dashed curve in Fig. 11.13. They also showed that the protective ozone shield can occur for much smaller oxygen levels ($<10^{-2}$ PAL).

A recent model by KASTING and DONAHUE (1980) gave a result in contrast with most of the previous models. As is seen by the solid curve in Fig. 11.13 the total ozone predicted with this model decreases monotonically with decreasing O_2 content and the critical O_2 level at which an effective UV shield is produced is larger than 10^{-2} PAL; these results are rather similar to those of the earlier simple models by BERKNER and MARSHALL (1965).

KASTING and DONAHUE (1980) argue that the difference in their result from that of LEVINE *et al.* (1979) may be caused by the fact that their model includes the surface loss of O_3 and uses a larger eddy diffusion coefficient; as a result, O_x can be transported faster to the surface where it is lost. The insignificance of O_3 transport in the model of Levine *et al.* may also be due to the fact that the model treats O_3 as an individual constituent rather than as a member of the O_x family, something that is criticized by K-D. In any case the model results are influenced by uncertainties in many input parameters and assumptions, and more studies are certainly needed in order to arrive at any final conclusion on the problem.

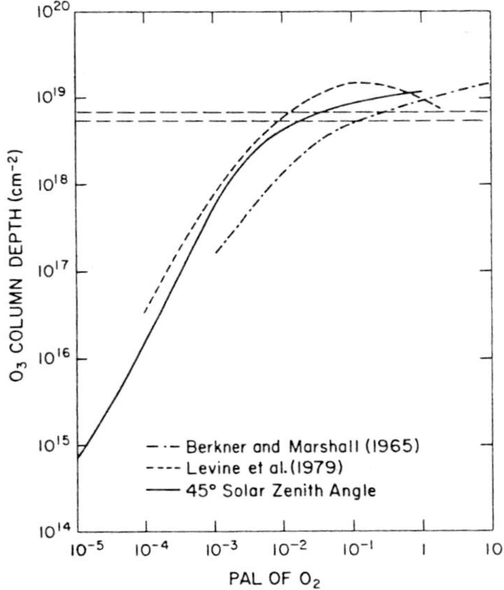

FIG. 11.13. Total zone versus oxygen content. The horizontal dashed lines represent estimates for the amount of total ozone required to produce an effective UV shield according to Berkner and Marshall (lower line) and Ratner and Walker (upper line). (KASTING and DONAHUE, 1980).

Figure 11.14 presents the result of a model calculation using the photochemistry and modeling technique discussed in this book; however, the model now assumes a depositing velocity -0.2 cm sec^{-1} for O_x and the lower boundary values for N_2O, NO_x, Cl_x and freons are reduced by a factor of square root of the O_2 PAL value. Table 11.3 summarizes the total ozone predicted for various oxygen levels. These results seem basically to agree with those of KASTING and DONAHUE (1980), although our model uses the updated values (WMO, 1981) for chemical reaction rate constants.

REID *et al.* (1978) studied the effect the passage of our solar system through a supernova remnant (SNR) shell might have had on the ancient atmosphere. Among various potential effects, they were particularly concerned with the decreased concentration of O_3 due to the increased concentrations of NO_x from high levels of ionizing radiation caused by intense cosmic ray flux within the SNR (this is an event basically similar to the SPE discussed in Section 10.2.1 but its intensity is much stronger). If such an event had occurred during a reversal of the earth's magnetic field, the magnetic field's ability to deviate the intruding charged particles (cosmic ray) to high latitudes would have been greatly reduced and stratospheric ozone might have been reduced significantly for the entire globe.

Some studies of rock magnetism indicate that the reversal of the

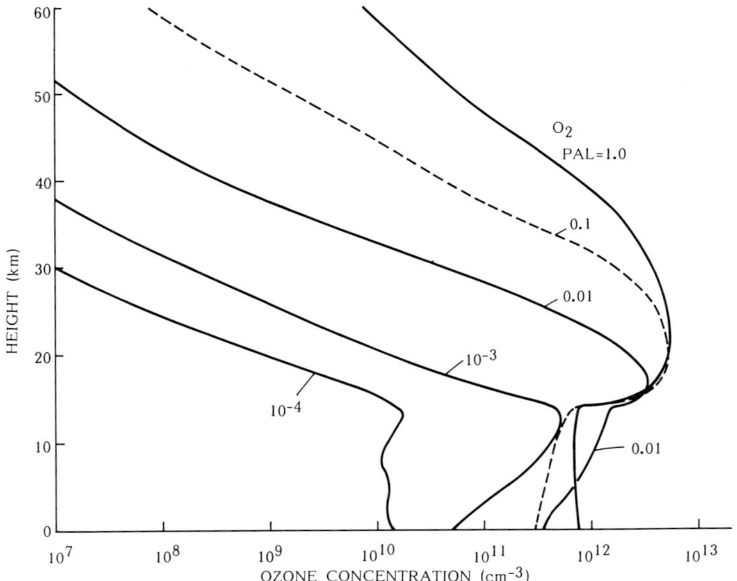

FIG. 11.14. Ozone density profiles predicted for various levels of atmospheric oxygen.

TABLE 11.3 Column density of O_3 predicted for different oxygen levels. Ratios in parentheses indicate the values predicted by KASTING and DONAHUE (1980).

O_2 level (PAL)	Ozone column density (cm^{-2})	(cm atm.)	ratio to PAL=1.0
1	1.067×10^{19}	0.397	1.000 (1.000)
10^{-1}	7.486×10^{18}	0.279	0.702 (0.760)
10^{-2}	3.640×10^{18}	0.135	0.341 (0.371)
10^{-3}	4.966×10^{17}	0.018	0.047 (0.050)
10^{-4}	2.260×10^{16}	8.4×10^{-3}	2.1×10^{-3} (1.3×10^{-3})

geomagnetic field occurred very rarely some 75–100 million years ago. This period corresponds to the Cretaceous period when many kinds of huge dinosaurs flourished. By the end of the Cretaceous period, however, the occurrence of reversal of the geomagnetic field increased suddently (see, for instance, COX, 1975). It is speculated that the passage of the earth through an SNR towards the end of the Cretaceous period may have caused a prolonged period of harsh environmental conditions for the biosphere including a large world-wide ozone depletion and a high level of damaging UV radiation on the earth's surface, which would have led to the extinction of the dinosaurs.

A new hypothesis, however, has been proposed recently to explain the extinction of the dinosaurs. It seems to be more likely now that the extinction at the end of the Cretaceous period might have resulted from a collision between the Earth and a large asteroid (or meteor). The collision may have generated a huge cloud of dust and debris, large enough to have darkened the Earth for several years. As a result, photosynthesis nearly ceased, and the Earth's food chain collapsed. Dinosaurs were particularly vulnerable to such a disaster because of their physical size. New evidence suggesting the occurrence of such a collision between the Earth and a large meteor was obtained recently when a high concentration of iridium was discovered in the Cretaceous geological strata (RUSSELL, 1979; HSU, 1980; ALVAREZ *et al.*, 1980).

11.4 Ozone in the Martian Atmosphere

Mars is the only planet other than Earth, where ozone has been observed in the atmosphere. The Martian ozone was first detected in 1969 near the south polar region by a UV spectrometer on board the Mariner 7 flyby spacecraft (BARTH and HORD 1971). Later extensive measurements by the Mariner 9 orbiter (BARTH *et al.*, 1973) and observations of O_2 ($^1\Delta_g$) dayglow emission at 1.27 μm by Earth-based telescopes (NOXON *et al.*, 1976) revealed a large seasonal difference (by a factor of $\sim$100) in the ozone column density observed at high latitudes. The maximum occurs in late winter or early spring which is similar to the seasonal variation in the earth's stratospheric ozone. However, the percentage change is much larger on Mars than on Earth (the latter has only ~50% seasonal variation).

The large seasonal variation in the Martian ozone has been interpreted mainly in terms of a large seasonal variation in water vapor associated with a large seasonal variation in atmospheric temperature. This is in contrast to the earth's ozone, whose seasonal variation occurs by the transport effect. The Martian atmosphere is thin (the surface pressure is ~10 mb, which correspond to that at an altitude of ~30 km on earth) and there are no oceans on the surface; thus, the atmospheric temperature of Mars is affected directly by the incoming solar energy flux and undergoes large seasonal and diurnal variations. This was confirmed by observations from the Viking Landers (HESS *et al.*, 1977). Viking orbiter data also show large variations in water vapor with latitudal, as well as seasonal and diurnal variations (FARMER *et al.*, 1977).

Seasonal models have been published concerning Martian atmospheric ozone (PARKINSON and HUNTEN, 1972; MCELROY and DONAHUE, 1972; LIU and DONAHUE, 1976; KONG and MCELROY, 1977; KRASTITSKY, 1978; SHIMAZAKI and SHIMIZU, 1979). All basically agree in showing that water vapor controls the amount of ozone present in the Martian atmosphere. Water vapor is the ultimate source of HO_x, which directly affects the ozone concentration via catalytic reactions that reduce the odd-oxygen density. Since water vapor is almost entirely condensed out at very low temperatures in winter on Mars (~150 K), little HO_x is produced and the ozone density should be strongly enhanced from the summer value.

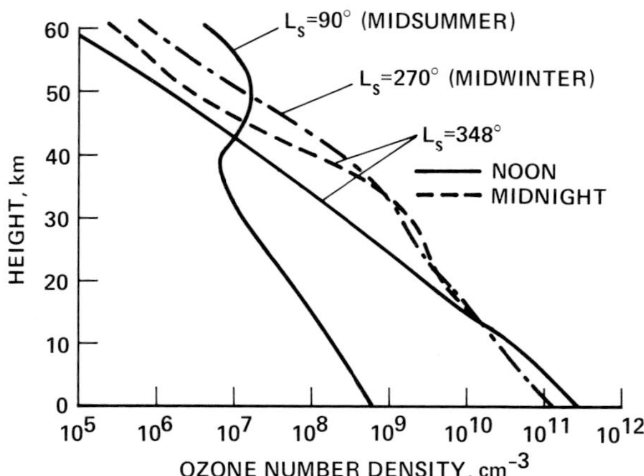

FIG. 11.15. Height profiles of ozone density on Mars for different seasons, predicted with a seasonal/diurnal model (SHIMAZAKI, 1981).

The ozone density profiles predicted with a seasonal/diurnal model are illustrated for different seasons in Fig. 11.15 (SHIMAZAKI, 1981). Below ~ 40–50 km the ozone density in midwinter is more than two orders of magnitude larger than that in midsummer; above these heights the summer density is larger than the winter density. This implies that a large O_x reduction by HO_x catalytic reactions in summer occurs only below ~40–50 km.

There is a small diurnal variation for both midsummer and midwinter; this is because the midsummer is sunlit and the midwinter is in complete darkness throughout the entire day. A considerable diurnal variation can be seen, however, in other seasons, for instance, in late winter as is shown for $L_s=348°$ in Fig. 11.15 (L_s is the aerocentric longitude, whose zero value corresponds to the spring equinox in the Northern Hemisphere). Above ~ 10 km the midnight $[O_3]$ is substantially larger than the noon value. Below this height the amount of O available for production of O_3 is too small to make the nighttime increase in $[O_3]$ appreciable.

Figure 11.16 shows the variation in total column ozone density from fall to early summer calculated with a seasonal/diurnal model. During the midsummer season the predicted column density is too small to be shown here.

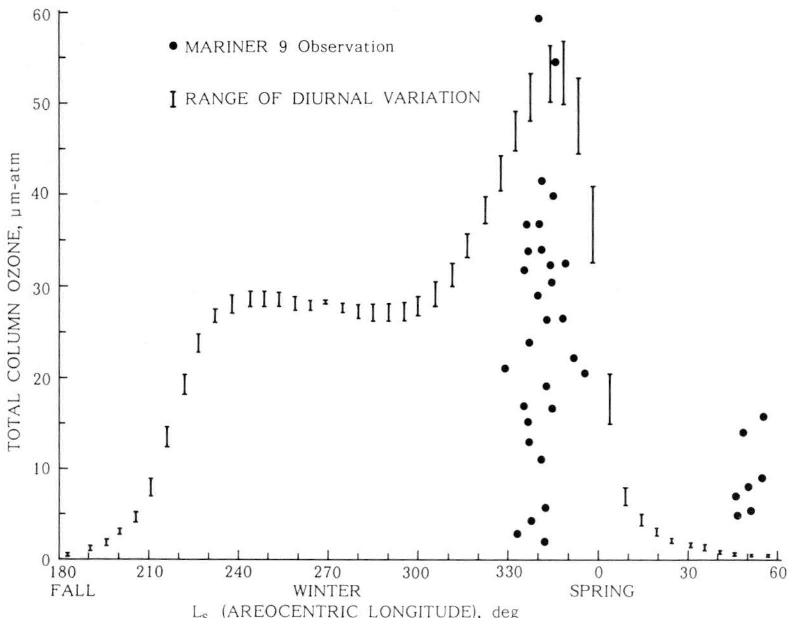

FIG. 11.16. Seasonal variations in the column ozone density of the Martian atmosphere predicted with a seasonal/diurnal model, are compared with Mariner 9 observations.

The model includes the effects of both seasonal and diurnal variations in atmospheric temperature, water vapor, and solar zenith angle. The model predicts the time of enhanced ozone density and its magnitude to be close to those observed by Mariner 9, but the observational data are so scattered that a detailed comparison of the model with observations would be a difficult task.

Most observations of Martian ozone have been made in late winter. During the summer the ozone column density is less than 3 μm atm, the detection limit of the Mariner 9 experiments. The large variability of the observed ozone column density on Mars may have been caused by a variable condition near the surface, where most of the ozone exists. Rapid changes in the amount of water vapor near the surface could occur, for instance, by changes in the cloud morphology and by the motion of cold air masses containing less water vapor. These could be reasons for the large, rapid variations in ozone from day to day and from place to palce (OWEN *et al.*, 1977; WEHRBEIN *et al.*, 1979). The surface loss of O_x due to the chemically active Martian surface (HUGUENIN, 1973) could be another factor affecting the ozone density.

Most Martian ozone has been observed at high latitudes. The Mars-5 satellite, however, first detected low-latitude ozone indicating a maximum density of $\sim 10^{10}$ cm^{-3} at an altitude of $\sim$38 km (KRASNOPOLSKY*et al.*, 1977). The existence of such a high ozone density may suggest that the Martian atmosphere is extremely dry and cold (<140 K) at $\sim$40 km at low latitudes (SHIMAZAKI, 1981).

Because of the small amount of ozone (at most $\sim$60 μm atm) Martian ozone generally should not play a role in atmospheric heating. However, it may play a significant role in determining the rate of deposition of CO_2 over the polar caps in winter, thus affecting the surface pressure (KUHN *et al.*, 1979).

Chapter 12

RADIATION IN THE MIDDLE ATMOSPHERE

Radiative energy transfer plays important roles in the middle atmosphere. Absorption of the solar UV radiation and emission of the infrared (IR) atmospheric radiation are the most important net heat source which maintains the climatological stationary, three-dimensional temperature structure of the middle atmosphere. Difference of heating, e.g. excess heating in lower latitudes over high latitudes in equinoxes and in the summer hemisphere over the winter hemisphere in solstices, is the primary driving force for the global mean circulation of the middle atmosphere. These large-scale atmospheric motions transport the trace gases and influence their distributions, which affect the solar heating and atmospheric emission, and therefore the net radiative heating and cooling.

Solar UV radiation also activates the atmospheric molecules in the middle atmosphere and produces ozone and other trace gases. These photochemical products are, in turn, responsible for the radiative energy budget in the middle atmosphere. Thus, the middle atmosphere forms a complex system in which atmospheric motion, radiative process and photochemical process are strongly interacting.

The radiative transfer in the middle atmosphere is by itself a complex process due to the wide range of the vertical distributions of pressure, temperature and absorbing gases. While a general discussion of the radiative transfer should be instructive for understanding the physical foundations (Goody, 1964; Kondratyev, 1969), the discussion in this chapter is focused mainly on the technical aspect in implementing the radiative heating and cooling calculations for the climatological and dynamical studies of the middle atmosphere. A brief discussion is then given on radiative-photochemical-dynamical interaction in the middle atmosphere.

12.1 Heating by Short Wave (Solar) Radiation

The heating of the atmosphere by solar radiation of wavelengths shorter than ~4 μm is the ultimate energy source driving atmospheric motion. About 99% of the solar radiation energy is contained in this spectral range with a maximum intensity at ~0.4828 μm which corresponds approximately to 6000 K black body radiation. The main radiative processes of solar radiation within this spectral range in the earth's atmosphere are absorption, reflection, and scattering; emission is entirely negligible because the atmospheric temperature is too low compared to the solar temperature.

A large part of the absorbed energy is used for dissociation, excitation and ionization of atmospheric molecules or atoms; among them dissociations of O_3 and O_2 are particularly important in the middle atmosphere. The energy of direct heating is therefore the total energy absorbed minus dissociation energy. Figure 12.1 compares the total energy absorbed by O_2 and O_3 and the energy of direct heating predicted for the winter mesosphere and lower thermosphere (Fukuyama, 1974). It is evident that only 10–20% of the total absorbed energy is used for direct heating.

The energy stored by the dissociation products, however, will be released quickly as heat through chemical reactions and also through deactivation by quenching if the products are in excited states. As is seen in Fig. 12.2, heating by chemical reactions and quenching minus cooling by airglow emission compensates for most of the energy spent for dissociation. As a result, the entire absorbed solar energy eventually would be used for atmospheric

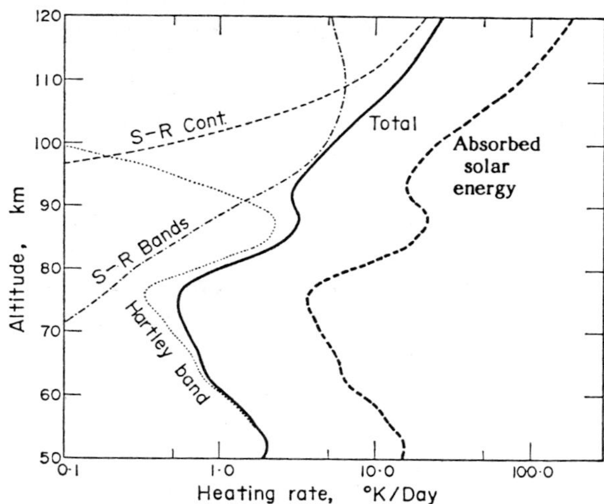

Fig. 12.1. Comparison of direct solar UV heating due to O_3 and O_2 absorption and total heating including dissociation and excitation energies (Fukuyama, 1974).

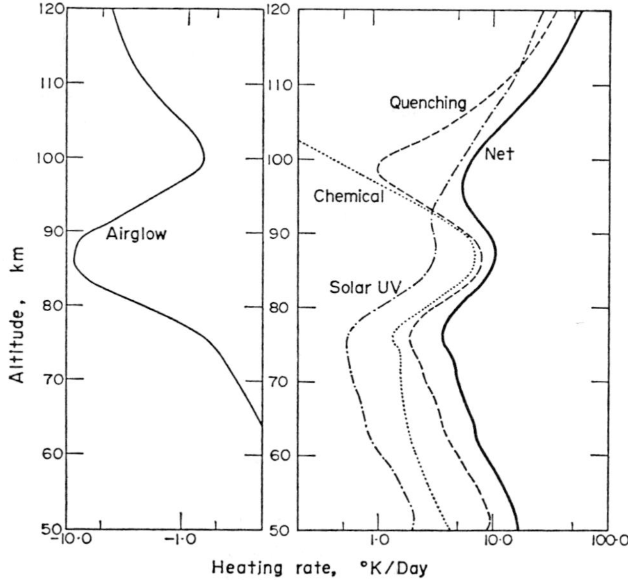

FIG. 12.2. Heating rates due to direct solar UV heating, chemical reactions, and quenching (deactivation of excited species), and cooling rates by airglow emission. The total is close to the total absorbed solar energy shown in Fig. 12.1 (FUKUYAMA, 1974).

heating in the stratosphere and mesosphere.

The radiative heating rate by individual molecule is tranditionally calculated by the rate of temperature change per unit time defined by

$$\frac{\partial T}{\partial t} = \frac{Q}{\rho c_P} = -\frac{1}{\rho c_P}\frac{\partial S}{\partial s} \tag{12.1}$$

where Q is the radiative heating (absorbed energy) per unit volume per unit time, ρ is the air density, and c_p is the specific heat at constant pressure. The heat capacity of the air is given by

$$\rho c_P = k \sum_i \left(1 + \frac{N_i}{2}\right) n_i \tag{12.2}$$

where k is Boltzmann's constant, and N_i is 5 for N_2 and O_2. S in (12.1) is the radiative flux of the absorbed energy and is expressed as

$$S(u) = \sum_\lambda I_\infty(\lambda) \cdot A(\lambda, u) \tag{12.3}$$

where I_∞ is the incident solar energy flux at the top of the atmosphere, u is the amount of absorber in the slant column of unit area given by

$$u = \int_s^\infty n \, ds = \int_z^\infty n \sec \chi \, dz \tag{12.4}$$

and A is absorptivity, which is the ratio of absorbed to incident radiation and is a function of u and wavelength λ.

The total energy flux is also calculated from (12.1) as follows:

$$S = \sum_\lambda S_\lambda = -\int_\infty^s \sum_\lambda Q_\lambda \, ds = \sum_\lambda \int_0^u \frac{Q_\lambda}{n} \, du \tag{12.5}$$

If A is given by an exponential function

$$A = 1 - e^{-\sigma(\lambda) \cdot u} \tag{12.6}$$

where σ is the absorption cross section, we can calculate a specific heating, the integrand in the rightmost equation of (12.5), as a function only of u (not including n) as follows:

$$q_\lambda = \frac{Q_\lambda}{n} = I_\infty(\lambda) \cdot \sigma(\lambda) e^{-\sigma(\lambda) \cdot u}. \tag{12.7}$$

The main absorbing molecules are O_3 in the stratosphere and mesosphere, H_2O in the troposphere and O_2 in the thermosphere. CO_2 is of secondary importance as an absorber throughout the region from the troposphere to the mesosphere. The absorption of each constituent is highly selective and strongly dependent on wavelength, and calculations of the heating rate taking into account the detailed wavelength dependence are formidable and very time consuming if not impossible. In order to avoid such inconvenience and to facilitate the heating rate calculations, empirical formulae have been developed for each trace gas by various investigators.

In what follows, while surveying the existing formulae for heating rate calculations for each constituent, we will try to develop a simple analytical formula that is applicable to heating rate calculations for all constituents. A polynomial is chosen for this purpose; since it was used in Chapter 4 for calculations of photodissociation rates of various molecules in the Schumann-Runge band system, one can calculate the heating rate and the photolysis rate by the same simple formula, thereby simplifying the numerical scheme and saving computation time for model calculations to investigate the effect of coupling between radiation and photochemistry. Later, a similar approach will also be used for cooling rate calculations.

12.1.1 Heating Due to O_3 and O_2 absorption

Since the absorption cross section σ of ozone has been well determined for the entire wavelength range, the accurate (exact) heating rate can be calculated straightforwardly by (12.3) or (12.5), using the absorptivity or the specific heating expressed by an exponential function as are given in (12.6) and (12.7). The result is summarized in Fig. 12.3 for the three O_3 bands i.e. the Hartley, the Huggins, and the Chappuis bands and their sum. It is seen in the figure that the solar energy is absorbed mainly in the Hartley band for u less than ~0.1 cm, whereas the absorption occurs mainly in the Chappuis band for u greater than ~1 cm. In the transition region there is a depression for the sum of the three bands.

A similar calculation can also be done for O_2 absorption for the Schumann-Runge continuum and the Herzberg continuum. For the Schumann-Runge band, however, the exponential term in (12.6) and (12.7) must be replaced by the transmission P calculated in (4.17). Heating due to O_2 absorption becomes increasingly important in the thermosphere. Photoabsorption in the Schumann-Runge bands provides the largest heat source up to ~ 110 km, above which the Schumann-Runge continuum is the dominant heat source (see Fig. 12.1). The effective heating efficiency of O_2 photodissociation in the thermosphere is evaluated to be ~0.3 (IZAKOV

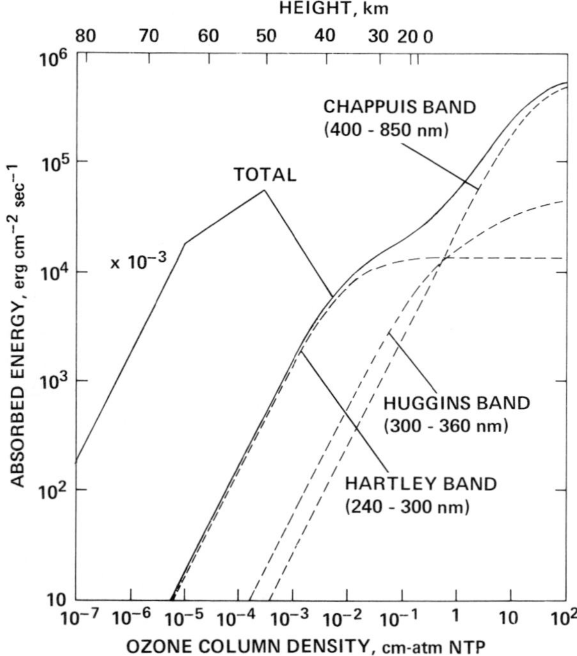

FIG. 12.3. Absorption by O_3 at the Hartley, Huggins and Chappuis bands as functions of ozone column density. The corresponding height measure under overhead sun condition at the equator, equinoxes is shown at the top abscissa.

and MOROV, 1971; CHANDRA and SINHA, 1973).

The solar energy deposited above ~100 km by O_2 photodissociation is released as heat at lower heights by the three-body recombination $O+O+M \rightarrow O_2+M$ after the dissociation products, O atoms, are transported down into the lower thermosphere by molecular and eddy diffusion. Similarly, the chemical energy can be transported horizontally through the mesosphere and lower thermosphere from the summer to the winter hemisphere with the transport of O atoms (KELLOGG, 1961).

The absorptivity of O_3 for the entire spectral range is approximated in the following form by SASAMORI *et al.* (1972) based on calculations of MANABE and STRICKLER (1964):

$$A = 0.045(u + 8.34 \times 10^{-4})^{0.38} - 3.1 \times 10^{-3} \tag{12.8}$$

where u is the thickness in cm atm NTP of the O_3 column density. Multiplying this by the total incident solar energy (~2 cal cm^{-2} min^{-1} or ~ 1.39×10^6 erg cm^{-2} sec^{-1}) gives the total flux of energy absorbed by O_3.

LACIS and HANSEN (1974) expressed the O_3 absorptivity in two different spectral ranges; one is for UV radiation

$$A = \frac{1.082u}{(1 + 138.6u)^{0.805}} + \frac{0.0658u}{1 + (103.6u)^3} \quad (10^{-4} \leq u \leq 1) \tag{12.9}$$

and the other is for visible radiation

$$A = \frac{0.02118u}{1 + 0.042u + 0.000323u^2} \quad (10^{-4} \leq u \leq 10). \tag{12.10}$$

The incident solar energy for the two ranges are $\sim 5.846\times10^4$ and ~ 5.49×10^5 erg cm^{-2}sec^{-1}, respectively.

In calculating the specific heating rate for O_3 absorption given by (12.7) LINDZEN and WILL (1973) proposed using the average values of I_∞ and σ for both the Hartley and the Chappuis bands. For the Huggins band they have given the following formula:

$$q_H = \frac{(I_\infty)_H}{Mu}\{e^{-\sigma_H u e^{-M\lambda_L}} - e^{-\sigma_H u e^{-M\lambda_S}}\} \tag{12.11}$$

where the suffix H refers to the values in the Huggins band, λ_L and λ_S are the wavelength limits for the band, and $M(=0.0126)$ is a constant.

Using equations similar to those used by LINDZEN and WILL (1973), STROBEL (1978) has determined new coefficients for calculating the heating rates in the three O_3 absorption bands. He has also developed a similar method for calculating the heating rate due to O_2 absorption.

We now develop a new method that can calculate the ozone absorption in *any* absorption band by a *same simple analytical formula* similar to (4.17), i.e.

$$\log S = \sum_{i=1}^{K} C_i (\log u)^i. \tag{12.12}$$

The coefficients C_i are determined by the least squares method so that the exact calculation of S by (12.5) as a function of u may be fitted with the equation (12.12). Resultant coefficients for the seventh-order polynomial are given in Table 12.1 for each of three O_3 absorption bands, and in Table 12.2 for each of three O_2 absorption, i.e. the Schumann-Runge continuum, the Schuman-Runge band, and the Herzberg band. The O_3 absorption calculated using our method is compared in Fig. 12.4 with the results by other formulae discussed earlier. Agreement is generally good, although our formula gives slightly larger absorption than most other formulae.

In our approach it is easy to recalculate a set of new coefficients if the input solar energy changes by, for instance, solar cycle variations or the change in sun-earth distance (the earth is about 3.28% closer to the sun in early January than in early July), or if the values are revised by new observations. For example, if the incident energy changes from I_∞ to a new value I'_∞, our method simply requires us to modify the coefficient C_0 to the new value

$$C'_0 = C_0 + \log (I'_\gamma / I_\gamma) \tag{12.13}$$

whereas all other coefficients remain unchanged.

Our formula can also calculate the heating by *diffuse radiation* reflected or scattered from the ground or from clouds. Using a known (or assumed)

TABLE 12.1 Coefficients of the seventh-order polynomial (12.12) to be used for calculations of heating rate due to O_3 absorption.

Spectrum range	Hartley band (240–300 nm)	Huggins band (300–360 nm)	Chappuis band (400–850 nm)
I_∞^*	1.406(4)**	5.915(4)	7.365(5)
C_0	4.14292(0)	4.23361(0)	4.43347(0)
C_1	−5.33910(−2)	4.12314(−1)	9.50435(−1)
C_2	1.38771(−2)	−1.60939(−1)	−6.23836(−2)
C_3	5.36080(−2)	1.26357(−2)	−3.20015(−2)
C_4	−8.90747(−3)	9.98976(−3)	−7.37424(−3)
C_5	−6.07296(−3)	6.52087(−4)	−6.81799(−4)
C_6	−8.42669(−4)	−1.25889(−4)	−1.89154(−6)
C_7	−3.75006(−5)	−1.27023(−5)	2.17012(−6)

*I_∞ is the incident solar energy (erg cm^{-2} sec^{-1}).
**$A(b)$ should read as $A \times 10^b$.

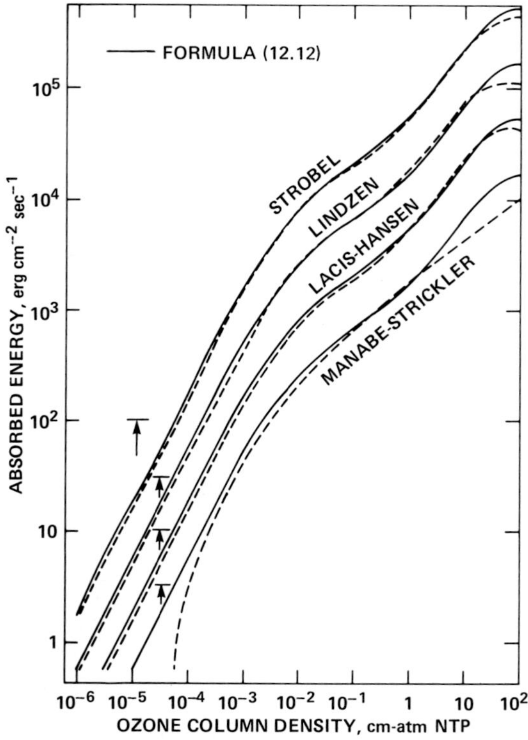

FIG. 12.4. Comparison of ozone absorption calculated by our method (shown by solid curves) with those calculated by various other formulae. Arrows indicate the ordinate at 10^2 erg cm^{-2} sec^{-1} for each pair of curves under comparison.

TABLE 12.2 Coefficients of the seventh-order polynomial (12.12) to be used for calculations of heating rate due to O_2 absorption.

Sepctrum range	Schumann-Runge continuum (135–175 nm)	Schumann-Runge band (175–205 nm)	Herzberg band (205–240 nm)
I_∞^*	7.0853(0)**	1.1408(2)	1.6873(3)
C_0	8.61249(−1)	6.82461(−1)	−6.58235(−1)
C_1	−1.19593(−2)	4.69617(−1)	9.97771(−1)
C_2	−2.55548(−2)	−9.29862(−2)	−8.92023(−4)
C_3	2.45084(−2)	1.90149(−2)	9.84498(−4)
C_4	−4.70005(−3)	4.88367(−3)	3.68662(−4)
C_5	−5.26400(−4)	−1.46366(−3)	−1.15262(−4)
C_6	1.47471(−4)	−1.01445(−4)	−5.95621(−5)
C_7	4.35660(−7)	2.54485(−5)	−6.10814(−6)

*I_∞ is the incident solar energy (erg cm$^{>872}$ sec^{-1}).
**$A(b)$ should read as $A \times 10^b$.

albedo, the incident energy flux for the reflected (upward) radiation is calculated by multiplying the albedo to the intensity of the downward radiation flux arriving at the reflection heights. The coefficient C_0 is then adjusted to that incident energy flux by means of (12.13). The angular integration for the diffuse radiation is taken into account approximately by applying a diffusive or air mass factor of 1.66 to u (see Appendix I).

12.1.2 Heating Due to H_2O, CO_2 and NO_2 absorption

HOWARD *et al.* (1956 a and b) have summarized their experimental results for the integrated absorption function for each absorption band of H_2O and CO_2 into the following formulae:

$$A = \int_{\Delta\lambda} A_\lambda \, d\lambda = cu^d (P + e)^k \tag{12.14}$$

for weaker absorption (smaller u), and

$$A = \int_{\Delta\lambda} A_\lambda \, d\lambda = C + D \log u + K \log (P + e) \tag{12.15}$$

for stronger absorption (larger u), where u is the column density of H_2O in g cm^{-2} or of CO_2 in cm atm NTP, and P and e represent the total pressure and the partial pressure of H_2O (or CO_2), respectively. Constants c, d, k, C, D, and K determined for each band based on experimental data on absorption are given in Table 12.3 (HOWARD *et al.*, 1956 a and b; BURCH *et al.*, 1962). The weaker absorption (12.14) should apply when A is smaller than a value A_c that is also given in Table 12.3, whereas the stronger absorption (12.15) should apply when A is larger than A_c.

The absorbed solar energy fluxes calculated using these equations are shown in Fig. 12.5 for each of seven H_2O bands (0.94, 1.1, 1.38, 1.87, 2.7, 3.2 and 6.3 μm) and their sum. Discontinuities are seen on some curves in the transition region where $A \approx A_c$, due to a small difference between the calculated values by (12.14) and (12.15). However, there is virtually no discontinuity, if the seven absorption bands are integrated. This variation of the sum is fitted to equation (12.12) for the range of u between 10^{-7} and 10^2 g cm^{-2}; the coefficients thus determined are given in Table 12.4.

Many formulae have been proposed to express the solar energy absorbed by H_2O. The classical formula of MÜGGE and MÖLLER (1932)

$$S = 0.172\,(mu)^{0.3} \quad \text{cal cm}^{-2}\,\text{min}^{-1} \tag{12.16}$$

is essentially a first order polynomial of the type of equation (12.12); changing the unit to erg cm^{-2}sec^{-1}, the equation can be written as

$$\log S = 5.079 + 0.303 \log mu \tag{12.17}$$

TABLE 12.3 Parameters of the formulae (12.14) and (12.15) for calculating the absorption by H_2O and CO_2. Sets of values with $d=0.5$ are from HOWARD *et al.* (1956 a and b), and those with $d \neq 0.5$ from BURCH *et al.* (1962).

H_2O band μ	spectral range cm^{-1}	c	d	k	C	D	K	A_c	k/d	K/D
6.3	1150–2050	356	0.5	0.30	302	218	157	160	0.60	0.72
3.2	2800–3300	40.2	0.5	0.30				500	0.60	
2.7	3340–4400	316	0.5	0.32	337	246	150	200	0.64	0.61
1.87	4800–5200	152	0.5	0.30	127	232	144	275	0.60	0.62
1.38	6500–8000	163	0.5	0.30	202	460	198	350	0.60	0.43
1.1	8300–9300	31	0.5	0.26				200	0.52	
0.94	10100–11500	38	0.5	0.27				200	0.54	
CO_2 band										
15	550–800	3.16	0.5	0.44	−68	55	47	50	0.88	0.85
10.4	750–1000	0.016	0.78	0.20				35		
9.4	1000–1110	0.023	0.75	0.22				40		
5.2	1870–1980	0.024	0.5	0.4				30	0.80	
4.8	1980–2160	0.12	0.5	0.37				60	0.74	
4.3	2160–2500	15.0	0.54	0.41	27.5	34	31.5	50	0.76	0.93
2.7	3480–3800	3.15	0.5	0.43	−137	77	68	50	0.86	0.88
		3.5	0.58	0.38					0.66	
2.0	4750–5200	0.492	0.5	0.39	−536	138	114	80	0.78	0.82
1.6	6000–6550	0.063	0.5	0.38				80	0.76	
1.4	6650–7250	0.058	0.5	0.41				80	0.82	

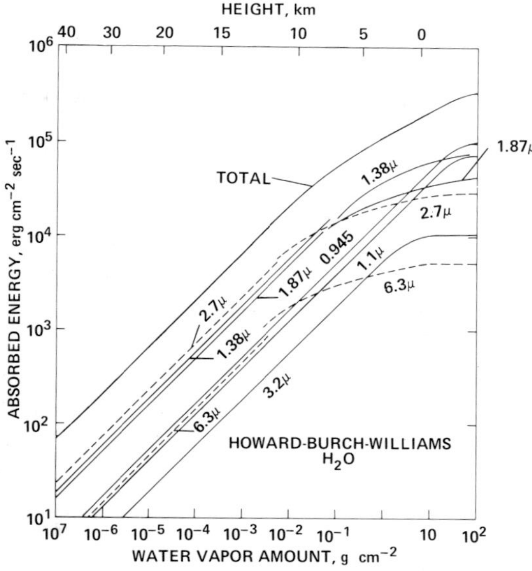

FIG. 12.5. Absorption by H_2O at seven bands in the range of 0.945–6.3 μm predicted with the formulae (12.14) and (12.15). The height scale is for $\chi=0°$.

TABLE 12.4 Coefficients of the seventh-order polynomial (12.12) to be used for calculations of heating rate due to absorption by H_2O, CO_2 and NO_2.

Spectrum range	H_2O (0.945–6.3 μm)	CO_2 (1.4–14.7 μm)	NO_2 (300–710 nm)
I_∞^*	3.292(5)**	1.162(5)	6.275(5)
C_0	5.05448(0)	3.53769(0)	5.79479(0)
C_1	2.91793(−1)	3.70827(−1)	1.74092(−1)
C_2	−8.46612(−3)	−5.91767(−2)	−1.60064(−1)
C_3	1.37175(−2)	9.35607(−4)	3.82587(−2)
C_4	−4.03685(−3)	3.04431(−3)	1.02528(−2)
C_5	−3.04987(−3)	2.00068(−4)	−1.61624(−3)
C_6	−5.10273(−4)	−5.18424(−5)	−5.63677(−4)
C_7	−2.72098(−5)	−5.24546(−6)	−3.71916(−5)

*I_∞ is the incident solar energy (erg cm^{-2} sec^{-1}).
**$A(b)$ should read as $A \times 10^b$.

where m is the air mass factor. Our corresponding equation is

$$\log S = 5.007 + 0.429 \log mu \tag{12.18}$$

KORB *et al.* (1957) have improved (12.17) and obtained a third order polynomial

$$\log S = 5.1036 + 0.347 \log mu - 0.056(\log mu)^2 - 0.006(\log mu)^3. \tag{12.19}$$

which can be compared with our third order polynomial

$$\log S = 5.0791 + 0.3150 \log mu - 0.0416(\log mu)^2 - 0.0029(\log mu)^3. \tag{12.20}$$

The result from the Korb formula agrees better with the data of Howard *et al.* than does the result of Mügge and Möller. Our seventh-order polynomial, however, indicates 2.4 times smaller mean error than the third-order polynomial (SHIMAZAKI and HELMLE, 1979). Our result also agrees favorably with those of SASAMORI *et al.* (1972) and LACIS and HANSEN (1974); both have developed formulae for the H_2O absorption based on results evaluated with the radiation chart of YAMAMOTO (1952). Sasamori's formula is given by

$$A = 0.110(mu + 6.31 \times 10^{-4})^{0.3} - 0.0121 \tag{12.21}$$

whereas Lacis and Hansen give the following formula:

$$A = \frac{2.9mu}{(1 + 141.5mu)^{0.635} + 5.925mu}. \tag{12.22}$$

The absorption in each of the eight CO_2 bands (1.4, 1.6, 2.0, 2.7, 4.3, 4.8, 5.2 and 14.7μm) and their sum calculated by Howard's equations (12.14) and (12.15) is illustrated in Fig. 12.6. Again, there are discontinuities on each curve in the transition region where $A \approx A_c$; in case of CO_2 the discontinuity is still significant even for the curve of the sum of eight bands. After smoothing out this discontinuity we have curve fitted this variation to equation (12.12) for the entire range of u. The coefficients determined for the seventh-order polynomial are tabulated in Table 12.4. The result calculated from our formula compares fairly well with those by other formula, for instance, by the CO_2 absorptivity developed by SASAMORI *et al.* (1972)

$$A = 2.35 \times 10^{-3}(u + 0.0129)^{0.26} - 7.5 \times 10^{-4} \tag{12.23}$$

which approximates the data of BURCH *et al.* (1960).

Table 12.4 also contains the coefficients applicable to the calculation of heating due to NO_2 absorption; these coefficients have been determined by curve fitting (12.12) to the absorbed solar energies calculated by the formulae of LUTHER (1976), who developed a formula similar to (12.7) for 300–475 nm and a formula similar to (12.11) for 475–710. nm.

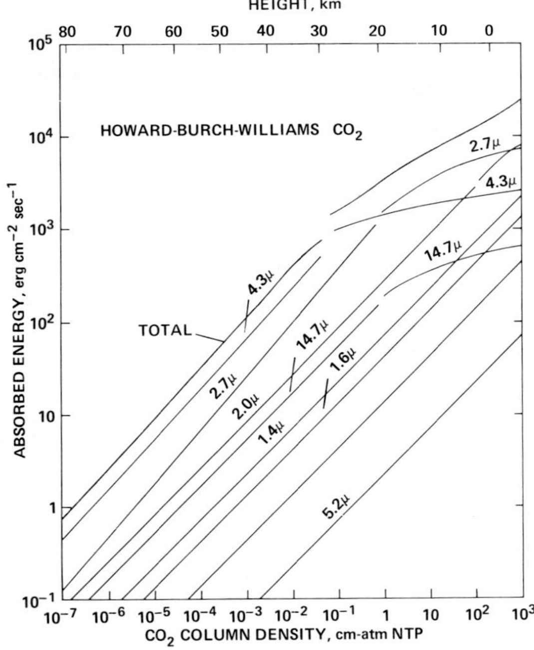

FIG. 12.6. Absorption by CO_2 eight bands in the range of 1.4–14.7 μm predicted with the formulae (12.14) and (12.15). The height scale is for $\chi=0°$.

12.1.3 Global distribution of heating rate

In this section are presented the global distributions of the solar heating rate computed using our method for representative distributions of the absorbing gases from the surface up to 90 km. We have taken the ozone distribution from Fig. 8.3(a) and (b). For H_2O we assume a mixing ratio of 3.5×10^{-6} above the tropopause, below which the relative humidity is assumed to vary linearly from 0.8 on the ground surface to the value corresponding to the mixing ratio 3.5×10^{-6} at the tropopause. A constant mixing ratio of 3.3×10^{-4} is assumed for CO_2 for the entire atmosphere. The distribution of NO_2 is taken from the Ames 2-D model (W. J. Borucki, private communication); the vertical distribution is essentially similar to the curve of NO_2 in Fig. 7.3 (a) and has a maximum mixing ratio at ~35–40 km.

The absorption rate is primarily a function of the density of absorbing gases, but it is also influenced by the atmospheric condition in which the absorbing gas is embeded. The latter effect arises from the change of the absorption cross section with atmospheric pressure and temperature. The effect is particularly significant in the near-infrared absorption of H_2O and

CO_2, since the bandwidth of rotational absorption lines of these molecules varies in proportion to the collisional frequencies of these molecules with N_2 and O_2. In addition to the collisional broadening, the bandwidth of infrared absorption lines changes through the Doppler effect associated with the random thermal motion of absorbing gases. However, the Doppler broadening becomes important, relative to the collisional broadening, only above ~30 km where the water vapor absorption is insignificant. Thus, here we consider only the effect of collisional broadening in an approximate manner for H_2O and CO_2 absorption.

Since the collision frequency is proportional to the pressure and inversely proportional to the square root of temperature, and since the absorption cross section appears in the radiative calculation scheme always as a multiplication factor to the density of absorbing gas, it is possible to make an appropriate correction for the absorbing gas density instead of making corrections for the absorption cross section. Thus, when H_2O and CO_2 distribute as a function of pressure and temperature, their effective absorption path lengths may be defined by

$$u_{\mathrm{eff}} = \int \left(\frac{p}{p_0}\right)^r \left(\frac{T_0}{T}\right)^{\frac{1}{2}} \mathrm{d}u \qquad (12.24)$$

where P_0 and T_0 represent normal standard pressure and temperature. For H_2O and CO_2 the power r can be evaluated from experimentally determined values for k/d in (12.13) and K/D in (12.14); the average values for all bands under consideration is ~0.6 for H_2O and ~0.8 for CO_2 (see Table 12.3). In this way the absorption data obtained at the standard pressure and temperature in the laboratory can be applied to calculations of IR absorption of H_2O and CO_2 at different pressure and temperature of the atmosphere. For UV absorption of O_3, O_2 and NO_2 there is no need to consider the effect of pressure and temperature changes; therefore, r can be set to zero.

The spherically-averaged global mean height profiles of the heating rate due to O_3 absorption calculated for March are shown in Fig. 12.7 (a) for three different O_3 bands. The heating rate is expressed by in unit of K day^{-1}. The ozone heating occurs predominantly in the Hartley band above ~30 km, below which the heating is mainly due to the Chappuis band. In the transition region, the heating in the Huggins band is also important. Heating rates due to absorption by CO_2, H_2O and NO_2 and the total heating including the O_3 absorption are shown in Fig. 12.7 (b). The result indicates that the O_3 heating is far more dominant over other components in the middle atmosphere, whereas heating is mainly due to the H_2O absorption in the troposphere.

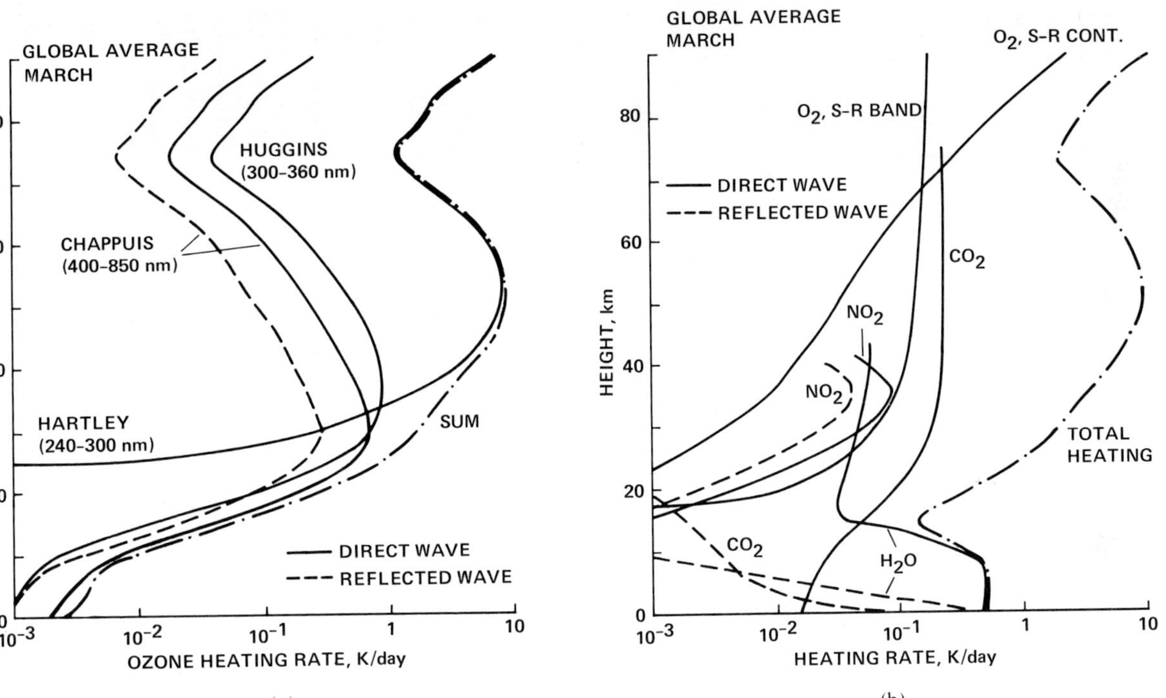

FIG. 12.7. (a) Height variations of the heating rate due to O_3 absorption at Hartley, Huggins, and Chappuis bands and their sum, representing the spherically-averaged global mean for March. (b) The same as Fig. 12.7(a) but showing the heating rate due to absorptions by H_2O, CO_2, NO_2, and O_2, and the total heating rate including the O_3 absorption.

Heating by reflected solar radiation is calculated assuming the global average of the effective albedo to be 0.25. The *effective albedo* takes into account the combined effect of reflection from clouds and the ground and is useful in calculating the heating by reflected radiation above the clouds. The approximation using effective albedo should give a reasonable heating rate in the middle atmosphere. It is essential for the troposphere, however, to take into account the cloud albedo and the surface albedo separately and to calculate the radiative heat fluxes between the cloud and the ground and also within the clouds.

The calculated heating rates by reflected radiation due to H_2O and CO_2 absorption are very small in the stratosphere (see Fig. 12.7 (b)). Furthermore, since the albedo of clouds is relatively small in the infrared and also since cloud water droplets absorb radiation almost at the same wavelengths as water vapor, the heating by reflected waves due to H_2O and CO_2 absorption should be even smaller than those calculated with the albedo 0.25; thus, the heating by reflected radiation may be ignored compared with the heating by direct radiation. On the other hand, the albedo of clouds is high in the visible and ultraviolet; therefore, heating by reflected radiation due to O_3 absorption is significant. In fact, this heating is comparable to the CO_2 heating by direct radiation in the lower stratosphere, and is about 22 percent of the O_3 heating by direct radiation at 25 km (see also LACIS and HANSEN, 1974).

It must be noted here that a great danger is involved in using our formula beyond the range of absorber amount not included for the analyses. Applicability of the formula (12.12) is limited to the range of u greater than 10^{-7} g cm^{-2} for H_2O and 10^{-7} cm atm NTP for O_3, CO_2 and NO_2; these critical values are met somewhere near 40 km for H_2O and ~80 km for CO_2 and O_3 (see the top abscissa of Figs. 12.3, 5, and 6), although these heights depend upon season and latitude and should be much higher near sunrise and sunset when the solar zenith angle is large. The heating rates in the regions above these critical heights are not shown in Fig. 12.7 (b) and are ignored in our calculation of atmospheric heating. Fortunately, however, contributions from this heating are generally small and their exclusion should not affect the global distribution of the *total* heating rate. The upper limits of the formula's applicability are 10^2 g cm^{-2} for H_2O, 10^2 cm atm NTP for O_3 and NO_2 and 10^3 cm atm NTP for CO_2. The column densities of these gases in the atmosphere are usually within these limits.

The heating rate due to CO_2 and H_2O absorption represented in terms of °K per day is nearly constant above certain heights. This phenomenon is caused by the assumption of a constant mixing ratio, that is, these constituents' concentrations change with height in proportion to the mass density of the atmosphere. Since the heat capacity also decreases with height in proportion to the mass density of the atmosphere, the rate of change of temperature should remain nearly constant with height, although the actual absorbed energy decreases with increasing height.

The latitudinal distribution of the diurnal heating rate is illustrated in Fig. 12.8 (a) and (b) for March and June, respectively, in terms of temperature change per day. Note that there is no solar radiation at night; therefore, the calculated values are the total heating during the sunlit hours of a day. For March, the global maximum of heating rate occurs at around 50 km in lower latitudes. The distribution of heating rate is almost symmetrical around the equator at levels below ~65 km, but a marked asymmetry appears in higher regions. This is certainly caused by the irregular distribution of ozone in that same area (see Fig. 8.3(a)). However, there is a general tendency that the heating rate takes the minimum at ~75 km where the ozone density is minimum, and that it increases above that height as the ozone density increases. The O_2 heating should also contribute to the increased heating rate at highest regions above ~80 km. It is also noted that the general pattern of the distribution of heating rate in the mesosphere is more closely related to the distribution of the ozone mixing ratio rather than the O_3 density itself as is learned by comparison between Fig. 12.8(a) and Fig. 8.3(c).

In June, the heating rate is generally larger over the summer hemisphere where the solar radiation is more intense. A maximum heating rate of ~15 K day^{-1} occurs at ~50 km near the north (summer) pole (see Fig. 12.8(b)). This is certainly due to the longer sunlit hours of a day at high latitudes in summer. Actually, in summer solstice the area at latitudes higher than 66.5° is sunlit all day long. A very large heating rate ($>$30 K day^{-1}) appears above ~85 km again near the north pole. This should be related partly to the very large O_3 density (mixing ratio) observed in that region (see Fig. 8.3(b) and (d)). Thus, the diurnal heating rate is affected strongly by the amount of atmospheric ozone (mixing ratio) as well as the intensity of solar radiation and the duration of sunlit hours.

The global distributions of the heating rate presented here are essentially similar to those of previous studies (e.g. PARK and LONDON, 1974, WANG *et al.*, 1982); however, the result for the mesosphere is much improved because we use the global O_3 distribution recently observed by the Solar Mesosphere Explorer satellite (see Section 8.3.1).

12.2 Radiative Transfer of Long Wave (Terrestrial) Radiation and Atmospheric Cooling

The maximum terrestrial radiation at atmospheric temperatures (~300 K) occurs near 10 μm (Wien's displacement law, see Appendix B), which is

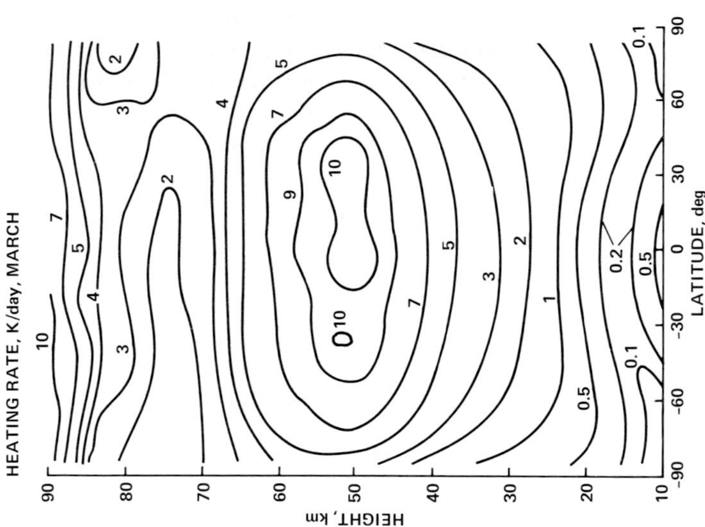

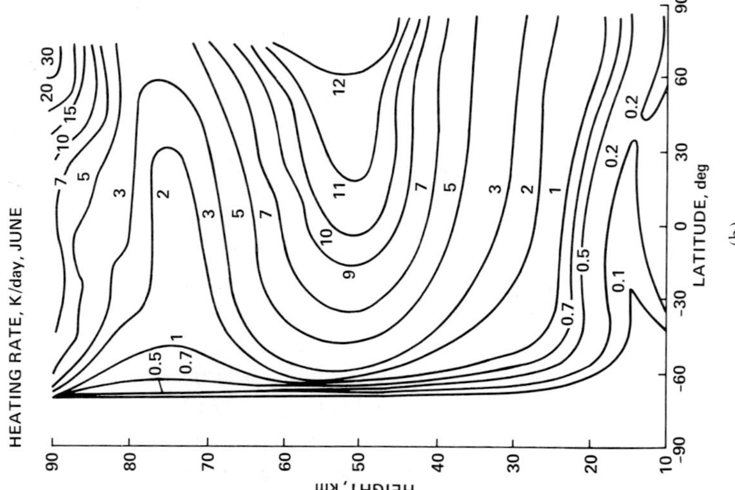

FIG. 12.8. (a) Contour plots in the meridional plane of the zonal-mean diurnal heating rate for March. (b) The same as Fig. 12.8(a) but for June.

far longer than the wavelength of the maximum incoming solar radiation ($\sim$0.48 μm). Since atmospheric molecules, particularly CO_2, H_2O, and O_3, have strong absorption bands in the 5–15 μm spectral range (see Fig. 12.9), re-emission of energy from these molecules is fundamental in this part of spectrum of the IR radiation.

12.2.1 Radiative transfer equation

The radiation loses its energy by gaseous absorption as it passes through the atmospheric layers and gains energy by emission from gases at the same wavelength (Kirchhoff's law). The net change in the intensity of radiation I is governed by the equation of radiative transfer (or Schwarzschilds' equation) given by (3.20). For our present purposes, we write it here in the form

$$\mu \frac{dI_\lambda(u, \mu)}{du} = k_\lambda(I_\lambda(u, \mu) - J_\lambda(T)) \tag{12.25}$$

where $\mu = \cos\theta$ (θ is the zenith angle of the ray path), u is the column abundance of absorbers in cm STP, K_λ is the absorption coefficient, and J is a source function. J can be represented rather accurately by the Planck balckbody function below $\sim$60 km, where a gas is in local thermodynamic equilibrium (LTE). The Planck function is a function of temperature and wavelength; the integral with wavelength over the band can be calculated with sufficient accuracy by (B.11) in Appendix B.

Equation (12.25) is usually solved for the upward propagating radiation and for the downward propagating radiation separately. Applying an angular integration with respect to θ to the solution of (12.25), one can calculate the upward and downward fluxes for diffuse radiation (see Appendix I). Then, the temperature change due to radiative transfer of terrestrial radiation may be calculated by an equation similar to (12.1) using the total of upward and downward fluxes for S. Normally, the process results in a cooling of the middle atmosphere. The absence of clouds in the middle atmosphere allows us to perform the radiative transfer calculation by neglec-

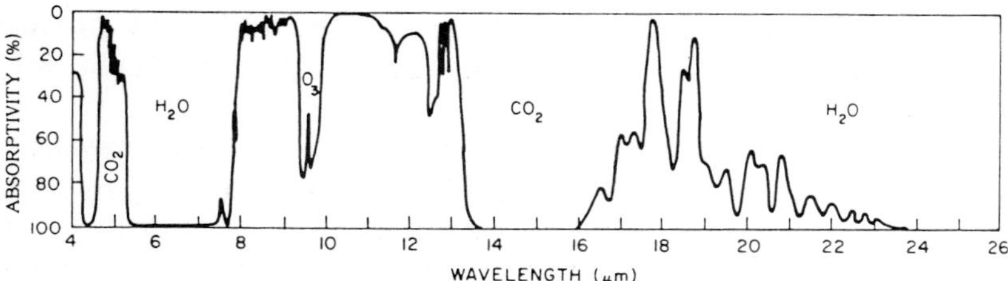

FIG. 12.9. Spectral absorptivity of the Earth's atmosphere at wavelengths in the terrestrial radiation regions; gases responsible for the main atmospheric absorption bands are indicated (COULSON, 1975).

ting the effect of scattering.

12.2.2 Models of absorption lines and bands

Solving the radiative transfer equation requires the transmission function of absorbers between the two heights, $T_\lambda\ (z,z')$. We now define the absorption function as a complement of T_λ as follows (cf. with (12.6)):

$$A_\lambda(z,z') = 1 - T_\lambda(z,z') = 1 - e^{-\frac{k_\lambda}{\mu L}\left|\int_z^{z'} n\,dz\right|} \tag{12.26}$$

where L is the Loschmidt number.The absorption function is a functional form of absorptivity.

In order to compute $A_\lambda\ (z,z')$ for a given spectral line in the atmosphere, it is necessary to describe k_λ as a function of wavelength λ (or frequency ν) for each line. Below ~35 km collisional processes determine the line shape and it is well represented by the Lorentz shape

$$k_\nu = \frac{s\alpha_L}{\pi\{(\nu - \nu_0)^2 + \alpha_L^2\}} \tag{12.27}$$

where s is the line intensity, α_L is the half-width of the line at one-half of the maximum of k_ν, and ν_0 is the frequency of the line center. For the pressure and temperature dependences of α_L, see (12.32) below.

In the upper layers of the atmosphere where the influence of collisions on line broadening decreases, the line shape is affected by the Doppler effect and has the form

$$k_\nu = \frac{s}{\alpha_D}\left(\frac{\ln 2}{\pi}\right)^{\frac{1}{2}} \exp\left\{-\left(\frac{\nu - \nu_0}{\alpha_D}\right)^2 \ln 2\right\} \tag{12.28}$$

where α_D is the doppler width of the line given by

$$\alpha_D = \sqrt{\frac{kT\nu_0^2 \ln 2}{m_i c^2}} \tag{12.29}$$

where k is the Boltzmann constant and m_i is the molecular mass of the absorbing gas. For comparable intensity and half-width the Doppler broadened line has more absorption near the center and less absorption in the wings than does a Lorentz line.

The most direct method to calculate the integral of A_λ over a wavelength interval is the *line-by-line* calculation, using parameters (ν_0, α_L, α_D and s) experimentally determined for each absorption line (McCLATCHEY *et al.*, 1972, 1973). Although the method should be most accurate, it is time consuming and suffers from the effect of overlapping of neighboring lines as well as uncertainties in the parameters.

The practical alternative to line-by-line integration is to use *band models*.

The simplest band model, developed by ELSASSER (1938), assumes an infinite array of equally spaced Lorentz lines, all of equal intensity. This model simulates the 15 μm CO_2 band reasonably well, since the lines are somewhat regularly spaced in this case. GOODY (1952) developed a statistical band model in which a random position and spacing of lines and an exponential (or Gaussian) distribution of line strengths are assumed. Goody's model fits H_2O and O_3 spectra, which lack regularity, as well as the CO_2 band.

The mean absorption in the Goody band model is expressed as

$$\bar{A} = \frac{\int A_\lambda \, d\lambda}{\Delta\lambda} = 1 - \exp\left\{-\frac{sm}{\delta}\left(1 + \frac{sm}{\pi\alpha_L}\right)^{-\frac{1}{2}}\right\} \tag{12.30}$$

for lines with the Lorentz profile, where s is the mean line intensity, m is the mass of the absorber, and δ is the mean line spacing.

In the actual non-homogeneous atmosphere k_λ is no longer constant with altitude and varies largely with changes in pressure and temperature. If we deal with stratospheric radiative processes using data obtained at a tropospheric condition the error is so large that the result of a heat transfer calculation may become meaningless (GOODY, 1964). A convenient way to simplify the calculations is a *scaling method* which modifies m and α_L by the following two-parameter scaling approximation:

$$m = \int \Phi(T) \, dm \tag{12.31}$$

and

$$\alpha_L = (\alpha_L)_0 \int \Psi(T)\left(\frac{p}{p_0}\right)^r dm/m. \tag{12.32}$$

RODGERS and WALSHAW (1966) have shown that Φ and Ψ can be written empirically with adequate accuracy as

$$\ln \Phi = f(T - 260) + g(T - 260)^2 \tag{12.33}$$

and

$$\ln \Psi = f'(T - 260) + g'(T - 260)^2. \tag{12.34}$$

The following values are applicable to the parameters in Goody's band model for the case involving the 15μm CO_2 band: $s/\delta = 718.7$ cm^{-1}, $(\alpha_L)_0 = 5.7\times10^{-2}$ cm^{-1}, $\nu_0 = 666.67$ cm^{-1}, $\pi(\alpha_L)_0/\delta = 0.448$, $f = 3.49\times10^{-3}$, $f' = -1.28\times10^{-6}$, and $f' = g' = 0$.

The Lorentz line profile is quite adequate in representing the line shape below ~30 km, whereas the Doppler profile holds above ~50 km (these critical heights are different for different constituents, see RODGERS and

WALSHAW, 1966). For radiative transfer analyses dealing with the entire middle atmosphere it is imperative to incorporate the combined influence of Lorentz and Doppler broadening effects. The mixed line shape (Voigt profile) may be given by

$$k_\nu = \frac{s}{\alpha_D\sqrt{\pi}}\frac{y}{\pi}\int_{-\infty}^{\infty}\frac{e^{-t^2}}{y^2+(x-t)^2}\,dt \tag{12.35}$$

where $x=(\nu-\nu_0/\alpha_D)/\alpha_D$ and $y=\alpha_L/\alpha_D$. If $y \gg 1$ the Voigt profile approaches the Lorentzian shape, whereas if $y \ll 1$ it reduces to the Doppler shape for small x but to the Lorentzian shape for large x. The critical condition ($y=1$) occurs at ~ 35 km for the 15 μm CO_2 band (SHVED and TSARITSYNA, 1963).

For a single Voigt line TIWARI (1978) has calculated a quantity

$$W^* = \bar{W}/[2\alpha_D/\sqrt{\ln 2}] \tag{12.36}$$

as a function of

$$x_D = \frac{s}{\alpha_D}\sqrt{\frac{\ln 2}{\pi}}\,m \tag{12.37}$$

for different values of

$$a = \frac{\alpha_L}{\alpha_D}\sqrt{\ln 2}. \tag{12.38}$$

$\bar{W}$ in (12.36) is the so-called *equivalent width* of a line given by $\int_{-\infty}^{\infty} A_\nu d\nu$ and is related to the mean absorption function by

$$\bar{A} = 1 - e^{-\bar{W}/\delta} \tag{12.39}$$

where δ is the average spacing between lines.

Tiwari's results are shown in Fig. 12.10. He also fitted these results to the analytical forms

$$W^* = \frac{\sqrt{\pi}}{2}[F_L\{1-e^{-\sqrt{x_D}} + x_D^2 e^{-\sqrt{x}}\}]^{\frac{1}{2}} \quad 0.1 \le x_D < 20 \tag{12.40}$$

and

$$W^* = \left[\frac{\pi}{4}F_L + \ln x_D\left(1-\frac{F_L}{x_D^2}\right)\right]^{\frac{1}{2}} \quad x_D \ge 20 \tag{12.41}$$

where

$$x = \frac{sm}{2\pi\alpha_L} \tag{12.42}$$

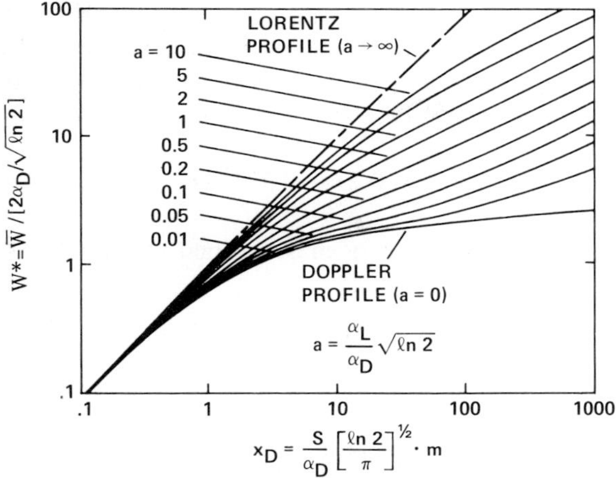

FIG. 12.10. Absorption of a single line of Voigt profile (TIWARI, 1978).

and

$$F_L = x_D^2\left[1 + \left(x_D\frac{\sqrt{\pi}}{4}a\right)^{5/4}\right]^{4/5}. \tag{12.43}$$

The variation of W^* calculated by above equations has a discontinuity at $x_D=20$ which occurs at ~55 km, because of a difference between the two equations (12.40) and (12.41).

12.2.3 Cooling rates due to IR radiation from atmospheric trace gases

The most important molecules in the transfer of terrestrial IR radiation are CO_2, H_2O and O_3 (See Fig. 12.9). We will now briefly review the various analytical methods for evaluating absorption functions for these molecules. A simple analytical form is then developed for the absorption function, using a form identical to that used for calculations of the absorbed solar energy in the previous section.

Carbon Dioxide Radiation from CO_2 molecules, particularly at the 14–16 μm band, is the most important for cooling of the middle atmosphere. Many empirical formulae have been developed in order to relate the mean absorption function to the amount of CO_2. The simplest analytical form is

$$\bar{A} = 1 - e^{-au^b_{CO_2}}. \tag{12.44}^*$$

A set of parameters has been determined to be $a=0.32$ and $b=0.4$ by

* Note that a parameter a in this equation (12.44) is not related to a defined in (12.38).

SHEKHTER (1950) and a=0.256 and b=0.41 by KONDRATYEV and NEDOVESOVA (1958). The latter agrees particularly well with the experimental data of HOWARD *et al.* (1956a) expressed in the general form (12.14), which takes the form in this case (for the 15 μm band) of

$$\bar{A} = \frac{1}{250}\int A_\nu \, d\nu = -0.27 + 0.22 \log [u(p + p_{CO_2})^{0.855}]. \qquad (12.45)$$

A more complex formula has been developed by RAMANATHAN (1976),

$$\bar{A} = 2A_0 \ln\left(1 + \sum_{i=1}^{10} \xi_i^{\frac{1}{2}}\right) \qquad (12.46)$$

where

$$\xi_i = 1.66\left(\frac{4\nu_0}{A_0\delta_i}\right)\int s_i q_i p_{CO_2} p \, dz \qquad (12.47)$$

A_0 (=17.3 cm^{-1}) is the bandwidth, ν_0(=0.064 cm^{-1} atm^{-1}) is the mean-half-width at 1 atm pressure, and q_i is the ratio of the abundance of individual isotopes to the total CO_2 abundance. Ten different bands including various isotopic bands contribute to the total absorption. Values of s_i, δ_i and q_i for each isotopic band are given in the paper of RAMANATHAN (1976), who states that application of this form to the Venus atmosphere gives results in excellent agreement with the more detailed line-by-line calculations by DICKINSON (1972).

We now try to fit the variation of W^* with x_D shown in Fig. 12.10 to a single analytical formula applicable to the entire range of x_D, i.e.

$$\log W^* = \sum_{i=0}^{K} C_i(\log x_D - 2)^i. \qquad (12.48)$$

Since W^* is also a function of a defined in (12.38) we need an additional formula to represent C_i as a function of a as follows:

$$C_i(a) = \sum_{j=0}^{K'} B_{ij}(\log a)^j. \qquad (12.49)$$

The coefficients B_{ij} are determined for the polynomials with K=4 and K' =7 as shown in Table 12.5. By this method 40 coefficients are sufficient to calculate the absorption function between any two heights of the atmosphere.

Cooling rates due to radiative transfer in the 15 μm CO_2 band calculated by our method for 5° N in March are shown in Fig. 12.11 for three different values or r, the pressure-scaling parameter introduced in (12.14). In

TABLE 12.5 Coefficients B_{ij} in (12.49) to be used for calculation of the absorption function for the 15 μm CO_2 band.

j \ i	0	1	2	3	4
0	1.1449(0)*	4.95197(−1)	−2.69216(−2)	1.58512(−2)	−2.30788(−3)
1	4.74644(−1)	1.20280(−1)	−5.39608(−2)	−4.33103(−3)	2.84428(−3)
2	−1.39471(−3)	3.51354(−3)	−8.52836(−4)	−3.56893(−4)	9.80791(−5)
3	−3.36581(−2)	1.30974(−2)	5.72778(−3)	−1.27612(−3)	−1.57184(−4)
4	1.62149(−4)	2.14834(−3)	−5.62323(−4)	−2.36154(−4)	5.80905(−5)
5	1.52256(−3)	−1.56919(−3)	−8.07880(−5)	1.58024(−4)	−1.28190(−5)
6	6.34865(−5)	−4.50270(−4)	1.20046(−4)	4.95364(−5)	−1.30224(−5)
7	−1.36820(−5)	−3.22590(−5)	1.69682(−5)	3.79406(−6)	−1.51590(−6)

*$A(b)$ should read as $A \times 10^b$.

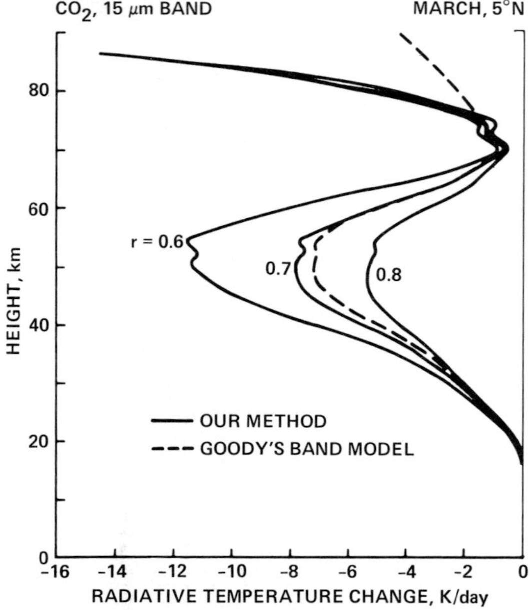

FIG. 12.11. Radiative cooling rate due to the 15 μm CO_2 band predicted with our method for equatorial region in March for three different values of r. The result for r=0.7 is compared with the corresponding case of Goody's band model.

general, radiative cooling follows the temperature profile, which is taken for the present analyses from the variation near the equator shown in Fig. 2.2(a). The cooling rate always takes a maximum at ~50 km where the temperature profile has positive curvature, whereas a minimum cooling is found at ~70 km near the level of negative curvature of the temperature profile.

The result from our method is in good agreement with that of Goody's band model except for the region above ~75 km, if the same value is used for r (see Fig. 12.11). The effect of r is very significant particularly near the region of maximum cooling around 50 km. A value of 0.7 has been chosen for r in all of the following calculations, since the cooling rate predicted with this value of r agrees well with the results of more accurate calculations; i.e., the predicted cooling rate of ~7.8 K day^{-1} at 50 km agrees reasonably well with the value of ~7.4 K day^{-1} calculated by FELS and SCHWARZKEPT (1981) and the value of ~6.7 K day^{-1} by APRUZERE *et al.* (1982).

The large increase in cooling rate above ~70 km predicted by our method is a characteristic peculiar to the Doppler line profile at these heights. It gives much larger cooling rate than those calculated with Goody's band model. The reason for this difference may lie in the fact that the half-width of the Lorentz profile on which the Goody's model is based becomes very small at low pressure (see (12.32)), whereas the Voigt profile, which actually represents the Doppler profile above ~70 km, has little effect of atmospheric pressure (see Eq. 12.29).

At heights above ~80 km, the assumption of LTE is not valid, and the source function is no longer represented by the Plank blackbody function. Complicated and detailed calculations are necessary for the source function at these heights, taking into account molecular processes such as transitions among the various energy levels of the CO_2 molecule including the isotopes (CURTIS and GOODY, 1956; KUHN and LONDON, 1969; DICKINSON, 1973; KUTEPOV and SHVED, 1978). Results of these non-LTE calculations indicate that the source function for the 15 μm CO_2 band are significantly less than the blackbody intensity above ~80 km; therefore, the cooling rate will be much less than those predicted in Fig. 12.11.

Figure 12.12 shows, separately, the radiative temperature change due to the downward flux and the upward flux. The figure indicates that cooling in the stratosphere is caused mainly by the downward flux and that the upward flux produces atmospheric heating above ~50 km and in the troposphere.

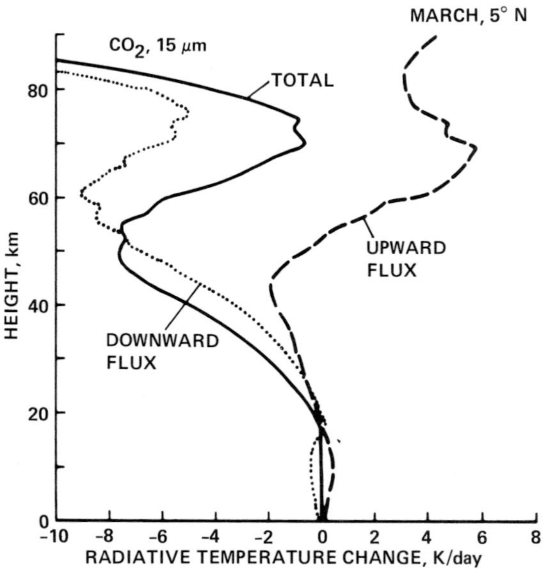

FIG. 12.12. Upward and downward radiative fluxes due to the 15 μm CO_2 band predicted with our method for equatorial region in March (r=0.7).

Ozone Three fundamental vibrational bands (ν_1, ν_2 and ν_3) of O_3 occur at wavelengths of 9.066, 14.27 and 9.597μm, respectively. The very strong ν_3 and moderately strong ν_1 fundamentals constitute the well-known 9.6 μm O_3 band. Since this band occurs in the atmospheric window (8–12 μm) and is located near the maximum of the blackbody radiation curve for atmospheric temperatures, it exerts a non-negligible influence on the IR energy budget. This is particularly true in the middle atmosphere, where a large amount of ozone exists.

Various analytical formulae have been worked out for expressing the absorption function of the 9.6 μm O_3 band by WALSHAW (1957), AIDA (1975), RAMANATHAN (1976) and others. We present here the formula by Ramanathan;

$$A=\frac{A_0}{\Delta\nu}\ln\left\{1+\frac{u_c}{\left[4+u_c\left(1+\frac{1}{\beta}\right)\right]^{\frac{1}{2}}}\right\} \tag{12.50}$$

where $A_0(=39\ \mathrm{cm}^{-1})$ is the bandwidth parameter, $\Delta\nu(=260\ \mathrm{cm}^{-1})$ is the entire band width, u_c is the dimensionless optical path length given by

$$u_c=\int\frac{S}{A_0}\rho_{O_3}\,\mathrm{d}z \tag{12.51}$$

and β is the half-width parameter given by

$$\beta=\frac{4\nu_0}{u_c\delta}\int p\,\mathrm{d}u_c. \tag{12.52}$$

The parameters in (12.51) and (12.52) are as follows: $S(=387\ \mathrm{cm}^{-2}\ \mathrm{atm}^{-1}$ STP) is the band intensity, $\nu_0\ (=0.076\ \mathrm{cm}^{-1})$ is the mean line half-width at 1 atm pressure and $\delta(=0.1\ \mathrm{cm}^{-1})$ is the mean line spacing. Equation (12.50) gives appreciably larger A than Walshaw's values for ozone amounts greater than 0.3 cm; this may be due to the omission of the ν_1 band in Walshaw's analysis. The result from Eq. (12.50) is in agreement with that from Aida's more detailed formulation except for ozone amounts between 0.2 and 0.4 cm where the present formulation results is an approximately 10% smaller A. Since the amount of O_3 is usually smaller than this range in the middle atmosphere, the absorption function represented by (12.50) should be sufficient for our present purposes.

We now fit the result from (12.50) into the polynomial function

$$\log A=\sum_{i=0}^{6}C_i(\log u_i)^i \tag{12.53}$$

where u_i is the amount of ozone in cm. The coefficients C_i are given in Table

12.6. The radiative temperature changes due to the 9.6 μm O_3 band predicted by our method with $r=0$ are illustrated in Fig. 12.13 for three latitudes (5°N, 65°N, and 65°S) in March. Since the effect of pressure dependences of the line half-width and the band intensity is already includ-

TABLE 12.6 Coefficients C_i's in (12.53) to be used for calculations of absorption functions for the 9.6 μm O_3 band, and three spectral ranges for H_2O; the vibration-rotation band centering at 6.3 μm, the continuum at 8–13 μm, and the purely rotation band at 12.5–250 μm.

	O_3 9.6 μm		H_2O 8–13 μm	
		5–8 μm		12.5–250 μm
C_0	−1.58429(−1)*	−8.93469(−2)	−6.73589(−1)	−1.60003(−1)
C_1	3.70719(−1)	8.84785(−2)	6.55012(−1)	1.23203(−1)
C_2	−1.47030(−1)	−1.12991(−2)	−1.27751(−1)	−1.03768(−2)
C_3	3.28693(−2)	1.32278(−2)	−1.64740(−2)	7.39209(−4)
C_4	8.76234(−3)	−6.38374(−3)	4.23706(−4)	−3.65905(−3)
C_5	−2.61758(−3)	−1.71655(−3)	2.40811(−4)	−5.50930(−4)
C_6	−5.81872(−4)	−1.06018(−4)	1.29986(−5)	−2.21885(−5)

*$A(b)$ should read as $A \times 10^b$.

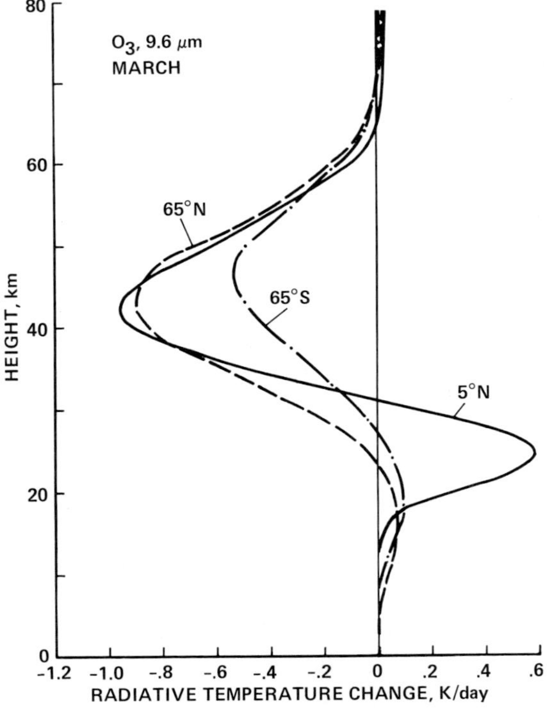

FIG. 12.13. Infrared heating and cooling rates due to the 9.6 μm O_3 band predicted with our method for three latitudes in March ($r=0$).

ed in parameters β and u_c in (12.50), the pressure-scaling factor r is not necessary to consider in this case; thus we set $r=0$. Maximum cooling of ~ 1.0 K day^{-1} occurs at 45–50 km at high latitudes and at slightly lower heights (40–45 km) at low latitudes. The predicted maximum cooling rate is smaller by a factor of ~2 than those calculated earlier by DICKINSON (1973) and RAMANATHAN (1976). This difference may be attributed mainly to the different ozone density profiles used rather than the different methods of analyses. The effect of ozone profiles is really significant as is judged from the difference between the two results at 65°N and 65°S shown in Fig. 12.13. The figure also shows that heating occurs in the stratosphere. The heating is particularly large (~0.6 K day^{-1}) at low latitudes where the lower stratosphere is extremely cold. There is also a tendency for slight heating above 65–70 km.

Water Vapor Water vapor has important absorption bands in both the solar and terrestrial wavelength ranges. It has a complex vibration-rotation spectrum in the infrared. The ν_1 and ν_3 fundamental vibrations are at wavelengths too short (2.74 and 2.66μm, respectively) to be of much importance in the terrestrial regime, although they absorb significant amounts of solar radiation (see Section 12.1.2). The ν_2 fundamental, however, is responsible for a very strong vibration-rotation band which is centered at ~6.3 μm extending from 5 to 8μm. A pure rotation band at wavelengths greater than 12μm also exerts a strong influence on the radiative energy balance of the atmosphere. There is also a weak continuum band in the 8.3–12.5 μm spectral region.

HOWARD (1965) has shown that their experimental data fit remarkably well to the equation

$$A_\lambda = 1 - \exp \frac{-1.97 W/W_0}{[1 + 6.57(W/W_0)]^{\frac{1}{2}}} \tag{12.54}$$

where W is the mass of water vapor and W_0 is the value of W at which $A\lambda=0.5$. This equation is of essentially the same form as Goody's band model (12.30).

RODGERS and WALSHAW (1966) listed the parameters s/δ and $\pi\alpha/\delta$ in (12.30) for 9 wavelength intervals of the vibration-rotation (6.3μm) band and 10 intervals of the rotation band (>12.5 μm). We have calculated the total absorption function for each band using these parameters and the result is fitted to equation (12.53), where u_i now represents the water vapor amount in g cm^{-2} or thickness in cm of the "precipitated water" layer (these units are essentially identical). For the continuum regime (8–13 μm), we have used the absorption function calculated using the absorption coefficients given for every 1 μm interval by KONDRATYEV (see Table 3.9 of his book, 1969). The values of the coefficients C_i determined for the three different wavelength ranges are shown in Table 12.6.

The radiative temperature changes due to H_2O emissions predicted by our method using $r=0.6$ for the equatorial region in March are shown in Fig. 12.14. Cooling in the middle atmosphere occurs mainly in the rotation band ($>12.5\mu$m). Although the vibration-rotation band (6.3μm) is strong, its contribution to the radiative cooling is much smaller than that of the rotation band; this is because of the generally low temperature in the middle atmosphere (the maximum blackbody radiation at 240 K is at $\sim 12\ \mu$m). In the troposphere, particularly below ~ 5 km, the contribution of the continuum (8–13 μm) dominates the other two bands. The absorption in this spectral range is weak, but since this region coincides with the atmospheric window (8–12 μm) and also since water vapor is more abundant in the

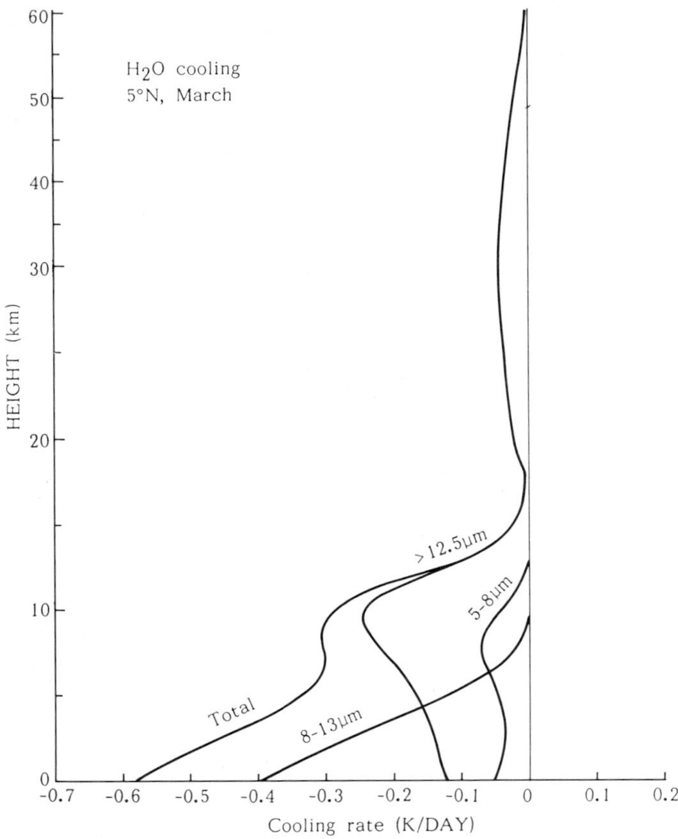

FIG. 12.14. Infrared heating and cooling rates due to emissions from H_2O at three different wavelengths ranges predicted with our method for equatorial region in March ($r=0.6$).

troposphere and the temperature is a little higher (the maximum blackbody radiation at 300 K is at ~9.65 μm), the radiative cooling in this spectral range becomes significant in the lower troposphere.

12.2.4 Global distributions of cooling rate

The total rates of radiative temperature change for the 15μm CO_2 band, the 9.6μm O_3 band, and the three H_2O bands (or continuum) have been calculated by our method for the meridional domain for March and June. The distributions of O_3, CO_2 and H_2O are the same as those used for the heating rate calculations in Section 12.1.3, and the temperature distributions are taken from Figs. 2.2 (a) and (b). The results are shown in Figs. 12.15 (a) and (b).

There is generally a negative temperature change (cooling) except for the mesosphere at high latitude in summer and near the tropopause at low latitudes, where infrared heating occurs. This heating is caused by the very low temperature there; while the local cooling by emission is small because of the low temperature, the radiative transfer from the surrounding warmer region can heat that area. The general pattern of radiative heating and cooling calculated here is reasonable and agrees well with the result of other calculations. However, the details are different because of the difference in the assumed distributions of temperature and concentrations of gaseous components as well as the different approaches used for calculating the absorption function.

12.3 Interactions among Radiation, Photochemistry and Dynamics

Atmospheric phenomena are so complicated that a change in one particular parameter (e.g. incoming solar flux, chemical composition, circulation etc.) affects many other physical and chemical properties, which eventually could feed back to the original parameter. The feedback can occur in either direction; a positive (or negative) feedback accelerates (or deaccelerates) the change in the original parameter. Coupling among the three processes, i.e. radiation, chemistry and dynamics, are so complex that it is almost impossible to predict quantitatively the impact of the coupling including feedback effects. In some cases even the qualitative effect is not known well.

Some important interactions among the three processes are illustrated schematically in Fig. 12.16. We now discuss briefly each of the interaction processes and feedback mechanisms.

12.3.1 Radiative-photochemical coupling

Temperature changes are affected by themselves due to a self-feedback

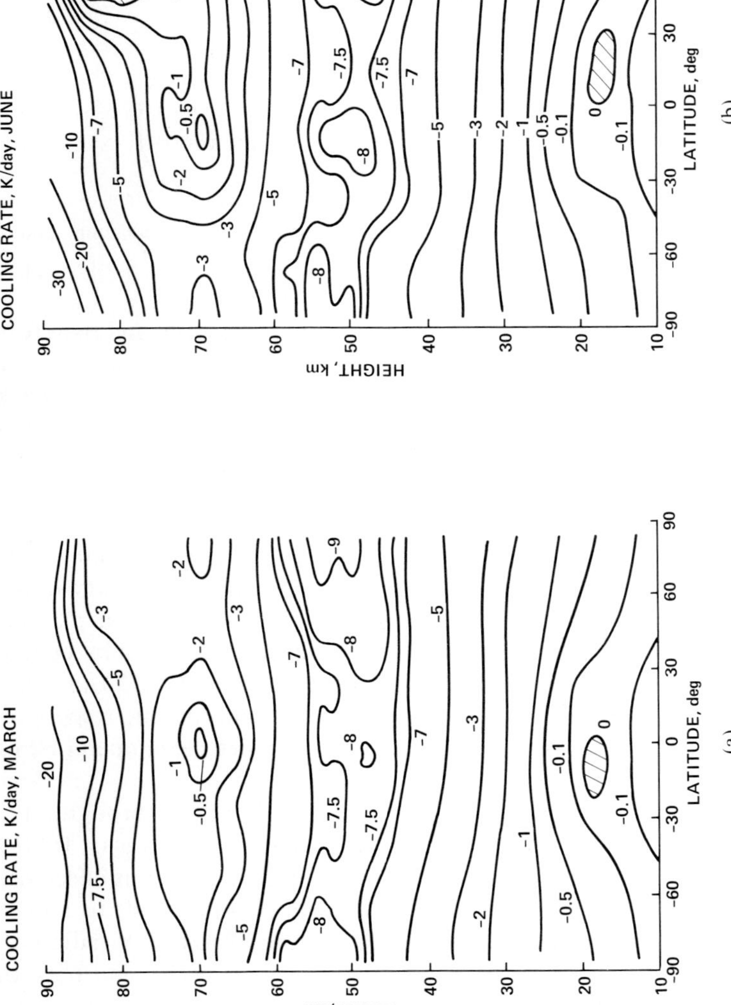

FIG. 12.15. (a) Contour plots in the meridional plane of total IR heating and cooling due to 15 μm CO_2, 9.6 μm O_3 and three H_2O bands predicted for March. The regions of heating are shaded. (b) The same as Fig. 12.15(a) but for June.

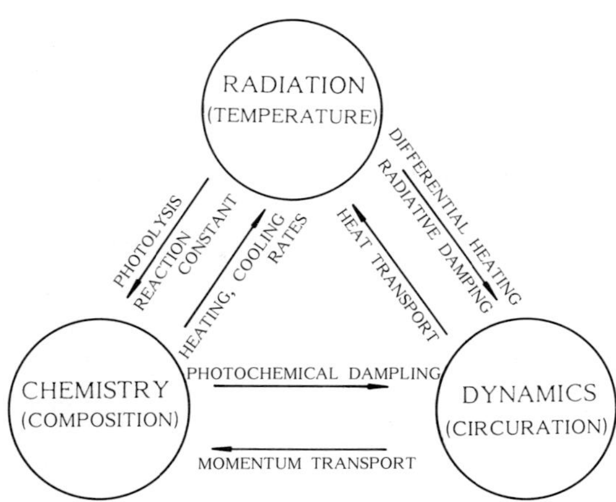

FIG. 12.16. Schematic diagram illustrating coupling among radiation, photochemistry and dynamics.

process. If the temperature increases, the rate of IR emission and therefore the cooling rate increases, which would result in a decrease in the temperature. This radiative damping of temperature is a negative feedback mechanism for temperature perturbations. If the cooling rate is assumed to be proportional to the increase in temperature, the process is called *Newtonian cooling*. It represents a good approximation to the cooling of the middle atmosphere (DICKINSON, 1973).

The photolyses of various molecules by solar UV radiation and subsequent chemical reactions are essential for inducing photochemistry in the middle atmosphere, which determines the chemical composition. As was discussed in previous sections, absorption of solar UV radiation by O_3 and O_2 and emission of IR radiation from atmospheric molecules (CO_2, O_3 and H_2O) are the ultimate causes of atmospheric heating and cooling, respectively. The net radiative heating or cooling rate determines the temperature of the middle atmosphere.

The temperature change affects photochemistry through the temperature-dependent rate constants of chemical reactions. If the temperature increases, the ozone concentration decreases by this effect (see Section 11.1.4). The decreased ozone concentration then tends to reduce the heating rate by weakening the ozone absorption of solar UV radiation, resulting in a decrease in the temperature. Thus, ozone photochemistry produces negative feedback and acts as a strong buffer to temperature perturbations (photochemical damping of temperature).

The effect of photochemical damping is to reduce the relaxation time (or accelerate the relaxation rate) of the atmospheric temperature. A detailed study of photochemical-radiative relaxation processes was carried out by LINDZEN and GOODY (1965) for an oxygen only atmosphere. Later, the study was extended by BLAKE and LINDZEN (1973) to include catalytic chemistry of HO_x and NO_x. The latter study has found that the photochemistry accelerates the thermal relaxation rate by a factor of 2 - 3 at the stratospause. These analyses, however, should be restricted to local perturbations with vertical scales sufficiently small so that equilibrium conditions could be assumed at all other heights. Taking into account the opacity effects, STROBEL (1977) showed that at the stratospause the photochemical acceleration of the thermal relaxation rate was substantially reduced and that large vertical scale temperature disturbances relaxed at approximately the Newtonian cooling rate. Large vertical scale perturbations in ozone density and temperature may even grow at larger optical depths due to opacity changes in the O_3 dissociation and heating rates.

If the ozone concentration in the upper stratosphere is decreased by, for instance, the increased CFM due to human activities, solar radiation can reach deeper into the middle and lower stratosphere. Thus, O_x production increases in these lower regions, and may compensate somewhat for the decreased O_3 in the upper region. As a result, the change in the total ozone could be minimal.

Perturbations of the stratospheric composition affect the stratospheric temperature profile via the changes in balance between heating and cooling rates. Changes in temperature affect chemical reaction rates, which in turn feed back on stratospheric composition. LUTHER *et al.* (1977) have shown that temperature feedback has approximately a 10% restoring effect on total ozone for a small ozone reduction caused by either the stratospheric injection of NO_x or the release of CFM.

A clear inverse correlation has been observed between the temperature and the ozone concentration in the upper stratosphere. Figure 12.17 shows the Nimbus 4 data on temporal variations of the 2 mb temperature and the 1 mb ozone mass mixing ratio observed at ~60°N latitude. There is a general tendency for the ozone density to be larger in winter when the upper stratospheric temperature is lower. It is interesting to see, however, that a large decrease of O_3 occurred during the period of a strong winter stratospheric warming in January, 1971.

The temperature distribution in radiative equilibrium was calculated by LEOVY (1964) taking into account the effect of radiative-photochemical coupling. The result is shown in Fig. 12.18. The model generally predicts well the observed temperature distribution in the 30–70 km region (cf. Fig. 2.2); this implies that the temperature in the middle atmosphere is determined to a first order approximation by the condition of radiative equilibrium. However, there are some important differences between the predicted and

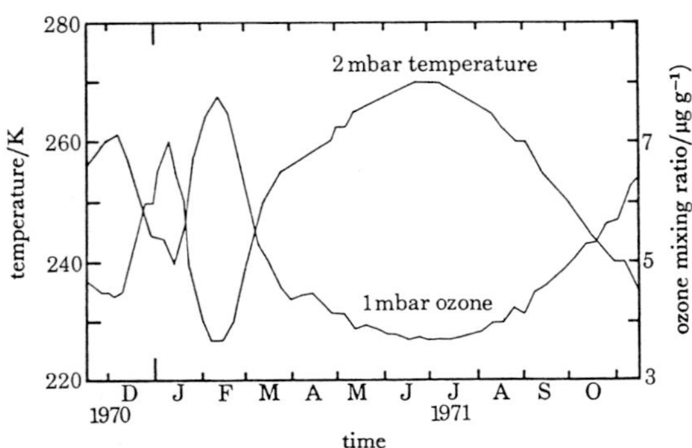

FIG. 12.17. Temporal variations of the 2 mb temperature and the 1 mb ozone mass mixing ratio in a latitutde band centered at 60°N from one year data of Nimbus 4 observations (KRUEGER *et al.*, 1980).

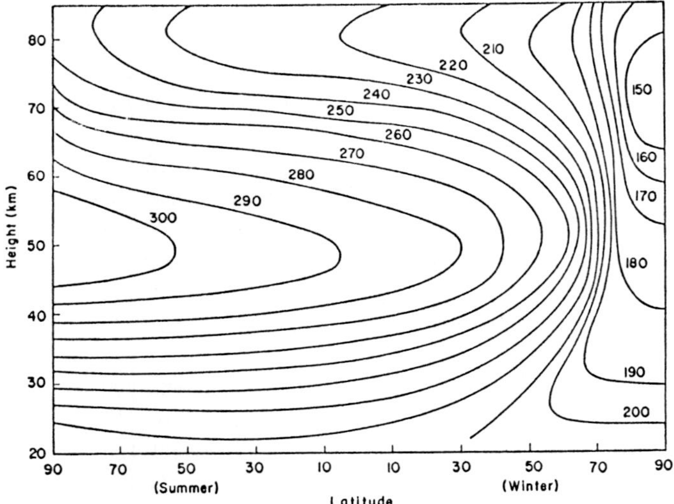

FIG. 12.18. Equilibrium temperature in the meridional plane in solstices determined by the solar UV heating and cooling by IR emission (LEOVY, 1964).

observed temperature distributions. Firstly, in winter polar regions where there is no solar insolation, the calculated temperature is too low. Secondly, over sunlit latitudes the predicted temperature in the mesosphere is generally too high; this may be caused by the fact that the model uses the ozone photochemistry of an *oxygen-only* atmosphere. The ozone concentration in the mesosphere in such a model is much larger than the observed values (see Fig. 5.2) and therefore the heating rate is too large. The final but most important discrepancy between the model and observations concerns the seasonal variation in mesospheric temperature. Observational evidence indicates a very cold summer mesospheric temperature, but the radiative equilibrium model predicts much higher temperatures in summer than in winter. This discrepancy is resolved only by taking into account the effect of radiative-dynamical coupling as will be discussed in the next Subsection (12.3.2).

It is well recognized that the thermal structure of the atmosphere is influenced by the presence of small quantities of CO_2, H_2O, O_3 and aerosols. The main radiative effect of these gases in the lower atmosphere is through the absorption of upwelling radiation and reradiation at the local temperature; this blanketing effect leads to an increase in the surface temperature, the so-called *greenhouse effect*. The greenhouse effect due to increased CO_2 from fossil fuel consumption will have the most important impact on climate and human activities (note, however, that increased CO_2 would cause cooling in and above the middle stratosphere as was discussed in Section 11.1.4).

The atmosphere also contains a large number of trace gases (both natural and man-made) having strong IR absorption bands. The greenhouse effect from chlorofluorocarbons was first evaluated by RAMANATHAN (1975). The greenhouse effect due to N_2O was studied by YUNG *et al.* (1976). Extensive calculations of greenhouse effects from man-made perturbations of various trace gases have been carried out by WANG *et al.* (1976). Table 12.7 summarizes the result of these calculations. Since the model cannot predict cloud feedback effects, cloud-top temperature or cloud-top height was assumed to be fixed. The former assumption seems to be more plausible and gives larger greenhouse effects than the latter assumption. Note that in these predictions the combined effects are calculated for CF_2Cl_2 and $CFCl_3$ and also for CH_3Cl and CCl_4.

It is now generally accepted that the global surface temperature has risen $\sim$0.4°C in the past century through the greenhouse effect resulting primarily from an increase in atmospheric CO_2. It is also recognized, however, that the temperature has dropped sharply, usually over 1 to 3 year periods, due to cooling effects from large volcanic eruptions. These volcanic events produce massive natural pollution of the stratosphere, resulting in backscattering of solar radiation and a cooling of the earth (HANSEN *et al.*, 1981). It is also known that stratospheric sulfuric acid aerosols, such as that produced by a reaction between atmospheric H_2O and volcanic SO_2, absorb incoming solar radiation, warming the stratosphere, but causing a proportional cooling at the earth's surface (TOON and POLLACK, 1980).

TABLE 12.7 Infrared bands of various trace gases and greenhouse effects due to changes in their concentrations as indicated (WANG *et al.*, 1976)

Trace gases	Band Center	Band designation	Assumed concentration on surface	Factor modifying concentration	Greenhouse Effect fixed cloud-top temperature	fixed cloud-top height
H_2O	7.78 μm	ν_1	0.28 ppmv			
	17.0	ν_2		2	0.68°K	0.44°K
	4.5	ν_3				
CH_4	7.66	ν_4	1.6 ppmv	2	0.28	0.20
NH_3	10.53	ν_2	6 ppbv	2	0.12	0.09
HNO_3	5.9	ν_2	1 ppbv			
	7.5	$\nu_3+\nu_4$		2	0.08	0.06
	21.8	ν_9				
	11.3	$\nu_5+2\nu_9$				
C_2H_4	10.5	ν_7	0.2 ppbv	2	0.01	0.01
SO_2	8.69	ν_1	2 ppbv	2	0.03	0.02
	7.35	ν_3				
CF_2Cl_2	9.13	ν_1	0.1 ppbv	20		
	8.68	ν_6				
	10.93	ν_8			0.54	0.36
$CFCl_3$	9.22	ν_1	0.1 ppbv	20		
	11.88	ν_4				
CH_3Cl	13.66	ν_2	0.5 ppbv	2		
	9.85	ν_6				
	7.14	$\nu_2+\nu_5$			0.02	0.01
CCl_4	12.99	ν_3	0.1 ppbv	2		
H_2O	6.25	ν_2				
	10	continuum		2*	1.03	0.65
	20	continuum				
	10−∞	rotational				
CO_2	15.0	ν_2	330 ppmv	1.25	0.79	0.53
O_3	9.6	ν_3	**	0.75	−0.47	−0.34

*Change in H_2O abundance is assumed to be a factor of 2 above 11 km; below this height, it is determined by assuming the condition of fixed relative humidity.

**The ozone density profile in the U.S. Standard Atmosphere (Appendix A) was used.

12.3.2 Radiative-dynamical coupling

The ultimate drive for large scale atmospheric motions is—at least in the region between 30 and 70 km—differencial heating, i.e. the difference between the heat input by absorption of solar insolation and the cooling due to the flux divergence of IR emission. The dynamics, however, also suffers from damping by radiation; the increased cooling rate due to an increase in temperature enhances damping for hydrodynamic waves.

Dynamical processes affect the temperature field of the atmosphere in a different fashion. Adiabatic processes, i.e. expansion and contration without loss or gain of heat, are responsible for the observed lapse rate of temperature in the troposphere (see Chapter 2). A similar process also causes a large seasonal difference in the mesospheric temperature.In particular, the extremely cold temperature in the summer mesosphere is caused by the effect of adiabatic cooling due to the upward motion and expansion associated with the global meridional circulation produced by a large seasonal difference in diabatic heating between the summer and winter hemispheres.

The most significant effect of dynamics on the temperature field appears during *sudden stratospheric warmings* which usually occur in the northern winter hemisphere. During these events the temperature rises by as much as 40°C for several days. The event occurs over most of the high latitude area including the polar regions and is accompanied by small temperature decreases in the lower latitude areas. The sudden temperature increase usually starts in the upper stratosphere and propagates down to the middle and lower stratosphere. It is also associated with a general tendency for temperature to decrease in the mesosphere. Figure 12.19 illustrates latitudinal variations in the zonally averaged temperature at 30 mb during and after a strong and sudden warming that occurred in January, 1963 (JULIAN and LABITZKE, 1965).

The sudden stratospheric warming in winter is known to be caused by the interaction between planetary waves and the mean flow. The upward propagating, growing waves extract energy from the mean flow and decelerates the westerly wind, but also accelerates the easterly wind due to absorption of the wave energy near a critical level, and the flow eventually becomes easterly. Above this critical level the planetary wave cannot propagate, and if the propagation of the wave from below still continues, the easterly is greatly enhanced. From *thermal wind** considerations, development of this new wind shear structure is consistent with the development of a meridional

* The thermal wind is the mean-shear vector in geostrophic balance with the gradient of mean temperature of a layer bounded by two isobaric surfaces. It is directed along the isotherms, with cold air to the left in the Northern Hemisphere.

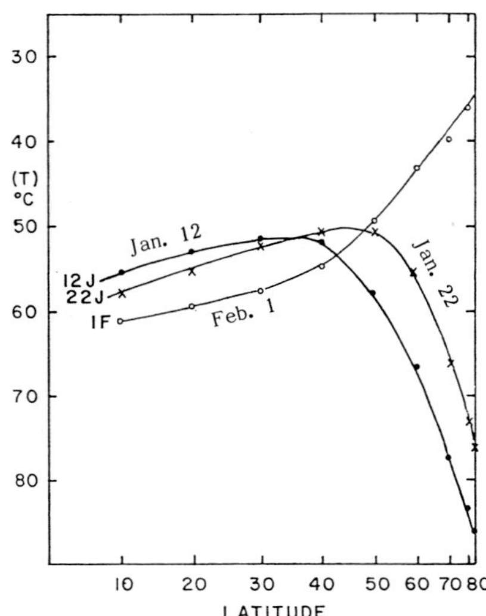

FIG. 12.19. The zonally averaged temperature at 30 mb as a function of latitude during and after a strong, sudden warming (JULIAN and LABITZKE, 1965).

temperature gradient opposite to the normal gradient, i.e. wide-spread warming of the polar air. Thus, the observed features associated with a sudden warming would follow. Details of the theory of sudden warmings are discussed in MATSUNO (1971), HOLTON (1975) and MATSUNO and NAKAMURA (1979).

12.3.3 Photochemical-dynamical coupling

It has been well recognized that large scale dynamics (the mean circulation and eddy motions) are important for the global distribution of stratospheric ozone. However, changes in ozone density adversely affect the dynamics through changes in heating rate due to ozone absorption of solar UV radiation (photochemical damping). SCHOEBEL and STROBEL (1978 b) evaluated the effects of various ozone density reductions on the zonally averaged circulation with a numerical quasi-geostrophic model. For global ozone perturbations by potential halocarbon pollution, they calculated about a 10% reduction in the zonal jet strength and less than a 5% change in the global mean stratospheric temperature. They also predict that large, uniform ozone reductions (~50%) would produce significant effects on the mean circulation, i.e. a substantial collapse of the stratosphere due to cooler temperatures and a weak polar night jet.

Chapter 13

IONIC CONSTITUENTS IN THE MIDDLE ATMOSPHERE

The production of various ions in the middle atmosphere is interesting in itself. Since ions interact closely with neutral constituents, measurements and analyses of ionic compositions also give us valuable information on some neutral constituents which otherwise is hard to obtain. We will first discuss various ionization sources and then discuss the ion chemistry in the mesosphere and lower thermosphere which is a significant source of NO. Observations of hydrated ions of proton (H^+) and nitrate (NO_3^-) in the *D*-region and their production mechanisms are relatively well known. After discussing these subjects, the most recent observations of non-proton hydrated positive ions and large acid cluster negative ions in the stratosphere are reviewed. The measurements of ion composition in the stratosphere are also useful to estimate the abundance of some important neutral constituents in the stratosphere, such as sulfuric acid.

13.1 Ionization Sources

The major agencies producing ionization in the middle atmosphere are solar electromagnetic radiations including EUV, X-ray and Lyman-α above ~60 km, and charged particle components of the galactic cosmic rays (GCR) below that height. These constitute the main quiet-time (background) ionization. Their ionization rates are shown by solid curves in Fig. 13.1. Other sources such as ionizations by meteors and scattered Lyman-α are of secondary importance in comparison to the avove mentioned sources.

As is seen from Fig. 3.9, the solar radiations that can penetrate below 100 km and ionize the atmospheric constituents are those of EUV with wavelengths around 120 nm and X-rays (below 10 nm). Each neutral constituent in the atmosphere can be ionized only at wavelengths below the ioniza-

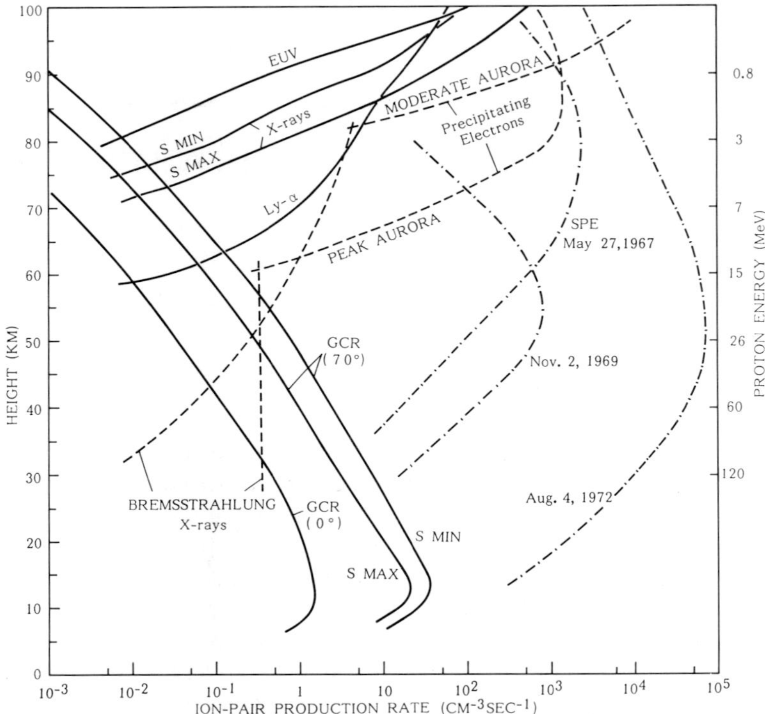

FIG. 13.1. Ion-pair production rates by various ionization sources. Solid curves are rates for background (quiet-time) ionization; broken curves are for special events such as auroral activities and solar proton events (see text for references for each calculation).

tion threshold of that constituent. Since the thresholds for the major constituents N_2 and O_2 are ~79.58 nm and ~102.78 nm, respectively, these molecules are not ionized below ~90 km. NO molecule has a longer threshold wavelength (~133.78 nm) and the solar Lyman-α radiation at ~ 121.6 nm can penetrate down through most of the mesosphere because the Ly-α line coincides, by chance, with a window in the O_2 absorption spectrum. Thus, ionization of NO by Ly-α can occur down to ~60 km and is the main source for ionospheric *D*-layer. The ionization rate by Ly-α illustrated in Fig. 13.1 is calculated using the NO distribution of MEIRA (1971) shown in Fig. 8.12.

The metastable O_2 ($^1\Delta_g$) molecule can be ionized by solar radiation in the wavelength range from 102.7 to 111.8 nm but the ground-state O_2 cannot be ionized. The ionization rate of O_2 ($^1\Delta_g$) predicted by HUNTEN and MCELROY (1968) is slightly less than the NO ionization rate by Ly-α at 75–

85 km, but the prediction by PAULSEN *et al.* (1972) shows more than an order of magnitude less than the NO ionization rate. Part of the reason for this dicrepahncy may lie in the different $O_2(^1\Delta_g)$ density used in the calculations.

Solar X-ray can ionize any atmospheric species. Its flux changes with the solar cycle; at the sunspot maximum the ionization due to solar X-ray exceeds that due to Ly-α ionization above ~85 km (see Fig. 13.1).

The ion production rates due to GCR shown in Fig. 13.1 are based on calculations by WEBBER (1962) above ~30 km, below which they are based on balloon measurements reported by NEHER (1967). The GCR ionization rate is affected by the solar cycle modulation as well as the latitudinal (or geomagnetic cutoff) effect. The ionization rate is larger at sunspot minimum than at maximum; this is caused by the effect of the solar magnetic field that tends to deflect the incoming GCR more strongly at sunspot maximum. The solar cycle modulation effect is significant at high latitudes but minimal at low latitudes. The ionization due to GCR occurs throughout the day, whereas all ionization due to solar flux occurs only during the daytime.

The ionization rate is greatly enhanced during special events such as auroral activities and solar proton events (SPE). Energetic electrons precipitated from the radiation belt are the main cause of the enhanced *D*-region ionization at high latitudes during auroral events associated with magnetic storms. Precipitation of energetic electrons also produces the bremsstrahlung X-ray, which can penetrate much deeper than the primary electrons. In Fig. 13.1 are illustrated profiles of the ionization rate due to energetic electrons and bremsstrahlung X-rays originating from them for the moderate visual aurora at night and average peak aurara during the daytime. These profiles are taken from ROSENBERG and LANZOROTTI (1979) who used exponential electron energy spectra and isotropic pitch angle distributions calculated by BERGER and SELTZER(1972) and BERGER *et al.* (1974). It is interesting to note that during moderate aurora at night the bremsstrahlung X-ray contribution could dominate the ionization from all other background sources above ~55 km (Note that there is no ionization due to solar radiaiton at night).

Solar cosmic ray particles usually have relativley little impact on the ionization of the middle atmosphere. During the solar proton event (SPE), however, soalr flares eject high energy protons (>10 MeV) that can dominate the ionization levels of polar regions over the entire range of the middle atmosphere for several days. Figure 13.1 includes illustrations of the ion-pair produciton rates for three SPE events in May, 1967, Nov. 1969 and Aug. 1972 calculated by ZMUDA and POTEMRA (1972) and POTEMRA (1974). The ion production rate at the severe SPE in August, 1972 is particularly large. These enhanced ionizations produce a large amount of NO in the middle atmosphere by the ion chemistry, as will be discussed in the next section.

The scale on the right ordinate of Fig. 13.1 shows the minimum energy a proton must have in order to penetrate to the related altitude shown on the left scale. Thus, protons of 60 MeV can penetrate to 40 km, whereas protons of 120 MeV are needed to penetrate down to 30 km.

13.2 Ion Chemistry and Odd-Nitrogen Production in the Mesosphere and Lower Thermosphere

The primary ions produced by various ionization processes at noon and midnight are shown in Fig. 13.2.These are taken from a diurnal model of OGAWA and SHIMAZAKI (1975). The nighttime ionizations are mainly due to resonant scattered EUV such as H I Lyman-α (121.6 nm), H II Lyman-β (102.7 nm), He I (584 Å) and He II (304 Å). The ionizations of NO and O_2 by geocoronal Ly-α and Ly-β, respectively, are responsible for maintaining the nighttime E region ionization at ~100 km.

The ions primarily produced are N_2^+, O_2^+, NO^+ and O^+, but among them, N_2^+ and O^+ are quickly converted to NO^+ and O_2^+ through reactions

$$N_2^+ + O \rightarrow NO^+ + N^{(*)}, \; k = 1.4 \times 10^{-10}(T_i/300)^{-0.44} \text{ cm}^3 \text{ sec}^{-1} \quad (13.1)$$

$$N_2^+ + O_2 \rightarrow O_2^+ + N_2, \; k = 5.1 \times 10^{-11}(T_i/300)^{-0.8} \text{ cm}^3 \text{ sec}^{-1} \quad (13.2)$$

$$O^+ + N_2 \rightarrow NO^+ + N^{(*)}, \; k = 6 \times 10^{-13} \text{ cm}^3 \text{ sec}^{-1} \quad (13.3)^{\S}$$

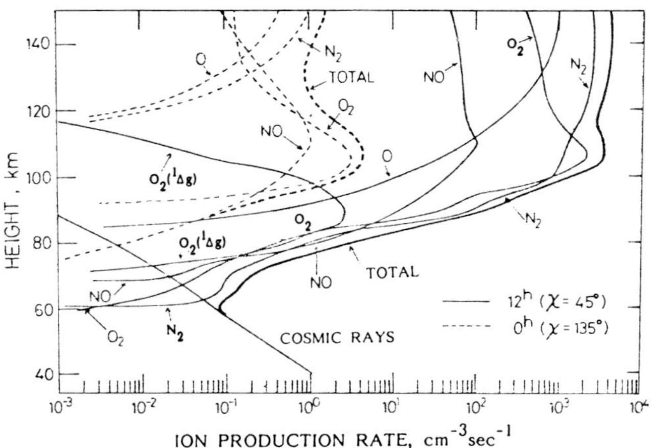

FIG. 13.2. Production rates of various ions due to various ionization processes at noon and midnight (OGAWA and SHIMAZAKI, 1975).

and

$$O^+ + O_2 \rightarrow O_2^+ + O, k = 2 \times 10^{-11} (T_i/300)^{-0.4} \text{ cm}^3 \text{ sec}^{-1}. \quad (13.4)^{\S}$$

N_2^+ ions are also lost by dissociative recombination with electrons

$$N_2^+ + e \rightarrow N^{(*)} + N^{(*)}, k = 2.2 \times 10^{-7} (T_e/300)^{-0.39} \text{ cm}^3 \text{ sec}^{-1} \quad (13.5)$$

However, the dissociation recombination of O_2^+ ions

$$O_2^+ + e \rightarrow O + O, k = 1.9 \times 10^{-7} (T_e/300)^{-0.65} \text{ cm}^3 \text{ sec}^{-1} \quad (13.6)$$

should not cause a large reduction of O_2^+ density, since there is a significant produciton of O_2^+ through reactions (13.2) and (13.4). Thus, the major ions present in the mesosphere and lower thermospher must be NO^+ and O_2^+; these results are confirmed by observations and also by model calculations shown, for example, in Fig. 13.3.

Reactions (13.1), (13.3) and (13.5) produce atomic nitrogens, denoted by $N^{(*)}$; they could be either in ground state $N(^4S)$ or in excited metastable state $N(^2D)$. There are also other reactions producing $N^{(*)}$ such as

$$NO^+ + e \rightarrow N^{(*)} + O, k = 4.0 \times 10^{-7} (T_e/300)^{-0.83} \text{ cm}^3 \text{ sec}^{-1} \quad (13.7)$$

$$N_2 + h\nu(800\text{–}1000 \text{ Å}) \rightarrow N^{(*)} + N^{(*)}, J_\infty = 1.1 \times 10^{-7} \text{ sec}^{-1} \quad (13.8)$$

and

$$N_2 + \text{photoelectron} \rightarrow N^{(*)} + N + e. \quad (13.9)$$

The cross seciton for the dissociative excitation by electron impact (13.9) has been measured by ZIPF and MCLAUGHLIN (1978); it increases from ~ 1.3×10^{-18} cm^2 at an incident electron energy of 10 eV to ~2.3×10^{-16} cm^2 at 100 eV.

The reaction of $N(^2D)$ with O_2 is a important source of NO in and above the mesosphere (NORTON and BARTH, 1970; NICOLET, 1970a), i.e.

$$N(^2D) + O_2 \rightarrow NO + O(^1D), k = 6 \times 10^{-12} \text{ cm}^3 \text{ sec}^{-1}. \quad (13.10)$$

Since the reaction of $N(^4S)$ with O_2 is too slow to produce NO (k~10^{-16} at T=300°), the *branching ratio* of production of $N(^2D)$ to that of $N(^4S)$ is an important factor in determining the rate of production of NO in the mesosphere and thermosphere. Model calculations of odd nitrogen in these regions have been made by STROBEL (1971), OGAWA and SHIMAZAKI (1975), ORAN *et al.* (1975), RUSCH *et al.* (1975), SWIDER and NARCISI

§If O^+ is in the excited state $O^+(^2D)$, the charge exchange reaction of $O^+(^2D)$ with N_2 would produce $N_2{}^+$ (instead of NO^+), which eventually goes into $N^{(*)}$ by reaction (13.5). The rate constant for $O^+(^2D)+N_2 \rightarrow O+N_2{}^+$ is 8×10^{-10} cm^3 sec^{-1}; similarly the rate constant for $O^+(^2D)+O_2 \rightarrow O+O_2{}^+$ is 7×10^{-10} cm^3 sec^{-1}.

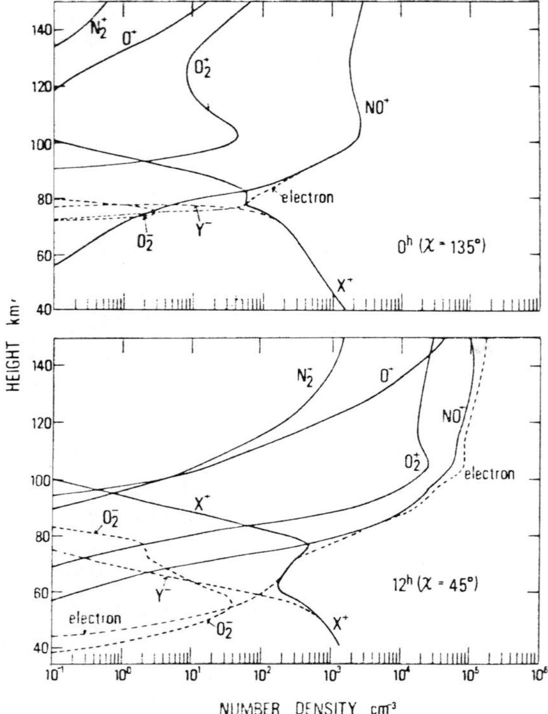

FIG. 13.3. Ion compositions at noon and midnight calculated in a time-dependent diurnal model (OGAWA and SHIMAZAKI, 1975).

(1977), KONDO and OGAWA (1977), and others; most works found that a large branching ratio (>50%) was necessary in order to explain the observed NO density in the mesosphere. KLEY *et al.* (1977) have measured the $N(^2D)$ yield by reaction (13.7) to be 0.76 ± 0.06. This reaction is a dominant $N(^2D)$ source in the thermosphere.

Main loss reactions for odd nitrogens in the mesosphere and thermosphere are

$$N(^2D) + NO \rightarrow N_2 + O, k = 6.1 \times 10^{-11}\ \text{cm}^3\,\text{sec}^{-1} \qquad (13.11)$$

and

$$N(^4S) + NO \rightarrow N_2 + O, \quad k = 3.4 \times 10^{-11}\ \text{cm}^3\,\text{sec}^{-1}. \qquad (13.12)$$

Distributions of various odd nitrogen components calculated in a time-dependent diurnal model taking into account all reactions discussed above and many others are shown in Fig. 13.4 for noon and mid-night conditions (see OGAWA and SHIMAZAKI, 1975 for details). Since concentrations of $N(^4S)$ and $N(^2D)$ are much less than that of NO in the region of our in-

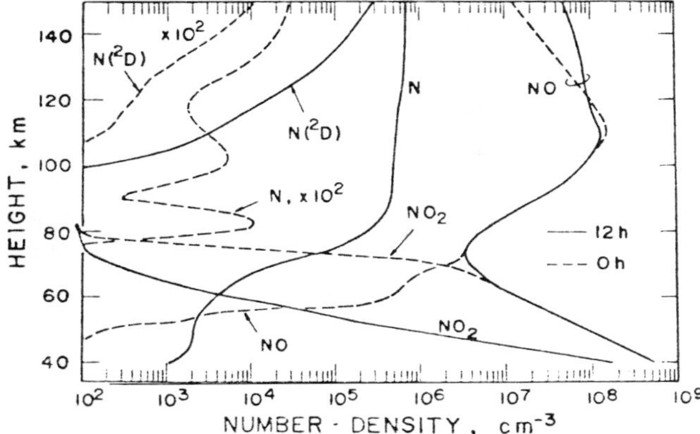

FIG. 13.4. Height distributions of odd-nitrogens in the mesosphere and lower thermosphere at noon and midnight calculated in a time-dependent diurnal model (OGAWA and SHIMAZAKI, 1975).

terest, the chemical time constant of NO, which is inversely proportional to the concentration of N, should be large. Thus, NO produced largely in the thermosphere is transported down into the mesosphere by vertical diffusion. This resembles the ozone situation in the lower stratosphere; ozones are produced mainly by photochemical processes in the upper stratosphere and their downward transport is the main source of the long-lived ozone in the lower stratosphere. The falling NO hardly enter the stratosphere, however, since they are decomposed largely in the mesosphere by predissociation in the $\delta(0.0)$ band; a dissociaiton product $N(^4S)$ further destroys another NO by reaction (13.12).

During the intense SPE's, substantial fluxes of secondary fast electrons e* with energies in the range of tens to hundres of electron volts are produced (see Section 10.3). These fast electrons initiate the ion and neutral odd-nitrogen chemistry by its action on the major neutral components of the middle atmosphere and produce various ions and $N^{(*)}$ (CRUTZEN *et al.*, 1975; RUSCH *et al.*, 1981),

$$e^* + N_2 \rightarrow N_2^+ + 2e \tag{13.13}$$

$$e^* + N_2 \rightarrow N^{(*)} + N^+ + 2e \tag{13.14}$$

$$e^* + N_2 \rightarrow N^{(*)} + N^{(*)} + e \tag{13.15}$$

$$e^* + O_2 \rightarrow O_2^+ + 2e \tag{13.16}$$

$$e^* + O_2 \rightarrow O + O^+ + 2e. \tag{13.17}$$

These reactions are followed by a series of ion-atom interchange and ion-electron recombination reactions including (13.1), (13.3), (13.5) and

$$N^+ + O_2 \rightarrow O^+ + NO, k = 2.4 \times 10^{-11} \text{ cm}^3 \text{ sec}^{-1} \quad (13.18)$$

$$\rightarrow NO^+ + O, k = 3.0 \times 10^{-10} \quad \text{"} \quad (13.19)$$

$$\rightarrow O_2^+ + N^{(*)}, k = 3.0 \times 10^{-10} \quad \text{"} \quad (13.20)$$

which produce additional atomic nitrogen and eventually nitric oxide by the reaction (13.10). The increased NO_x should then cause the decrease in ozone density in the middle atmosphere as was discussed in Section 10.2.1.

13.3 Positive Cluster Ions

13.3.1 Proton water cluster ions in the D-region

Ionic components in the middle atmosphere are measured by mass spectrometers aboard rockets and balloons. The first such observations were made for positive ions in the *D*-region (mostly upper mesosphere and lower thermosphere) by NARCISI and BAILEY (1965). They found, as expected, NO^+ (mass number 30) and $O_2{}^+$ (32) as being the most abundant ions above ~80 km. Below this height, however, they observed that these ions were replaced by ions with mass numbers 19, 37 and 55. Observations have been repeated by NARCISI (1967), GOLDBERG and BLUMLE (1970),

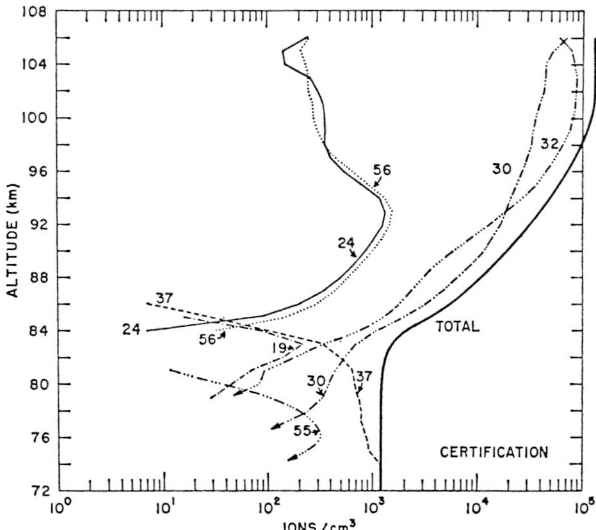

FIG. 13.5. Height profiles of major positive ions observed in the D- and E-regions at about 20° solar zenith angle. Numerals indicate the mass numbers of ions (NARCISI *et al.*, 1972).

GOLDBERG and AIKIN, (1971), JOHANNESSEN and KRANKOWSKY (1972), ZBINDEN *et al.* (1975) and others. Figure 13.5 shows a result obtained by NARCISI *et al.* (1972). By increasing the sensitivity of the mass spectrometer at larger mass numbers, ARNOLD and JOOS (1979) observed ions with larger mass numbers 73, 91, 109, 127 and 145 in the cold mesopause at high latitudes during summer, as is shown in Fig. 13.6.

The positive ions with mass numbers 19, 37, 55, 73, ..., 145 are interpreted as the proton water cluster (or hydrated) ions whose general form is $H^+ (H_2O)_n$ (or perhaps more properly $H_3O^+(H_2O)_{n-1}$), where $n=1$, 2, 3, ..., 8. In general, the clustering bonds become weaker for larger n values, and the ions with larger mass numbers are unstable particularly at higher temperatures; therefore, they can be produced more easily in the cold mesosphere developed at high latitudes during summer.

The conversion of the primarily produced O_2^+ ions into the proton hydrated ions is understood quite satisfactority through the following reaction series (FEHSENFELD and FERGUSON, 1969; GOOD *et al.*, 1970): The associaiton of O_2^+ with O_2 first yields $O_2^+ \cdot O_2$ (or O_4^+), ions, which change to the O_2^+ water cluster ions by the *switching reaction* with H_2O

$$O_2^+ \cdot O_2 + H_2O \rightarrow O_2^+ \cdot H_2O + O_2. \tag{13.21}$$

The further reaction with H_2O produces the first proton hydrates (19^+) by

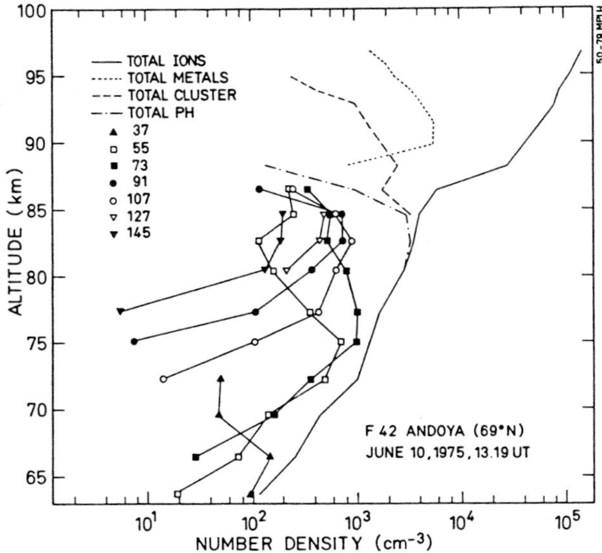

FIG. 13.6. Height profiles of number densities of proton water cluster ions in the cold mesosphere at high latitudes in summer (ARNOLD and JOOS, 1979).

$$O_2^+ \cdot H_2O + H_2O \rightarrow H^+ \cdot (H_2O) + OH + O_2. \qquad (13.22)$$

The second (37^+), third (55^+) and higher order proton hydrates are then produced by successive clustering and thermal breakup (reverse) reactions,

$$H^+ \cdot (H_2O)_{n-1} + H_2O + M \rightleftarrows H^+(H_2O)_n + M. \qquad (13.23)$$

O_4^+ ions are lost mainly by reaction with O. Since the concentration of O increases rapidly above ~80 km, the above theory accounts for the sudden cessation in the production of water cluster ions above ~80 km. An additional reason is perhaps the cold temperature at the mesopause; most water vapor is condensed out when it passes through the cold mesopause and the atmosphere above that height is extremely dry. This is the *cold trap* effect which is similar to but much stronger than the effect due to cold temperatures at the tropopause. The water cluster ions may serve as the condensation nuclei for the production of *noctilucent clouds* in the summer polar mesosphere, particularly when there are no other nucleus such as meteoric dust (REID, 1975).

An understanding of the conversion of NO^+ ions into the proton hydrates is a more difficult problem, but the following mechanism is most likely (DUNKIN *et al.*, 1971; THOMAS, 1976; REID, 1977; and others): Firstly, hydrates of NO^+ are produced either by the direct three-body association of NO^+ with H_2O

$$NO^+ + H_2O + M \rightarrow NO^+ \cdot H_2O + M \qquad (13.24)$$

or by the two-step process of

$$NO^+ + CO_2 + M \rightarrow NO^+ \cdot CO_2 + M \qquad (13.25)$$

and

$$NO^+ \cdot CO_2 + H_2O \rightarrow NO^+ \cdot H_2O + CO_2. \qquad (13.26)$$

Similar two step reactions may also take place through clustering of NO^+ with N_2, followed by switching reaction with H_2O. Successive hydrations of $NO^+ \cdot H_2O$ ions by three-body association with H_2O produces $NO^+ \cdot (H_2O)_n$. When the third hydrate $NO^+ \cdot (H_2O)_3$ is produced, it can be converted to the proton hydrate of the third order by the fast reaction

$$NO^+ \cdot (H_2O)_3 + H_2O \rightarrow H^+(H_2O)_3 + HNO_2. \qquad (13.27)$$

Proton hydrates of other orders are then produced by successive clustering and thermal breakup reactions (13.23).

Cluster ions are lost mainly by recombination with electrons. Recombination coefficients of water-cluster ions have been measured at various temperatures by LEU *et al.*, (1973) who reported values in the range from

10^{-6} to 10^{-5} cm^3 sec^{-1}. These values are considerably larger than the dissociative recombination coefficients of simple molecular ions N_2^+, O_2^+ and NO^+ given in (13.5)–(13.7). The larger recombination coefficients for water cluster ions seem to favor an interpretation of the observed electron density in the *D*-region in terms of electron-positive ion recombination.

Water cluster ions and their relative abundances among various orders of clustering (*n* values) depend mainly upon the temperature and the amount of water vapor available. REID (1977) calculated four cases by combining the two water-vapor profiles with each of the two temperature profiles. Figure 13.7 (a) and (b) illustrate, respectively, two extreme cases of the warm dry atmosphere (~210 K at the mesopause and 1.2 ppm H_2O at 60 km) and the cold wet atmosphere (~140K at the mesopause and 9 ppm H_2O at 60 km). The colder temperature makes the breakup harder and stabilizes clustering at larger mass numbers. Thus, the colder and the wetter the atmosphere, the more water cluster ions are produced to the extended height range and the more fractional abundance are associated with the higher order (larger *n*) hydrates. Recently, proton hydrates having up to 20 water molecules have been observed at the arctic summer mesosphere by BJÖRN and ARNOLD (1981).

13.3.2 Non-proton hydrated ions in the stratosphere

Measurements of stratospheric positive ions have been made in recent years by *in situ* mass spectrometers using rockets (ARNOLD *et al.*, 1977) and balloons (ARIJS *et al*, 1978; ARNOLD *et al.*, 1978). The balloon observations are probably better since they are not affected by composition changes due to shock wave-induced collisional breakup during sampling. While ions with mass number 55, 73 and 91 were observed and identified as proton hydrates (PH) with $n=3$, 4, and 5, all measurements indicate a clear tendency that these ions disappear quickly below ~40 km, and are replaced by non-proton hydrates (NPH). Figure 13.8 illustrates the variation in fractional abundance of NPH with height. A high accuracy measurement in the balloon-borne experiment indicates the fractional abundance of NPH being ~0.43 at 37 km.

Early observations gave conflicting results for the mass numbers of NPH. The first mass spectrometric measurements by ARNOLD *et al.* (1977) reported the mass numbers 29 ± 2, 42 ± 2, 60 ± 2 and 80 ± 2. These ions were tentatively identified as hydrated protonated formaldehyde ions $H^+\cdot(CH_2O)_m\cdot(H_2O)_n$, but these ion cores were quickly ruled out by FEHSENFED *et al.* (1978) on the basis that those ions react with H_2O to form proton hydrates and CH_2O for small *n*. ARIJS *et al.* (1978) observed ions 60^+, 78^+ and 96^+ and tentatively assigned them to hydrates of N_3^+. FERGUSON (1978) proposed to interpret the observed ions as protonated

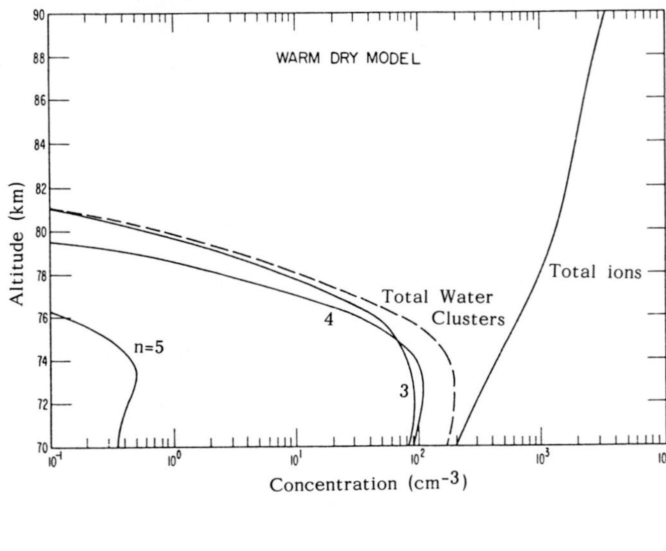

(a)

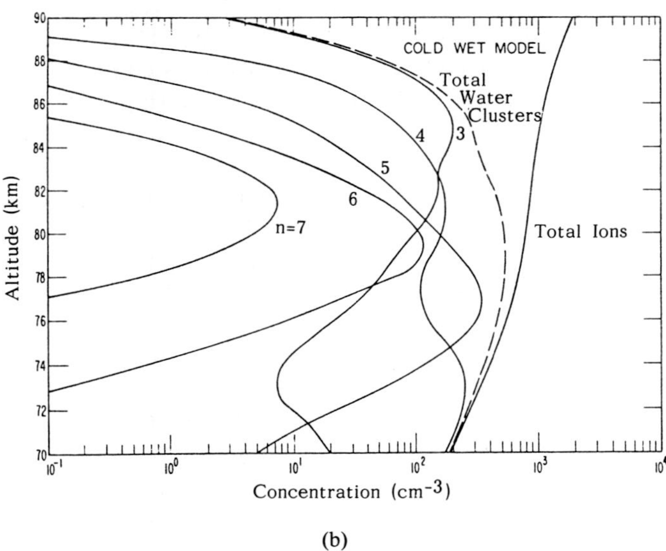

(b)

FIG. 13.7. Profiles of proton water cluster ions $H^+(H_2O)_n$ calculated in a model for a warm dry atmosphere (a), and for a cold wet atmosphere (b) (REID, 1977).

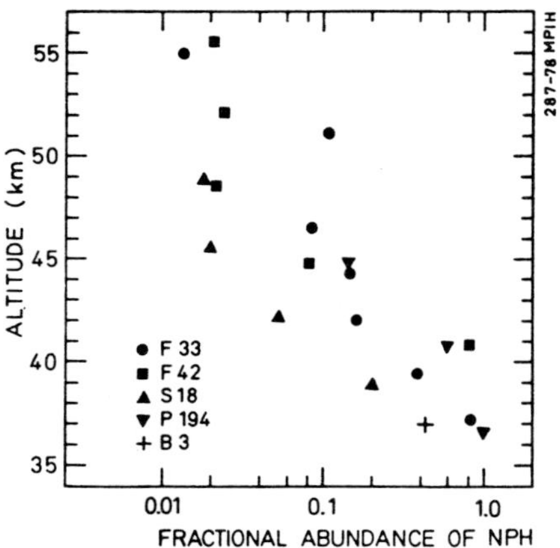

FIG. 13.8. Fractional abundances of "Non-Proton Hydrates" as measured in 4 rocket-borne experiments and in a balloon-borne experiment denoted by + (ARNOLD *et al.*, 1978).

sodium hydroxide ions $H^+NaOH \cdot (H_2O)_n$, which gives 41^+, 59^+, 77^+ and 95^+ for $n=0,1,2$ and 3, respectively. Later, however, ARNOLD *et al.*. (1981a) have found an evidence contrary to FERGUSON's interpretation (see below, p. 342).

In a recent article, ARNOLD *et al.* (1981a) identify the NPH as $H X^+ X_L (H_2O)_M$ and propose that NPH are formed from PH by proton transfer reactions involving a stratospheric trace gas X that has a mass 41 ± 1, a proton affinity larger than that of H_2O and a number density of $\sim 10^5$ cm^{-3} around 36 km. Observed mass numbers and tentative identifications of stratospheric positive ions are shown in Table 13.1.

Composition measurements of stratospheric positive ions have recently

TABLE 13.1 Mass numbers, tentative identifications and fractional abundance of the positive ions measured at ~36.5 km. X is most likely CH_3CN (ARNOLD *et al.*, 1981 a).

Mass (amu)	Ion		Abundance (%)
55 ± 1	$H^+(H_2O)_3$	HP	3.4
73 ± 1	$H^+(H_2O)_4$	HP	26.8
78 ± 1	$H X^+(H_2O)_2$	NHP	7.2
91 ± 1	$H^+(H_2O)_5$	HP	8.9
96 ± 1	$H X^+(H_2O)_3$	NHP	19.6
101 ± 1	$H X^+X(H_2O)$	NHP	10.7
119 ± 1	$H X^+X(H_2O)_2$	NHP	16.1
136 ± 1	$H X^+X(H_2O)_3$	NHP	4.4
141 ± 1	$H X^+X_2(H_2O)$	NHP	2.9

been extended down to ~24 km by HENSCHEN and ARNOLD (1981), whose result indicates that the fractional abundances for NPH's increase from ~ 0.45 at 42 km to ~0.8 at 24 km. At the steady state equilibrium the ratio of abundances of NPH and PH may be given by

$$\frac{[\mathrm{NPH}]}{[\mathrm{PH}]} = \frac{k[X]}{\alpha[M]n_-} \tag{13.28}$$

where k is the rate constant of the proton transfer reactions of PH with X

$$\mathrm{H}^+ \cdot (\mathrm{H_2O})_n + X \rightarrow \mathrm{H}X^+(\mathrm{H_2O})_{n-m} + (\mathrm{H_2O})_m \tag{13.29}$$

and α is the coefficient of ion-ion recombination reaction, which is a 3-body process below ~35 km.

Since k is constant as 3×10^{-9} cm^3 sec^{-1} (SMITH D. *et al.*, 1981) and $\alpha(\sim 2\times10^{-25}$ cm^6 sec$^{-1})$ and n_-(negative ion density) change little with height, (13.28) indicates that the mixing ratio of X is almost proportional to the ratio of [NPH] to [PH]. Thus, it can be inferred from the measured variations of [NPH]/[PH] that the mixing ratios of X is $\sim 2\times10^{-12}$ below 30 km. It decreases slowly above this height to $\sim 5\times10^{-13}$ at 42 km, and drops to 10^{-14}—10^{-13} at the stratospause. This result suggests that the trace gas X probably originates from below ~30 km and that it is photochemically destroyed while it is mixed upwards into the upper stratosphere.

After examining various possibilities, ARNOLD *et al.* (1981a) concluded that X most likely cannot be identified as NaOH or MgOH as was suggested by FERGUSON (1978). Acetonitrile (CH_3CN) may be the best candidate, since it has a proper mass (41$^+$), smaller latent heat of evaporation, smaller bond energies, and originates from lower regions (HENSCHEN and ARNOLD, 1981; ARIJS *et al.*, 1983). A model study of C_3CN was made by BRASSEUR *et al.* (1983b).

The absence of positive cluster ions containing Na^+- or Mg^+- cores may imply that most of these metallic ions produced in the higher regions (~90 km, see Fig. 13.6) by the ablation of meteorites may be converted to aerosols as they are transported down into the lower regions. These aerosols undergo coagulation and sedimentation into the middle stratosphere where it may play a role in inducing sulfuric acid vapor condensation and thereby may contribute to the formation of the "Junge layer" of stratospheric aerosols (see Appendix J).

13.4 Negative Ions

13.4.1 NO_3^-, CO_3^- and HCO_3^- and their hydration in the D-region

When electrons are produced by ionization, a negative ion chemistry is initiated by their attachment to neutral constituents,

$$e + O_2 + M \rightarrow O_2^- + M \tag{13.30}$$

and

$$e + O_3 \rightarrow O^- + O_2. \tag{13.31}$$

The relatively unstable O_2^- and O^- ions are converted eventually to stable ions NO_3^- and HCO_3^- predominately through the following chain reactions (FERGUSON, 1974): First, O_3^- ions are produced mainly by the reaction

$$O_2^- + O_3 \rightarrow O_3^- + O_2 \tag{13.32}$$

and partly by the reaction

$$O^- + O_3 \rightarrow O_3^- + O \tag{13.33}$$

or

$$O^- + O_2 + M \rightarrow O_3^- + M \tag{13.34}$$

which follow a reaction series producing NO_3^-

$$O_3^- + CO_2 \rightarrow CO_3^- + O_2 \tag{13.35}$$
$$CO_3^- + NO \rightarrow NO_2^- + CO_2 \tag{13.36}$$
$$NO_2^- + O_3 \rightarrow NO_3^- + O_2. \tag{13.37}$$

The reactions of NO_2^- with H and of O^- with CH_4 produced OH^-, whose three-body reaction with CO_2 is the source for HCO_3^- ions.

The terminal ions NO_3^- and HCO_3^- are sufficiently stable against attacks from atmospheric species. They are lost only by recombination with positive ions, which has a rate coefficient as small as 6×10^{-8} cm^3 sec^{-1} (SMITH *et al.*, 1976). The reason for the predominance of NO_3^- is its large electron affinity, which has been measured as 3.91 ± 0.24 eV (DAVIDSON *et al.*, 1977).

The negative ion chemistry is controlled to a large extent by the concentrations of neutral constituents, particularly O_3, O, NO, CO_2 and H, as are seen from the above reaction scheme. Because the reaction rate constant for reaction (13.36) is relatively small and also because the NO concentration is low, this reaction is a rate-determining reaction for the production of NO_3^-. A part of CO_3^- should also return to O_2^- by reaction with O.

Hydrations of CO_3^-, NO_3^- and HCO_3^- are discussed by KEESEE *et al.* (1979) based on laboratory results. They conclude that in the normal nighttime lower ionosphere, monohydrates or dihydrates of these negative ions should be prevalent; however, the extent of hydration is sensitive to atmospheric temperature.

The first rocket-borne *in situ* measurements of the *D*-region negative ion composition were made by NARCISI *et al.* (1971, 1972) and ARNOLD *et al.* (1971). These two measurements are apparently contradictory in rela-

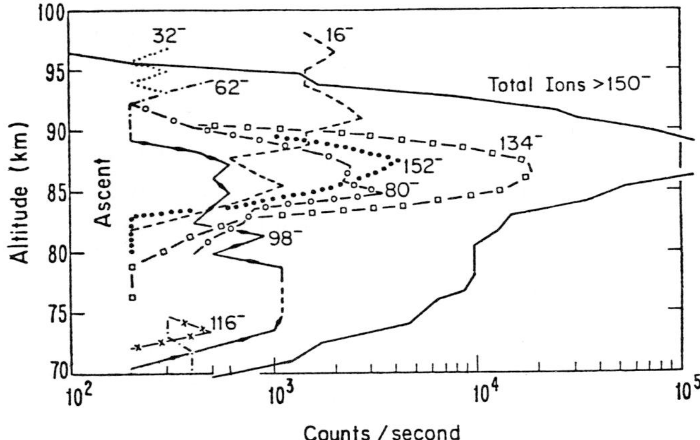

FIG. 13.9. Negative ion composition in the D-region measured during total solar eclipse (NARCISI *et al.*, 1972).

tion to the extent of hydration of negative ions. As is shown in Fig. 13.9, Narcisi *et al.* observed hydrates of up to five waters, i.e. $NO_3^- \cdot (H_2O)_n$, $n=0, 1, 2, \cdots, 5$ during a total eclipse. This results in the mass number 62, 80, 98, 116, 134 and 152. Arnold *et al.*, on the other hand, observed little or no hydrated negative ions. However, KEESEE *et al.* (1979) argues that the differences in time and place, and concomitant variation of atmospheric conditions may account for many of the differences in the observational results. While the results of Narcisi *et al.* agree reasonably well with the hydration on the basis of laboratory results, the results of Arnold *et al.* could be consistent with a somewhat warmer and/or drier mesosphere. Measurements of the H_2O emission lines at 22.2 GHz by GIBBINS *et al.* (1982) show that the concentration in the mesospheric water vapor can be highly variable. Large variabilities have been observed with periods of 2–5 day duration, suggesting that they may be originated by dynamical effects due to planetary wave propagation.

13.4.2 Acid cluster ions in the stratosphere

The mass numbers of most abundant negative ions are large (>150) even in the *D*-region as is seen in Fig. 13.9. The first negative ion observations in the stratosphere were reported by ARNOLD and HENSCHEN (1978), who observed negative ions in the mass range from 125 to 295 amu. The mass numbers of stratospheric negative ions are generally much larger than those of stratospheric positive ions, which are in the range from 55 to 140.

Many observations have been made since then with higher mass resolutions mostly at the floating levels of balloons ($\sim$30–35 km) by ARIJS *et al.*

(1981, 1982b), ARNOLD *et al.* (1981b), VIGGIANO and ARNOLD (1981a). In the most recent experiments VIGGIANO *et al.* (1983b) extended the mass range of the instrument to over 600 amu and conducted measurements in the height range from 15 to 35 km on both ascent and descent as well as float altitude. The mass numbers of observed negative ions and their tentative identification are given in Table 13.2. The fractional abundances of various ions are shown in Fig. 13.10.

The high mass numbers of stratospheric negative ions indicate cluster ions with core ions and attached ligand molecules. $NO_3^-(62)$ is considered a favoured core ion because of the strong electron affinity of NO_3^-. Thus, the ion of the mass number 125 and the most abundant ion with the mass number 188 are naturally identified as $NO_3^- \cdot HNO_3$ and $NO_3^- \cdot (HNO_3)_2$ respectively. Some of the observed negative ions, therefore have the general form $NO_3^- \cdot (HNO_3)_n$. These ions are formed by switching reactions of $NO_3^- \cdot (H_2O)_n$ with NHO_3.

Most of the other ions are grouped as a family in the general form $HSO_4^- \cdot (H_2SO_4)_m \cdot (HNO_3)_n$. These ions are produced by reactions of the NO_3^- core ions with H_2SO_4.

$$NO_3^- \cdot (HNO_3)_n + H_2SO_4 \rightarrow HSO_4^- \cdot (HNO_3)_n + HNO_3. \quad (13.38)$$

The laboratory measured rate constant for these processes for $n=0$, 1 and 2 are fast (VIGGIANO *et al.*, 1980); the values determined in the recent ex-

TABLE 13.2 Mass numbers and tentative identifications of the negative ions observed at 20–34 km (VIGGIANO *et al.*, 1983 b).

Mass (amu)	Ion
62±2	NO_3^-
80±2	$NO_3^-(H_2O)$
125±2	$NO_3^-(HNO_3)$
143±2	$NO_3^-(NHO_3)\,(H_2O)$
160±2	$HSO_4^-(HNO_3)$
188±2	$NO_3^-(HNO_3)_2$
195±2	$HSO_4^-(H_2SO_4)$
206±2	$NO_3^-(HNO_3)_2(H_2O)$
223±2	$HSO_4^-(HNO_3)_2$
241±2	$HSO_4^-(HNO_3)_2(H_2O)$
251±2	$NO_3^-(HNO_3)_3$
257/258±2	$HSO_4^-(H_2SO_4)\,(HNO_3)$
278±3	$HSO_4^-(H_2SO_4)\,(HSO_3)$
286±3	$HSO_4^-(HNO_3)_3$
293±2	$HSO_4^-(H_2SO_4)_2$
391±2	$HSO_4^-(H_2SO_4)_3$
409±3	$HSO_4^-(H_2SO_4)_3(H_2O)$
489±2	$HSO_4^-(H_2SO_4)_4$

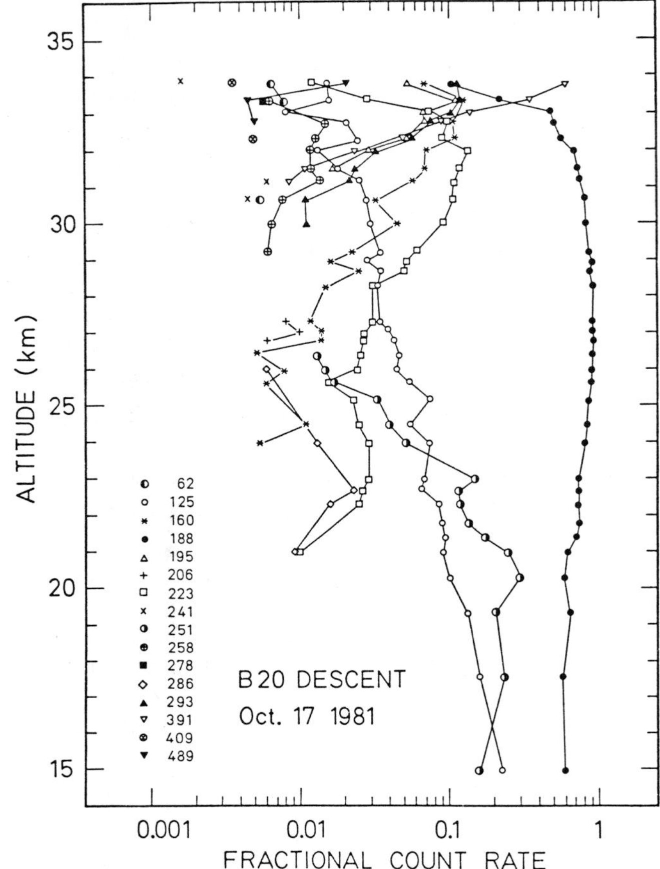

FIG. 13.10. The negative ion height profile in the stratosphere measured during the balloon's descent (VIGGIANO *et al.*, 1983b).

periments are 2.6×10^{-9}, 2.3×10^{-9}, and 1.1×10^{-9} cm^3 sec^{-1} for the three reactions, respectively (VIGGIANO *et al.*, 1982).

As is seen in Fig. 13.10, $NO_3^- \cdot (HNO_3)_2$ ions dominate at all heights except near the very uppermost heights, where HSO_4^- core ions, particularly $HSO_4^- \cdot (H_2SO_4)_3$, tend to dominate. The changeover in the ion composition occurs because the sulfuric acid vapor concentration becomes large above ~30 km due to the temperature increase which increase the H_2SO_4 vapor pressures over the stratospheric aerosol.

Measurements of ion composition in the stratosphere are useful for evaluating the concentrations of sulfuric acid vapor, which is thought to be the most important nucleating agent for stratospheric aerosol particles (see

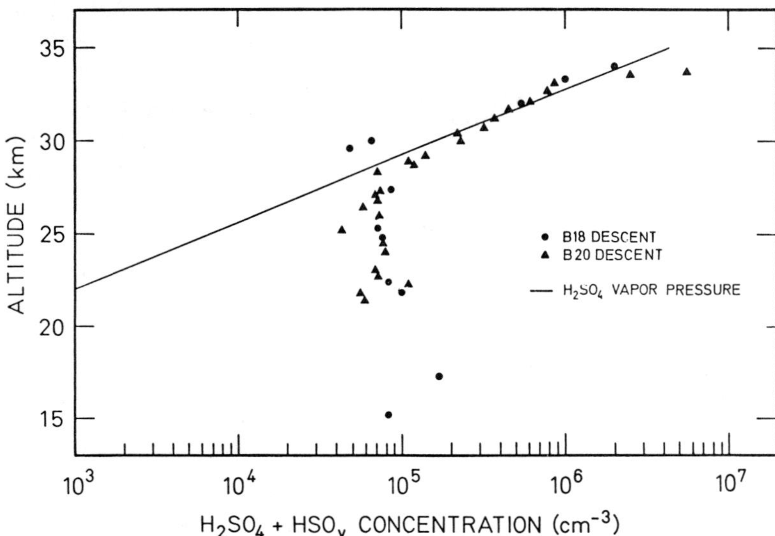

FIG. 13.11 Concentrations for H_2SO_4+HSO_y derived from ion composition measurements on two balloon descent flights compared with the H_2SO_4 vapor pressure calculated for average fall temperatures (VIGGIANO and ARNOLD, 1983a).

Appendix J). The technique makes measurements of this important constituent in the stratosphere possible; its concentration is too low to be measured by other methods. ARNOLD and FABIAN (1980) first applied the method, which relates the concentrations of HSO_4^- cored ions n(S) and NO_3^- cored ions n(N) by

$$k \cdot n(N) \cdot [H_2SO_4] = \alpha \cdot n(S) \cdot n_+ \tag{13.39}$$

where k is the rate constant of the chemical reaction of the type (13.38), α is the ion-ion recombination coefficient, and n_+ is the total positive ion concentration.

The technique is also applied to the newer data by ARNOLD *et al.* (1981c) and VIGGIANO and ARNOLD (1981b, 1983a). In particular, VIGGIANO and ARNOLD (1983a) use a mass spectrometer with a large mass range and better measurements of the parameters involved in the derivation of the H_2SO_4 concentration. They further add the following reaction for the conversion from nitrate cored ions to sulfate cored ions

$$NO_3^- \cdot (HNO_3)_n + HSO_3 \rightarrow HSO_4^- \cdot (HNO_3)_n + NO_2. \tag{13.40}$$

The results from their data analysis for H_2SO_4 plus HSO_y are compared in Fig. 13.11 with the H_2SO_4 vapor pressure calculated over a 75% H_2SO_4/25% H_2O solution for average fall temperatures. The measured concentrations by two balloon flights agree well with each other and follow exceptionally well the H_2SO_4 vapor pressure curve above ~28 km. Below this height supersaturation occurs. This is consistent with the view that aerosols are formed at heights where sulfuric acid is supersaturated. The supersaturation rate reaches ~40 at 23 km, from where aerosols diffuse upwards and H_2SO_4 evaporates until equilibrium is established at higher regions.

APPENDIX

Appendix A (Chapter 2, 5, 8)

U. S. Standard atmosphere, 1976 (see 350 page)

height z(km)	temperature T(°K)	pressure P(mb)	mass density $\rho(\text{g}\cdot\text{cm}^{-3})$	gravitational acceleration $g(\text{cm}\cdot\text{sec}^{-2})$	number density $n_i(\text{cm}^{-3})$	molecular weight m	scale height H(Km)	ozone number density $n(\text{cm}^{-3})$
0	288.15	1.013(3)*	1.225(−3)	9.807(2)	2.547(19)	28.964	8.435	7.50(11)
5	255.68	5.405(2)	7.364(−4)	9.791(2)	1.531(19)	28.964	7.496	5.68(11)
10	223.25	2.650(2)	4.135(−4)	9.776(2)	8.598(18)	28.964	6.555	1.12(12)
15	216.65	1.211(2)	1.948(−4)	9.761(2)	4.049(18)	28.964	6.372	2.63(12)
20	216.65	5.529(1)	8.891(−5)	9.745(2)	1.849(18)	28.964	6.382	4.75(12)
25	221.55	2.549(1)	4.008(−5)	9.730(2)	8.334(17)	28.964	6.536	4.27(12)
30	226.51	1.197(1)	1.841(−5)	9.715(2)	3.828(17)	28.964	6.693	2.51(12)
35	236.51	5.746(0)	8.463(−6)	9.700(2)	1.760(17)	28.964	7.000	1.39(12)
40	250.35	2.871(0)	3.996(−6)	9.684(2)	8.308(16)	28.964	7.421	6.04(11)
45	264.16	1.491(0)	1.966(−6)	9.669(2)	4.088(16)	28.964	7.842	2.20(11)
50	270.65	7.978(−1)	1.027(−6)	9.654(2)	2.135(16)	28.964	8.047	6.60(10)
60	247.02	2.196(−1)	3.097(−7)	9.624(2)	6.439(15)	28.964	7.368	7.30(9)
70	219.59	5.221(−2)	8.283(−8)	9.594(2)	1.722(15)	28.964	6.570	5.36(8)
80	198.64	1.052(−2)	1.846(−8)	9.564(2)	3.848(14)	28.964	5.962	
90	186.87	1.836(−3)	3.416(−9)	9.535(2)	7.116(13)	28.91	5.636	
100	195.08	3.201(−4)	5.60(−10)	9.505(2)	1.189(13)	28.40	6.009	
110	240.00	7.104(−5)	9.71(−11)	9.476(2)	2.144(12)	27.27	7.723	
120	360.00	2.538(−5)	2.22(−11)	9.447(2)	5.107(11)	26.20	12.09	
150	634.39	4.542(−6)	2.08(−12)	9.360(2)	5.186(10)	24.10	23.38	
200	854.56	8.474(−7)	2.54(−13)	9.218(2)	7.182(9)	21.30	36.18	
300	976.01	8.770(−8)	1.92(−14)	8.943(2)	6.509(8)	17.73	51.19	
400	995.83	1.452(−8)	2.80(−15)	8.680(2)	1.056(8)	15.98	59.68	
600	999.85	8.21(−10)	1.14(−16)	8.188(2)	5.950(6)	11.51	88.24	
1000	1000.0	7.51(−11)	3.56(−18)	7.322(2)	5.442(5)	3.94	288.2	

*$A(b)$ Should read as $A\times10^{b}$.

Appendix B (Chapter 3, 12)

Black body radiation

The energy density of radiation emitted from a black body of the temperature T in the wavelength range λ to $\lambda+\mathrm{d}\lambda$ is expressed by the Planck radiation formula

$$u_\lambda \cdot \mathrm{d}\lambda = \frac{8\pi hc}{\lambda^5} \frac{\mathrm{d}\lambda}{e^{hc/k\lambda T}-1} \text{ erg cm}^{-3} \tag{B.1}$$

where h is the Planck constant, c is the velocity of light and k is the Boltzmann constant. The energy flux in unit solid angle is then calculated by

$$E_\lambda \cdot \mathrm{d}\lambda = \frac{c}{4\pi} u_\lambda \mathrm{d}\lambda = \frac{2hc^2}{\lambda^5} \frac{\mathrm{d}\lambda}{e^{hc/k\lambda T}-1} \text{ erg cm}^{-2} \text{ sec}^{-1} \tag{B.2}$$

Inserting numerical values for the physical constants in (B.2), we obtain

$$E_\lambda \cdot \mathrm{d}\lambda = \frac{1.19\times 10^{-5}}{\lambda^5} \frac{\mathrm{d}\lambda}{e^{1.439/\lambda T}-1} \text{ erg cm}^{-2} \text{ sec}^{-1} \tag{B.3}$$

This expression is usually used as the *source function* in calculations of IR radiative transfer in planetary atmospheres.

The solar radiation flux I_∞ received at the top of the atmosphere may be calculated by multiplying the solid angle of the sun $\Delta\omega$ by (B.3) as follows:

$$I_\infty \cdot \mathrm{d}\lambda = E_\lambda \cdot \mathrm{d}\lambda \cdot \Delta\omega = E_\lambda \cdot \frac{\pi a^2}{r^2} \mathrm{d}\lambda \tag{B.4}$$

where a is the radius of the sun ($\sim 6.9555\times 10^5$ km), and r is the distance between the sun and the earth. For the average sun-earth distance ($\sim 1.496\times 10^8$ km), (B.4) becomes by virtue of (B.3)

$$I_\infty \cdot \mathrm{d}\lambda = \frac{8.09\times 10^{-10}}{\lambda^5} \frac{\mathrm{d}\lambda}{e^{1.439/\lambda T}-1} \text{ erg cm}^{-2} \text{ sec}^{-1} \tag{B.5}$$

The maximum intensity of radiation occurs under the condition $\partial E_\lambda/\partial\lambda = 0$, which gives

$$\lambda_{\mathrm{m}} \cdot T = 0.2897 \text{ cm deg} \tag{B.6}$$

This equation is called Wien's displacement law. Thus, the maximum intensity occurs at the wavelength $\lambda_{\mathrm{m}} = 4\ 82.8$ nm (blue visible light) for solar radiation of 6,000 K, and at $\lambda_{\mathrm{m}} = 9.66\ \mu$m (IR range) for terrestrial radiation of 300 K.

The total energy emitted from the black body surface into the hemisphere at *all* wavelengths is calculated by

$$E(T)=\int_0^\infty\int_0^{\frac{\pi}{2}}\frac{2\pi r\sin\theta\cdot\mathrm{d}(r\cdot\sin\theta)}{r^2}E_\lambda\cdot\mathrm{d}\theta\cdot\mathrm{d}\lambda=\pi\int_0^\infty E_\lambda\mathrm{d}\lambda=\sigma\cdot T^4 \tag{B.7}$$

where σ is the Stefan-Boltzmann constant (5.672×10^{-5} erg cm^{-2} sec^{-1} deg^{-4}).

It is sometimes necessary to calculate the energy flux in some particular wavelength range. In particular, for calculations of the atmospheric cooling rate due to terrestrial radiation, the source function has to be calculated by integrating (B.2) for the specific wavelength range of the absorption band of particular molecules such as CO_2, H_2O and O_3. A convenient and practical method of integration has been developed by Widger and Woodall (1976) as is described below.

Integration of (B.2) for the wavelength range 0 to λ would give

$$I_\infty(0,\lambda)=\frac{2k^4}{h^3c^2}T^4\int_x^\infty\frac{x^3}{e^x-1}\mathrm{d}x \tag{B.8}$$

where

$$x=\frac{hc}{\lambda kT}. \tag{B.9}$$

The integral on the right hand side of (B.8) may be calculated by the following expansion series:

$$I(x)=\int_x^\infty\frac{x^3}{e^x-1}\mathrm{d}x=\sum_{n=1}^\infty e^{-nx}\left(\frac{x^3}{n}+\frac{3x^2}{n^2}+\frac{6x}{n^3}+\frac{6}{n^4}\right) \tag{B.10}$$

Over most ranges of temperature and wavelength of meteorological concern, (B.10) gives a rapidly converging solution. For most practical purposes, the first 2 or 3 terms of the series are enough for cooling rate calculations. Using (B.8) and (B.10), we can calculate the source function for wavelength between λ_1 and λ_2 by

$$S(\lambda_1,\lambda_2)=I_\infty(0,\lambda_2)-I_\infty(0,\lambda_1)=2.776\times10^{-6}T^4\{I(X_2)-I(X_1)\} \tag{B.11}$$

where

$$X_i=\frac{hc}{\lambda_i kT}\,(i=1\text{ or }2). \tag{B.12}$$

Appendix C (Chapter 3, 4)

Solar fluxes and absorption cross-sections of various molecules

The solar photon fluxes have been compiled from the data of BROADFOOT(1972), THEKAEKARA(1973), SAMAIN and SIMON(1976), BRUECKNER *et al.* (1976) and SIMON (1978). Absorption cross-sections are taken from various sources, as discussed in Chapter 4 for each molecule. The wavelength divisions and data in the table (see 354–359 pages) are used in the model calculations discussed in Chapter 7.

DIV	WAVELENGTH (nm)	SOLAR FLUX ($CM^{-2} \cdot SEC^{-1}$)	ABSORPTION CROSS SECTION (CM^2) O2	O3	N2O	NO2	N2O5	HNO3	H2O	HNO4
Ø	LYMAN-ALPHA (121.57)	3.ØØE+11	1.ØØE-2Ø						1.5ØE-17	
1	135.Ø - 14Ø.Ø	1.ØØE+1Ø	6.7ØE-18	1.6ØE-17	4.8ØE-18	9.28E-18	-I	-I	3.4ØE-18	-I
2	14Ø.Ø - 145.Ø	1.4ØE+1Ø	1.38E-17	7.7ØE-18	1.Ø8E-18	1.3ØE-17	-I	-I	8.4ØE-19	-I
3	145.Ø - 15Ø.Ø	2.25E+1Ø	1.43E-17	5.1ØE-18	6.6ØE-18	1.11E-17	-I	-I	5.1ØE-19	-I
4	15Ø.Ø - 155.Ø	3.5ØE+1Ø	1.12E-17	3.9ØE-18	1.9ØE-18	1.3ØE-17	-I	-I	1.Ø7E-18	-I
5	155.Ø - 16Ø.Ø	5.5ØE+1Ø	7.9ØE-18	2.ØØE-18	1.Ø2E-19	1.48E-17	-I	-I	2.3ØE-18	-I
6	16Ø.Ø - 165.Ø	8.5ØE+1Ø	4.8ØE-18	1.15E-18	3.7ØE-2Ø	1.52E-17	-I	-I	3.6ØE-18	-I
7	165.Ø - 17Ø.Ø	1.35E+11	2.2ØE-18	8.3ØE-19	5.8ØE-2Ø	1.52E-17	-I	7.1ØE-18	4.6ØE-18	-I
8	17Ø.Ø - 175.Ø	2.3ØE+11	6.7ØE-19	8.1ØE-19	1.Ø2E-19	1.3ØE-17	-I	9.4ØE-18	3.6ØE-18	-I
9	175.Ø - 176.3	7.4ØE+1Ø		8.23E-19	1.22E-19	9.3ØE-18	-I	1.16E-17	2.33E-18	-I
1Ø	176.3 - 176.9	3.66E+1Ø		8.26E-19	1.26E-19	8.37E-18	-I	1.22E-17	2.Ø9E-18	-I
11	176.9 - 177.5	3.84E+1Ø		8.29E-19	1.29E-19	7.81E-18	-I	1.26E-17	1.94E-18	-I
12	177.5 - 178.3	5.6ØE+1Ø		8.19E-19	1.31E-19	7.Ø7E-18	-I	1.29E-17	1.59E-18	-I
13	178.3 - 179.3	7.6ØE+1Ø		7.92E-19	1.34E-19	6.7ØE-18	-I	1.29E-17	1.Ø8E-18	-I
14	179.3 - 18Ø.4	9.35E+1Ø		7.62E-19	1.36E-19	5.58E-18	-I	1.3ØE-17	6.92E-19	-I
15	18Ø.4 - 181.6	1.14E+11		7.29E-19	1.37E-19	4.46E-18	-I	1.3ØE-17	4.17E-19	-I
16	181.6 - 183.1	1.65E+11		6.94E-19	1.36E-19	3.76E-18	-I	1.3ØE-17	2.35E-19	-I
17	183.1 - 184.6	1.95E+11		6.65E-19	1.32E-19	2.79E-18	-I	1.26E-17	1.Ø7E-19	-I
18	184.6 - 186.3	2.55E+11	Schumann-Runge bands	6.35E-19	1.26E-19	2.23E-18	-I	1.21E-17	4.44E-2Ø	-I
19	186.3 - 188.2	3.42E+11		6.Ø4E-19	1.16E-19	1.49E-18	-I	1.16E-17	1.69E-2Ø	-I
2Ø	188.2 - 19Ø.2	4.6ØE+11		5.43E-19	1.Ø3E-19	1.12E-18	-I	1.14E-17	6.92E-21	-I
21	19Ø.2 - 192.4	6.38E+11		4.82E-19	8.77E-2Ø	7.44E-19	-I	1.14E-17	2.71E-21	9.7ØE-18
22	192.4 - 194.7	8.97E+11		4.29E-19	7.Ø9E-2Ø	5.95E-19	-I	1.Ø3E-17	1.Ø1E-21	8.8ØE-18
23	194.7 - 197.2	1.25E+12		3.86E-19	5.4ØE-2Ø	4.46E-19	-I	8.29E-18	3.49E-22	7.7ØE-18
24	197.2 - 198.5	8.45E+11		3.47E-19	4.21E-2Ø	4.Ø9E-19	-I	6.23E-18	1.Ø2E-22	6.3ØE-18
25	198.5 - 2ØØ.Ø	1.19E+12		3.47E-19	3.45E-2Ø	4.Ø9E-19	-I	6.23E-18	1.Ø2E-22	5.6ØE-18
26	2ØØ.Ø - 2Ø2.5	2.25E+12		3.18E-19	2.52E-2Ø	4.46E-19	-I	4.37E-18	2.77E-23	4.9ØE-18
27	2Ø2.5 - 2Ø5.Ø	2.3ØE+12		3.21E-19	1.63E-2Ø	4.84E-19	-I	3.21E-18	7.62E-24	4.ØØE-18
28	2Ø5.Ø - 21Ø.Ø	6.ØØE+12	1.18E-23	3.75E-19	7.85E-21	3.5ØE-19	6.6ØE-18	2.1ØE-18	-I	2.9ØE-18
29	21Ø.Ø - 215.Ø	1.15E+13	1.Ø8E-23	5.3ØE-19	2.56E-21	4.ØØE-19	4.7ØE-18	9.7ØE-19	-I	1.9ØE-18
3Ø	215.Ø - 22Ø.Ø	2.ØØE+13	9.3ØE-24	9.5ØE-19	7.36E-22	4.ØØE-19	2.9ØE-18	3.28E-19	-I	1.4ØE-18
31	22Ø.Ø - 225.Ø	3.ØØE+13	7.3ØE-24	1.8ØE-18	1.96E-22	3.6ØE-19	1.85E-18	1.44E-19	-I	1.Ø5E-18
32	225.Ø - 23Ø.Ø	3.55E+13	5.ØØE-24	2.9ØE-18	5.25E-23	2.6ØE-19	1.2ØE-18	8.51E-2Ø	-I	8.6ØE-19
33	23Ø.Ø - 235.Ø	3.ØØE+13	2.9ØE-24	4.4ØE-18	1.53E-23	1.7ØE-19	8.ØØE-19	5.63E-2Ø	-I	7.4ØE-19
34	235.Ø - 24Ø.Ø	2.8ØE+13	1.4ØE-24	6.2ØE-18	5.2ØE-24	9.ØØE-2Ø	6.ØØE-19	3.74E-2Ø	-I	6.3ØE-19
35	24Ø.Ø - 245.Ø	3.3ØE+13	3.1ØE-25	8.1ØE-18	2.18E-24	4.5ØE-2Ø	5.3ØE-19	2.6ØE-2Ø	-I	5.5ØE-19
36	245.Ø - 25Ø.Ø	4.ØØE+13	-I	9.8ØE-18	-I	3.ØØE-2Ø	3.8ØE-19	2.1ØE-2Ø	-I	4.5ØE-19
37	25Ø.Ø - 255.Ø	5.ØØE+13	-I	1.1ØE-17	-I	2.3ØE-2Ø	3.ØØE-19	1.95E-2Ø	-I	3.85E-19
38	255.Ø - 26Ø.Ø	6.ØØE+13	-I	1.1ØE-17	-I	1.9ØE-2Ø	2.3ØE-19	1.94E-2Ø	-I	3.1ØE-19
39	26Ø.Ø - 265.Ø	8.ØØE+13	-I	1.Ø7E-17	-I	2.2ØE-2Ø	2.2ØE-19	1.9ØE-2Ø	-I	2.55E-19
4Ø	265.Ø - 27Ø.Ø	1.45E+14	-I	9.8ØE-18	-I	2.8ØE-2Ø	1.65E-19	1.8ØE-2Ø	-I	2.ØØE-19
41	27Ø.Ø - 275.Ø	1.45E+14	-I	7.7ØE-18	-I	4.ØØE-2Ø	1.28E-19	1.63E-2Ø	-I	1.55E-19
42	275.Ø - 28Ø.Ø	5.25E+13	-I	5.8ØE-18	-I	5.ØØE-2Ø	9.4ØE-2Ø	1.4ØE-2Ø	-I	1.Ø5E-19

43	28Ø.Ø	-	285.Ø	6.5ØE+13	-I	3.8ØE-18	-I	6.6ØE-2Ø	6.9ØE-2Ø	1.14E-2Ø	-I	7.6ØE-2Ø
44	285.Ø	-	29Ø.Ø	1.5ØE+14	-I	2.4ØE-18	-I	7.7ØE-2Ø	5.3ØE-2Ø	8.77E-21	-I	5.ØØE-2Ø
45	29Ø.Ø	-	295.Ø	4.5ØE+14	-I	1.43E-18	-I	9.5ØE-2Ø	4.1ØE-2Ø	6.34E-21	-I	3.2ØE-2Ø
46	295.Ø	-	3ØØ.Ø	3.5ØE+14	-I	7.5ØE-19	-I	1.1ØE-19	3.1ØE-2Ø	4.26E-21	-I	2.Ø5E-2Ø
47	3ØØ.Ø	-	3Ø2.5	1.53E+14	-I	3.22E-19	-I	1.25E-19	2.ØØE-2Ø	2.4ØE-21	-I	1.45E-2Ø
48	3Ø2.5	-	3Ø5.Ø	1.5ØE+14	-I	2.27E-19	-I	1.32E-19	1.8ØE-2Ø	1.8ØE-21	-I	1.2ØE-2Ø
49	3Ø5.Ø	-	3Ø7.5	1.58E+14	-I	1.63E-19	-I	1.55E-19	1.6ØE-2Ø	1.4ØE-21	-I	9.6ØE-21
5Ø	3Ø7.5	-	31Ø.Ø	1.7ØE+14	-I	1.11E-19	-I	1.7ØE-19	1.4ØE-2Ø	1.1ØE-21	-I	7.6ØE-21
51	31Ø.Ø	-	312.5	1.78E+14	-I	8.57E-2Ø	-I	1.8ØE-19	1.25E-2Ø	8.ØØE-22	-I	6.2ØE-21
52	312.5	-	315.Ø	1.95E+14	-I	6.17E-2Ø	-I	1.9ØE-19	1.18E-2Ø	5.6ØE-22	-I	4.8ØE-21
53	315.Ø	-	317.5	2.1ØE+14	-I	4.72E-2Ø	-I	2.ØØE-19	9.5ØE-21	3.4ØE-22	-I	3.8ØE-21
54	317.5	-	32Ø.Ø	2.5ØE+14	-I	3.25E-2Ø	-I	2.3ØE-19	8.ØØE-21	2.1ØE-22	-I	3.1ØE-21
55	32Ø.Ø	-	325.Ø	7.2ØE+14	-I	1.63E-2Ø	-I	2.6ØE-19	6.6ØE-21	1.2ØE-22	-I	2.3ØE-21
56	325.Ø	-	33Ø.Ø	7.ØØE+14	-I	8.57E-21	-I	2.9ØE-19	5.ØØE-21	2.ØØE-23	-I	1.5ØE-21
57	33Ø.Ø	-	335.Ø	8.ØØE+14	-I	6.4ØE-21	-I	3.ØØE-19	4.21E-21	-I	-I	-I
58	335.Ø	-	34Ø.Ø	8.5ØE+14	-I	3.2ØE-21	-I	3.1ØE-19	3.ØØE-21	-I	-I	-I
59	34Ø.Ø	-	345.Ø	9.ØØE+14	-I	1.47E-21	-I	3.25E-19	2.68E-21	-I	-I	-I
6Ø	345.Ø	-	35Ø.Ø	1.ØØE+15	-I	6.7ØE-22	-I	3.4ØE-19	2.15E-21	-I	-I	-I
61	35Ø.Ø	-	355.Ø	1.Ø3E+15	-I	3.ØØE-22	-I	3.6ØE-19	1.75E-21	-I	-I	-I
62	355.Ø	-	36Ø.Ø	1.Ø5E+15	-I	1.3ØE-22	-I	3.8ØE-19	1.33E-21	-I	-I	-I
63	36Ø.Ø	-	365.Ø	1.Ø5E+15	-I	4.ØØE-23	-I	4.1ØE-19	1.ØØE-21	-I	-I	-I
64	365.Ø	-	37Ø.Ø	1.18E+15	-I	-I	-I	4.4ØE-19	6.3ØE-22	-I	-I	-I
65	37Ø.Ø	-	375.Ø	1.23E+15	-I	-I	-I	4.6ØE-19	4.ØØE-22	-I	-I	-I
66	375.Ø	-	38Ø.Ø	1.23E+15	-I	-I	-I	4.6ØE-19	2.55E-22	-I	-I	-I
67	38Ø.Ø	-	385.Ø	1.15E+15	-I	-I	-I	4.3ØE-19	1.34E-22	-I	-I	-I
68	385.Ø	-	39Ø.Ø	1.1ØE+15	-I	-I	-I	4.1ØE-19	-I	-I	-I	-I
69	39Ø.Ø	-	395.Ø	1.2ØE+15	-I	-I	-I	4.5ØE-19	-I	-I	-I	-I
7Ø	395.Ø	-	4ØØ.Ø	1.35E+15	-I	-I	-I	4.5ØE-19	-I	-I	-I	-I

DIV	WAVELENGTH (nm)	CH4	CO2	H2O2	HO2	(CH2O)1	(CH2O)2	HCL	CL2	HOCL
		ABSORPTION CROSS SECTION (CM^2)								
Ø	LYMAN-ALPHA (121.57)	1.8ØE-17	5.ØØE-2Ø							
1	135.Ø - 14Ø.Ø	7.22E-18	7.ØØE-19	7.8ØE-18	-I	-I	-I	1.7ØE-18	-I	-I
2	14Ø.Ø - 145.Ø	8.ØØE-19	5.2ØE-19	5.6ØE-18	-I	-I	-I	2.51E-18	-I	-I
3	145.Ø - 15Ø.Ø	2.Ø8E-2Ø	6.3ØE-19	4.1ØE-18	-I	-I	-I	3.24E-18	-I	-I
4	15Ø.Ø - 155.Ø	6.Ø6E-21	5.2ØE-19	3.6ØE-18	-I	-I	-I	3.72E-18	-I	-I
5	155.Ø - 16Ø.Ø	3.Ø9E-21	3.7ØE-19	3.3ØE-18	-I	-I	-I	3.47E-18	-I	-I
6	16Ø.Ø - 165.Ø	1.23E-21	1.7ØE-19	3.7ØE-18	-I	-I	-I	2.97E-18	-I	-I
7	165.Ø - 17Ø.Ø	-I	7.2ØE-2Ø	3.9ØE-18	-I	-I	-I	2.Ø4E-18	-I	-I
8	17Ø.Ø - 175.Ø	-I	1.3ØE-2Ø	3.2ØE-18	-I	-I	-I	1.31E-18	-I	-I
9	175.Ø - 176.3	-I	4.89E-21	2.13E-18	-I	-I	-I	9.8ØE-19	-I	-I
1Ø	176.3 - 176.9	-I	3.81E-21	1.92E-18	-I	-I	-I	8.7ØE-19	-I	-I
11	176.9 - 177.5	-I	3.21E-21	1.78E-18	-I	-I	-I	8.1ØE-19	-I	-I
12	177.5 - 178.3	-I	2.58E-21	1.67E-18	-I	-I	-I	7.4ØE-19	-I	-I
13	178.3 - 179.3	-I	1.93E-21	1.56E-18	-I	-I	-I	6.7ØE-19	-I	-I
14	179.3 - 18Ø.4	-I	1.38E-21	1.44E-18	-I	-I	-I	5.9ØE-19	-I	-I
15	18Ø.4 - 181.6	-I	9.4ØE-22	1.32E-18	-I	-I	-I	5.1ØE-19	-I	-I
16	181.6 - 183.1	-I	6.Ø9E-22	1.19E-18	-I	-I	-I	4.4ØE-19	-I	-I
17	183.1 - 184.6	-I	3.92E-22	1.14E-18	5.72E-18	-I	-I	3.6ØE-19	-I	-I
18	184.6 - 186.3	-I	2.43E-22	1.1ØE-18	5.85E-18	-I	-I	2.8ØE-19	-I	-I
19	186.3 - 188.2	-I	1.44E-22	1.Ø6E-18	5.98E-18	-I	-I	2.2ØE-19	-I	-I
2Ø	188.2 - 19Ø.2	-I	7.74E-23	7.2ØE-19	6.14E-18	-I	-I	1.6ØE-19	-I	-I
21	19Ø.2 - 192.4	-I	3.95E-23	6.8ØE-19	6.3ØE-18	-I	-I	1.15E-19	-I	-I
22	192.4 - 194.7	-I	2.Ø9E-23	6.3ØE-19	6.44E-18	-I	-I	8.ØØE-2Ø	-I	-I
23	194.7 - 197.2	-I	1.17E-23	5.8ØE-19	6.54E-18	-I	-I	5.ØØE-2Ø	-I	-I
24	197.2 - 198.5	-I	6.76E-24	5.5ØE-19	6.64E-18	-I	-I	3.6ØE-2Ø	-I	-I
25	198.5 - 2ØØ.Ø	-I	6.76E-24	5.2ØE-19	6.64E-18	-I	-I	2.8ØE-2Ø	-I	-I
26	2ØØ.Ø - 2Ø2.5	-I	4.49E-24	5.ØØE-19	6.75E-18	-I	-I	1.9ØE-2Ø	-I	5.2ØE-2Ø
27	2Ø2.5 - 2Ø5.Ø	-I	3.54E-24	4.6ØE-19	6.81E-18	-I	-I	1.2ØE-2Ø	-I	5.4ØE-2Ø
28	2Ø5.Ø - 21Ø.Ø	-I	3.1ØE-24	4.ØØE-19	6.85E-18	-I	-I	6.2ØE-21	-I	5.8ØE-2Ø
29	21Ø.Ø - 215.Ø	-I	2.8ØE-24	3.3ØE-19	6.8ØE-18	-I	-I	2.2ØE-21	-I	7.2ØE-2Ø
3Ø	215.Ø - 22Ø.Ø	-I	2.2ØE-24	2.8ØE-19	6.55E-18	-I	-I	8.ØØE-22	-I	9.5ØE-2Ø
31	22Ø.Ø - 225.Ø	-I	-I	2.4ØE-19	5.85E-18	-I	-I	-I	-I	1.35E-19
32	225.Ø - 23Ø.Ø	-I	-I	2.1ØE-19	4.9ØE-18	-I	-I	-I	-I	1.7ØE-19
33	23Ø.Ø - 235.Ø	-I	-I	1.7ØE-19	4.ØØE-18	-I	-I	-I	-I	2.ØØE-19
34	235.Ø - 24Ø.Ø	-I	-I	1.4ØE-19	3.ØØE-18	-I	-I	-I	-I	2.2ØE-19
35	24Ø.Ø - 245.Ø	-I	-I	1.15E-19	1.98E-18	3.4ØE-21	-I	-I	8.4ØE-22	2.2ØE-19
36	245.Ø - 25Ø.Ø	-I	-I	9.2ØE-2Ø	1.29E-18	3.7ØE-21	-I	-I	1.Ø2E-21	2.ØØE-19
37	25Ø.Ø - 255.Ø	-I	-I	7.5ØE-2Ø	7.7ØE-19	4.6ØE-21	-I	-I	1.3ØE-21	1.6ØE-19
38	255.Ø - 26Ø.Ø	-I	-I	6.2ØE-2Ø	4.6ØE-19	6.1ØE-21	-I	-I	1.8ØE-21	1.15E-19
39	26Ø.Ø - 265.Ø	-I	-I	4.8ØE-2Ø	3.ØØE-19	7.9ØE-21	-I	-I	3.ØØE-21	8.7ØE-2Ø
4Ø	265.Ø - 27Ø.Ø	-I	-I	3.8ØE-2Ø	1.4ØE-19	1.Ø4E-2Ø	-I	-I	8.1ØE-21	6.6ØE-2Ø
41	27Ø.Ø - 275.Ø	-I	-I	3.ØØE-2Ø	7.6ØE-2Ø	1.31E-2Ø	-I	-I	1.2ØE-2Ø	5.5ØE-2Ø
42	275.Ø - 28Ø.Ø	-I	-I	2.2ØE-2Ø	-I	1.55E-2Ø	-I	-I	2.3ØE-2Ø	5.ØØE-2Ø

43	28Ø.Ø - 285.Ø	-I	-I	1.7ØE-2Ø	-I	1.8ØE-2Ø	8.5ØE-21	-I	3.5ØE-2Ø	4.9ØE-2Ø
44	285.Ø - 29Ø.Ø	-I	-I	1.25E-2Ø	-I	2.2ØE-2Ø	8.4ØE-21	-I	5.3ØE-2Ø	5.1ØE-2Ø
45	29Ø.Ø - 295.Ø	-I	-I	8.7ØE-21	-I	2.4ØE-2Ø	8.ØØE-21	-I	7.7ØE-2Ø	5.5ØE-2Ø
46	295.Ø - 3ØØ.Ø	-I	-I	7.7ØE-21	-I	2.5ØE-2Ø	7.7ØE-21	-I	1.Ø5E-19	5.9ØE-2Ø
47	3ØØ.Ø - 3Ø2.5	-I	-I	6.2ØE-21	-I	2.5ØE-2Ø	7.4ØE-21	-I	1.28E-19	6.1ØE-2Ø
48	3Ø2.5 - 3Ø5.Ø	-I	-I	5.7ØE-21	-I	2.5ØE-2Ø	7.3ØE-21	-I	1.45E-19	6.2ØE-2Ø
49	3Ø5.Ø - 3Ø7.5	-I	-I	5.ØØE-21	-I	2.45E-2Ø	7.3ØE-21	-I	1.62E-19	6.3ØE-2Ø
5Ø	3Ø7.5 - 31Ø.Ø	-I	-I	4.3ØE-21	-I	2.4ØE-2Ø	7.3ØE-21	-I	1.74E-19	6.2ØE-2Ø
51	31Ø.Ø - 312.5	-I	-I	3.5ØE-21	-I	2.25E-2Ø	7.4ØE-21	-I	1.9ØE-19	6.1ØE-2Ø
52	312.5 - 315.Ø	-I	-I	3.1ØE-21	-I	2.1ØE-2Ø	7.6ØE-21	-I	2.ØØE-19	5.8ØE-2Ø
53	315.Ø - 317.5	-I	-I	2.7ØE-21	-I	1.85E-2Ø	7.9ØE-21	-I	2.15E-19	5.5ØE-2Ø
54	317.5 - 32Ø.Ø	-I	-I	2.4ØE-21	-I	1.6ØE-2Ø	8.4ØE-21	-I	2.3ØE-19	5.2ØE-2Ø
55	32Ø.Ø - 325.Ø	-I	-I	2.Ø5E-21	-I	1.25E-2Ø	9.7ØE-21	-I	2.45E-19	4.7ØE-2Ø
56	325.Ø - 33Ø.Ø	-I	-I	1.6ØE-21	-I	9.ØØE-21	1.3ØE-2Ø	-I	2.52E-19	4.ØØE-2Ø
57	33Ø.Ø - 335.Ø	-I	-I	1.2ØE-21	-I	5.ØØE-21	1.55E-2Ø	-I	2.5ØE-19	3.3ØE-2Ø
58	335.Ø - 34Ø.Ø	-I	-I	9.6ØE-22	-I	1.8ØE-21	1.4ØE-2Ø	-I	2.45E-19	2.65E-2Ø
59	34Ø.Ø - 345.Ø	-I	-I	7.4ØE-22	-I	4.ØØE-22	1.ØØE-2Ø	-I	2.25E-19	2.1ØE-2Ø
6Ø	345.Ø - 35Ø.Ø	-I	-I	5.6ØE-22	-I	-I	4.5ØE-21	-I	2.Ø5E-19	1.6ØE-2Ø
61	35Ø.Ø - 355.Ø	-I	-I	-I	-I	-I	1.7ØE-21	-I	1.75E-19	1.25E-2Ø
62	355.Ø - 36Ø.Ø	-I	-I	-I	-I	-I	4.5ØE-22	-I	1.45E-19	9.2ØE-21
63	36Ø.Ø - 365.Ø	-I	-I	-I	-I	-I	-I	-I	1.2ØE-19	6.8ØE-21
64	365.Ø - 37Ø.Ø	-I	-I	-I	-I	-I	-I	-I	9.5ØE-2Ø	5.1ØE-21
65	37Ø.Ø - 375.Ø	-I	-I	-I	-I	-I	-I	-I	7.2ØE-2Ø	3.8ØE-21
66	375.Ø - 38Ø.Ø	-I	-I	-I	-I	-I	-I	-I	5.5ØE-2Ø	2.8ØE-21
67	38Ø.Ø - 385.Ø	-I	-I	-I	-I	-I	-I	-I	4.4ØE-2Ø	2.1ØE-21
68	385.Ø - 39Ø.Ø	-I	-I	-I	-I	-I	-I	-I	3.7ØE-2Ø	1.5ØE-21
69	39Ø.Ø - 395.Ø	-I	-I	-I	-I	-I	-I	-I	2.9ØE-2Ø	1.1ØE-21
7Ø	395.Ø - 4ØØ.Ø	-I	-I	-I	-I	-I	-I	-I	2.2ØE-2Ø	8.4ØE-22

DIV	WAVELENGTH (nm)	ABSORPTION CROSS SECTION (CM^2)							
		CLO	CFCL3	CF2CL2	CCL4	CLONO2	CH3CL	SO2	COS
1	135.0 - 140.0	-I	3.40E-17	1.20E-18	-I	-I	-I	-I	-I
2	140.0 - 145.0	-I	1.30E-17	3.80E-19	-I	-I	-I	3.16E-18	-I
3	145.0 - 150.0	-I	5.20E-18	1.90E-19	-I	-I	-I	4.22E-18	-I
4	150.0 - 155.0	-I	3.60E-18	2.10E-19	-I	-I	-I	5.62E-18	-I
5	155.0 - 160.0	-I	3.50E-18	2.90E-19	-I	-I	-I	3.65E-18	-I
6	160.0 - 165.0	-I	5.20E-18	1.80E-19	-I	-I	-I	1.33E-18	-I
7	165.0 - 170.0	-I	6.20E-18	2.00E-19	-I	-I	-I	4.22E-19	-I
8	170.0 - 175.0	-I	5.20E-18	3.20E-19	-I	-I	-I	4.87E-19	-I
9	175.0 - 176.3	-I	4.80E-18	4.50E-19	1.02E-17	-I	9.80E-19	7.80E-19	-I
10	176.3 - 176.9	-I	4.60E-18	5.10E-19	1.01E-17	-I	8.70E-19	8.77E-19	-I
11	176.9 - 177.5	-I	4.50E-18	5.50E-19	1.00E-17	-I	8.10E-19	9.52E-19	-I
12	177.5 - 178.3	-I	4.20E-18	6.00E-19	9.60E-18	-I	7.40E-19	1.04E-18	-I
13	178.3 - 179.3	-I	3.90E-18	6.80E-19	8.90E-18	-I	6.70E-19	1.16E-18	-I
14	179.3 - 180.4	-I	3.70E-18	7.50E-19	7.90E-18	-I	5.90E-19	1.31E-18	-I
15	180.4 - 181.6	-I	3.40E-18	8.40E-19	6.80E-18	-I	5.10E-19	1.50E-18	-I
16	181.6 - 183.1	-I	3.10E-18	9.30E-19	5.60E-18	-I	4.40E-19	1.75E-18	-I
17	183.1 - 184.6	-I	2.80E-18	9.90E-19	4.50E-18	-I	3.60E-19	2.24E-18	-I
18	184.6 - 186.3	-I	2.50E-18	9.50E-19	3.80E-18	-I	2.80E-19	3.00E-18	-I
19	186.3 - 188.2	-I	2.30E-18	8.50E-19	2.50E-18	-I	2.20E-19	4.06E-18	-I
20	188.2 - 190.2	-I	1.80E-18	6.20E-19	1.70E-18	-I	1.60E-19	4.70E-18	-I
21	190.2 - 192.4	-I	1.60E-18	4.50E-19	1.10E-18	4.80E-18	1.00E-19	4.40E-18	-I
22	192.4 - 194.7	-I	1.30E-18	3.20E-19	7.80E-19	3.85E-18	6.80E-20	3.97E-18	-I
23	194.7 - 197.2	-I	1.05E-18	2.00E-19	6.80E-19	3.30E-18	3.80E-20	3.46E-18	2.70E-20
24	197.2 - 198.5	-I	8.50E-19	1.40E-19	6.50E-19	3.10E-18	2.60E-20	2.97E-18	3.00E-20
25	198.5 - 200.0	-I	7.30E-19	1.00E-19	6.40E-19	2.90E-18	2.00E-20	2.97E-18	3.36E-20
26	200.0 - 202.5	-I	5.70E-19	6.00E-20	6.20E-19	2.90E-18	1.20E-20	2.55E-18	5.00E-20
27	202.5 - 205.0	-I	4.30E-19	3.50E-20	6.00E-19	2.90E-18	7.20E-21	2.21E-18	6.27E-20
28	205.0 - 210.0	-I	2.40E-19	1.30E-20	5.30E-19	3.12E-18	3.30E-21	1.78E-18	1.02E-19
29	210.0 - 215.0	-I	9.00E-20	2.90E-21	3.80E-19	3.46E-18	1.10E-21	1.15E-18	1.82E-19
30	215.0 - 220.0	-I	4.00E-20	1.20E-21	2.40E-19	3.55E-18	4.00E-22	5.62E-19	2.71E-19
31	220.0 - 225.0	-I	1.40E-20	4.00E-22	1.15E-19	3.15E-18	-I	2.37E-19	2.91E-19
32	225.0 - 230.0	7.40E-19	6.00E-21	1.00E-22	5.80E-20	2.44E-18	-I	1.33E-19	2.70E-19
33	230.0 - 235.0	1.10E-18	-I	-I	2.70E-20	1.74E-18	-I	2.05E-20	2.02E-19
34	235.0 - 240.0	1.55E-18	-I	-I	1.30E-20	1.20E-18	-I	-I	1.17E-19
35	240.0 - 245.0	2.20E-18	-I	-I	-I	8.46E-19	-I	-I	5.49E-20
36	245.0 - 250.0	3.10E-18	-I	-I	-I	6.10E-19	-I	-I	2.51E-20
37	250.0 - 255.0	4.00E-18	-I	-I	-I	4.62E-19	-I	-I	1.06E-20
38	255.0 - 260.0	5.00E-18	-I	-I	-I	3.53E-19	-I	-I	4.73E-21
39	260.0 - 265.0	5.90E-18	-I	-I	-I	2.70E-19	-I	-I	1.89E-21
40	265.0 - 270.0	6.30E-18	-I	-I	-I	2.08E-19	-I	-I	7.33E-22
41	270.0 - 275.0	6.50E-18	-I	-I	-I	1.61E-19	-I	-I	-I
42	275.0 - 280.0	6.00E-18	-I	-I	-I	1.22E-19	-I	-I	-I

43	28Ø.Ø	-	285.Ø	4.9ØE-18	-I	-I	-I	8.95E-2Ø	-I	-I	-I
44	285.Ø	-	29Ø.Ø	3.5ØE-18	-I	-I	-I	6.48E-2Ø	-I	-I	-I
45	29Ø.Ø	-	295.Ø	2.3ØE-18	-I	-I	-I	4.6ØE-2Ø	-I	-I	-I
46	295.Ø	-	3ØØ.Ø	1.4ØE-18	-I	-I	-I	3.13E-2Ø	-I	-I	-I
47	3ØØ.Ø	-	3Ø2.5	9.7ØE-19	-I	-I	-I	2.4ØE-2Ø	-I	-I	-I
48	3Ø2.5	-	3Ø5.Ø	7.5ØE-19	-I	-I	-I	2.ØØE-2Ø	-I	-I	-I
49	3Ø5.Ø	-	3Ø7.5	-I	-I	-I	-I	1.6ØE-2Ø	-I	-I	-I
5Ø	3Ø7.5	-	31Ø.Ø	-I	-I	-I	-I	1.3ØE-2Ø	-I	-I	-I
51	31Ø.Ø	-	312.5	-I	-I	-I	-I	1.1ØE-2Ø	-I	-I	-I
52	312.5	-	315.Ø	-I	-I	-I	-I	9.3ØE-21	-I	-I	-I
53	315.Ø	-	317.5	-I	-I	-I	-I	7.8ØE-21	-I	-I	-I
54	317.5	-	32Ø.Ø	-I	-I	-I	-I	6.6ØE-21	-I	-I	-I
55	32Ø.Ø	-	325.Ø	-I	-I	-I	-I	5.2ØE-21	-I	-I	-I
56	325.Ø	-	33Ø.Ø	-I	-I	-I	-I	4.ØØE-21	-I	-I	-I
57	33Ø.Ø	-	335.Ø	-I	-I	-I	-I	3.ØØE-21	-I	-I	-I
58	335.Ø	-	34Ø.Ø	-I	-I	-I	-I	2.6ØE-21	-I	-I	-I
59	34Ø.Ø	-	345.Ø	-I	-I	-I	-I	2.25E-21	-I	-I	-I
6Ø	345.Ø	-	35Ø.Ø	-I	-I	-I	-I	2.Ø5E-21	-I	-I	-I
61	35Ø.Ø	-	355.Ø	-I	-I	-I	-I	1.9ØE-21	-I	-I	-I
62	355.Ø	-	36Ø.Ø	-I	-I	-I	-I	1.75E-21	-I	-I	-I
63	36Ø.Ø	-	365.Ø	-I	-I	-I	-I	1.6ØE-21	-I	-I	-I
64	365.Ø	-	37Ø.Ø	-I	-I	-I	-I	1.45E-21	-I	-I	-I
65	37Ø.Ø	-	375.Ø	-I	-I	-I	-I	1.3ØE-21	-I	-I	-I
66	375.Ø	-	38Ø.Ø	-I	-I	-I	-I	1.2ØE-21	-I	-I	-I
67	38Ø.Ø	-	385.Ø	-I	-I	-I	-I	1.Ø5E-21	-I	-I	-I
68	385.Ø	-	39Ø.Ø	-I	-I	-I	-I	9.ØØE-22	-I	-I	-I
69	39Ø.Ø	-	395.Ø	-I	-I	-I	-I	7.5ØE-22	-I	-I	-I
7Ø	395.Ø	-	4ØØ.Ø	-I	-I	-I	-I	6.2ØE-22	-I	-I	-I

Appendix D (Chapter 3, 4)

Absorption calculations at grazing incidence

When solar radiation enters the earth's atmosphere at a large incidence angle, as occurs near sunrise, sunset or at very high latitudes, or when absorption or emission occurs in long tangential paths as for the limb occultation measurements, the effects of the earth's curvature on the calculation of transmission can no longer be ignored. The columnar content of the material along the slant path from the sun to the observational point P (see Fig. D.1) is given by

$$N = \int_{\infty}^{P} n(\chi_i, z_i) \mathrm{d}s \tag{D.1}$$

where n is the number density at any point P_i on the path at altitude z_i and solar zenith angle χ_i. (D.1) may be used to calculate the optical depth using (3.5) and the absorption of solar radiation using (3.8).

CHAPMAN(1931) has shown that if n varies exponentially with altitude, N could be expressed by

$$N_P = n_P H Ch(\eta_P, \chi_P) \tag{D.2}$$

where H is the scale height and

$$\eta = (R+z)/H \tag{D.3}$$

The suffix P indicates the values at the observational point P.

The Chapman function Ch in (D.2) is a fairly complex function. CHAP-

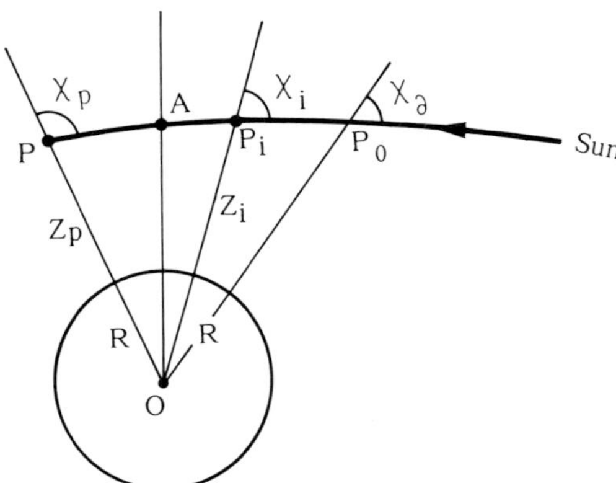

FIG. D.1. A ray path at grazing incidence; P is the observational point and A is the grazing point.

MAN(1953) and WILKES(1954) presented its value in the form of tables. Since then, various convenient approximate formulae have been derived (e.g. SWIDER 1964; FITZMAURICE 1964; SWIDER and GARDNER 1969; RISHBETH and GARRIOTT 1969; SMITH and SMITH 1972; WANG *et al.*, 1981).

Taking into account the fact that N is most sensitive to the scale height and the density near the lowest point of the path, i.e., near the observational point (P) for zenith angles $\leq 90°$ and near the grazing height (A) for zenith angles $>90°$, SMITH and SMITH (1972) have derived the following approximation formulae:

$$N = n_P H_P \left(\frac{\pi}{2}\eta_P\right)^{1/2} e^{y^2}(1-\mathrm{erf}\,(y)) \quad \text{for} \quad \chi_P \leq 90° \tag{D.4}$$

and

$$N = \left(\frac{\pi}{2}\eta_A\right)^{1/2} H_A\{2n_A - n_P e^{y^2}(1-\mathrm{erf}\,(y))\} \quad \text{for} \quad \chi_P > 90° \tag{D.5}$$

where

$$y = \left(\frac{\eta_P}{2}\right)^{1/2} |\cos \chi_P| \tag{D.6}$$

and

$$\mathrm{erf}\,(y) = \frac{2}{\sqrt{\pi}} \int_0^y e^{-u^2}\,\mathrm{d}u \tag{D.7}$$

The error function (D.7) can be calculated within an accuracy of $\pm 2.5 \times 10^{-5}$ by (ABRAMOWITZ and STEGUN 1964)

$$\mathrm{erf}\,(y) \fallingdotseq 1 - (a_1 t + a_2 t^2 + a_3 t^3) \cdot e^{-y^2} \tag{D.8}$$

with

$$t = \frac{1}{1 + 0.47047y}$$

$$a_1 = 0.3480242, a_2 = -0.0958798, \text{and } a_3 = 0.7478556$$

The formula (D.2) and its approximation (D.4) and (D.5) are good so long as n decreases with altitude so that an effective scale height can be defined. However, they cannot be applied if n has a peak value in the middle atmosphere, such as is the case for O_3. In these cases, N has to be calculated from (D.1) by a numerical method.

Fig. D.2 illustrates the geometry of the ray path near P_i, exaggerating the effect of refraction; the ray path is refracted at P_i with an angle of refraction α_i. Applying Snell's law at P_i, we have

$$r_i \sin \chi_i = r_{i+1} \sin \alpha_i \tag{D.9}$$

where r_i and r_{i+1} are the mean refractive indices of the layers above P_i and P_{i+1}, respectively. r_i is a function of atmospheric density and wavelength as given by (3.30) or (3.31).

The path length $\Delta s_i (= P_i P_{i+1})$ at P_i is related to the angle $\Delta\chi_i$ subtended at the center of the earth by the simple geometrical relation

$$\frac{\Delta s_i}{\sin(\Delta\chi_i)} = \frac{R + z_i}{\sin \chi_{i+1}} = \frac{R + z_{i+1}}{\sin \alpha_i} \tag{D.10}$$

where R is the earth's radius. Using (D.9), it follows from the relationship between the last two quantities of (D.10) that

$$r_i(R + z_i) \sin \chi_i = r_{i+1}(R + z_{i+1}) \sin \chi_{i+1} = \text{const } (C) \tag{D.11}$$

The constant C can be evaluated at P by

$$C = r_P(R + z_P) \sin \chi_P \tag{D.12}$$

We now can derive the following expression for Δs from (D.10)

$$\Delta s_i = \frac{C}{r_i} \frac{\sin(\Delta\chi_i)}{\sin \chi_i \sin \chi_{i+1}} \tag{D.13}$$

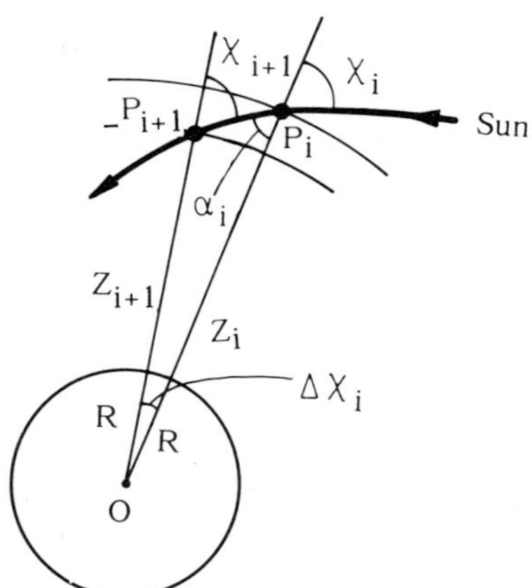

FIG. D.2. Geometry of a refracted path near P_i.

In order to calculate N from (D.1) by numerical integration for a non-exponential atmosphere, the value of n in (D.1) must be evaluated by interpolation as the value at the height

$$z_i = \frac{C}{r_i \cdot \sin \chi_i} - R \tag{D.14}$$

In an actual calculation we start the integration from some point P_0 instead of ∞ (see Fig. D.1). It may be sufficient to choose P_0 at the point where $\chi_0 = 75°$. Thus, the integration is carried out by dividing the entire path length P_0P into small segments Δs_i, each of which subtends a small angle $\Delta\chi_i$.

The effect of the earth's curvature on the photodissociation rate has been considered by ISAKSEN (1973) and TURCO (1975). The curvature effect on the atmospheric heating rate by O_3 absorption has been calculated by WANG *et al.* (1982). The latter study shows that for the summer solstice the percentage difference of the heating rate caused by the effect of grazing incidence has a peak around 62.5° latitude, above which it drops rapidly in the polar region primarily due to the nearly 24 hr duration of the day. The peak percentage differences are ~12.2%, ~7.4% and ~2.1% at 15, 30, and 45 km, respectively. For the equinoxes the percentage difference at 30 km, where the peak of the ozone mixing ratio is located, is ~5.4% and ~ 9% at 60° and 70°, respectively, from which it increases almost exponentially toward the poles. The result of the winter solstice is similar to that for the equinoxes, but the effect occurs at much lower latitudes. Thus, a systematic error is introduced by using a plane-parallel model; it depends upon the altitude, latitude and season. For instance, at 30 km errors greater than 10 % occur in latitudes above ~72.5° during equinoxes but they occur above ~50° in the winter hemisphere.

Appendix E (Chapter 4)

Coefficients of the polynomials to be used in the calculation of O_2 photodissociation and atmospheric transmission in the schumann-runge band system.

Table E.1 shows the subdivisions of the Schumann-Runge band system for which HUDSON and MAHLE (1972b) tabulated the O_2 photodissociaiton cross-section (R) and the atmospheric transmission (P) for selected values of the O_2 column density, N(O_2). Solar flux data are taken from NASA (1979).

Table E.2 lists the coefficients R_m in Eq. (4.16) and P_m in Eq. (4.17) determined for each subdivision for the range of *large* N(O_2). This range of N(O_2) depends upon the subdivision and is shown in the first column. The

TABLE E.1 Subdivisions of the Schumann-Runge band system by HUDSON and MAHLE (1972 b) and integrated solar fluxes from NASA (1979).

Subdivision	Band	Wavelength range (nm)	Wavenumber range (cm^{-1})	Solar flux (photons cm^{-2} sec^{-1})
1	19–0	175.34–175.58	57,032–56,954	1.58×10^{10}
	18–0	175.58–175.89	56,954–56,854	2.23×10^{10}
	17–0	175.89–176.32	56,854–56,715	2.80×10^{10}
2	16–0	176.32–176.86	56,715–56,542	3.66×10^{10}
3	15–0	176.86–177.46	56,542–56,351	5.18×10^{10}
4	14–0	177.46–178.26	56,351–56,098	6.60×10^{10}
5	13–0	178.26–179.26	56,098–55,785	8.79×10^{10}
6	12–0	179.26–180.36	55,785–55,445	1.03×10^{11}
7	11–0	180.36–181.64	55,445–55,054	1.43×10^{11}
8	10–0	181.64–183.06	55,054–54,627	2.07×10^{11}
9	9–0	183.06–184.62	54,627–54,165	2.09×10^{11}
10	8–0	184.62–186.34	54,165–53,665	2.56×10^{11}
11	7–0	186.34–188.22	53,665–53,129	3.96×10^{11}
12	6–0	188.22–190.24	53,129–52,565	4.63×10^{11}
13	5–0	190.24–192.40	52,565–51,975	6.38×10^{11}
14	4–0	192.40–194.70	51,975–51,361	7.16×10^{11}
15	3–0	194.70–197.18	51,361–50,715	1.14×10^{12}
16	2–0	197.18–198.50	50,715–50,378	1.54×10^{12}
17	2–0	198.50–200.00	50,378–50,000	
18	1–0	200.00–202.50	50,000–49,383	
19	0–0	202.50–205.00	49,383–48,780	

ranges are mainly applicable to the stratosphere. If $N(O_2)$ is larger than these ranges, both the O_2 dissociation and the atmospheric transmission are essentially zero.

Table E.3 is similar to Table E.2 but for the ranges of *small* $N(O_2)$, which is mainly applicable to the mesosphere and lower thermosphere. If $N(O_2)$ is smaller than these ranges, the O_2 dissociation coefficient and atmospheric transmission are essentially height-independent, and can be calculated by using the smallest values of $N(O_2)$ in each range.

TABLE E.2 Polynomial coefficients R'_ms and P'_ms to be used in Eqs. (4.16) and (4.17), respectively, for larger N(O_2), applicable mainly to stratosphere.

M	RANGE OF N(O2)	RØ	R1	R2	R3	R4	R5	R6	R7
1	1.E+18 - 5.E+2Ø	-2.Ø48Ø6E+Ø1	-2.76268E+ØØ	-1.67295E+ØØ	-2.55342E-Ø1	-2.4986ØE-Ø1	-1.1346ØE+ØØ	-1.Ø4919E+ØØ	-2.97345E-Ø1
2	1.E+18 - 5.E+2Ø	-2.Ø1189E+Ø1	-1.53585E+ØØ	-1.Ø3382E+ØØ	-6.59266E-Ø1	-2.51588E-Ø1	-6.98111E-Ø2	-3.51642E-Ø2	-1.Ø59Ø2E-Ø2
3	1.E+18 - 1.E+21	-2.ØØ59ØE+Ø1	-1.Ø6359E+ØØ	-5.22563E-Ø1	-2.36Ø46E-Ø1	-7.35Ø65E-Ø2	-9.69734E-Ø2	-6.Ø8623E-Ø2	-7.89959E-Ø3
4	1.E+18 - 5.E+21	-2.Ø16Ø9E+Ø1	-6.29715E-Ø1	-1.92698E-Ø1	-1.68Ø47E-Ø1	-9.27665E-Ø2	-1.91286E-Ø2	8.8Ø4ØØE-Ø4	-2.5891ØE-Ø3
5	1.E+18 - 5.E+21	-2.Ø3117E+Ø1	-5.377Ø2E-Ø1	-3.966Ø1E-Ø2	-9.13848E-Ø2	-6.31ØØ5E-Ø2	-1.62742E-Ø2	-2.88987E-Ø3	-3.875Ø5E-Ø3
6	1.E+18 - 1.E+22	-2.Ø3Ø79E+Ø1	-6.68644E-Ø1	-5.397Ø7E-Ø2	-6.395Ø7E-Ø2	-1.3Ø7Ø6E-Ø2	-2.44645E-Ø3	-1.27985E-Ø2	-1.13413E-Ø3
7	1.E+19 - 2.E+22	-2.Ø2581E+Ø1	-8.31471E-Ø1	-1.44Ø56E-Ø2	8.13544E-Ø3	-1.9Ø714E-Ø1	1.3247ØE-Ø1	-5.9694ØE-Ø2	7.62761E-Ø3
8	1.E+19 - 2.E+22	-2.Ø4267E+Ø1	-6.6398ØE-Ø1	-1.63893E-Ø1	-9.73765E-Ø2	4.6216ØE-Ø1	-4.52384E-Ø1	1.64551E-Ø1	-2.22698E-Ø2
9	1.E+19 - 5.E+22	-2.Ø5254E+Ø1	-6.42623E-Ø1	-3.74217E-Ø2	-5.53253E-Ø2	1.44693E-Ø1	-1.61912E-Ø1	6.47285E-Ø2	-9.42813E-Ø3
1Ø	4.E+19 - 1.E+23	-2.Ø6331E+Ø1	-2.24Ø53E-Ø1	-1.317Ø8E+ØØ	1.69Ø61E+ØØ	-9.82365E-Ø1	2.17962E-Ø1	-3.285Ø2E-Ø3	-3.75722E-Ø3
11	4.E+19 - 2.E+23	-2.Ø7144E+Ø1	-2.744Ø3E-Ø1	-1.22766E+ØØ	1.831Ø8E+ØØ	-1.33447E+ØØ	4.69217E-Ø1	-7.48568E-Ø2	3.63494E-Ø3
12	1.E+2Ø - 5.E+23	-2.1Ø451E+Ø1	3.67732E-Ø1	-1.95769E+ØØ	1.98872E+ØØ	-1.Ø736ØE+ØØ	2.92118E-Ø1	-3.52564E-Ø2	1.Ø2275E-Ø3
13	1.E+2Ø - 5.E+23	-2.17516E+Ø1	3.Ø3Ø47E+ØØ	-6.8Ø78ØE+ØØ	6.53Ø71E+ØØ	-3.42ØØ9E+ØØ	9.73889E-Ø1	-1.39668E-Ø1	7.6Ø662E-Ø3
14	1.E+21 - 1.E+24	7.66467E+Ø1	-2.683Ø6E+Ø2	3.Ø7813E+Ø2	-1.92534E+Ø2	7.Ø8221E+Ø1	-1.53517E+Ø1	1.82Ø92E+ØØ	-9.15774E-Ø2
15	1.E+21 - 1.E+24	3.81973E+Ø1	-1.62638E+Ø2	1.84522E+Ø2	-1.14178E+Ø2	4.15986E+Ø1	-8.95181E+ØØ	1.Ø5754E+ØØ	-5.31984E-Ø2
16	1.E+21 - 1.E+24	1.62465E+Ø1	-1.Ø2638E+Ø2	1.15585E+Ø2	-7.13163E+Ø1	2.6ØØØ2E+Ø1	-5.61752E+ØØ	6.68662E-Ø1	-3.4Ø6Ø5E-Ø2
17	1.E+21 - 1.E+24	3.726ØØE+ØØ	-7.12992E+Ø1	8.Ø4788E+Ø1	-4.95663E+Ø1	1.8Ø45ØE+Ø1	-3.9Ø125E+ØØ	4.663Ø8E-Ø1	-2.39875E-Ø2
18	4.E+21 - 1.E+24	-1.Ø72Ø8E+Ø2	1.81679E+Ø2	-1.648Ø5E+Ø2	8.13447E+Ø1	-2.34729E+Ø1	3.92228E+ØØ	-3.44815E-Ø1	1.17162E-Ø2
19	4.E+21 - 1.E+24	7.79173E+Ø1	-2.29584E+Ø2	2.22728E+Ø2	-1.1951ØE+Ø2	3.83753E+Ø1	-7.39294E+ØØ	7.94ØØ3E-Ø1	-3.69148E-Ø2

M	RANGE OF N(O2)	PØ	P1	P2	P3	P4	P5	P6	P7
1	1.E+18 - 5.E+2Ø	-1.42376E+ØØ	-2.34831E+ØØ	-1.63124E+ØØ	-7.49668E-Ø1	-6.5Ø856E-Ø1	-6.28299E-Ø1	-2.95676E-Ø1	-4.98395E-Ø2
2	1.E+18 - 5.E+2Ø	-7.88571E-Ø1	-1.26114E+ØØ	-1.Ø6286E+ØØ	-6.3Ø674E-Ø1	-2.74757E-Ø1	-9.54411E-Ø2	-3.Ø3966E-Ø2	-6.14974E-Ø3
3	1.E+18 - 1.E+21	-4.72423E-Ø1	-7.Ø5128E-Ø1	-5.38672E-Ø1	-2.85Ø17E-Ø1	-1.32979E-Ø1	-7.876Ø8E-Ø2	-4.24757E-Ø2	-1.Ø3882E-Ø2
4	1.E+18 - 5.E+21	-2.Ø6972E-Ø1	-3.Ø8335E-Ø1	-2.628Ø8E-Ø1	-1.74ØØØE-Ø1	-6.95555E-Ø2	-8.42733E-Ø3	-6.24646E-Ø3	-5.Ø1186E-Ø3
5	1.E+18 - 5.E+21	-1.412Ø3E-Ø1	-1.85182E-Ø1	-1.45294E-Ø1	-1.Ø1287E-Ø1	-4.96Ø69E-Ø2	-1.16Ø34E-Ø2	-6.149Ø7E-Ø3	-3.63282E-Ø3
6	1.E+18 - 1.E+22	-1.6Ø814E-Ø1	-1.889Ø9E-Ø1	-1.11587E-Ø1	-8.17415E-Ø2	-4.281Ø4E-Ø2	4.592Ø5E-Ø3	-2.Ø9365E-Ø4	-5.4472ØE-Ø3
7	1.E+19 - 2.E+22	-1.88331E-Ø1	-2.255Ø5E-Ø1	-1.69766E-Ø1	-4.99569E-Ø2	1.62467E-Ø1	-3.11Ø58E-Ø1	1.59597E-Ø1	-3.Ø15Ø6E-Ø2
8	1.E+19 - 2.E+22	-1.18695E-Ø1	-1.Ø2827E-Ø1	-1.77928E-Ø1	-1.1678ØE-Ø1	4.54637E-Ø1	-4.725Ø7E-Ø1	1.81513E-Ø1	-2.54336E-Ø2
9	1.E+19 - 5.E+22	-7.71293E-Ø2	-7.9246ØE-Ø2	-1.11Ø55E-Ø1	-7.58633E-Ø2	2.44768E-Ø1	-2.57512E-Ø1	1.ØØ79ØE-Ø1	-1.42188E-Ø2
1Ø	4.E+19 - 1.E+23	-1.42972E-Ø1	8.2192ØE-Ø1	-3.Ø6262E+ØØ	4.53979E+ØØ	-3.48874E+ØØ	1.36735E+ØØ	-2.628Ø7E-Ø1	1.9Ø252E-Ø2
11	4.E+19 - 2.E+23	-3.85551E-Ø2	-4.51282E-Ø2	-1.22822E-Ø1	1.662ØØE-Ø1	-1.5Ø74ØE-Ø1	3.51198E-Ø2	4.28979E-Ø3	-2.Ø7169E-Ø3
12	1.E+2Ø - 5.E+23	3.73Ø71E-Ø2	-2.66814E-Ø1	2.73349E-Ø1	-1.35995E-Ø1	-1.365Ø7E-Ø2	1.18728E-Ø2	1.32679E-Ø3	-8.Ø6338E-Ø4
13	1.E+2Ø - 5.E+23	-1.57234E-Ø1	7.54227E-Ø1	-1.64949E+ØØ	1.74561E+ØØ	-1.Ø28Ø9E+ØØ	3.2Ø547E-Ø1	-4.83664E-Ø2	2.5Ø999E-Ø3
14	1.E+21 - 1.E+24	9.3921ØE+ØØ	-2.911Ø3E+Ø1	3.77361E+Ø1	-2.65674E+Ø1	1.Ø9467E+Ø1	-2.65233E+ØØ	3.52Ø48E-Ø1	-1.99858E-Ø2
15	1.E+21 - 1.E+24	1.19433E+Ø1	-3.42168E+Ø1	4.13165E+Ø1	-2.72894E+Ø1	1.Ø6553E+Ø1	-2.46921E+ØØ	3.15935E-Ø1	-1.73869E-Ø2
16	1.E+21 - 1.E+24	9.87Ø29E+ØØ	-2.81575E+Ø1	3.39217E+Ø1	-2.24166E+Ø1	8.78839E+ØØ	-2.Ø52Ø3E+ØØ	2.65391E-Ø1	-1.48156E-Ø2
17	1.E+21 - 1.E+24	8.25926E+ØØ	-2.37494E+Ø1	2.88525E+Ø1	-1.92263E+Ø1	7.6Ø468E+ØØ	-1.79213E+ØØ	2.34141E-Ø1	-1.32233E-Ø2
18	4.E+21 - 1.E+24	1.79299E+Ø1	-4.7357ØE+Ø1	5.3Ø918E+Ø1	-3.281Ø5E+Ø1	1.2Ø973E+Ø1	-2.66936E+ØØ	3.2778ØE-Ø1	-1.74336E-Ø2
19	4.E+21 - 1.E+24	2.Ø7817E+Ø1	-5.29Ø32E+Ø1	5.74819E+Ø1	-3.45977E+Ø1	1.24778E+Ø1	-2.7Ø342E+ØØ	3.26996E-Ø1	-1.71763E-Ø2

TABLE E.3 Polynomial coefficients R'_ms and P'_ms to be used in Eqs. (4.16) and (4.17), respectively, for smaller N(O_2), applicable mainly to mesosphere and lower thermosphere.

M	RANGE OF N(O2)	R0	R1	R2	R3	R4	R5	R6	R7
1	1.E+15 - 1.E+18	-2.46032E+01	-1.31785E+01	-1.26651E+01	-6.71776E+00	-2.10190E+00	-3.86653E-01	-3.86762E-02	-1.62222E-03
2	1.E+15 - 1.E+18	-2.52558E+01	-1.43187E+01	-1.41313E+01	-7.98286E+00	-2.73309E+00	-5.60602E-01	-6.33488E-02	-3.03019E-03
3	1.E+15 - 1.E+18	-2.27331E+01	-7.84675E+00	-7.81741E+00	-4.75334E+00	-1.79382E+00	-4.06677E-01	-5.04787E-02	-2.62790E-03
4	1.E+15 - 1.E+18	-2.09828E+01	-4.19774E+00	-6.47870E+00	-5.75346E+00	-2.75846E+00	-7.20358E-01	-9.69702E-02	-5.27910E-03
5	1.E+15 - 1.E+18	-1.96206E+01	1.04428E+00	8.15731E-01	-5.21930E-01	-6.38024E-01	-2.29428E-01	-3.63934E-02	-2.18813E-03
6	1.E+15 - 1.E+18	-2.62305E+01	-1.69386E+01	-1.83855E+01	-1.12327E+01	-4.08273E+00	-8.74826E-01	-1.01987E-01	-4.98703E-03
7	1.E+15 - 1.E+19	-2.02367E+01	-6.01527E-01	7.46044E-01	1.17969E+00	6.60310E-01	1.86373E-01	2.65457E-02	1.51727E-03
8	1.E+15 - 1.E+19	-2.01390E+01	7.05725E-01	2.21479E+00	1.50756E+00	4.41123E-01	5.20374E-02	-6.11841E-05	-3.27766E-04
9	1.E+15 - 1.E+19	-2.04593E+01	-1.59215E-01	1.33005E+00	1.57179E+00	8.10553E-01	2.19383E-01	3.04051E-02	1.70330E-03
10	1.E+15 - 4.E+19	-2.05796E+01	-6.89409E-01	-1.91305E-01	1.97736E-01	1.85162E-01	6.38784E-02	1.02337E-02	6.33300E-04
11	4.E+16 - 4.E+19	-2.06841E+01	-5.77041E-01	-2.33187E-01	1.06403E-01	1.36037E-01	5.08980E-02	8.49614E-03	5.39431E-04
12	4.E+16 - 1.E+20	-2.08826E+01	-4.81480E-01	-2.20505E-01	7.63882E-02	1.15885E-01	4.54115E-02	7.80383E-03	5.06629E-04
13	1.E+17 - 1.E+20	-2.11039E+01	-2.65537E-01	-2.10783E-01	-6.35894E-02	1.06192E-02	1.12702E-02	2.50650E-03	1.85463E-04
14	4.E+17 - 1.E+21	-2.14057E+01	-8.39951E-02	-9.66625E-02	-5.21516E-02	-8.12161E-03	2.86132E-03	1.14686E-03	1.08490E-04
15	7.E+17 - 1.E+21	-2.19908E+01	-2.82908E-02	-3.00377E-02	-1.81368E-02	-6.50268E-03	-1.34759E-03	-1.46842E-04	-6.40320E-06
16	7.E+17 - 1.E+21	-2.21106E+01	-3.20983E-02	-3.58789E-02	-2.43569E-02	-6.93775E-03	1.70546E-05	3.30022E-04	3.97401E-05
17	4.E+18 - 1.E+21	-2.27823E+01	-1.13507E-03	-3.66184E-03	-3.31054E-03	-1.23975E-03	-1.78828E-04	4.90656E-07	1.56967E-06
18	7.E+18 - 4.E+21	-2.28857E+01	1.06289E-03	-8.63400E-04	-1.91416E-03	-6.70409E-04	7.00961E-05	6.32402E-05	7.33272E-06
19	1.E+19 - 4.E+21	-2.29208E+01	4.29477E-04	4.28129E-04	-2.12221E-04	-3.36076E-04	-1.24075E-04	-1.80463E-05	-8.71585E-07

M	RANGE OF N(O2)	P0	P1	P2	P3	P4	P5	P6	P7
1	1.E+15 - 1.E+18	1.66768E+00	6.09207E+00	7.84896E+00	5.10914E+00	1.87788E+00	3.96087E-01	4.48058E-02	2.10988E-03
2	1.E+15 - 1.E+18	-2.73220E-01	2.40370E-01	7.74402E-01	6.00236E-01	2.27416E-01	4.68358E-02	5.04076E-03	2.22582E-04
3	1.E+15 - 1.E+18	-8.82601E-03	6.38453E-01	1.06719E+00	7.48552E-01	2.82013E-01	5.99695E-02	6.80332E-03	3.20946E-04
4	1.E+15 - 1.E+18	3.54460E-01	1.23255E+00	1.50072E+00	9.34023E-01	3.32671E-01	6.87809E-02	7.69548E-03	3.60810E-04
5	1.E+15 - 1.E+18	1.52102E-01	6.16201E-01	7.67031E-01	4.73380E-01	1.65554E-01	3.35235E-02	3.67546E-03	1.69198E-04
6	1.E+15 - 1.E+18	7.35285E-02	4.72677E-01	6.69566E-01	4.39548E-01	1.58225E-01	3.22889E-02	3.51425E-03	1.58792E-04
7	1.E+15 - 1.E+19	-1.86506E-01	-2.05020E-01	-3.29157E-02	5.61326E-02	3.82935E-02	1.07312E-02	1.45305E-03	7.81811E-05
8	1.E+15 - 1.E+19	-1.22368E-01	-1.43924E-01	-1.05826E-01	-7.64255E-02	-3.96551E-02	-1.16577E-02	-1.73845E-03	-1.02717E-04
9	1.E+15 - 1.E+19	-7.60281E-02	-7.89696E-02	-9.40945E-03	2.27239E-02	1.43291E-02	3.83589E-03	5.01717E-04	2.62806E-05
10	1.E+15 - 4.E+19	-5.56984E-02	-8.04998E-02	-3.64416E-02	3.90584E-03	9.51296E-03	3.55349E-03	5.72060E-04	3.49732E-05
11	4.E+16 - 4.E+19	-3.78908E-02	-6.06320E-02	-3.53598E-02	-5.14949E-03	3.47237E-03	1.78840E-03	3.21156E-04	2.08438E-05
12	4.E+16 - 1.E+20	-2.16054E-02	-3.68148E-02	-2.48409E-02	-7.12558E-03	-3.20913E-05	4.80951E-04	1.07435E-04	7.54280E-06
13	1.E+17 - 1.E+20	-1.08268E-02	-2.15116E-02	-1.84108E-02	-8.20470E-03	-1.83104E-03	-1.31811E-04	1.65650E-05	2.42997E-06
14	4.E+17 - 1.E+21	-4.60371E-03	-1.08231E-02	-1.22654E-02	-7.46278E-03	-2.30284E-03	-2.87088E-04	5.59496E-06	2.88806E-06
15	7.E+17 - 1.E+21	-1.20045E-03	-2.65200E-03	-3.06153E-03	-2.45150E-03	-1.29978E-03	-4.07393E-04	-6.68070E-05	-4.38681E-06
16	7.E+17 - 1.E+21	-9.08389E-04	-2.04025E-03	-2.39738E-03	-1.80993E-03	-8.59103E-04	-2.40350E-04	-3.57612E-05	-2.17006E-06
17	4.E+18 - 1.E+21	-2.35939E-04	-4.19732E-04	-3.85550E-04	-4.42780E-04	-3.71148E-04	-1.56723E-04	-3.09049E-05	-2.28771E-06
18	7.E+18 - 4.E+21	-2.39635E-04	-5.12774E-04	-1.98765E-04	-1.27220E-04	-2.85606E-04	-1.85424E-04	-4.56973E-05	-3.87931E-06
19	1.E+19 - 4.E+21	-2.13687E-04	-4.79376E-04	-1.36789E-04	-5.16679E-05	-2.43231E-04	-1.73966E-04	-4.42493E-05	-3.81239E-06

Appendix F (Chapter 6)

Solution of the chemical rate equation

The chemical rate equation (7.3)

$$\frac{dn}{dt}=Q-\beta n-\alpha n^2 \tag{F.1}$$

is a generalized Riccati differential equation. Putting

$$n=y'/y \tag{F.2}$$

(F.1) is transferred to a second order homoganeous linear differential equation

$$\frac{d^2y}{dt^2}=-\beta\frac{dy}{dt}+\alpha Qy \tag{F.3}$$

The general solutin of (F.3) is given by

$$y=c_i e^{\lambda_1 t}+c_2 e^{\lambda_2 t} \tag{F.4}$$

where

$$\lambda_{1,2}=\frac{-\beta\pm\sqrt{\beta^2+4\alpha Q}}{2} \tag{F.5}$$

The suffices 1 and 2 correspond to the + and − sign in the right hand side of the equation, respectively. Thus,

$$\lambda_1=-\frac{\beta}{2}+\gamma,\ \lambda_2=-\frac{\beta}{2}-\gamma \tag{F.6}$$

where

$$\gamma=\frac{\sqrt{\beta^2+4\alpha Q}}{2} \tag{F.7}$$

Substituting (F.4) into (F.2) gives

$$n=-\frac{\beta}{2\alpha}+\frac{\gamma}{\alpha}\cdot\frac{c_1e^{\gamma t}-c_2e^{-\gamma t}}{c_1e^{\gamma t}+c_2e^{-\gamma t}} \tag{F.8}$$

At $t=0$, the initial value n_0 is obtained from (F.8) as

$$n_0=-\frac{\beta}{2\alpha}+\frac{\gamma}{\alpha}\frac{c_1-c_2}{c_1+c_2} \tag{F.9}$$

and if $t\to\infty$ n approaches the equilibrium value n_e given by

$$n_e = -\frac{\beta}{2\alpha} + \frac{\gamma}{\alpha} \tag{F.10}$$

Using (F.10) we obtain from (F.9)

$$\frac{c_2}{c_1} = \frac{n_e - n_0}{n_0 + n_e + \dfrac{\beta}{\alpha}} \tag{F.11}$$

(F.8) is now written by using (F.10) and (F.11)

$$n = -\frac{\beta}{2\alpha} + \left(n_e + \frac{\beta}{2\alpha}\right) \frac{e^{\gamma t} - \dfrac{n_e - n_0}{n_0 + n_e + \dfrac{\beta}{\alpha}} e^{-\gamma t}}{e^{\gamma t} + \dfrac{n_e - n_0}{n_0 + n_e + \dfrac{\beta}{\alpha}} e^{-\gamma t}} \tag{F.12}$$

$$= -\frac{\beta}{2\alpha} + \left(n_e + \frac{\beta}{2\alpha}\right) \frac{n_0 + \dfrac{\beta}{2\alpha} + \left(n_e + \dfrac{\beta}{2\alpha}\right) \tanh \gamma t}{n_e + \dfrac{\beta}{2\alpha} + \left(n_0 + \dfrac{\beta}{2\alpha}\right) \tanh \gamma t} \tag{F.13}$$

Appendix G (Chapter 7)

Limiting diffusive flux

The main source of atmospheric hydrogen is through photochemical reactions in the middle atmosphere ($H_2O + h\nu$, $H_2O + O(^1D)$ etc.). If the time of hydrogen escape from the exosphere to space is shorter than the time for transport from the lower regions, the escape flux of exospheric hydrogen is actually limited by the rate at which the hydrogen is transported upwards from the middle atmosphere.

The vertical flux for the ith constituent given by (7.14) can be rewritten as

$$\phi_i = -(D_i + K)\left\{\frac{\partial n_i}{\partial z} + \left(\frac{1}{T}\frac{\partial T}{\partial z} + \frac{1}{\overline{H}}\right) n_i\right\} + D_i n_i \left(\frac{1}{\overline{H}} - \frac{1}{H_i}\right) \tag{G.1}$$

The first term on the right-hand side is proportional to $\partial f_i / \partial z$ (f_i is the mixing ratio) and will vanish if f_i=constant (or n_i is completely mixed); then the flux is given by the second term

$$\phi_l = D_i n_i \left(\frac{1}{\overline{H}} - \frac{1}{H_i}\right) = b_i \frac{f_i}{1 + f_i}\left(\frac{1}{\overline{H}} - \frac{1}{H_i}\right) \tag{G.2}$$

where b_i is the *binary diffusion parameter* defined by

$$b_i = D_i(n + n_i) \tag{G.3}$$

The molecular diffusion coefficient D_i is given by (2.18). The flux ϕ_l is called the *limiting diffusive flux*, since the upward flux of a light, non-reactive constituent can not exceed ϕ_l; if the upward flux ϕ_i exceeds ϕ_l, the first term of (G.1) must be positive or $-\partial f_i/\partial z > 0$, and consequently f_i must decrease with height. Thus, if this situation extended to great height f_i would becomes negative above a certain height.

For minor, light gases, $f_i \ll 1$ and $H_i \gg \bar{H}$; therefore ϕ_l is always positive and is given by

$$\phi_l = b_i f_i \frac{1}{\bar{H}} \tag{G.4}$$

This implies that the light gas has an upward limiting flux in order to maintain the distribution of complete mixing below the turbopause. Adopting $b_i \approx 6.5 \times 10^{17} T^{0.7}$ (HUNTEN 1973) for a hydrogen atom and $\bar{H} \approx 6.4$ km, we obtain

$$\phi_l \approx 10^{12} T^{0.7} f_i \tag{G.5}$$

The most commonly used escape flux is the Jeans flux given by

$$J_c = \frac{1}{2} n_c \exp\left(-\frac{r_c}{H_c}\right)\left(1 + \frac{r_c}{H_c}\right)\sqrt{\frac{2kT}{\pi m}}, \tag{G.6}$$

where r is the geocentric distance and the suffix c indicates the value at the exobase (which is a height from which a fraction e^{-l} of molecules traveling vertically upward with enough energy to escape will do so without further collisions). Actually, the escape flux is not sensitive to the precise value chosen for r_c (HUNTEN 1973).

The Jeans expansion velocity J_c/n_c is very sensitive to the exospheric temperature. It is $\sim 7.26 \times 10^3$ cm sec^{-1} for $T_c = 1{,}500$ K, $\sim 7.92 \times 10^2$ cm sec^{-1} for $T = 1{,}000$ K and ~ 0.96 cm sec^{-1} for $T_c = 500$ K. The limiting diffusive velocity ϕ_l/n_i at 100 km is calculated as ~ 4 cm sec^{-1} for $T = 200$ K and $n = 10^{13}$ cm^{-3}. Thus, the limiting flux should actually control the escape flux if the thermospheric temperature is greater than ~ 500 K (WALKER 1977).

Measurements of the density of hydrogen and temperature at the exobase have shown that the escape flux for hydrogen is $\sim 10^8$ cm^{-2} sec^{-1} (DONAHUE 1966; JOSEPH 1967; BRINTON and MAYR 1971; VIDAL-MAJAR *et al.*, 1973; HUNTEN and STROBEL 1974; TINSLEY 1974). This escape flux is about the same as the limiting flux, if f_i of hydrogen is $\sim 2 \times 10^{-6}$ in the lower thermosphere.

Appendix H (Chapter 7)

A numerical method for a two dimensional photochemical model

The continuity equation that describes the rate of temporal change in the zonally averaged concentration n in the meridional plane can be written as

$$\frac{\partial n}{\partial t}=Q-Ln-\left(\frac{\partial}{\partial y}-\frac{\tan\theta}{R}\right)F_y-\frac{\partial}{\partial z}F_z \tag{H.1}$$

where Q is the chemical production rate, L the chemical loss coefficient, R the earth's radius, θ the latitude, and F_y and F_z the northward (or y) and upward (or z) components of the flux, respectively. These flux components are assumed to be composed of the mean meridional motions and the large scale eddy-diffusion; they may be written in the following way if the latter is expressed in terms of the eddy-diffusion tensor (see Section 7.2 and Section 9.3):

$$F_y=u-K_{yy}\frac{\partial\mu}{\partial y}-K_{zy}\frac{\partial\mu}{\partial_z} \tag{H.2}$$

and

$$F_z=w-K_{yz}\frac{\partial\mu}{\partial y}-K_{zz}\frac{\partial\mu}{\partial z} \tag{H.3}$$

where μ is the volume mixing ratio.

In solving (H.1) by the finite difference method it is useful to use the *alternating direction implicit method*, which consists of two half line iterations, i.e. one for the row (latitudinal or y) direction and the other for the column (vertical or z) direction. In each line iteration the loss term and the transport terms in the corresponding direction are considered to be implicit, whereas the production term and the transport terms in the other direction are considered explicit. Thus, we first determine the intermediate solution by applying a line iteration in the y idrection as follows:

$$\begin{aligned} n^{l+\frac{1}{2}}(y,z)= {} & n^l(y,z)+r\cdot\Delta t\{A_r n^{l+\frac{1}{2}}(y+\Delta y,z)-B_r n^{l+\frac{1}{2}}(y,z) \\ & +C_r n^{l+\frac{1}{2}}(y-\Delta y,z)\}+r\cdot\Delta t\{A_c n^l(y,z+\Delta z) \\ & -B_c n^l(y,z)+C_c n^l(y,z-\Delta z)+Q\} \end{aligned} \tag{H.4}$$

where l indicates the time step, and coefficients A, B, and C can be derived from (H.1) in expressions similar to (7.22)–(7.24); the suffix r and c represent, respectively, transport terms in the y (row) direction and z (column) direction. The loss term is included in B_r. (H.4) includes three unknowns, i.e. $n^{l+\frac{1}{2}}(y+\Delta y, z)$, $n^{l+\frac{1}{2}}(y, z)$ and $n^{l+\frac{1}{2}}(y-\Delta y, z)$, and all such equations written for various y's along a meridional circle are solved simultaneously for N unknowns $n^{l+\frac{1}{2}}(y_i, z)$, $i=1, 2, 3, \ldots . N$, where N is the total number of grid points along the meridional circle. The end boundary conditions at (or near) the poles are usually given by zero flux across the vertical boundaries.

From the intermediate solution $n^{l+\frac{1}{2}}(y, z)$ we then calculate the final solution by applying an iteration in the z direction as follows:

$$\begin{aligned} n^{l+1}(y,z) = n^{l+\frac{1}{2}}(y,z) + r\cdot\Delta t\{A_c^{l+1}(y, z+\Delta z) - B_c n^{l+1}(y,z) \\ + C_c n^{l+1}(y, z-\Delta z)\} + r\cdot\Delta t\{A_r n^{l+\frac{1}{2}}(y+\Delta y, z) \\ - B_r n^{l+\frac{1}{2}}(y,z) + C_r n^{l+\frac{1}{2}}(y-\Delta y, z) + Q\} \end{aligned} \quad \text{(H.5)}$$

The loss term is now included in B_c. The upper and lower boundary conditions are essentially the same as those used in 1-D models (see Section 7.1.2.3).

A parameter r in (H.4) and (H.5) is a positive constant called the *iteration parameter*. Although r need not be fixed, it must be the same for both parts of the iteration, i.e. in both (H.4) and (H.5) (see, for instance, YOUNG 1962). The author has verified that an iteration parameter of 1/4, which is chosen somewhat arbitrarily, causes the solution to converge satisfactorily in most cases.

Appendix I (Chapter 3, 12)

Diffusive fluxes of terrestrial IR radiation

The fundamental equation of radiative transfer is the Schwarzchild equation (see (12.25))

$$\mu \frac{\mathrm{d}I_\lambda(u,\mu)}{\mathrm{d}u} = k_\lambda(I_\lambda(u,\mu) - J_\lambda) \quad \text{(I.1)}$$

where

$$\mu = \cos\theta, \quad \text{(I.2)}$$

$$u = \int_z^\infty n\,\mathrm{d}z/L, \quad \text{(I.3)}$$

θ is the zenith angle of the ray path, L is the Loschmidt number, k_λ is the absorption coefficient and J_λ is the source function represented by the Planck blackbody function.

We now define the transmission function of the absorbers between z and z' as

$$T_\lambda(u, u') = e^{-\frac{k_\lambda}{\mu}\int_{z'}^{z} n\mathrm{d}z/L} = e^{\frac{k_\lambda}{\mu}(u-u')} \tag{I.4}$$

Multiplying (I.4) with (I.1) and integrating the result between u_0 and u we obtain the following solution for $I_\lambda(u, \mu)$:

$$I_\lambda(u, \mu) - I_\lambda(u_0, \mu) T_\lambda(u, u_0) = -\int_{u_0}^{u} \frac{k_\lambda}{\mu} J_\lambda \cdot T_\lambda(u, u')\, \mathrm{d}u' \tag{I.5}$$

If J_λ is given by a spatial integral of I_λ as is in case of scattering calculations, (I.5) represents an integral form of the transfer equation. In contrast, (I.1) is a differential form of the transfer equation.

The integrand of the right-hand side of (I.5) is equal to $J_\lambda\ (\partial T_\lambda(u, u')/\partial u')$ by virtue of (I.4). For the purposes of numerical integration the Planck function profile may be easier to differentiate than the transmission function. Thus, through integration by part (I.5) can be expressed as

$$I_\lambda(u, \mu) = (I_\lambda(u_0, \mu) - J_\lambda(u_0))\, T_\lambda(u, u_0) + J_\lambda(u) - \int_{u_0}^{u} \frac{\partial J_\lambda}{\partial u'} T_\lambda(u, u')\mathrm{d}u' \tag{I.6}$$

J_λ is a function of u (or z) through the variation of temperature with height. If u_0 is the total column density of the absorber at the ground (I.6) represents the intensity of the upward propagating radiation, where $\mu > 0$ and $u < u_0$.

For the downwardly propagating radiation $\mu < 0$, $u_0 = 0$ and the incoming IR radiation from space is negligibly small. Thus, the intensity of downward radiation is expressed as

$$I_\lambda(u, -\mu) = J_\lambda(0) T_\lambda(u, 0) - J_\lambda(u) + \int_{0}^{u} \frac{\partial J_\lambda}{\partial u'} T_\lambda(u, u')\, \mathrm{d}u' \tag{I.7}$$

where $J_\lambda(0)$ represents the Planck function of the temperature at the height where u is essentially equal to zero.

Applying angular integration to (I.6) and (I.7) we obtain the equation for the upward and downward flux, respectively, as follows:

$$F_\lambda^\uparrow(u) = (F_\lambda^\uparrow(u_0) - J_\lambda(u_0))\, T_\lambda^*(u, u_0) + \pi\, J_\lambda(u) - \int_{u_0}^{u} \frac{\partial J_\lambda}{\partial u'} T_\lambda^*(u, u')\mathrm{d}u' \tag{I.8}$$

and

$$F_\lambda^{\downarrow}(u) = J_\lambda(0)\, T_\lambda^*(u,0) - \pi\, J_\lambda(u) + \int_0^u \frac{\partial J_\lambda}{\partial u'} T_\lambda^*(u,u')\mathrm{d}u' \tag{I.9}$$

where $T_\lambda{}^*$ is the modified transmission function given by

$$T_\lambda^*(u,u') = \int_0^{2\pi} \int_0^{\pi/2} T_\lambda(u,u')\sin\theta\cos\theta\, \mathrm{d}\theta\, \mathrm{d}\varphi = 2\pi E_3(k_\lambda(u-u')) \tag{I.10}$$

The general form of the exponential integral of order 3 is

$$E_3(x) = \int_1^\infty e^{-xt}/t^3\, \mathrm{d}t \tag{I.11}$$

Numerical values of E_3 can be found, for instance, in Table A.8.1 of GOODY(1964).
If T_λ^* is related to T_λ by

$$T_\lambda^*(u,\ u') = \pi\ T_\lambda(\beta_u,\ \beta_u') \tag{I.12}$$

it is found that $\beta = 1.66$ with a fair degree of accuracy (see, for instance, KONDRATYEV 1969). β is called the air mass factor.

Appendix J (Chapter 9, 13)

Stratospheric aerosol

Aerosol is a mixed particle of fluid and solid floating in the atmosphere. Its radius (r) ranges from $\sim 10^{-3}$ to $\sim 20\ \mu$m; particles smaller than $\sim 5 \times 10^{-3}\ \mu$m change to larger particles quickly by *coagulation* and particles larger than $\sim 20\ \mu$m is lost quickly from the atmosphere by *sedimentation*.

Aerosols are usually classified into three categories according to their sizes; *i.e.* Aikin particles ($r < 0.1\ \mu$m), large particles ($r = 0.1$–$1.0\ \mu$m) and giant particles ($r > 1.0\ \mu$m). The stratospheric aerosols are large particles. Since only Aikin particles can rise into the stratosphere from the troposphere, stratospheric aerosols must have been produced within the stratosphere. Aikin particles could be important as condensation nuclei for stratospheric aerosols. The size distributions of stratospheric particles observed at various latitudes are shown in Fig. 9.3.

Physical processes which affect stratospheric aerosol particles include *nucleation*, *condensation*, *coagulation*, *evaporation* and *sedimentation* in the stratosphere, and *washout* below the tropopause (HAMILL *et al.*, 1977; TOON *et al.*, 1979). Heterogeneous nucleation is believed to be the dominant nucleation mechanism, in which the stratospheric solution droplets are formed on solid particles. Coagulation and condensation are important processes for producing large particles in the lower stratosphere. JUNGE *et al.* (1961) have observed that the aerosol concentration increases with height in the lower stratosphere, reaching a maximum at a height around 20 km. Above this height the aerosol concentration decreases mainly through loss by evaporation. The evaporation rate increases with height because of the increase of the atmospheric temperature and the decrease of the atmospheric pressure. The peak of stratospheric aerosol is called the *Junge layer*. The vertical profiles of the stratospheric aerosol concentration observed at various latitudes are shown in Fig. 9.4.

JUNGE *et al.* (1961) first determined that suflur—mostly in the form of suflate—was the major chemical element in stratospheric aerosol under normal conditions. The majority of stratospheric particles are composed mainly of sulfuric acid (H_2SO_4) and varying amount of ammonium sulfate ($(NH_4)_2SO_4$), ammonium persulfate ($(NH_4)_2S_2O_8$), and silicone (Si). Sources for the sulfuric acid vapor in the stratosphere include sulfur dioxide (SO_2), carbonyl sulfide (COS) and other sulfur gases that reach the stratosphere. The refining and combustion of fossil fuels release SO_2 to the atmosphere. SO_2 is also a major chemical element in the volcanic gas. The chemical processes of producing H_2SO_4 from SO_2 may initiate with the production of a precursor SO_3 either by the reaction of SO_2 with hydroxyle radical

$$SO_2+OH+M \rightarrow HSO_3+M, \quad k_0=3.0\times10^{-31}(T/300)^{-3.4}\ \text{cm}^6\ \text{sec}^{-}(*)$$
$$k_\infty=2.0\times10^{-12}\ \text{cm}^3\ \text{sec}^{-1}(*)$$

followed by

$$HSO_3+OH \rightarrow SO_3+H_2O, \quad k=1\times10^{-11}\ \text{cm}^3\ \text{sec}^{-1}$$

or by the reaction of SO_2 with atomic oxygen

$$SO_2+O+M \rightarrow SO_3+M, \quad k=3.4\times10^{-32}\ e^{-1130/T}\ \text{cm}^6\ \text{sec}^{-1}$$

The sulfuric acid vapor is then produced by the reaction

$$SO_3+H_2O \rightarrow H_2SO_4, \quad k=9.1\times10^{-13}\ \text{cm}^3\ \text{sec}^{-1}.$$

COS is a very stable component, but it can be decomposed into S and CO by the photolysis (see J_{27} in Chapter 4) in the stratosphere. The sulfur atom then can be converted to SO_2 by the following reactions:

*See (5.31) for k_0 and k_∞; these will be used to calculate k by (5.34).

$$S+O_2 \rightarrow SO+O, \ k=2.3\times 10^{-12}\ cm^3\ sec^{-1}$$

followed by

$$SO+O_2 \rightarrow SO_2+O, \ \ k=2.4\times 10^{-13}e^{-2370/T}\ cm^3\ sec^{-1}$$

$$SO+O_3 \rightarrow SO_2+O_2, \ k=3.6\times 10^{-12}e^{-1100/T}\ cm^3\ sec^{-1}$$

or

$$SO+NO_2 \rightarrow SO_2+NO, \ k=1.4\times 10^{-11}\ cm^3\ sec^{-1}$$

COS is formed in the troposphere by the reaction of CS_2 and OH. Sources for atmospheric CS_2 could include the oceans, marshes and industrial processes. Roles of COS for stratospheric aerosol formation have been discussed by CRUTZEN(1976) and TURCO *et al.* (1980). Vertical transport is important for mixing up from the troposphere chemical components such as SO_2 and COS and Aikin particles, that are essential for producing aerosols in the stratosphere.

Stratospheric aerosol can affect the atmosphere's radiation balance, and therefore the climate, by scattering and absorbing the radiation from the sun and earth (CADLE and GRAMS 1975). Scattering by particles of the size smaller than the wavelength of radiation is called the *Rayleigh* scattering, whereas scattering by larger particles the *Mie* scattering. Mie scattering is most important in the stratosphere. The energy of radiation scattered or absorbed by stratospheric aerosols is ~20% of the incoming solar radiation.

A large amount of sulfur containing gases would be injected directly into the stratosphere by active volcano eruptions. The events should cause a large increase of aerosol concentration in the stratosphere. The increased stratospheric aerosol would scatter more solar radiation backward, which would cause the increase of the temperature in the atmosphere above the Junge layer, but tend to cool the atmosphere below it and the groung (TOON and POLLACK 1980).

REFERENCES

ABRAMOWITZ, M. and I. A. STEGUN (eds.), Handbook of mathematical functions with formulae, graphs, and mathematical tables, National Bureau of Standards, *App. Math. Series*, **55**, 1046 pp., U. S. Dept. of Commerce, 1964.

ACKERMAN, M., Ultraviolet solar radiation related to mesospheric processes, in *Mesospheric Models and Related Experiments*, ed. by G. Fiocco, pp. 149–159, D. Raidel Pub. Co., Dordrecht-Holland, 1971.

ACKERMAN, M., Stratospheric water vapor from high resolution infrared spectra, *Planet. Space Sci.*, **22**, 1265–1267, 1974.

ACKERMAN, M., *In situ* measurements of middle atmosphere composition, *J. Atmosph. Terr. Phys.*, **41**, 723–733, 1979.

ACKERMAN, M. and F. BIAUME, Structure of the Schumann-Runge bands from the 0-0 to the 13-0 band, *J. Molec. Spectra*, **35**, 73–82, 1970.

ACKERMAN, M., F. BIAUME, and G. KOCKARTS, Absorption cross sections of the Schumann-Runge bands of molecular oxygen, *Planet. Space Sci.*, **18**, 1639–1651, 1970.

ACKERMAN, M. and C. MULLER, Stratospheric nitrogen dioxide from infrared absorption spectra, *Nature*, **240**, 300–301, 1972.

ACKERMAN, M. and C. MULLER, Stratospheric methane and nitrogen dioxide from infrared absorption spectra, *Pure Appl. Geophys.*, **106–108**, 1325–1335, 1973.

ACKERMAN, M. and P. C. SIMON, Rocket measurements of solar fluxes at 1216 Å, 1450 Å and 1710 Å, *Solar Phys.*, **30**, 345–350, 1973.

ACKERMAN, M., D. FRIMONT, C. MULLER, D. NEVEJANS, J-C. FONTANELLA, A. GIRARD, and N. LOUISNARD, Stratospheric nitric oxide from infrared spectra, *Nature*, **245**, 205–206, 1973.

ACKERMAN, M., J-C. FONTANELLA, D. FRIMONT, A. GIRARD., N. LOUISNARD, and C. MULLER, Simultaneous measurements of NO and NO_2 in the stratosphere, *Planet. Space Sci.*, **23**, 651–660, 1975.

ACKERMAN, M., D. FRIMONT, A. GIRARD, M. GOTTINGNIES, and C. MULLER, Stratospheric HCl from infrared spectra, *Geophys. Res. Lett.*, **3**, 81–83, 1976.

AIDA, M., A theoretical examination of absorption in the 9.6 micron ozone band, *J. Quant. Spectros. Radia. Transfer*, **15**, 389–403, 1975.

AIMEDIEU, P., Measurement of the vertical ozone distribution by means of an *in situ* gas phase chemiluminescence ozonometer during the intercomparison ozone campaign, GAP, France, June, 1981, *Planet. Space Sci.*, **31**, 743–748, 1983.

AIMEDIEU, P. and J. BARAT, Instrument to measure stratospheric ozone with high resolution, *Rev. Sci. Instrum.*, **52**, 432–437, 1981.

ALLAN, R. J., K. S. GROVES, and A. F. TUCK, Global OH distribution derived from general circulation model fields of ozone and water vapor, *J. Geophys. Res.*, **86**, 5303–5320, 1981.

ALLISON, A. C., A. DALGARNO, and N. W. PASACHOFF, Absorption by vibrationally excited molecular oxygen in the Schumann-Runge continuum, *Planet. Space Sci.*, **19**, 1463–1473, 1971.

ALVAREZ, L. W., W. ALVAREZ, F. ASARO, and H. V. MICHEL, Extraterrestrial cause for the Cretaceous-Tertiary extinction, *Science*, **208**, 1095–1108, 1980.

AMES, W. F., *Numerical Methods for Partial Differential Equations*, Barnes and Noble, Inc., 365 pp., 1977.

ANDERSON, D. E., P. D. FELDMAN, E. P. GENTIEN, and R. R. MEIER, The UV dayglow 2, Ly-α and Ly-β emission and the H distribution in the mesosphere and thermosphere, *Geophys. Res. Lett.*, **7**, 529–532, 1980.

ANDERSON, J. G., Rocket measurement of OH in the atmosphere, *J. Geophys. Res.*, **76**, 7820–7824, 1971.

ANDERSON, J. G., The absolute concentration of $O(^3P)$ in the earth's stratosphere, *Geophys. Res. Lett.*, **2**, 231–234, 1975.

ANDERSON, J. G., The absolute concentration of OH ($X^2\Pi$) in the earth's stratosphere, *Geophys. Res. Lett.*, **3**, 165–168, 1976.

ANDERSON, J. G., H. J. GROSSL, R. E. SHETTER, and J. J. MARGITAN, Stratospheric free chlorine measured by balloon-borne *in situ* resonance fluorescence, *J. Geophys. Res.*, **85**, 2869–2887, 1980.

ANDERSON, J. G., H. J. GROSSL, R. E. SHETTER, and J. J. MARGITAN, HO_2 in the stratosphere: three *in situ* measurments, *Geophys. Res. Lett.*, **8**, 289–292, 1981.

ANDREWS, D. G. and M. E. MCINTYRE, An exact theory of nonlinear waves on a Lagrangian-mean flow, *J. Fluid Mech.*, **89**, 609–646, 1978.

ANGELL, J. K. and J. KORSHOVER, Quasi-biennial and long-term fluctuations in total ozone, *Monthly Weather Rev.*, **101**, 426–443, 1973.

ANGELL, J. K. and J. KORSHOVER, Global analysis of recent total ozone fluctuations, *Ibid*, **104**, 63–75, 1976.

ANGELL, J. K. and J. KORSHOVER, Global ozone variations and update into 1976, *Ibid*, **106**, 725–737, 1978a.

ANGELL, J. K. and J. KORSHOVER, Recent rocketsonde-derived temperature variations in the western hemisphere, *J. Atmos. Sci.*, **35**, 1758–1764, 1978b.

APRUZESE, J., M. R. SCHOEBERL, and D. F. STROBEL, Parameterization of IR cooling in a middle atmosphere dynamics model 1. Effects of the zonally averaged circulation, *J. Geophys. Res.*, **87**, 8951–8966, 1982.

ARIJS, E., J. INGELS, and D. NEVEJANS, Mass spectrometric measurement of the positive ion composition in the stratosphere, *Nature*, **271**, 642–644, 1978.

ARIJS, E., D. NEVEJANS, P. FREDERICK, and J. INGELS, Stratospheric negative ion composition measurements in the stratosphere, *Geophys. Res. Lett.*, **8**, 121–124, 1981.

ARIJS, E., D. NEVEJANS, P. FREDERICK, and J. INGELS, Stratospheric positive ion composition measurements, ion abundance and related trace gas detection, *J. Atmos. Terr. Phys.*, **44**, 43–54, 1982a.

ARIJS, E., D. NEVEJANS, P. FREDERICK, and J. INGELS, Stratospheric negative ion composition measurements, ion abundances and related trace gas detection, *J. Atmos. Terr. Phys.*, **44**, 681–693, 1982b.

ARIJS, E., D. NEVEJANS, and J. INGELS, Positive ion composition measurement and acetonitrile in the upper stratosphere, *Nature*, **303**, 314–316, 1983.

ARNOLD, F., J. KISSEL, D. KRANKOWSKY, H. WIEDER, and J. ZAHRINGER, Negative ions in the lower ionosphere, A mass spectrometric measurement, *J. Atmos. Terr. Phys.*, **33**, 1169–1175, 1971.

ARNOLD, F., D. KRANKOWSKY, and K. H. MARIEN, First mass spectrometric measurements of positive ions in the stratosphere, *Nature*, **267**, 30–32, 1977.

ARNOLD, F., H. BOHRINGER, and G. HENSCHEN, Composition measurements of stratospheric positive ions, *Geophys. Res. Lett.*, **8**, 653–656, 1978.

ARNOLD, F. and G. HENSCHEN, First mass analysis of stratospheric negative ions, *Nature*, **275**, 521–522, 1978.

ARNOLD, F. and W. JOOS, Rapid growth of atmospheric cluster ions at the cold mesopause, *Geophys. Res. Lett.*, **6**, 763–769, 1979.

ARNOLD, F. and P. FABIAN, First measurements of gas phase sulphuric acid in the stratosphere, *Nature*, **283**, 55–57, 1980.

ARNOLD, F., G. HENSCHEN, and E. E. FERGUSON, Mass spectrometric measurements of fractional ion abundances in the stratosphere—positive ions, *Planet. Space Sci.*, **29**, 185–193, 1981a.

ARNOLD, F., P. FABIAN, E. E. FERGUSON, and W. JOOS, Mass spectrometric measurements of fractional ion abundances in the stratosphere — negative ions, *Planet. Space Sci.*, **29**, 195–203, 1981b.

ARNOLD, F., P. FABIAN, and W. JOOS, Measurements of the height variation of sulfuric acid vapor concentrations in the stratosphere, *Geophys. Res. Lett.*, **8**, 293–296, 1981c.

ARVESEN, J. C., R. N. GRIFFIN, Jr., and B. D. PEARSON, Jr., Determination of extraterrestrial solar spectral irradiance from a research aircraft, *App. Opt.*, **8**, 2215–2232, 1969.

ASHBY, R. W., A numerical study of the chemical composition of the stratosphere, A thesis of PhD, the University of Wisconsin, Madison, 302 pp., 1976.

ATKINSON, R., G. M. BREUER, J. N. PITTS, Jr., and H. L. SANDOVAL, Tropospheric and stratospheric sinks for halocarbons: Photooxidation, $O(^1D)$ atoms, and OH radical reactions, *J. Geophys. Res.*, **81**, 5765–5770, 1976.

AUSLOOPS, P., R. E. REHBERT, and L. GLASGOW, Photodecomposition of chloromethanes absorbed on silica surfaces, *J. Res. (NBS)*, **82**, 1–8, 1977.

BAILEY, D. K., Abnormal ionization in the lower ionosphere associated with cosmic ray flux enhancements, *Proc. IRE*, **47**, 255–266, 1959.

BAKER, K. D., R. H. BISHOP, and L. R. MEGILL, Rocket measurements of $O_2(^1\Delta_g)$, emissions in the auroral zone, *J. Geophys. Res.*, **79**, 243–248, 1974.

BAKER, K. D., A. F. NAGY, R. O. OLSEN, E. S. ORAN, J. RANDHAWN, D. F. STROBEL, and T. TOHMATSU, Measurement of the nitric oxide altitude distribution in the mid-latitude mesosphere, *J. Geophys. Res.*, **82**, 3281–3286, 1977.

BALUTEAU, J. P. and E. BUSSELETTI, High resolution spectra of the stratosphere between 30 and 200 cm^{-1}, *Nature Phys. Sci.*, **241**, 113–114, 1973.

BANKS, P. M. and G. KOCKARTS, *Aeronomy*, Part A and B, Academic Press, 1973.

BARBE, A., P. HARCHE, C. SECROUN, and P. JOUVE, Measurements of tropospheric and stratospheric H_2CO by an infrared high resolution technique, *Geophys. Res. Lett.*, **6**, 463–465, 1979.

BARNES, I., V. BASTIAN, K. H. BECKER, E. H. FINK, and F. ZABEL, Rate constant of the reaction of OH with HO_2NO_2, *Chem. Phys. Lett.*, **83**, 459–464, 1981.

BARTH, C. A. Rocket measurement of nitric oxide in the upper atmosphere, *Planet. Space Sci.*, **14**, 623–630, 1966.

BARTH, C. A. and C. W. HORD, Mariner ultraviolet spectrometer: Topography and polar cap, *Science*, **173**, 107–201, 1971.

BARTH, C. A., C. W. HORD, A. I. STEWART, A. L. LANE, M. L. DICK, and G. P. ANDERSON, Mariner 9 ultraviolet spectrometer experiment: Seasonal variation of ozone on Mars, *Science*, **179**, 795–796, 1973.

BARTH, C. A., D. W. RUSCH, R. J. THOMAS, G. H. MOUNT, G. J. ROTTMAN, G. E. THOMAS, R.

W. Sanders, and G. M. Lawrence, Solar mesosphere explorer: scientific objectives and results, *Geophys. Res. Lett.*, **10**, 237–240, 1983.

Bates, D. R. and P. B. Hays, Atmospheric nitrous oxide, *Planet. Space Sci.*, **15**, 189–197, 1967.

Baum, W. A., F. S. Johnson, J. J. Oberly, C. C. Rockwood, C. V. Strain, and R. Tousey, Solar ultraviolet spectrum to 88 kilometers, *Phys. Rev.*, **70**, 781–782, 1946.

Berger, M. J. and S. M. Seltzer, Bremsstrahlung in the atmosphere, *J. Atmos. Terr. Phys.*, **34**, 85–108, 1972.

Berger, M. J., S. M. Seltzer, and K. Maeda, Some new results on electron transport in the atmosphere, *J. Atmos. Terr. Phys.*, **36**, 591–617, 1974.

Berkner, L. V. and L. C. Marshall, The history of oxygen concentration in the earth's atmosphere, *Faraday Soc. Diss.*, **37**, 122–141, 1964.

Berkner, L. V. and L. C. Marshall, On the origin and rise of oxygen concentration in the earth's atmosphere, *J. Atmos. Sci.*, **22**, 225–261, 1965.

Berkner, L. V. and L. C. Marshall, Limitation on oxygen concentration in a primitive planetary atmosphere, *J. Atmos. Sci.*, **23**, 132–143, 1966.

Bethke, G. W., Oscillator strengths in the far ultraviolet, I. Nitric oxide, *J. Chem. Phys.*, **31**, 662–668, 1959a.

Bethke, G. W. Oscillator strengths in the far ultraviolet, II. Oxygen Schumann-Runge band; *Ibid*, **31**, 669–673, 1959b.

Björn, L. G. and F. Arnold, Mass spectrometric detection of precondensation nuclei at the arctic summer mesosphere, *Geophys. Res. Lett.*, **8**, 1167–1170, 1981.

Bischof, W., P. Fabian, and R. Borchers, Decrease in CO_2 mixing ratio observed in the stratosphere, *Nature*, **288**, 347–348, 1980.

Blackshear, W. T. and R. H. Tolson, High correlation between variations in monthly averages of solar activity and total atmospheric ozone, *Geophys. Res. Lett.*, **5**, 921–924, 1978.

Blake, A. J., An atmospheric absorption model for the Schumann-Runge bands of oxygen, *J. Geophys. Res.*, **84**, 3272–3282, 1979.

Blake, A. J. and J. H. Carver, The evolutionary role of atmospheric ozone, *J. Atmos. Sci.*, **34**, 720–728, 1977.

Blake, D. and R. S. Lindzen, Effect of photochemical models on calculated equilibrium and cooling rates in the stratosphere, *Mon. Weather Rev.*, **101**, 783–802, 1973.

Blake, D. R., E. W. Mayer, S. C. Tyler, Y. Makide, D. C. Montague, and F. S. Rowland, Global increase in atmospheric methane concentrations between 1978 and 1980, *Geophys. Res. Lett.*, **9**, 477–480, 1982.

Blamont, J. E., M. L. Chanin, and G. Megie, Vertical distribution and temperature profile of the nighttime atmospheric sodium layer obtained by laser backscatter, *Ann. Geophys.*, **28**, 833–838, 1972.

Blatherwick, R. D., A. Goldman, D. G. Murcray, F. J. Murcray, G. R. Cook, and J. W. Van Allen, Simultaneous mixing ratio profiles of stratospheric NO and NO_2 as derived from balloon-borne infrared solar spectra, *Geophys. Res. Lett.*, **7**, 471–473, 1980.

Bloxam, R. M., A. W. Brewer, and C. T. McElroy, NO_2 measurements by absorption spectrophotometer observations from the ground and high altitude balloon, Churchill, Manitoba, July, 1974, *Proc. Fourth Conference on CIAP, DOT-TSC-OST-75-38*, pp. 454–457, U. S. Dept. of Transp., Washington, D.C., 1975.

Bojkov, R. D., Differences in Dobson spectrometer and filter ozonometer measurements of total ozone, *J. Appl. Meteor.*, **8**, 362–368, 1969.

Borucki, W. J., D. S. Colburn, R. C. Whitten, L. A. Capone, and M. Covert, Model analysis of the ozone depletion due to the August 1972 solar proton event, *EOS Trans.*,

AGU, **59**, 284, 1978.

BORUCKI, W. J., R. C. WHITTEN, V. R. WATSON, H. T. WOODWARD, C. A. RIEGEL, L. A. CAPONE, and T. BECKER, Model predictions of latitude-dependent ozone depletion due to supersonic transport operations, *AIAA Journal*, **14**, 1738–1745, 1976.

BORUCKI, W. J., R. C. WHITTEN, H. T. WOODWARD, L. A. CAPONE, C. A. RIEGEL, and S. GAINES, Stratospheric ozone decrease due to chlorofluoromethane photolysis: predictions of latitudinal dependence, *J. Atmos. Sci.*, **37**, 686–697, 1980.

BRASSEUR, G., Un modele bidimensionnel du comportement de l'ozone dans la stratosphere, *Planet. Space Sci.*, **26**, 139–159, 1978.

BRASSEUR, G., A. DERUDDER, and P. C. SIMON, Implication for stratospheric composition of a reduced absorption cross section in the Herzberg continuum of molecular oxygen, *Geophys. Res. Lett.*, **10**, 20–23, 1983 a.

BRASSEUR, G., E. ARIJS, A. DE RUDDER, D. NEVEJANS, and J. INGELS, Acetonitrile in the atmosphere, *Geophys. Res. Lett.*, **10**, 725–728, 1983b.

BRECKENBRIDGE, W. H. and H. TANBE, Ultraviolet absorption spectrum of carbonyl sulfide, *J. Chem. Phys.*, **52**, 1713–1715, 1970.

BREMNER, J. M., S. G. ROBBINS, and A. M. BLACKMER, Seasonal variability in emission of nitrous oxide from soil, *Geophys. Res. Lett.*, **7**, 641–644, 1980.

BREWER, A. W., Evidence for a world cicrulation provided by the measurements of helium and water vapor distribution in the stratosphere, *Quart. J. Roy. Meteor Soc.*, **75**, 351-363, 1949.

BREWER, A. W. and J. R. MILFORD, The Oxford Kew ozonesonde, *Proc. Roy. Soc. London, A,* **256**, 470–495, 1960.

BREWER, A. W. and K. P. B. THOMSON, A radiometer-sonde for observing stratospheric emission due to water vapor in its rotation band, *Quart. J. Roy. Met. Soc.*, **98**, 187–192, 1972.

BRINTON, H. C. and H. G. MAYR, Temporal variations of thermospheric hydrogen derived from *in situ* measurements, *J. Geophys. Res.*, **76**, 6198-6201, 1971.

BROADFOOT, A. L., The solar spectrum 2100–3200 Å, *Astrophys. J.*, **173**, 681–689, 1972.

BRUECKNER, G. E., J.-D. F. BARTOE, M. O. KJELDSETH, and M. E. VAN HOOSIER, Absolute solar ultraviolet intensities and their variations with solar activity I. The wavelength region 1750–2100 Å, *Astrophys. J.*, **209**, 935–944, 1976.

BUIJS, H. L., G. L. VAIL, G. TRAMBLAY, and D. J. W. KENDALL, Simultaneous measurement of the volume mixing ratios of HCl and HF in the stratosphere, *Geophys. Res. Lett.*, **7**, 205–208, 1980.

BURCH, D. E., D. GRYVNAK and D. WILLIAMS, The infrared absorption by carbon dioxide, *Ohio State University Res. Foundation, Rep. on Project 778*, 88 pp., 1960.

BURCH, D. E., D. GRYVNAK, E. G. SINGLETON, W. L. FRANCE, and D. WILLIAMS , Infrared absorption by carbon dioxide, water vapor, and minor atmospheric constituents, Research Report, AFCRL-62-698, 316 pp. 1962.

BURKHARDT, E. G., C. A. LAMBERT, and C. K. N. PATEL, Stratsspheric nitric oxide: Measurements during daytime and sunset, *Science*, **188**, 1111-1113, 1975.

BURNETT, C. R. and E. B. BURNETT, Spectroscopic measurements of the vertical column abundance of hydroxyl (OH) in the earth's atmosphere, *J. Geophys. Res.*, **86**, 5185–5202, 1981.

BUSH, Y. A., A. L. SCHMELTEKOPF, F. C. FEHSENFELD, D. L. ALBRITTON, J. R. MCAFEE, P. D. GOLDAN, and E. E. FERGUSON, Stratospheric measurements of methane at several latitudes, *Geophys. Res. Lett.*, **5**, 1027–1029, 1978.

CADLE, R. D. and G. W. GRAMS, Stratospheric aerosol particles and their optical properties, *Rev. Geophys. Space Phys.*, **13**, 475–501, 1975.

CALLEAR, A. B. and M. J. PILLING, Fluorescence of nitric oxide-VI. Predissociation and cascade quenching of NO $D^2\Sigma^+$ ($v=0$) and NO $C^2\Pi(v=0)$ and oscillator strengths of the

$\delta(0,0)$ and $\epsilon(0,0)$ bands, *J. Chem. Soc. Fraday Trans.*, **66**, 1886–1906, 1970.

Callis, L. B., V. Ramanathan, R. E. Boughner and B. R. Barkstrom, The stratosphere: scattering effect, a coupled 1-D model and thermal balance effects, *Proc. 4-th Conf. on CIAP*, pp. 224–233, U. S. Dept. Transp. Washington, D. C., 1976.

Callis, L. B. and J. E. Nealy, Solar UV variability and its effect on stratospheric thermal structure and trace constituents, *Geophys. Res. Lett.*, **5**, 249–252, 1978.

Callis, L. B., M. Natarajan, and J. E. Nealy, Ozone and temperature trends associated with the 11-year solar cycle, *Science*, **204**, 1303–1306, 1979.

Callis, L. B. and M. Natarajan, Atmospheric carbon dioxide and chlorofluoromethanes: combined effects on stratospheric ozone, temperature, and surface temperature, *Geophys. Res. Lett.*, **8**, 587–590, 1981.

Calvert, J. G. and J. N. Pitts, Jr., *Photochemistry*, John Wiley and Sons, Inc., 899 pp., 1966.

Calvert, J. G., J. A. Kerr, K. L. Demerjian, and R. D. McQuigg, Photolysis of formaldephyde as a hydrogen atom source in the lower atmosphere, *Science*, **175**, 751–752, 1972.

Calvert, J. G., F. Su, J. W. Bottenheim, and O. P. Strauss, Mechanism of the homogenous oxidation of sulfur dioxide in the troposphere, *Atmos. Environ.*, **12**, 197–226, 1978.

Camy-Peyret, C., J.-M. Flaud, J. Laurent, and G. M. Stockes, First infrared measurement of atmospheric NO_2 from the ground, *Geophys. Res. Lett.*, **10**, 35–38, 1983.

Carver, J. H., B. H. Horton, G. W. A. Lockey, and B. Rofe, Ultraviolet ion chamber measurements of the solar minimum brightness temperature, *Solar Phys.*, **27**, 347–353, 1972a.

Carver, J. H., B. H. Horton, R. S. O'Brien, and B. Rofe, Ozone determination by lunar rocket photometry, *Planet. Space Sci.*, **20**, 217–223, 1972b.

Carver, J. H., H. P. Gies, T. I. Hobbs, B. R. Lewis, and D. G. McCoy, Temperature dependence of the molecular oxygen photoabsorption cross-section near the H Lyman-alpha line, *J. Geophys. Res.*, **82**, 1955–1960, 1977.

Chakrabarty, D. K. and P. Chakrabarty, The evolution of ozone with changing solar activity, *Geophys. Res. Lett.*, **9**, 76–78, 1982.

Chaloner, C. P., J. R. Drummond, J. T. Houghton, R. F. Jarnot, and H. K. Roscoe, Stratospheric measurements of H_2O and the diurnal change of NO and NO_2, *Nature*, **258**, 696–697, 1975.

Chaloner, C. P., J. R. Drummond, J. T. Houghton, R. F. Jarnot, and H. K. Roscoe, Infrared measurements of stratospheric composition I. The balloon instrument and water vapor measurements, *Proc. Roy. Soc. London A*, **364**, 145–159, 1978.

Chameides, W. L., S. Ciliu, and R. J. Cicerone, Possible variations in atmospheric methane, *J. Geophys. Res.*, **82**, 1795–1798, 1977.

Chance, K. V., J. C. Brasunas, and W. A. Traub, Far infrared measurement of stratospheric HCI, *Geophys. Res. Lett.*, **7**, 704–706, 1980.

Chandra, S. Energetics and thermal structure of the middle atmosphere, *Planet. Space Sci.*, **28** 585–593, 1980.

Chandra, S. and A. K. Sinha, The diurnal heat budget of the thermosphere, *Planet. Space Sci.*, **21**, 593–604, 1973.

Chang, J. S., Simulations, perturbations, and interpretations, *Proc. 3rd CIAP Conf., DOT-TSC-OST-74-15,* pp. 330–341, U. S. Dept. of Transp., Washington, D.C., 1974.

Chang, J. S., W. H. Duewer, and D. J. Wuebbles, The atmospheric nuclear tests of the 1950's and 1960's: a possible test of ozone depletion theories, *J. Geophys. Res.*, **84**, 1755–1765, 1979a.

CHANG, J. S., J. R. BERKER, J. E. DAVENPORT, and D. M. GOLDEN, Chlorine nitrate photolysis by a new technique: very low pressure photolysis, *Chem. Phys. Lett.*, **60**, 385–390, 1979b.

CHAPMAN, S., A theory of upper atmospheric ozone, *Mem. Roy. Meteorol. Soc.*, **3**, 103–125, 1930.

CHAPMAN, S., The absorption and dissociation on ionizing effect of monochromatic radiation in an atmosphere on a rotating earth. II, Grazing incidence, *Proc. Phys. Soc.,* **43**, 483–501, 1931.

CHAPMAN, S., Upper atmospheric nomenclature, *J. Atmos. Terr. Phys.*, **1**, 121–124, and 201, 1950.

CHAPMAN, S., Note on the grazing-incidence integral $Ch(X,\chi)$ for monochromatic absorption in an exponential atmosphere, *Proc. Phys. Soc., London, Sec. B*, **66**, 710–712, 1953.

CHAPMAN S. and T. G. COWLING, *The Mathematical Theory of Non-Uniform Gases*, 2nd edition, 415 pp. Cambridge Univ. Press, 1953.

CHARNEY, J. G. and P. G. DRAZIN, Propagation of planetary-scale disturbances from the lower into the upper atmosphere, *J. Geophys. Res.*, **66**, 83–109, 1961.

CHEMICAL MANUFACTURES ASSOCIATION, World production and release of chlorofluorocabons 11 and 12 through 1980, Chemical Manufactures Association Fluorocarbon program panel, July, 29, 1981.

CHRISTIE, A. D., Secular or cyclic change in ozone, *Pure Appl. Geophys.*, **106–108**, 1000–1009, 1973.

CHRISTIE, A. D., Atmospheric ozone depletion by nuclear weapons testing, *J. Geophys. Res.*, **81**, 2583–2594, 1976.

CIAP, The effects of stratospheric pollution by aircraft, *Climate Impact Assessment Program Report of Findings*, Report DOT-TST-75-50, ed. by A.J. Grobecker, S.C. Coroniti and R.H. Cannon Jr., U. S. Dept. of Trans., Washington, D.C., 1974.

CIAP, *The natural stratosphere of 1974*, CIAP monograph 1, ed. by A. J. Grobecker, U. S. Dept. of Transp., Washington, D. C., 1975.

CICERONE, R. J., Halogens in the atmosphere, *Rev. Geophys. Space Phys.*, **19**, 123–139. 1981.

CICERONE, R. J., R. S. STOLARSKI, and S. WALTERS, Stratospheric ozone destruction by man-made chlorofluoromethanes, *Science,* **185**, 1165–1167, 1974.

CICERONE, R. J., J. D. SHETLER, D. H. STEDMAN, T. J. KELLY, and S. C. LIU, Atmospheric N_2O: measurements to determine its sources, sinks, and variations, *J. Geophys. Res.*, **83**, 3042–3050, 1978.

CICERONE, R. J., S. WALTERS, and S. C. LIU, Nonlinear response of stratospheric ozone column to chlorine injections, *J. Geophys. Res.*, **88**, 3647–3661, 1983.

CIESLIK, S., Determination experimentale des forces d'oscillateur des bands β, γ, δ et ε de la molecule NO, *Bull. Cl. Sci. Acad. R. Belg.*, **63**, 886, 1977.

CIESLIK, S. and M. NICOLET, The aeronomic dissociation of nitric oxide, *Planet. Space Sci.*, **21**, 925–938, 1973.

CLARK, T. A. and D. J. W. KENDALL, Far infrared emission spectrum of the stratosphere from balloons, *Nature*, **260**, 31–32, 1976.

CLYNE, M. A. A., P. B. MONKHOUSE, and L. W. TOWNSEND, Reactions of $O(^3P_j)$ atoms with halogens: the rate constants for the elementary reactions $O+BrO$, $O+Br_2$, $O+Cl_2$, *Int. J. Chem. Kinet.*, **8**, 425–449, 1976.

COFFEY, M. T., W. G. MANKIN, and A. GOLDMAN, Simultaneous spectroscopic determination of the latitudinal, seasonal, and diurnal variability of stratospheric N_2O, NO, NO_2, and HNO_3, *J. Geophys. Res.*, **86**, 7331–7341, 1981a.

COFFEY, M. T., W. G. MANKIN, and R. J. CICERONE, Spectroscopic detection of stratospheric hydrogen cyanide, *Science,* **214**, 333–335, 1981b.

Cogley, A. C. and W. J. Borucki, Exponential approximation for daily average solar heating or photolysis, *J. Atmos. Sci.*, **33**, 1347–1356, 1976.

Colegrove, F. D., W. B. Hanson, and F. S. Johnson, Eddy diffusion and oxygen transport in the lower thermosphere, *J. Geophys. Res.*, **70**, 4931–4941, 1965.

Conrad, R. and W. Seiler, Acid soils as a source of atmospheric carbon monoxide, *Geophys. Res. Lett.*, **9**, 1353–1356, 1982.

Coulson, K. L., *Solar and Terrestrial Radiation*, 317 pp., Academic Press, 1975.

Cox, A., The frequency of geomagnetic reversals and the symmetry of the nondipole field, *Rev. Geophys. Space Phys.*, **13**, No. 3, 35–51, 1975.

Cox, R. A., and R. G. Derwent, The ultra-violet absorption spectrum of gaseous nitrous acid, *J. Photochem.*, **6**, 23–34, 1976.

Cox, R. A. and K. Patrick, Kinetics at the reaction of HO_2+NO_2 (+ M)→HO_2NO_2 using molecular modulation spectrometer, *Int. J. Chem. Kinet.*, **11**, 635–648, 1979.

Cox, R. A., J. P. Burrows, and T. J. Wallington, Rate coefficient for the reaction $OH+HO_2=H_2O+O_2$ at 1 atmosphere pressure and 308°K, *Chem. Phys. Lett.*, **84**, 217–221, 1981.

Craig, R. A., *The Upper Atmosphere, Meteorology and Physics*, 509 pp., Academic Press, 1965.

Creel, C. L. and J. Ross, Photodissociation of NO_2 in the region 458–630 nm, *J. Chem. Phys.*, **64**, 3560–3566, 1976.

Cronn, D. R., R. A. Rasmussen, E. Robinson, and D. E. Harsch, Halogenated compound identification and measurement in the troposphere and lower stratosphere, *J. Geophys. Res.*, **82**, 5535–5544, 1977.

Crutzen, P. J., The influence of nitrogen oxides on the atmospheric ozone content, *Quart. J. Roy. Met. Soc.*, **96**, 320–325, 1970.

Crutzen, P., A discussion of the chemistry of some minor constituents in the stratosphere and troposphere, *Pure Appl. Geophys.*, **106–108**, 1385–1399, 1973.

Crutzen, P., A review of upper atmosphere photochemistry, *Can. J. Chem.*, **52**, 1560–1581, 1974.

Crutzen, P. J., The possible importance of CSO for the sulfate layer of the stratosphere, *Geophys. Res. Lett.*, **3**, 73–76, 1976.

Crutzen, P. J., I. S. A. Isaksen, and G. C. Reid, Solar proton events: stratospheric sources of nitric oxide, *Science*, **189**, 457–459, 1975.

Crutzen, P. J., I. S. A. Isaksen, and J. R. McAfee, The impact of the chlorocarbon industry on the ozone layer, *J. Geophys. Res.*, **83**, 345–363, 1978.

Cumming, C., and R. P. Lowe, Balloon-borne spectroscopic measurement of stratospheric methane, *J. Geophys. Res.*, **78**, 5259–5264, 1973.

Cunnold, D., F. Alyea, N. Phipplips, and R. Prinn, A three-dimensional dynamical-chemical model of atmospheric ozone, *J. Atmos. Sci.*, **32**, 170–194, 1975.

Curtis, A. R. and R. M. Goody, Thermal radiation in the upper atmosphere, *Proc. Roy. Soc. London, A*, **236**, 193–206, 1956.

Danielsen, E. F., An objective method for determining the generalized transport tensor for two-dimensional Eulerian models, *J. Atmos. Sci.*, **38**, 1319–1339, 1981.

Danielsen, E. F., Statistics of cold cumulonimbus anvils based on enhanced infrared photographs, *Geophys. Res. Lett.*, **9**, 601–604, 1982a.

Danielsen, E. F., A dehydration mechanism for the stratosphere, *Geophys. Res. Lett.*, **9**, 605–608, 1982b.

Danielsen, E., R. Black, J. Shedlovsky, A. Wartburg, P. Haagenson, and W. Pollock, Observed disribution of radioactivity, ozone, and potential vorticity associated with tropopause folding, *J. Geophys. Res.*, **75**, 2353–2361, 1970.

DAVIDSON, J. A., F. C. FEHSENFELD, and C. J. HOWARD, The heats of formation of NO_3^- and NO_3^- association complexes with HNO_3 and HBr, *Int. J. Chem. Kinet.*, **9**, 17–29, 1977.

DAVIS, D. D., W. HEAPS, and T. MCGEE, Direct measurements of natural tropospheric levels of OH via an aircraft borne tunable dye laser, *Geophys. Res. Lett.*, **3**, 331–333, 1976.

DEGUCHI, S. and D. O. MUHLEMAN, Mesospheric water vapor, *J. Geophys. Res.*, **87**, 1343–1346, 1982.

DELUISI, J. J., Umkehr vertical-ozone profile errors caused by the presence of stratospheric aerosols, *J. Geophys. Res.*, **84**, 1766–1770, 1979.

DELUISI, J. J., Disparities in the determination of atmospheric ozone from solar ultraviolet transmission measurements 298.1–319.5 nm, *Geophys. Res. Lett.*, **7**, 1102–1104, 1980.

DELUISI, J. J. and J. NIMIRA, Preliminary comparison of satellite BUV observations of the vertical distribution of ozone in the upper stratosphere, *J. Geophys. Res.*, **83**, 379–384, 1978.

DELUISI, J. J., C. L. MATEER, and D. F. HEATH, Comparison of seasonal variations of upper stratospheric ozone concentration revealed by umkehr and Nimbus 4 BUV observations, *J. Geophys. Res.*, **84**, 3728–3732, 1979.

DEMORE, W. B., Rate constant and possible pressure dependence of the reaction $OH+HO_2$, *J. Phys. Chem.*, **86**, 121–126, 1982.

DEMORE, W. B. and M. PATAPOFF, Temperature and pressure dependence of CO_2 extinction coefficients, *J. Geophys. Res.*, **77**, 6291–6293, 1972.

DETWILER, C. R., D. L. HARRETT, J. D. PURCELL, and R. TOUSEY, The intensity distribution in the ultraviolet solar spectrum, *Ann. Geophys.*, **17**, 263–272, 1961.

DICKINSON, R. E., Infrared radiative heating and cooling in the Venusian mesosphere. 1: Global mean radiative equilibrium, *J. Atmos. Sci.*, **28**, 1531–1556, 1972.

DICKINSON, R. E., Method of parameterization for infrared cooling between altitudes of 30 and 70 km, *J. Geophys. Res.*, **78**, 4451–4457, 1973.

DICKINSON, P. H., R. C. BOLDEN, and R. A. YOUNG, Measurement of atomic oxygen in the lower ionosphere using a rocket-borne resonance lamp, *Nature*, **252**, 289–291, 1974.

DICKINSON, P. H. G., W. C. BAIN, L. THOMAS, E. R. WILLIAMS, D. B. JENKINS, and N. D. TWIDDY, The determination of the atomic oxygen concentration and associated parameters in the lower ionosphere, *Proc. Roy. Soc. London, A*, **369**, 379–408, 1980.

DICKSON, D., Congress faces decision on CFC, New ozone data from NASA, *Nature*, **293**, 3–4, 1981.

DOBSON, G. M. B., A photoelectric spectrophotometer for measuring the amount of atmospheric ozone, *Proc. Phy. Soc.*, **43**, 324–339, 1931.

DOBSON, G. M. B., Origin and distribution of polyatomic molecules in the atmosphere, *Proc. Roy. Soc., A*, **236**, 187–193, 1956.

DOBSON, G. M. B., A. W. BREWER, and B. CWILONG, Meteorology of the lower stratosphere, *Proc. Roy. Soc., London, A*, **189**, 144–147, 1946.

DONAHUE, T. M., The problem of atomic hydrogen, *Ann. Geophys.*, **22**, 175–188, 1966.

DRUMMOND, J. R. and R. F. JARNOT, Infrared measurements of stratospheric composition II Simultaneous NO and NO_2 measurements, *Proc. Roy. Soc., London, A*, **364**, 237–254, 1978.

DRUMMOND, J. W., J. M. ROSEN, and D. J. HOFMAN, Balloon borne chemiluminescent measurement of NO to 45 km, *Nature*, **265**, 319–320, 1977.

DUCE, R. A., W. H. ZOLLER, and J. L. MOYERS, Particulate and gaseous halogens in the antarctic atmosphere, *J. Geophys. Res.*, **78**, 7802–7811, 1973.

DUNKERTON, T., On the mean meridional mass motions of the stratosphere and mesosphere, *J. Atmos. Sci.*, **35**, 2325–2333, 1978.

DUNKIN, D. B., F. C. FEHSENFELD, A. L. SCHMELTEKOPF, and E. E. FERGUSON, Three-body association reactions of NO^+ with O_2, N_2 and CO_2, *J. Chem. Phys.*, **54**, 3817–3822, 1971.

DÜTSCH, H. U., Atmospheric ozone and ultraviolet radiation, *World Survey of Climatology*, **4**, 383–432, 1969.

DÜTSCH, H. U., Photochemistry of atmospheric ozone, *Adv. Geophys.*, **15**, 219–322, 1971.

DÜTSCH, H. U., The search for solar cycle-ozone relationships, *J. Atmosph. Terr. Phys.*, **41**, 771–785, 1979.

EHHALT, D. H., *In situ* observations, *Phil. Trans. Roy. Soc., London, A*, **296**, 175–189, 1980.

EHHALT, D. H. and L. E. HEIDT, The concentration of molecular H_2 and CH_4 in the stratosphere, *Pure Appl. Geophys.*, **106–108**, 1340–1352, 1973.

EHHALT, D. H., N. ROPER, and H. E. MOORE, Vertical profiles of nitrous oxide in the troposphere, *J. Geophys. Res.*, **80**, 1653–1655, 1975a.

EHHALT, D. H., L. E. HEIDT, R. A. LUEB, and E. A. MARTELL, Concentrations of CH_4, CO, CO_2, H_2, H_2O and N_2O in the upper stratosphere, *J. Atmos. Sci.*, **32**, 163–169, 1975b.

EHHALT, D. H., U. SCHMIDT, and L. E. HEIDT, Vertical profiles of molecular hydrogen in the troposphere and stratosphere, *J. Geophys. Res.*, **82**, 5907–5911, 1977.

ELLSAESSER, H. W., J. E. HARRIES, D. KLEY, and R. PENNDORF, Stratospheric H_2O, *Planet. Space. Sci.*, **28**, 827–835, 1980.

ELSASSER, W. M., Mean absorption and equivalent absorption coefficient of a band spectrum, *Phys. Rev.*, **54**, 126–129, 1938.

EVANS, W. F. J., D. M. HUNTEN, E. J. LLEWELLYN, and A. VALLANCE JONES, Altitude profile of the infrared atmospheric system of oxygen in the dayglow, *J. Geophys. Res.*, **73**, 2885–2896, 1968.

EVANS, W. F. J., E. J. LLEWELLYN, and A. VALLANCE JONES, Balloon observations of the temporal variation of the infrared atmospheric oxygen bands in the airglow, *Planet. Space Sci.*, **17**, 933–947, 1969.

EVANS, W. F. J. and E. J. LLEWELLYN, Molecular oxygen emissions in the airglow, *Ann. Geophys.*, **26**, 167–178, 1970.

EVANS, W. F. J. and E. J. LLEWELLYN, Measurements of mesosphere ozone from observations of the 1.27μ band, *Radio Sci.*, **7**, 45–50, 1972.

EVANS, W. F. J., J. B. KERR, D. I. WARDE, J. C. MCCONNELL, B. A. RIDLEY, and H. I. SCHIFF, Intercomparison of NO, NO_2 and HNO_3 measurements with photochemical theory, *Atmosphere (Can. Met. Soc.)*, **14**, 187–198, 1976.

EVANS, W. F. J., J. B. KERR, C. T. MCELROY, R. S. O'BRIEN, B. A. RIDLEY, and D. I. WARDE, The odd nitrogen mixing ratio in the stratosphere, *Geophys. Res. Lett.*, **4**, 235–238, 1977.

EVANS, W. F. J., C. T. MCELROY, and J. B. KERR, Simulation of nitrogen constituent measurements from the August 28, 1976, stratoprobe III flight, *J. Geophys. Res.*, **86**, 12,066–12,070, 1981.

EYRE, J. R. and H. K. ROSCOE, Radiometric measurements of HCl, *Nature*, **226**, 243–244, 1977.

FAA, Chemical kinetic and photochemical data sheets for atmospheric reactions, Report No. FAA-EE-80-17, ed. by R. F. Hampson, Federal Aviation Adm., U. S. Dept. of Transp., Washington D. C., 1981.

FABIAN, P. and C. E. JUNGE, Global rate of ozone destruction at the earth's surface, *Arch. Met. Geophy. Biokl., Ser. A.*, **19**, 161–172, 1970.

FABIAN, P., R. BORCHERS, K. H. WEILER, U. SCHMIDT, A. VOLZ, D. H. EHHALT, W. SEILER, and F. MULLER, Simultaneously measured vertical profiles of H_2, CH_4, CO, N_2O, $CFCl_3$, and CF_2CL_2 in the mid-latitude stratosphere and troposphere, *J. Geophys. Res.*, **84**, 3149–3154, 1979a.

FABIAN, P., J. A. PYLE, and R. J. WELLS, The August 1972 solar proton event and the atmospheric ozone layer, *Nature*, **277**, 458–460, 1979b.

FABIAN, P., R. BORCHERS, G. FLENTJE, W. A. MATTHEWS, W. SEILER, H. GIEHL, K. BUNSE, F. MULLER, U. SCHMIDT, A. VOLZ, A. KHEDIM, and F. J. JOHNEN, The vertical distribution of stable trace gases at mid latitudes, *J. Geophys. Res.*, **86**, 5179–5184, 1981.

FALCONER, P. D. and J. D. HOLDEMAN, Measurements of atmospheric ozone made from a GASP equipped 747 airliner: mid-March 1975, *Geophys. Res. Lett.*, **3**, 101–104, 1975.

FARLOW, N. H., G. V. FERRY, H. Y. LEM, and D. M. HAYES, Latitudinal variations of stratospheric aerosols, *J. Geophys. Res.*, **84**, 733–743, 1979.

FARMER, A. J. D., P. FABIAN, B. R. LEWIS, K. H. LONAN, and G. N. HADDAD, Experimental oscillator strengths for the Schumann-Runge band system in oxygen, *J. Quant. Spectrosc. Rad. Trans.*, **8**, 1739–1746, 1968.

FARMER, C. B., Infrared measurements of stratospheric composition, *Can. J. Chem.*, **52**, 1544–1549, 1974.

FARMER, C. B., O. F. RAPER, and R. H. NORTON, Spectroscopic detection and vertical distribution of HCl in the troposphere and stratosphere, *Geophys. Res. Lett.*, **3**, 13–16, 1976.

FARMER, C. B., D. W. DAVIS, A. L. HOLLAND, D. D. LAPORTE, and P. E. DOMS, Mars: water vapor observations from Viking orbiters, *J. Geophys. Res.*, **82**, 4225–4248, 1977.

FARMER, C. B., O. F. RAPER, B. D. ROBBINS, R. A. TOTH, and C. MULLER, Simultaneous spectroscopic measurements of stratospheric species: O_3, CH_4, CO, CO_2, N_2O, H_2O, HCl and HF at northern and southern mid-latitudes, *J. Geophys. Res.*, **85**, 1621–1632, 1980.

FEELY, H. W. and J. SPAR, Tungsten-185 from nuclear bomb tests as a tracer for stratospheric meteorology, *Nature*, **188**, 1062–1064, 1960.

FEHSENFELD, F. C. and E. E. FERGUSON, Origin of water cluster ion in the *D*-region, *J. Geophys. Res.*, **74**, 2217–2222, 1969.

FEHSENFELD, F. C., I. DOTAN, D. L. ALBRITTON, C. J. HOWARD, and E. E. FERGUSON, Stratospheric positive ion chemistry of formaldehyde and methanol, *J. Geophys. Res.*, **83**, 1333–1336, 1978.

FELDMAN, P. D. and P. Z. TAKACS, A search for molecular hydrogen fluorescence near 100 km, *J. Atmos. Sci.*, **32**, 2209–2212, 1975.

FELS, S. B. and M. D. SCHWARZKOPF, An efficient, accurate algorithm for calculating CO_2 15 μ band cooling rates, *J. Geophys. Res.*, **86**, 1205–1232, 1981.

FERGUSON, E. E., Laboratory measurements of ionospheric ion-molecule reaction rates, *Rev. Geophys. Space Phys.*, **12**, 703–713, 1974.

FERGUSON, E., Sodium hydroxide in the stratosphere, *Geophys. Res. Lett.*, **5**, 1035–1038, 1978.

FITZMAURICE, J. A., Simplification of the Chapman function for atmospheric attenuation, *App. Opt.*, **3**, 640, 1964.

FOLEY, H. M. and M. A. RUDERMAN, Stratospheric NO production from past nuclear explosions, *J. Geophys. Res.*, **78**, 4441–4450, 1973.

FONTANELLA, J. C., A. GIRARD, L. GRAMONT, and N. LOUISNARD, Vertical distribution of NO, NO_2 and HNO_3 as derived from stratospheric absorption infrared spectra, *App. Opt.*, **14**, 825–839, 1975.

FREDERICK, J. E. and R. D. HUDSON, Predissociation linewidths and oscillator strengths for the (2-0) to (13-0) Schumann-Runge bands of O_2, *J. Molec. Spectrosc.*, **74**, 247–256, 1979a.

FREDERICK, J. E. and R. D. HUDSON, Predissociation of nitric oxide in the mesosphere and stratosphere, *J. Atmos. Sci.*, **36**, 737–745, 1979b.

FREDERICK, J. E. and R. D. HUDSON, Atmospheric opacity in Schumann-Runge bands and the aeronomic dissociation of water vapor, *J. Atmos. Sci.*, **37**, 1088–1098, 1980a.

FREDERICK, J. E. and R. D. HUDSON, Dissociation of molecular oxygen in the Schumann-Runge bands, *J. Atmos. Sci.*, **37**, 1099–1106, 1980b.

FREDERICK, J. E. and J. E. MENTALL, Solar irradiance in the stratosphere: implications for the Herzberg continuum absorption of O_2, *Geophys. Res. Lett.*, **9**, 461–464, 1982.

FROIDEVAUX, L. and T. L. YUNG, Radiation and chemistry in the stratosphere: sensitivity to O_2 absorptioncross sections in the Herzberg continuum, *Geophys. Res. Lett.*, **9**, 854–857, 1982.

FUKUYAMA, K., Latitudinal distributions of photochemical heating rates in the winter mesosphere and lower thermosphere, *J. Atmos. Terr. Phys.*, **36**, 1321–1334, 1974.

GALBALLY, I. E. and C. R. ROY, Destruction of ozone at the earth's surface, *Quart. J. Roy. Met. Soc.*, **106**, 599–620, 1980.

GARCIA, R. R. and S. SOLOMON, A numerical model of the zonally averaged dynamical and chemical structure of the middle atmosphere, *J. Geophys. Res.*, **88**, 1379–1400, 1983.

GEAR, C. W., *Numerical Initial Value Problems in Ordinary Differential Equations*, 253 pp., Prentice-Hall Inc., Englewood Cliffs, New Jersey, 1971.

GEORGII, H. W. and F. X. MEIXNER, Measurement of the tropospheric and stratospheric SO_2 distribution, *J. Geophys. Res.*, **85**, 7433–7438, 1980.

GERSON, N. C. and J. KAPLAN, Nomenclature of the upper atmosphere, *J. Atmos. Terr. Phys.*, **1**, 200, 1951.

GIBBINS, C. J., P. R. SCHWARTZ, D. L. THACKER, and R. M. BEVILARQUA, The variability of mesospheric water vapor, *Geophys. Res. Lett.*, **9**, 131–134, 1982.

GIBSON, G. E. and N. S. BAYLISS, Variation with temperature of the continuous absorption spectrum of diatomic molecules: Part I Experimental, the absorption spectrum of chlorine, *Phys. Rev.*, **44**, 188–192, 1933.

GIBSON, A. J. and M. C. W. SANFORD, Daytime laser radar measurements of the atmosphere sodium layer, *Nature*, **239**, 509–511, 1972.

GIBSON, A. J. and L. THOMAS, Ultraviolet laser sounding of the troposphere and lower stratosphere, *Nature*, **250**, 561–563, 1975.

GIDEL, L. T., P. J. CRUTZEN, and J. FISHMAN, A two-dimensional photochemical model of the atmosphere, 1. Chlorocarbon emissions and their effect on stratospheric ozone, *J. Geophys. Res.*, **88**, 6622–6640, 1983.

GILLE, J. C., G. P. ANDERSON, and P. L. BAILEY, Comparison of near coincident LRIR and OSO-3 measurements of equatorial night ozone profiles, *Geophys. Res. Lett.*, **7**, 525–528, 1980a.

GILLE, J. C., P. L. BAILEY, and J. M. RUSSELL III, Temperature and composition measurements from the l.r.i.r. and l.i.m.s. experiments on Nimbus 6 and 7, *Phil. Trans. R. Soc., London, A*, **296**, 205–218, 1980b.

GILMORE, F. R., The production of nitrogen oxides by low altitude nuclear explosions, *J. Geophys. Res.*, **80**, 4553–4554, 1975.

GIRARD, A., Spectromètre à Grille, *App. Opt.*, **2**, 79–87, 1963.

GIRARD, A., L. GRAMONT, N. LOUISNARD, S. LE BOITEUX, and G. FERGANT, Latitudinal variation of HNO_3, HCl and HF vertical column density above 11.5 km, *Geophys. Res. Lett.*, **9**, 135–138, 1982.

GOLDAN, P. D., Y. A. BUSH, F. C. FEHSENFELD, D. L. ALBRITTON, P. J. CRUTZEN, A. L. SCHMELTEKOPF, and E. E. FERGUSON, Tropospheric N_2O mixing ratio measurements, *J. Geophys. Res.*, **83**, 935–939, 1978.

GOLDAN, P. D., W. C. KUSTER, D. L. ALBRITTON, and A. L. SCHMELTEKOPT, Stratospheric $CFCl_3$, CF_2Cl_2, and N_2O height profile measurements at several latitudes, *J. Geophys. Res.*, **85**, 413–423, 1980.

GOLDAN, P. D., W. C. KUSTER, A. L. SCHMELTEKOPT, F. C. FEHSENFELD, and D. L. ALBRITTON, Correction of atmospheric N_2O mixing ratio data, *J. Geophys. Res.*, **86**, 5385–5386, 1981.

GOLDBERG, R. A. and L. J. BLUMLE, Positive ion composition from a rocket-borne mass spectrometer, *J. Geophys. Res.*, **75**, 133–142, 1970.

GOLDBERG, R. A. and A. C. AIKIN, Studies of positive ion composition in the equatorial *D*-region ionosphere, *J. Geophys. Res.*, **76**, 8352–8464, 1971.

GOLDMAN, A., D. G. MURCRAY, F. H. MURCRAY, W. J. WILLIAMS, and F. S. BONOMO, Identification of the ν_3 NO_2 band in the solar spectrum observed from a balloon borne spectrometer, *Nature*, **225**, 443–444, 1970.

GOLDMAN, A., D. G. MURCRAY, F. H. MURCRAY, W. J. WILLIAMS, J. N. BROOKS, and C. M. BRADFORD, Vertical distribution of CO in the atmosphere, *J. Geophys. Res.*, **78**, 5273–5283, 1973a.

GOLDMAN, A., D. G. MURCRAY, F. H. MURCRAY, and W. J. WILLIAMS, Balloon-borne infrared measurements of the vertical distribution of N_2O in the atmosphere, *J. Opt. Soc. Amer.*, **63**, 843–846, 1973b.

GOLDMAN, A., D. G. MURCRAY, F. H. MURCRAY, W. J. WILLIAMS, and J. N. BROOKS, Distribution of water vapor in the stratosphere as determined from balloon measurements of atmospheric emission spectra in the 24–29 μm region, *App. Opt.*, **12**, 1045–1053, 1973c.

GOLDMAN, A., F. G. FERNOLD, W. J. WILLIAMS, and D. G. MURCRAY, Vertical distribution of NO_2 in the stratosphere as determined from balloon measurements of solar spectra in the 4500 Å region, *Geophys. Res. Lett.*, **5**, 257–260, 1978.

GOLDMAN, A., D. G. MURCRAY, F. J. MURCRAY, G. R. COOK, J. W. VAN ALLEN, F. S. BONOMO, and R. D. BLATHERWICK, identification of the 3 vibration rotation band of CF_4 in balloon-borne infrared solar spectra, *Geophys. Res. Lett.*, **6**, 609–612, 1979.

GOLDMAN, A., F. J. MURCRAY, R. D. BLATHERWICK, F. S. BONOMO, F. H. MURCRAY, and D. B. MURCRAY, Spectroscopic identification of $CHClF_2$ (F-22) in the lower stratosphere, *Geophys. Res. Lett.*, **8**, 1012–1014, 1981.

GOLOMB, D., K. WATANABE, and F. F. MARMO, Absorption coefficients of sulfur dioxide in the vacuum ultraviolet, *J. Chem. Phys.*, **36**, 958–960, 1962.

GOOD, A., D. A. DURDEN, and P. KEBARLO Mechanism and rate constants of ion-molecular reactions leading to formations of $H^+(H_2O)_n$ in moist oxygen and air, *J. Chem. Phys.*, **52**, 222–229, 1970.

GOODY, R. M., A statistical model for water-vapor absorption, *Quart. J. Roy. Met. Soc.*, **78**, 165–169, 1952.

GOODY, R. M., *The Physics of the Stratosphere*, 187 pp., Cambridge Univ. Press, London, 1954.

GOODY, R. M., *Atmospheric Radiation I. Theoretical Basis*, 436 pp., Oxford University Press, 1964.

GÖTZ, F. W. P., Zum Strahlungsklima des Spitzbergensoumers. Strahlungs und Ozonmessungen in der Königsbucht, *Gerl. Beitr. Geophys.*, **31**, 119–154, 1931.

GÖTZ, F. W. P., A. R. MEETHAM, and G. M. B. DOBSON, The vertical distribution of ozone in the atmosphere, *Proc. Roy. Soc., A*, **145**, 416–446, 1934.

GRAHAM, R. A. and H. S. JOHNSTON, The photochemistry of NO_3 and the kinetics of the N_2O_5-O_3 system, *J. Phys. Chem.*, **82**, 254–268, 1978.

GRAHAM, R. A., A. M. WINER, and J. N. PITTS, Ultraviolet and infrared absorption cross sections of gas phase HO_2NO_2, *Geophys. Res. Lett.*, **5**, 909–911, 1978.

GROVES, K. S. and A. F. TUCK, Stratospheric O_3 - CO_2 coupling in a photochemical-radiative column model. I: Without chlorine chemistry, *Quart. J. Roy. Met. Soc.*, **106**, 125–140, 1980a.

GROVES, K. S. and A. F. TUCK, Stratospheric O_3 - CO_2 coupling in a photochemical-radiative column model. II: With chlorine chemistry, *Quart. J. Roy. Met. Soc.*, **106**, 141–157, 1980b.

GUDIKSEN, P. H., A. W. FAIRHALL, and R. J. REED, Roles of mean meridional circulation and eddy diffusion in the transport of trace substances in the lower stratosphere, *J. Geophys. Res.*, **73**, 4461–4473, 1968.

GUENTHER, B., R. DASGUPTA, and D. HEATH, Twilight ozone measurement by solar occultation from AES, *Geophys. Res. Lett.*, **4**, 434–436, 1977.

HALMANN, M., Isotope effects on Franck-Condon factors –VI. Pressure-broadened absorption intensities of the Schumann-Runge bands of $^{16}O_2$ and $^{18}O_2$, *J. Chem. Phys.*, **44**, 2406–2408, 1966.

HAMILL, P., O. B. TOON, and C. S. KIANG, Microphysical processes affecting stratospheric aerosol particles, *J. Atmos. Sci.*, **34**, 1104–1119, 1977.

HAN, R. V., L. R. MEGILL, and C. L. WYATT, Recent observation of the equatorial $O_2(^1\Delta_g)$ emission after sunset, *J. Geophys. Res.*, **78**, 6140–6149, 1973.

HANSEN, J., D. JOHNSON, A. LOCIS, S. LEBEDEFT, P. LEE, D. RIND, and G. RUSSELL, Climate impact of increasing atmospheric carbon dioxide, *Science*, **213**, 957–966, 1981.

HANSER, F. A., B. SELLERS, and D. C. BRIEHL, Ultraviolet spectrophotometer for measuring columnar atmosphere ozone from aircraft, *App. Opt.*, **17**, 1649–1656, 1978.

HARKER, A. B., W. HO, and J. J. RATTO, Photodissociation quantum yield of NO_2 in the region 375 to 420 nm, *Chem. Phys. Lett.*, **50**, 394–397, 1977.

HARRIES, J. E., Measurement of stratospheric water vapor using far-infrared technique, *J. Atmos. Sci.*, **30**, 1691–1698, 1973.

HARRIES, J. E., The distribution of water vapor in the stratosphere, *Rev. Geophys. Space Phys.*, **14**, 565–575, 1976.

HARRIES, J. E., Ratio of HNO_3 to NO_2 concentrations in daytime stratosphere, *Nature*, **274**, 235, 1978.

HARRIES, J. E., Spectroscopic observations of middle atmosphere composition, *Phil. Trans. Roy. Soc. London, A*, **286**, 161–173, 1980a.

HARRIES, J. E., Atmospheric radiometry at submillimeter wavelengths, *App. Opt.*, **19**, 3075–3081, 1980b.

HARRIES, J. E. and W. J. BURROUGHS, Measurements of submillimeter wavelength radiation emitted by the stratosphere, *Quart. J. Roy. Met. Soc.*, **97**, 539–536, 1971.

HARRIES, J. E., D. G. MOSS, and N. R. W. SWANN, H_2O, O_3, N_2O and HNO_3 in the arctic stratosphere, *Nature*, **250**, 475–476, 1974a.

HARRIES, J. E., J. R. BIRCH, J. W. FLEMING, N. W. B. STONGE, D. G. MOSS, N. R. W. SWANN, and G. F. NEILL, Studies of stratospheric H_2O, O_3, HNO_3, N_2O and NO_2 from aircraft, *Proc. Third. Conf. on CIAP, DOT-TSC-OST-74-15*, pp. 197–212, U. S. Dept. of Transp., Washington, D. C., 1974b.

HARRIES, J. E., D. G. MOSS, N. R. W. SWANN, and G. F. NEIL, Simultaneous measuemen of H_2O, NO_2 and HNO_3 in the daytime stratosphere from 16 to 35 km, *Nature*, **259**, 300–302, 1976.

HARRISON, A. W., E. J. LLEWELLYN, and D. C. NICHOLLS, Night air-glow hydroxyl rotational temperatures, *Can. J. Phys.*, **48**, 1766–1768, 1970.

HARWOOD, R. S. and J. A. PYLE, A two-dimensional mean circulation model for the atmosphere below 80 km, *Quart. J. Roy. Met. Soc.* **101**, 723–747, 1975.

HARWOOD, R. S. and J. A. PYLE, Studies of the ozone budget using a zonal mean circulation model and linearized photochemistry, *Ibid.*, **103**, 319–343, 1977.

HASEBE, F., A global analysis of the fluctuation of total ozone I. Application of the optimum interpolation of the network data with random and systematic errors, *J. Meteorol. Soc., Japan,* **58**, 95–103, 1980a.

HASEBE, F., A global analysis of the fluctuation of total ozone II. Non-stationary annual oscillation, quasi-biennial oscillation, and long-term variations in total ozone, *Ibid.*, **58**,

104–117, 1980b.

Hassen, V., G. R. Hebert, and R. W. Nicholls, Measured transition probabilitites for bands of the Schumann-Runge ($B^3\Sigma_u^- - X^3\Sigma_g^-$) band system of molecular oxygen, *J. Phys. B: Atom. Molec. Phys.*, **3**, 1188–1194, 1970.

Hassen, V. and R. W. Nicholls, Absolute spectral absorption measurements on molecular oxygen from 2640–1920 Å: II. Continuum measurements 2430–1920 Å, *J. Phys. B: Atom. Molec. Phys.*, **4**, 1789–1797, 1971.

Heaps, W. S. and T. J. McGee, Balloon borne lidar measurements of stratospheric hydroxyl radical *J. Geophys. Res.*, **88**, 5281–5289, 1983.

Heath, D., Space observations of the variability of solar irradiance in the near and far ultraviolet, *J. Geophys. Res.*, **78**, 2779–2792, 1973.

Heath, D. F., Secular changes in stratospheric ozone from satellite observations (1970–1979), a paper presented at AGU fall meeting, San Francisco. December, 1981.

Heath, D. F., C. L. Mateer, and A. J. Krueger, The Nimbus-4 backscatter ultraviolet (BUV) atmospheric ozone experiment—two years operation, *Pure Appl. Geophys.*, **106–108**, 1278–1253, 1973.

Heath, D. F., A. J. Krueger, H. A. Roeder, and B. D. Henderson, The solar backscatter ultraviolet and total ozone mapping spectrometer, (SBUV/TOMS) for Nimbus G, *Opt. Eng.*, **14**, 323–331, 1975.

Heath, D. F. and M. P. Thekaekara, Measures of the solar spectral irradiance between 1200 and 3000 Å, NASA TM X-71172, 1976.

Heath, D. F., A. J. Krueger, and P. J. Crutzen, Solar proton events influence on stratospheric ozone, *Science*, **197**, 886–889, 1977.

Heidt, L. E., R. A. Lueb, W. Pollock, and D. H. Ehhalt, Stratospheric profiles of CCl_3F and CCl_2F_2, *Geophys. Res. Lett*, **2**, 445–447, 1975.

Heidt, L. E., J. P. Krasnec. R. A. Lueb, W. H. Pollock, B. E. Henry, and P. J. Crutzen, Latitudinal distributions of CO and CH_4 over the Pacific, *J. Geophys. Res.*, **85**, 7329–7336, 1980.

Henderson, W. R. and H. I. Schiff, A simple sensor for the measurement of atomic oxygen height profiles in the upper atmosphere, *Planet. Space Sci.*, **18**, 1527–1534. 1970.

Henschen G. and F. Arnold, Extended positive ion composition measurements in the stratosphere-Implications for neutral trace gases, *Geophys. Res. Lett.*, **8**, 999–1001, 1981.

Hering, W. S., Ozonsonde observations over North America, vol. 1, AFCRL Research report, AFCRL-64-30(I), 1964.

Hering, W. S., Ozone and atmospheric transport processes, *Tellus*, **18**, 329–336, 1966.

Hering, W. S. and T. R. Borden, Jr., ozonsonde observations over North America, vol. 2, Environmental research papers No. 38, AFCRL-64-30(II), 1964.

Hering, W. S. and T. R. Borden, Jr., Ozonsonde observations over North America, vol. 3, Environmental research papers No. 133, AFCRL-64-30(III), 1965.

Hering, W. S. and T. R. Borden, Jr., Ozonsonde observations over North America, vol. 4, Environmental research papers No. 279, AFCRL-64-30(IV), 1967.

Hering, W. S. and H. U. Dütsch, Comparison of chemiluminescent and electrochemical ozonesonde observations, *J. Geophys. Res.*, **70**, 5483–5490, 1965.

Herman, J. R., The response of stratospheric constituents to a solar eclipse, sunrise and sunset, *J. Geophys. Res.*, **84**, 3701–3710, 1979.

Herman, J. R. and J. E. Mentall, O_2 absorption cross section (190–225 nm) from stratospheric solar flux measurements, *J. Geophys. Res.*, **87**, 8967–8975, 1982.

Heroux, L. and R. A. Swirbalus, Full-disk solar fluxes between 1230 and 1940 Å, *J. Geophys. Res.*, **81**, 436–440, 1976.

Herzberg, G. and G. Scheibe, Uber die Absorptionspektra der dampfformgen Methylhalogenide und einiger andere Methlyl-verbindungen im Ultraviolett und im Schumann-Gebiet, *Z. Physik. Chem.*, **B 7**, 390–406, 1930.

HESS, S. L., *Introduction to Theoretical Meteorology*, 362 pp., pub. by Holt, Rainhart and Winston, New York, 1959.

HESS, S. L., R. M. HENRY, C. B. LEAVY, J. A. RYAN, and J. E. TILLMAN, Meteorological result from the surface of Mars: Viking 1 and 2, *J. Geophys. Res.*, **82**, 4559–4574, 1977.

HESSTVEDT, E., On the effect of vertical addy transport on atmospheric composition in the mesosphere and lower themosphere, *Geophys. Norv.*, **27**, 1–35, 1968.

HESSTVEDT, E., Comments on photochemical modeling, *Proc. 2nd CIAP Conference, DOT-TSC-OST-73-4*, pp. 285–290, U. S. Dept. of Transp., Washington, D.C., 1972.

HESSTVEDT, E., Reduction of stratospheric ozone from high-flying aircraft, studies in a two-dimensional photochemical model with transport, *Can. J. Chem.*, **52**, 1592–1598, 1974.

HESSTVEDT, E., S. HENRIKSON, and H. HJARTARSON, On the development of an aerobic atmosphere, a model experiment, *Geophys. Norv.*, **31**, 1–8, 1974.

HIDALGO, H. and P. J. CRUTZEN, The tropospheric and stratospheric composition perturbed by NO_x emissions of high-altitude aircraft, *J. Geophys. Res.*, **82**, 5833–5866, 1977.

HILSENRATH, E., B. GUENTHER and P. DUNN, Water vapor in the lower stratosphere measured from aircraft flight, *J. Geophys. Res.*, **82**, 5453–5458, 1977.

HILSENRATH, E., Rocket observations of the vertical distribution of ozone in the polar night and during a mid winter warming, *Geophys. Res. Lett.*, **7**, 581–584, 1980.

HILSENRATH, E. and D. F. HEATH, Seasonal and interannual variations in total ozone revealed by the Nimbus 4 backscattered ultraviolet experiment, *J. Geophys. Res.*, **84**, 6969–6979, 1979.

HILSENRATH, E. and P. T. KIRSCHNER, Recent assessment of the preformance and accuracy of a chemiluminescent rocket sonde for upper atmospheric ozone measurements, *Rev. Sci. Instrum.*, **51**, 1381–1389, 1981.

HILSENRATH, E. and B. M. SCHLESINGER, Total ozone seasonal and interannual variations derived from the 7 year Numbus-4 BUV datas set, *J. Geophys. Res.*, **86**, 12,087–12,096, 1981.

HINES, C. O., Eddy diffusion coefficients due to instabilities in internal gravity waves, *J. Geophys. Res.*, **75**, 3937–3939, 1970.

HOCHANADEL, C. J. and J. A. GHORMLEY, Absorption spectrum and reaction kinetics of the HO_2 radical in the gas phase, *J. Chem. Phys.*, **56**, 4426–4432, 1972.

HOCHANADEL, C. J., T. J. SWORSKI, and P. J. OGREN, Rate constants for the reactions of HO_2 with OH an HO_2, *J. Phys. Chem.*, **84**, 3274–3277, 1980.

HODGES, R. R., Eddy diffusion coefficients due to instabilities in internal gravity waves, *J. Geophys. Res.*, **74**, 4087–4090, 1969.

HOLT, R. B., C. K. MCLANE, and O. OLDENERG, Ultraviolet absorption spectrum of hydrogen peroxide, *J. Chem. Phys.*, **16**, 225–229 and 638, 1948.

HOLTON, J. R., The dynamic meteorology of the stratosphere and mesosphere, *Ame. Meteorol. Soc., Meteorol. Mon.*, **15**, 216 pp., 1975.

HOLTON, J. R., *An Introduction to Dynamic Meteorology*, 2nd edition, 391 pp., Academic Press, New york, San Francisco, and London, 1979.

HOLTON, J. R., An advective model for two-dimensional transport of stratospheric trace species, *J. Geophys. Res.*, **86**, 11,987–11,994, 1981.

HOLTON, J. R., The role of gravity wave induced drag and diffusion in the momentum budget of the mesosphere, *J. Atmos. Sci.*, **39**, 791–799, 1982.

HOLTON, J. R. and W. M. WEHRBEIN, A further study of the annual cycle of the zonal mean circulation in the middle atmosphere, *J. Atmos. Sci.*, **38**, 1504–1509, 1981.

HORVATH, J. J. and C. J. MASON, Nitric oxide mixing ratios near the stratopause measured by a rocket-borne chemiluminescent detector, *Geophys. Res. Lett.*, **5**, 1023–1026, 1978.

Houghton, J. T. and J. S. Seeley, Spectroscopic observation of the water vapor content of the stratosphere, *Quart. J. Roy. Met. Soc.*, **86**, 358–370, 1960.

Howard, C. J., Temperature dependence of the reaction $HO_2+NO \rightarrow OH+NO_2$, *J. Chem. Phys.*, **71**, 2352–2359, 1979.

Howard, C. J. and K. M. Evenson, Kinetics of the reaction of HO_2 with NO, *Geophys. Res. Lett.*, **4**, 437–440, 1977.

Howard, J. N., Transmission and detection of infrared radiation, Chap. 10 of Handbook of Geophysics and Space Environments, ed. by S. L. Valley, 36 pp., McGraw Hill Inc., New York, 1965.

Howard, J. N., D. E. Burch, and D. Williams, Infrared transmission of synthetic atmosphere II. Absorption of carbon dioxide, *J. Opt. Soc.*, **46**, 237–241, 1956a.

Howard, J. N., D. E. Burch, and D. Williams, Infrared transmission of synthetic atmosphere III. Absorption of water vapor, *J. Opt. Soc.*, **46**, 242–245, 1956b.

Hsu, K. J., Terrestrial catastrophe caused by cometary impact at the end of Cretaceous, *Nature*, **285**, 201–203, 1980.

Hudson, R. D., Critical review of ultraviolet photoabsorption cross sections for molecules of astrophysical and aeronomic interest, *Rev. Geophys. Space Phys.*, **9**, 305–406, 1971.

Hudson, R. D., Absorption cross sections of stratospheric molecules, *Can. J. Chem.*, **52**, 1465–1478, 1974.

Hudson, R. D. and S. H. Mahle, Photodissociation rates of molecular oxygen in the mesosphere and lower thermosphere, *J. Geophys. Res.*, **77**, 2902–2914, 1972a.

Hudson, R. D. and S. H. Mahle, Interpolation constants for calculation of transmittance and rate of dissociation of molecular oxygen in the mesosphere and lower thermosphere, NASA Technical Memo., NASA TM X-58084, 1972b.

Huebert, B. J., Nitric acid and aerosol nitrate measurements in the equatorial pacific region, *Geophys. Res. Lett.*, **7**, 325–328, 1980.

Huebert, B. J. and A. L. Lazrus, Global tropospheric measurements of nitric acid vapor and particulate nitrate, *Geophys. Res. Lett.*, **5**, 577–580, 1978.

Huebner, R. H., R. J. Cellotta, S. R. Mielezarek, and C. E. Kuyat, Apparent oscillator strength for molecular oxygen derived from electron energy loss measurements, *J. Chem. Phys.*, **63**, 241–248, 1975a.

Huebner, R. H., D. L. Bushnell, R. J. Celotta, S. R. Mielezarek, and O. E. Kuyatt, Ultraviolet photoabsorption by halocarbons 11 and 12 from electron impact measurements, *Nature*, **257**, 376–378, 1975b.

Huguenin, R. L., Photostimulated oxidation of magnetite, 2. Mechanism, *J. Geophys. Res.*, **78**, 8495–8506, 1973.

Hunt, B. G., Photochemistry of ozone in a moist atmosphere, *J. Geophys. Res.*, **71**, 1385–1289, 1966.

Hunt B. G., Experiments with a stratospheric general circulation model III. Large-scale diffusion of ozone including photochemistry, *Mon. Wea. Rev.*, **97**, 287–306, 1969.

Hunt, B. G. and S. Manabe, Experiments with a stratospheric general circulation model II. Large-svale diffusion of traces in the stratosphere, *Mon. Wea. Rev.*, **96**, 503–539, 1968.

Hunten, D. M., The escape of light gases from planetary atmosphere, *J. Atmos. Sci.*, **30**, 1481–1494, 1973.

Hunten, D. M., Energetics of thermospheric eddy transport, *J. Geophys. Res.*, **79**, 2533–2534, 1974.

Hunten, D. M., A second-order effect of stratospheric vertical motions, *Geophys. Res. Lett.*, **10**, 333–335, 1983.

Hunten, D. M. and M. B. McElroy, Metastable $O_2(^1\Delta_g)$ as a major source of ions in the *D*-

region, *J. Geophys. Res.*, **73**, 2421–2428, 1968.

HUNTEN, D. M. and D. F. STROBEL, Production and escape of terrestrial hydrogen, *J. Atmos. Sci.*, **31**, 305–317, 1974.

HYSON, P. and C. M. R. PLATT, Radiomatric measurements of stratospheric water vapor in the southern hemisphere, *J. Geophys. Res.*, **79**, 5001–5004, 1974.

HYSON, P., Stratospheric water vapor measurements over Australia 1973–1976, *Quart. J. Roy. Met.*, **104**, 225–228, 1978.

HYSON, P., P. J. FRASER, and G. T. PEARMAN, A two-dimensional transport simulation model for trace atmospheric constituents, *J. Geophys. Res.*, **85**, 4443–4455, 1980.

ILLIES, A. J. and G. A. TAKACS, Gas phase ultraviolet photoabsorption cross-section for nitrosyl chloride and nityl chloride, *J. Photochem.*, **6**, 35–42, 1976/77.

INN, E. C. Y., Absorption coefficient of HCl in the region 1400–2200 Å, *J. Atmos. Sci.*, **32**, 2375–2377, 1975.

INN, E. C. Y. and Y. TANAKA, Absorption coefficient of ozone in the ultraviolet and visible regions, *J. Opt. Soc. Amer.*, **43**, 870–873, 1953.

INN, E. C. Y., K. WATANABE, and M. ZELIKOFF, Absorption coefficients of gases in the vacuum ultraviolet, Part III CO_2, *J. Chem. Phys.*, **21**, 1648–1650, 1953.

INN, E. C. Y., J. F. VEDDER, B. J. TYSON, and D. O'HARA, COS in the stratosphere, *Geophys. Res. Lett.*, **6**, 191–193, 1979.

INN, E. C. Y., J. F. VEDDER, and D. O'HARA, Measurement of stratospheric sulfur constituents, *Geophys. Res. Lett.*, **8**, 5–8, 1981.

ISAKSEN, I. S. A., Diurnal variations of atmospheric constituents in an oxygen-hydrogen-carbon atmospheric model and the role of minor neutral constituents in the chemistry of the lower ionosphere, *Geophys. Pub.*, **30**, 1–63, 1973.

ISAKSEN, I. S. A., K. H. MILDBO, J. SUNDE, and P. J. CRUTZEN, A simplified method to include molecular scattering and reflection in calculations of photon fluxes and photodissociation rates, *Geophys. Norv.*, **31**, 11–26, 1977.

IWAGAMI, N. and T. OGAWA, Nitric oxide γ band airglow radiometer with a self-absorbing gas cell, *App. Opt.*, **20**, 2522–2526, 1981.

IZAKOV, M. N. and S. K. MAROV, The heating function of thermosphere by solar radiaon in Schumann-Runge continuum, *Geomoy. Aeron*, **10**, 495–499, 1971.

JAFFE, R. L. and S. R. LANGHOFF, Theoretical study of the photo-dissociation of HOCl, *J. Chem. Phys.*, **68**, 1638–1648, 1978.

JESSON, J. P., L. C. GLASGOW, D. L. FILKIN, and C. MILLER, The stratospheric abundance of peroxynitric acid, *Geophys. Res. Lett.*, **4**, 513–516, 1977.

JOHANNESSEN, A. and D. KRANKOWSKY, Positive-ion compostion measurement in the upper mesosphere and lower thermosphere at high latitude during summer *J. Geophys. Res.*, **77**, 2888–2901, 1972.

JOHN, D. S. St., S. P. BAILEY, W. H. FELLNER, J. M. MINOR, and R. D. SNEE, Time series search for trend in total ozone measurements, *J. Geophys. Res.*, **86**, 7299–7311, 1981.

JOHNSON, F. S., Transport processes in the upper atmosphere, *J. Atmos. Sci.*, **32**, 1658–1662, 1975.

JOHNSON, F. S. and E. M. WILKINS, Thermal upper limit on eddy diffusion in the mesosphere and lower thermosphere, *J. Geophys. Res.*, **70**, 1281–1284, 1965. (see also Correction in **70**, 4063, 1965).

JOHNSTON, H. S., Reduciton of stratospheric ozone by nitrogen oxide catalysts from SST exhaust, *Science*, **173** 517–522, 1971.

JOHNSTON, H. S., Catalytic reduction of stratospheric ozone by nitrogen oxides, *Adv. Environ. Sci. Tech.*, **4**, 263–380, 1974.

JOHNSTON, H. S., Expected short-term local effect of nuclear bombs on stratospheric ozone, *J.*

Geophys. Res., **82**, 3119–3124, 1977.

JOHNSTON, H. S., E. D. MORRIS, and J. VAN DEN BOGAERDE, Molecular modulation kinetic spectrometry, ClO and ClO_2 radicals in the photolysis of chlorine in oxygen, *J. Amer. Chem. Soc.*, **91**, 7712–7727, 1969.

JOHNSTON, H. S. and R. A. GRAHAM, Gas-phase ultraviolet absorption spectrum of nitric acid vapor. *J. Phys. Chem.*, **77**, 62–63, 1973.

JOHNSTON, H. S., G. WHITTEN, and J. BIRKS, Effect of nuclear explosions on stratospheric nitric oxide and ozone, *J. Geophys. Res.*, **78**, 6107–6135, 1973.

JOHNSTON, H. S. and R. A. GRAHAM, Photochemistry of NO_x and HNO_x compounds, *Can. J. Chem.*, **52**, 1415–1423, 1974.

JOHNSTON, H. S. and G. S. SELWYN, New cross sections for the absorption of near ultraviolet radiation by nitrous oxide, *Geophys. Res. Lett.*, **2**, 549–551, 1975.

JOHNSTON, H. S., D. KATTENHORN, and G. WHITTEN, Use of excess carbon 14 data to calibrate models of stratospheric ozone depletion by supersonic transports, *J. Geophys. Res.*, **81**, 368–380, 1976.

JONES, I. T. N. and R. D. WAYNE, The photolysis of ozone by ultraviolet radiation IV. Effect of photolysis wavelengths on primary step, *Proc. Roy. Soc. London, A*, **319**, 273–287, 1970.

JOSEPH, J. H., Diurnal and solar variations of neutral hydrogen in the thermosphere, *Ann. Geophys.*, **23**, 365–374, 1967.

JOURDAIN, J. L., G. LEBRAS, G-POULET, J. CONBOURIEU, P. RIGAUD, and B. LEROY, UV absorption spectrum of ClO($A^2\Pi$-$X^2\Pi$) up to (1,0) band, *Chem. Phys. Lett.*, **57**, 109–112, 1978.

JPL, Chemical kinetic and photochemical data for use in stratospheric modeling, Evaluation No. 4, NASA panel for data evaluation, JPL Pub. 81-3, 123 pp., Jet Prop. Labor., Pasadena, CA, 1981.

JPL, Chemical kinetic and photochemical data for use in stratospheric modeling, Evaluation No. 5, NASA panel for data evaluation, JPL Pub. 82-57, 186 pp., Jet Prop. Labor., CA. 1982.

JULIAN, P. R. and K. B. LABITZKE, A study of atmospheric energetics during the January-February 1963 stratospheric warming, *J. Atmos. Sci.*, **22**, 597–610, 1965.

JULIENNE, P. S., ${}^3\Sigma_u^- - {}^3\Sigma_u^+$ coupling in the $O_2B^3\Sigma_u^-$ predissociation, *J. Molec. Spectros.*, **63**, 60–79, 1976.

JUNGE, C. E., *Air Chemistry and Radioactivity*, 382 pp., Academic Press, New York and London, 1963.

JUNGE, C. E., C. W. CHAGNON, and I. E. MANSON, Stratospheric aerosols, *J. Met.*, **18**, 81–108, 1961.

KAGANN, R. H., J. W. ELKINS, and R. L. SAMS, Absolute band strengths of halocarbons F-11 and F-12 in the 8- to 16- μm region, *J. Geophys. Res.*, **88**, 1427–1432, 1983.

KALKSTEIN, M. I., Rhodium-102 high-altitude tracer experiment, *Science*, **137**, 645–652, 1962.

KASAHARA, A. and T. SASAMORI, Simulation experiments with a 12-layer stratospheric global circulation model. II. Momentum balance and energetics in the stratosphere, *J. Atmos. Sci.*, **31**, 408–421, 1974.

KASTING, J. F., S. C. LIU, and T. M. DONAHUE, Oxygen levels in the pre-biological atmosphere, *J. Geophys. Res.*, **84**, 3097–3107, 1979.

KASTING, J. F. and T. M. DONAHUE, The evolution of atmospheric ozone, *J. Geophys. Res.*, **85**, 3255–3263, 1980.

KATAMORI, M., Photochemical-radiative equilibrium of the earth's paleoatmospheres with various amounts of oxygen, *J. Met. Soc. Japan*, **57**, 243–252, 1979.

KEATING, G. M., Relation between monthly variations of global ozone and solar activity,

Nature, **274**, 873–874, 1978.

Keating, G. M., L. R. Lake, J. Y. Nicholson III, and M. Natarajan, Global ozone long-term trends from satellite measurements and the response to solar activity variations, *J. Geophys. Res.*, **86**, 9873–9880, 1981.

Keesee, R. G., N. Lee, and A. W. Castleman, Jr., Atmospheric negative ion hydration derived from laboratory results and comparison to rocket-borne measurements in the lower ionosphere, *J. Geophys. Res.*, **84**, 3719–3722, 1979.

Kellogg, W. W., Warming of the polar mesosphere and lower ionosphere in winter, *J. Meteor.*, **18**, 378–381, 1961.

Keneshea, T. J., A technique for solving the general reaction-rate equations in the atmosphere, AFCRL-67-0211, Air Force Cambridge Res. Labor., 1967.

Kerr, J. B. and C. T. McElroy, Measurement of stratospheric nitrogen dioxide from the AES stratospheric balloon program, *Atoms. Can. Met. Soc.*, **14**, 166–171, 1976.

Keyser, L. F., Absolute rate constant of the reaciton $OH+HO_2 \rightarrow H_2O+O_2$, *J. Phys. Chem.*, **85**, 3667–3673, 1981.

Keyser, L. F., Kinetics of the reaction $O+HO_2 \rightarrow OH+O_2$ from 229 to 372 K. *J. Phys. Chem.*, **86**, 3439–3446, 1982.

Khalil, M. A. K. and R. A. Rasmussen, Increase and seasonal cycles of nitrous oxide in the earth's atmosphere, *Tellus*, **35B**, 161–169, 1983.

Khrgian, A. K., *The Physics of Atmospheric Ozone*, Leningrad Gidrometeoizdat, 1973 (Translated by D. Lederman, 259 pp., Keter Publishing House, Jerusalem Ltd., 1975).

Kida, H., A numerical investigation of the atmospheric general circulation and stratospheric-tropospheric mass exchange. II. Lagrangian motion of the atmosphere, *J. Met. Soc. Japan*, **55**, 70–88, 1977.

Kida, H., General circulation of air parcels and transport characteristics derived from a hemispheric GCM, Part 1. A determination of advective mass flow in the lower stratosphere, *J. Met. Soc. Japan*, **61**, 171–187, 1983a.

Kida, H., General circulation of air parcels and transport characteristics derived from a hemispheric GCM, Part 2. Very long-term motions of air parcels in the troposphere and stratosphere, *J. Met. Soc. Japan*, **61**, 510–523, 1983b.

King, P., I. R. McKinnon, J. G. Mathieson, and I. R. Wilson, Upper limit to stratospheric N_2O_5 abundance, *J. Atmos. Sci.*, **33**, 1657–1659, 1976.

Kirchhoff, V. W. J. H., B. R. Clemensha, and D. M. Simonich, Sodium nightglow measurements and implications on the sodium photochemistry, *J. Geophys. Res.*, **84**, 1323–1327, 1979.

Klemm, R. B., S. Glicker, and L. J. Stief, Relative quantum yeild for the production of O-atom and S-atom from photodissociation of COS in the vacuum-UV, *Chem. Phys. Lett.*, **33**, 512–517, 1975.

Kley, D., G. M. Lawrence, and E. J. Stone, The yield of $N(^2D)$ atoms in the dissociative recombination of NO^+, *J. Chem. Phys.*, **66**, 4157–4165, 1977.

Kley, D. and E. J. Stone, Measurement of water vapor in the stratosphere by photodissociation with Ly-α(1216Å) light, *Rev. Sci. Instrum.*, **49**, 691–697, 1978.

Kley, D., E. J. Stone, W. R. Henderson, J. W. Drummond, W. J. Harrop, A. L. Schmeltekopf, T. L. Thompson, and R. H. Winkler, *In situ* measurements of the mixing ratio of water vapor in the stratosphere, *J. Atmos. Sci.*, **36**, 2513–2524, 1979.

Knauth, H.-D., Thermal decomposition of $ClONO_2$ in responce of NO, ClNO and N_2 (in German), *Ber. Bunsenges. Phys. Chem.*, **82**, 212–216, 1978.

Knauth, H.-D., H. Alberti, and H. Clausen, The equilibrium constant of the gas reaction $Cl_2O+H_2O \rightarrow 2HOCl$ and the UV spectrum of HOCl, *J. Phys. Chem.*, **83**, 1604–1612, 1979.

Ko, M. K. W. and N. D. Sze, Effect of recent rate data revisions on stratospheric modeling,

Geophys. Res. Lett., **10**, 341–344, 1983.

KOCHANSKI, A., Atmospheric motions from sodium cloud drifts, *J. Geophys. Res.*, **69**, 3651–3662, 1964.

KOCKARTS, G., Absorption par l'oxygene moleculaire dans les bandes de Schumann-Runge, *Aeronomica Acta*, No. 107, 1972.

KOCKARTS, G., Absorption and photodissociation in the Schumann-Runge bands of molecular oxygen in the terrestrial atmosphere, *Planet. Space Sci.*, **24**, 589–604, 1976.

KOMHYR, W. D., Electrochemical concentration cells for gas analysis, *Ann. Geophys.*, **25**, 203–210, 1969.

KOMHYR, W. D., Dobson spectrometer systematic total ozone measurement error, *Geophys. Res. Lett.*, **7**, 161–163, 1980.

KOMHYR, W. D., E. W. BARRETT, G. SLOCUM, and H. K. WELCKMANN, Atmospheric total ozone increase during the 1900s, *Nature*, **232**, 390–391, 1971.

KONDO, Y. and T. OGAWA, A temperature-dependent model of the thermospheric odd nitrogen, *J. Geomag. Geoelectr.*, **29**, 65–80, 1977.

KONDRATYEV, K. Ya., *Radiation in the Atmosphere*, 912 pp., Academic Press, New York and London, 1969.

KONDRATYEV, K. Ya. and L. L. NEDOVESOVA, On thermal emission of carbon dioxide in the atmosphere, *Proc. Acad. Sci., USSR, Geophys. Sect.* (English Transl.), No. 12, 1958.

KONG, T. Y. and M. B. MCELROY, Photochemistry of the Martian atmosphere, *Icarus*, **32**, 168–189, 1977.

KORB, G., J. MICHALOWSKY, and F. MÖLLER, Investigations in the heat balance of the troposphere, AFCRL, TN-58-238, 1957.

KOYAMA, T., Gaseous mehabolism in lake sediments and paddy soild and the production of atmospheric methane and hydrogen, *J. Geophys. Res.*, **68**, 3971–3973, 1963.

KRAMER, R. F. and G. F. WIDHOPF, Evaluations of daylight and diurnally averaged photolysis rate coefficients in atmospheric photochemical models, *J. Atmos. Sci.*, **35**, 1725–1734, 1978.

KRASNOPOLSKY, V. A., A. A. KRISCO, and V. N. ROGOCHEV, Ultraviolet photometry of Mars by the satellite Mars-5, *Kosmicheskie Issledovanya*, **15**, 255–259, 1977 (English translation: *Cosmic Res.*, **15**, 214–218).

KRASTITSKY, O. P., A model for the diurnal variation of the composition of the Martian atmosphere, *Kosmicheskie Issledovanya*, **16**, 432–442, 1978 (English translation: *Cosmic Res.*, **16**, 350–356).

KREY, R. W., R. J. LAGOMARSINO, and J. J. FREY, Stratospheric concentrations of CCl_3F in 1974, *J. Geophys. Res.*, **81**, 1557–1560, 1976.

KREY, R. W., R. J. LAGOMARSINO, and M. SCHONBERG, Stratospheric concentrations of N_2O in July, 1975, *Geophys. Res. Lett.*, **4**, 271–274, 1977.

KRUEGER, A. J., The mean ozone distribution from several series of rocket soundings to 52 km at latitude from 58°S to 64°N, *Pure Appl. Geophys.*, **106–108**, 1272–1280, 1973.

KRUEGER, A. J., Nimbus 7 total ozone mapping spectrometer (TOMS) data during the Gap, France, ozone intercomparisons of June 1981, *Planet. Space Sci.*, **31**, 773–777, 1983.

KRUEGER, A. and R. A. MINZNER, A mid-latitude ozone model for the 1976 U. S. standard atmosphere, *J. Geophys. Res.*, **81**, 4477–4481, 1976.

KRUEGER, A. J., B. GUENTHER, A. J. FLEIG, D. F. HEATH, E. HILSENRATH, R. MCPETERS, and C. PRABHAKARA, Satellite ozone measurements, *Phil. Trans. Roy.Soc., London, A*, **296**, 191–204, 1980.

KUHN, P. M. and L. P. STEARNS, Radiomatic observations of atmospheric water vapor injection by thunderstorms, *J. Atmos. Sci.*, **30**, 507–509, 1973.

KUHN, P. M., L. P. STEARNS, and M. S. LOJKO, Latitudinal profile of stratospheric water vapor, *Geophys. Res. Lett.*, **2**, 227–230, 1975.

KUHN, W. R. and J. LONDON, Infrared radiative cooling in the middle atmosphere (30–110 km), *J. Atmos. Sci.*, **26**, 189–204, 1969.

KUHN, W. R., S. K. ATREYA, and S. E. POSTAWKO, The influence of ozone on Martian atmospheric temperature, *J. Geophys. Res.*, **84**, 8341–8342, 1979.

KULKARNI, R. N., Measurements of NO_2 using the Dobson spectrophotometer, *J. Atmos. Sci.*, **32**, 1641–1643, 1975.

KUNZI, K. F. and E. R. CARLSON, Atmospheric CO volume mixing ratio profiles determined from ground based measurements of the $J=1\rightarrow 0$ and $J=2\rightarrow 1$ emission lines, *J. Geophys. Res.*, **87**, 7235–7241, 1982.

KURYLO, M. J., O. KLAIS, and A. H. LAUFER, Mechanistic investigation of the $HO+HO_2$ reaction, *J. Phys. Chem.*, **85**, 3674–3678, 1981.

KURYLO, M. J., K. D. CORNETT, and J. L. MURPHY, The temperature dependence of the rate constant for the reaction of hydroxyl radicals with nitric oxide, *J. Geophys. Res.*, **87**, 3081–3085, 1982.

KURZEJA, R. J., The diurnal variation of minor constituents in the stratosphere and its effect on the ozone concentration, *J. Atmos. Sci.*, **32**, 899–909, 1975.

KUTEPOV, A. A. and G. M. SHVED, Radiative transfer in the 15 μm CO_2 band with the breakdown of local thermodynamic equilibrium in the earth's atmosphere, *Izvestiya, Atmospheric and Oceanic Physics* (English Transl.), **14**, 16–30, 1978.

KYLE, T. G., D. G. MURCRAY, F. H. MURCRAY, and W. J. WILLIAMS, Abundance of methane in the atmosphere above 20 kilometers, *J. Geophys. Res.*, **74**, 3421–3425, 1969.

LACIS, A. A. and J. E. HANSEN, A parameterization for the absorption of solar radiation in the Earth's atmosphere, *J. Atmos. Sci.*, **31**, 118–133, 1974.

LANGHOFF, S. R., R. L. JAFFE, and J. Q. ARNOLD, Effective cross sections and rate constants for predissociation of ClO in the Earth's atmosphere, *J. Quant. Spectrosc. Rad. Transf.*, **18**, 227–235, 1977.

LAUFER, A. H. and J. R. MCNESBY, Denterium isotope effect in vacuum-ultraviolet absorption coefficients of water and methane, *Can. J. Chem.*, **43**, 3487–3490, 1965.

LAZRUS, A. L., and B. W. GANDRUD, Distribution of Stratospheric nitric acid vapor, *J. Atmos. Sci.*, **31**, 1107–1108, 1974.

LAZRUS, A. L., B. W. GANDRUD, R. N. WOODARD, and W. A. SEDLACEK, Direct measurements of stratospheric chlorine and bromine, *J. Geophys. Res.*, **81**, 1067–1070, 1976.

LAZRUS, A. L., B. W. GANDRUD, J. GREENBERG, J. BONELLI, E. MROZ, and W. A. SEDLACEK, Mid-latitude seasonal measurements of stratospheric acid chlorine vapor, *Geophys. Res. Lett.*, **4**, 587–589, 1977.

LAZRUS, A. L., R. D. CADLE, B. W. GANDRUD, J. P. GREENBERG, B. J. HUEBERT, and W. I. ROSE, Sulfur and halogen chemistry of the stratosphere and of volcanic cruption plumes, *J. Geophys. Res.*, **84**, 7869–7875, 1979.

LEAN, J. L., UV rocket spectroscopy measurement of the nighttime ozone distribution, *J. Geophys. Res.*, **88**, 1468–1474, 1983.

LEIFER, R., K. SOMMERS, and S. F. GUGGENHEIM, Atmospheric trace gas measurements with a new clean air sampling system, *Geophys. Res. Lett.*, **8**, 1079–1082, 1981.

LEOVY, C., Radiative equilibrium of the mesosphere, *J. Atmos. Sci.*, **21**, 238–248, 1964.

LEU, M. T., Rate constant for the reaction $HO_2+NO\rightarrow OH+NO_2$, *J. Chem. Phys.*, **74**, 1662–1666, 1979.

LEU, M. T., M. A. BIONDI, and R. JOHNSON, Measurements of the recombination of electrons with $H_3O^+(H_2O)_n$-series ions, *Phys. Rev.*, **7A**, 292–298, 1973.

LEVINE, J. S., P. B. HAYS, and J. C. G. WALKER, The evolution and variability of atmospheric ozone over geological time, *Icarus,* **39**, 295–309, 1979.

LEVY II, H., J. D. MAHLMAN, and W. J. MOXIM, A preliminary report on the numerical simulation of the three-dimensional structure and variability of atmospheric N_2O, *Geophys. Res. Lett.*, **6**, 155–158, 1979.

LEWIS, B. R., J. H. CARVER, T. I. HOBBS, D. G. MCCOY, and H. P. F. GIES, Experimentally determined oscillator strengths and line-widths for the Schumann-Runge band system of molecular oxygen—I. The (6-0) to (14-0) bands, *J. Quant. Spectrosc. Rad. Trans.*, **20**, 191–203, 1978.

LEWIS, B. R., J. H. CARVER, T. I. HOBBS, D. G. MCCOY, and H. P. F. GIES, Experimentally determined oscillator strengths and line-widths for the Schumann-Runge band system of molecular oxygen—II. The (2-0) to (5-0) bands, *Ibid*, **22**, 213–221, 1979.

LII, R. R., R. A. GORSE, M. C. SAUER, and S. GORDON, Rate constant for the reaction of OH with HO_2, *J. Phys. Chem.*, **84**, 819–821, 1980.

LIN, C. L., N. K. ROHATGI, and W. B. DEMORE, Ultraviolet absorption cross sections of hydrogen peroxide, *Geophys. Res. Lett.*, **5**, 113–115, 1978.

LINDSAY, J. C., The solar extreme UV radiation (l–400 Å), *Planet. Space Sci.*, **12**, 379–391, 1963.

LINDZEN, R. S., Turbulence and stress due to gravity wave and tidal breakdown, *J. Geophys. Res.*, **86**, 9707–9714, 1981.

LINDZEN, R. S. and R. GOODY, Radiative and photochemical processes in mesospheric dynamics: Part I, Models for radiative and photochemical processes, *J. Atmos. Sci.*, **22**, 341–348, 1965.

LINDZEN, R. S. and D. I. WILL, An analytical formula for heating due to ozone absorption, *J. Atmos. Sci.*, **30**, 513–515, 1973.

LITTLEJOHN, D. and H. S. JOHNSTON, Rate constant for the reaction of hydroxyl radicals and peroxynitric acid (abstract), *EOS Trans. AGU*, **61**, 966, 1980.

LIU, S. C. and T. M. DONAHUE, The regulation of hydrogen and oxygen escape from Mars, *Icarus*, **28**, 231–246, 1976.

LIU, S. C., T. M. DONAHUE, R. J. CICERONE, and W. L. CHAMEIDES, Effect of water vapor on the destruction of ozone in the stratosphere perturbed by Cl_x or NO_x pollutants, *J. Geophys. Res.*, **81**, 3111–3118, 1976.

LOEWENSTEIN, M., J. P. PADDOCK, I. G. POPPOFF, and H. F. SAVAGE, NO and O_3 measurements in the lower stratosphere from a U-2 aircraft, *Nature*, **249**, 817–818, 1974.

LOEWENSTEIN, M. and H. SAVAGE, Latitudinal measurements of NO and O_3 in the lower stratosphere from 5° to 82° North, *Geophys. Res. Lett.*, **2**, 448–450, 1975.

LOEWENSTEIN, M., W. J. BORUCKI, H. F. SAVAGE, J. G. BORUCKI, and R. C. WHITTEN, Geographical variations of NO and O_3 in the lower stratosphere, *J. Geophys. Res.*, **83**, 1875–1882, 1978.

LOGAN, J., M. J. PRATHER, S. C. WOFSY, and H. B. MCELROY, Atmospheric chemistry: response to human influence, *Phil, Trans. Roy. Soc. London, A*, **290**, 187–234, 1978.

LONDON, J. and S. OLTMANS, Further studies of ozone and sunspots, *Pure Appl. Geophys.*, **106–108**, 1302–1306, 1973.

LONDON, J., J. E. FREDERICK, and G. P. ANDERSON, Satellite observations of the global distribution of stratospheric ozone, *J. Geophys. Res.*, **82**, 2543–2556, 1977.

LONDON, J. and C. A. REBER, Solar activity and total atmospheric ozone, *Geophys. Res. Lett.*, **6**, 869–872, 1979.

LOUIS, J. F., A two-dimensional transport model of the atmosphere, Ph. D. Thesis, Univ. of Colo., Bouder, CO., 1974.

LOUISNARD, N., G. EICKEN, R. GIRAUDET and A. GIRARD, Preliminary results of the spectrometric balloon-borne experiment perforemed on May 15, 1978, Tech. Note ONERA, 1978.

LOVELOCK, J. E., Atmospheric fluorine compounds as indicators of air movements, *Nature*, **230**, 379, 1971.

LOVELOCK, J. E., Atmospheric halocarbons and stratospheric ozone, *Nature*, **252**, 292–294, 1974.

LOVELOCK, J. E., R. J. MAGGS, and R. J. WADE, Halogenated hydrocarbons in and over the Atlantic, *Nature*, **241**, 194–196, 1973.

LUTHER, F. M., Monthly values of eddy diffusion coefficients in the lower stratosphere, Laurence Livermore labor., UCPL-74616, 1973.

LUTHER, F. M., A parameterization of solar absorption by nitrogen dioxide, *J. App. Met.*, **15**, 479–481, 1976.

LUTHER, F. M. and R. J. GELINNS, Effect of molecular multiple scattering and surface albedo on atmospheric photodissociation rates, *J. Geophys. Res.*, **81**, 1125–1132, 1976.

LUTHER, F. M., D. J. WUEBBLES, and J. S. CHANG, Temperature feedback in a stratospheric model, *J. Geophys. Res.*, **82** 4935–4942, 1977.

LUTHER, F. M. and W. H. DUEWER, Effect of changes in stratospheric water vapor on ozone reduction estimates, *J. Geophys. Res.*, **83**, 2395–2402, 1978.

LUTHER, F. M., J. S. CHANG, W. H. DUEWER, J. E. PENNER, R. L. TARP, and D. J. WUEBBLES, Potential environmental effects of aircraft emissions, Lawrence Livermore Labor., UCRL-52861, 222 pp., 1979.

MAGNOTTA, F. and H. S. JOHNSTON, Photodissociation quantum yields for the NO_3 free radical, *Geophys. Res. Lett.*, **7**, 679–772, 1980.

MAHLMAN, J. D. and W. J. MOXIM, Tracer simulation using a global general circulation model: results from a midlatitude instrantaneous source experiment, *J. Atmos. Sci.*, **35**, 1370–1374, 1978.

MAIER, E. J., A. C. AIKIN, and J. E. AINSWORTH, Stratospheric nitric oxide and ozone measurements using photoionization mass spectrometry and UV absorption, *Geophys. Res. Lett.*, **5**, 37–40, 1978.

MANABE, S. and R. F. STRICKLER, Thermal equilibrium of the atmosphere with a convective adjustment, *J. Atmos. Sci.*, **21**, 361–365, 1964.

MANABE, S. and B. G. HUNT, Experiments with a stratospheric general circulation model. I, Radiative and dynamic aspects, *Mon. Wea. Rev.*, **96**, 477–502, 1968.

MANABE, S. and J. D. MAHLMAN, Simulation of seasonal and interhemispheric variations in the stratospheric circulation, *J. Atmos. Sci.*, **33**, 2185–2217, 1976.

MANKIN, W. G., M. T. COFFY, D. W. T. GRIFFITH, and S. R. DRAYSON, Spectroscopic measurement of carbonyl sulfide (OCS) in the stratosphere, *Geophys. Res. Lett.*, **6**, 853–856, 1979.

MANTIS, H., C. C. REPAPIS, and C. S. ZEREFOS, The summer maximum in total ozone over Northwest Europe, *Pure Appl. Geophys.*, **119**, 213–230, 1980/1981.

MARR, G. V., Photoionization and photodissociation results for molecules, in *Photoionization Processes in Gases*, Chap. 8, pp. 157–192, Acad. Press Inc., Ney York, 1967.

MASTENBROOK, H. J., The variability of water vapor in the stratosphere, *J. Atmos. Sci.*, **28**, 1495–1501, 1971.

MASTENBROOK, H. J., Water-vapor measurements in the lower stratosphere, *Can. J. Chem.*, **52**, 1527–1531, 1974.

MATEER, C. L., A review of some aspects of inferring the ozone profile by inversions of ultraviolet radiance measurement; in *The Mathematics of Profile Inversion*, ed. by L. Colin, NASA TMS-62, 2-25, 1972.

MATEER, C L., D. F. HEATH, and A. J. KRUEGER, Estimation of total ozone from satellite measurement of backscattered ultraviolet earth radiance. *J. Atmos. Sci.*, **28**, 1307–1311, 1971.

MATSUNO, T., Vertical propagation of stationary planetary waves in the winter northern hemisphere, *J. Atmos. Sci.*, **27**, 871–883, 1970.

MATSUNO, T., A dynamical model of the stratospheric sudden warming, *J. Atmos. Sci.*, **28**, 1479–1494, 1971.

MATSUNO, T., Lagragian motion of air parcels in the stratosphere in the presence of planetary waves, *Pure Appl. Geophys.*, **118**, 189–216, 1980.

MATSUNO, T. and K. NAKAMURA, The Eulerian- and Lagrangian-mean meridional circulations in the stratosphere at the time of a sudden warming *J. Atmos. Sci.*, **36**, 640–654, 1979.

MATSUNO, T. and T. SHIMAZAKI, The atmosphere in the stratosphere and mesosphere, *Lectures on Atmospheric Science*, Vol. 3, 279 pp., Univ. of Tokyo Press, 1981 (in Japanese).

MATTHIAS, A. D., A. M. BLACKMER, and J. M. BREMNER, Diurnal variability in the concentration of nitrous oxide in surface air, *Geophys. Res. Lett.*, **6**, 441–443, 1979.

MAUERSBERGER, K. and R. FINSTAD, Crabon dioxide measurements in the stratosphere, *Geophys. Res. Lett.*, **7**, 873–876, 1980.

MAUERSBERGER, K., R. FINSTAD, S. ANDERSON, and D. ROBBINS, A comparison of ozone measurements, *Geophys. Res. Lett.*, **8**, 361–364, 1981.

MAYNARD, W. C. (ed.), Middle atmosphere electrodynamics, 261 pp., NASA CP-2090, 1979.

MCCLATCHEY, R. A., R. W. FENN, J. E. A. SELBY, F. E. VOLZ, and J. S. GARING, Optical Properties of the Atmosphere, 3rd edition, 108 pp., *AFCRL. Environmental Res. Paper*, No. 411, 1972.

MCCLATCHEY, R. A., W. S. SENEDICT, S. A. CLOUGH, D. E. BURCH, R. F. CALFEE, K. FOX, L. S. ROTHMAN, and J. S. GARING, AFCRL Atmospheric absorption line parameters compilation, AFCRL TR-73-0096, Environmental Res. Papers, No. 434, 1973.

MCCONNELL, J. C. and M. B. MCELROY, Odd nitrogen in the atmosphere, *J. Atmos. Sci.*, **30**, 1465–1480, 1973.

MCELROY, M. B. and J. C. MCCONNELL, Nitrous oxide: A natural source of stratospheric NO, *J. Atmos. Sci.*, **28**, 1095–1098, 1971.

MCELROY, M. B. and T. M. DONAHUE, Stability of the Martian atmosphere, *Science*, **177**, 986–988, 1972.

MCELROY, M. B., S. C. WOFSY, J. E. PENNER and J. C. MCCONNELL, Atmospheric ozone: possible impact of stratospheric aviation, *J. Atmos. Sci.*, **31**, 287–303, 1974.

MCGHAN, M., A. SHAW, L. R. MEGILL, W. SEDLACEK, P. R. GUTHALS, and M. M. FOWLER, Measurements of nitric oxide after a nuclear burst, *J. Geophys. Res.*, **86**, 1167–1173, 1981.

MCINTYRE, M. E., Towards a Lagrangian-mean description of stratospheric circulations and chemical transports, *Phil. Trans. Roy. Soc. London, A*, **296**, 129–148, 1980.

MCKINNON, D. and H. W. MOREWOOD, Water vapor distribution in the lower stratosphere over North and South America, *J. Atmos. Sci.*, **27**, 483–493, 1970.

MCNESBY, J. R. and H. OKABE, Vacuum ultraviolet photochemistry, *Adv. Photochem.*, **3**, 157–240, 1964.

MCPETERS, R. D., The behavior of ozone near the stratopause from two years of BUV observations, *J. Geophys. Res.*, **85**, 4545–4550, 1980.

MCPETERS, R. D., C. H. JACKMAN, and E. G. STASSINOPOULOS, Observations of ozone depletion associated with solar proton events, *J. Geophys. Res.*, **86**, 12,071–12,081, 1981.

MCPETERS, R. D., D. F. HEATH, and P. K. BHARTIA, Average ozone profiles for 1979 from the Nimbus 7 SBUV instruments, *J. Geophys. Res.*, **89**, 5199–5214, 1984.

MCQUIGG, R. D. and J. G. CALVERT, The photodecomposition of CH_2O, CD_2O, CHDO, and CH_2O-CD_2O mixtures at xenon flash lamp intensities, *J. Amer. Chem. Soc.*, **91**, 1590–1599, 1969.

MCRAE, J. E. and T. E. GRAEDEL, Carbon dioxide in the urban atmosphere: dependencies and trends, *J. Geophys. Res.*, **84**, 5011–5017, 1979.

MEGIE, G. and J. E. BLAMONT, Laser sounding of atmospheric sodium: Interpretation in terms of global atmospheric parameters, *Planet Space Sci.*, **25**, 1093–1109, 1977.

MEGIE, G., J. Y. ALLAIN, M. L. CHANIN, and J. E. BLAMONT, Vertical profile of stratospheric ozone by lidar sounding from the ground, *Nature*, **270**, 329–331, 1977.

MEIER, R. R. and P. MANGE, Spatial and temporal variations of the Lyman-alpha airglow and related atomic hydrogen distributions, *Planet. Space Sci.*, **21**, 309–327, 1973.

MEIER, R. R., D. E. ANDERSON, Jr., and N. NICOLET, Radiation field in the troposphere and stratospher from 240–1000 nm, I. General analysis, *Planet. Space Sci.*, **30**, 923–933, 1982.

MEIRA, L. G. Jr., Rocket measurements of upper atmospheric nitric oxide and their consequences to the lower ionosphere, *J. Geophys. Res.*, **76**, 202–212, 1971.

MENZIES, R. T., Remote measurement of ClO in the stratosphere, *Geophys. Res. Lett.*, **6**, 151–154, 1979.

MENZIES, R. T., A re-evaluation of laser heterodyne radiometer ClO measurements, *Geophys. Res. Lett.*, **10**, 729–732, 1983.

MENZIES, R. T., C. W. RUTLEDGE, R. A. ZANTESON, and D. L. SPEARS, Balloon-borne laser heterodyne radiometer for measurements of stratospheric trace species, *App. Opt.*, **20**, 536–544, 1981.

MILLER, C., D. L. FILKIN, and J. P. JESSON, The fluorocarbon ozone theory, VI. Atmospheric modeling: calculation of the diurnal steady state, *Atmos. Environ.*, **13**, 381–394, 1979.

MILLER, C., D. L. FILKIN, and A. J. OWENS, A two-dimensional model of stratospheric chemistry and transport, *J. Geophys. Res.*, **86**, 12,039–12,065, 1981.

MILLER, F., A. VIDAL-MADJAR, J. GUIDON, and R. G. ROBLE, Ozone number density profiles in the lower mesosphere as determined by the French experiment on board OSO-8, *Geophys. Res. Lett.*, **6**, 863–865, 1979.

MOLINA, L. T., S. D. SCHINKE, and M. J. MOLINA, Ultraviolet absorption spectrum of hydrogen peroxide vapor, *Geophys. Res. Lett.*, **4**, 580–582, 1977.

MOLINA, L. T. and M. J. MOLINA, Ultraviolet spectrum of HOCl, *J. Phys. Chem.*, **82**, 2410–2412, 1978.

MOLINA, L. T. and M. J. MOLINA, Chlorine nitrate ultraviolet absorption spectrum at stratospheric temperatures, *J. Photochem.*, **11**, 139–144, 1979.

MOLINA, L. T., T. ISHIKAWA, and M. J. MOLINA, Quantum yield for production of OH in the photolysis of HOCl at 309 nm, *J. Chem. Phys.*, **84**, 821–826, 1980.

MOLINA, L. T., J. J. LAMB, and M. J. MOLINA, Temperature dependent UV absorption cross sections for carbonyl sulfide, *Geophys. Res. Lett.*, **8**, 1008–1011, 1981.

MOLINA, M. J. and F. S. ROWLAND, Stratospheric sink for chlorofluoromethanes: chlorine atom-catalysed destruction of ozone, *Nature*, **249**, 810–812, 1974.

MOORTGAT, G. K. and E. KUDSZUS, Mathematical expression for the $O(^1D)$ quantum yields from the O_3 photolysis as a funciton of temperature (230–320 K) and wavelength (295–320 nm), *Geophys. Res. Lett.*, **5**, 191–194, 1978.

MOORTGAT, G. K., and P. WARNECK, CO_2 and H_2 quantum yields in the photodecomposition of formaldehyde in air, *J. Chem. Phys.*, **70**, 3639–3651, 1979.

MOREELS, G., W. F. J. EVANS, J. E. BLAMONT, and A. VALLANCE JONES, A balloon-borne observation of the intensity variations of the OH emission in the evening twilight, *Planet. Space Sci.*, **18**, 637–640, 1970.

MORSS, D. A. and W. R. KUHN, Paleoatmospheric temperature structure, *Icarus*, **33**, 40–49, 1978.

MÜGGE, R. and F. MÖLLER, Zur Berechnung von Strahlunsströmen und Temperaturänderungen in Atmosphären von Beliebigen Aufbau, *Z. für Geophysik.*, **8**, 53–64, 1932.

MUGNAI, A., P. PETRONELLI, and G. FIOCCO, Sensitivity of the photodissociation of NO_2, NO_3, HNO_3, and H_2O_2 to the solar radiation diffused by the ground and by atmospheric particles, *J. Atmos. Terr. Phys.*, **41**, 351–359, 1979.

MURCRAY, D. G., F. H. MURCRAY, and W. J. WILLIAMS, Further data concerning the distribution of water vapor ion in the stratosphere, *Quart. J. Roy. Met. Soc.*, **92**, 159–161, 1966.

MURCRAY, D. G., F. H. MURCRAY, and W. J. WILLIAMS, A balloon-borne grating spectrometer, *App. Opt.*, **6**, 191–196, 1967.

MURCRAY, D. G., T. G. KYLE, F. H. MURCRAY, and W. J. WILLIAMS, Nitric acid and nitric oxide in the lower stratosphere, *Nature*, **218**, 78–79, 1968.

MURCRAY, D. G., F. H. MURCRAY, and W. J. WILLIAMS, Distribution of water vapor in the stratosphere as derived from setting sun absorption data, *J. Geophys. Res.*, **74**, 5369–5373, 1969.

MURCRAY, D. G., A. GOLDMAN, A. CSOEKE-POECKH, F. H. MURCRAY, W. J. WILLIAMS, and R. N. STOCKER, Nitric acid distribution in the stratosphere, *J. Geophys. Res.*, **78** 7033–7038, 1973.

MURCRAY, D. G., F. S. BONOMO, J. N. BROOKS, A. GOLDMAN, F. H. MURCRAY, and W. J. WILLIAMS, Detection of fluorocarbons in the stratosphere, *Geophys. Res. Lett.*, **2**, 108–112, 1975a.

MURCRAY, D. G., D. B. BARKER, J. N. BROOKS, A. GOLDMAN, and W. J. WILLIAMS, Seasonal and latitudinal variation of the stratospheric concentration of HNO_3, *Geophys. Res. Lett.*, **2**, 223–225, 1975b.

MURCRAY, D. G., A. GOLDMAN, W. J. WILLIAMS, F. H. MURCRAY, F. S. BONOMO, C. M. BRADFORD, G. R. COOK, P. L. HANST, and M. J. MOLINA, Upper limit for stratospheric $ClONO_2$ from balloon-borne infrared measurements, *Geophys. Res. Lett.*, **4**, 227–230, 1977.

MURCRAY, D. G., A. GOLDMAN, F. H. MURCRAY, F. J. MURCRAY, and W. J. WILLIAMS, Stratospheric distribution of $ClONO_2$, *Geophys. Res. Lett.*, **6**, 857–859, 1979.

MURGATROYD, R. J. and R. M. GOODY, Sources and sinks of radiative energy from 30 to 90 km, *Quart. J. Roy. Met. Soc.*, **84**, 225–234, 1958.

MURGATROYD, R. J. and F. SINGLETON, Possible meridional circulation in the stratosphere and mesosphere, *Quart. J. Roy. Met. Soc.*, **87**, 125–135, 1961.

MYER, J. A. and J. A. SAMSON, Vacuum-ultraviolet absorption cross sections of CO, HCl and ICN between 1050 and 2100 Å, *J. Chem. Phys.*, **52**, 266–271, 1970.

NAGATA, T., T. TOHMATSU, and T. OGAWA, Sounding rocket measurement of atmospheric ozone density, 1965–1970, *Space Res.*, **11**, 849–855, Akademie-Verlag, Berlin, 1971.

NAKATA, R. S., K. WATANABE, and F. M. MATSUNAGA, Absorption and photoionization coefficients of CO_2 in the region 580–1670 Å, *Sci. Light*, **14**, 54–71, 1965.

NAKAYAMA, T., M. Y. KITAMURA, and K. WATANABE, Ionization potential and absoption coefficients of nitrogen dioxide, *J. Chem. Phys.*, **30**, 1180–1186, 1959.

NARCISI, R. S., Ion composition of the mesosphere, *Space Res.*, **7**, 186–196, 1967.

NARCISI, R. S. and A. D. BAILEY, Mass spectrometric measurements of positive ion at altitudes from 64 to 112 kilometers, *J. Geophys. Res.*, **70**, 3687–3700, 1965.

NARCISI, R. S., A. D. BAILEY, L. DELLA LUCCA, C. SHERMAN, and D. M. THOMAS, Mass spectrometric measurements of negative ions in the *D- and lower E*-regions, *J. Atmos. Terr. Phys.*, **33**, 1147–1159, 1971.

NARCISI, R. S., A D. BAILEY, L. E. WLODYKA, and C. R. PHILBRICK, Ion composition measurements in the lower ionosphere during the November 1966 and March 1970 solar eclipses, *J. Atmos. Terr. Phys.*, **34**, 647–658, 1972.

NAS, *Environmental impact of stratospheric flight, Biological and climate effects of aircraft emissions in the stratosphere*, National Academy of Sciences report, 348 pp., 1975.

NAS, *Halocarbons, Environmental effects of chlorofluoromethane release*, National Academy of Sciences report, 125 pp., 1976.

NAS, *Stratospheric ozone depletion by Halocarbons: chemistry and transport*, National Academy of Sciences report, 238 pp., 1979.

NAS, *Causes and effects of stratospheric ozone reduction: an update*, National Academy of Sciences report, 339 pp., 1982.

NASA, *Chlorofluoromethanes and the Stratosphere*, ed. by R. D. Hudson, 266 pp., NASA Ref. Publ. 1010, 1977.

NASA, *The Stratosphere: Present and Future*, ed. by R. D. Hudson and E. I. Reed, 432 pp., NASA Ref. Publ. 1049, 1979.

NASTROM, G. D. and A. D. BELMONT, Periodic variations in stratospheric-mesospheric temperature from 20–65 km at 80°N to 30°S, *J. Atmos. Sci.*, **32**, 1715–1722, 1975.

NAUJOKAT, B., Long-term variations in the stratosphere in the northern hemisphere during the last two sunspot cycles, *J. Geophys. Res.*, **86**, 9811–9816, 1981.

NBS, *Chemical kinetic and Photochemical Data for Modeling Atmospheric Chemistry*, ed. by R. F. Hampton and D. Garvin, 113 pp., National Bureau of Standards, Tech. Note 866, 1975.

NBS, *Reaction Rate and Photochemical Data for Atmospheric Chemistry*, ed. by R. F. Hampton and D. Garvin, 106 pp., National Bureau of Standards, Special Pub., 513, 1977.

NEHER, H. V., Cosmic-ray particles that changed from 1954 to 1958 to 1965, *J. Geophys. Res.*, **72**, 1527–1539, 1967.

NELSON, H. H., W. J. MARINELLI, and H. S. JOHNSTON, The kinetics and product yield of the reaction of OH radicals with HNO_3, *Chem. Phys. Lett.*, **78**, 495–499, 1981.

NEWELL, R. E., Transfer through the tropopause and within the stratosphere, *Quart. J. Roy. Met. Soc.*, **89**, 167–204, 1963.

NEWELL, R. E. and S. GOULD-STEWART, A stratospheric fountain?, *J. Atmos. Sci.*, **38**, 2789–2796, 1982.

NICOLET, M., The properties and constitution of the upper atmosphere, in *Physics of the Upper Atmosphere*, ed. by J. A. Ratcliffe, pp. 17–71, Academic Press, New York, 1960.

NICOLET, M., The origin of nitric oxide in the terrestrial atmosphere, *Planet. Space Sci.*, **18**, 1111–1118, 1970a.

NICOLET, M., Aeronomic reactions of hydrogen and ozone, in *Mesospheric Models and Related Experiments*, ed. by A. Fiocco, pp. 1–51, D. Reidel, Dordrecht, Netherlands, 1970b.

NICOLET, M., Aeronomic chemistry of the stratosphere, *Proc. Survey Conf. CIAP.*, DOT-TSC-OST-72-13, pp. 44–70, 1972.

NICOLET, M., An overview of aeronomic processes in the stratosphere and mesosphere, *Can. J. Chem.*, **52**, 1381–1396, 1974.

NICOLET, M., Stratospheric ozone: An introduction to its study, *Rev. Geophys. Space Phys.*, **13**, 593–636, 1975.

NICOLET, M., Photodissociation of nitric oxide in the mesosphere and stratosphere: simplified numerical relations for atmospheric model calculations, *Geophys. Res. Lett.*, **6**, 866–868, 1979.

NICOLET, M., Solar UV radiation and its absoption in the mesosphere and stratosphere, *Pure Appl. Geophys.*, **118**, 3–19, 1980.

NICOLET, M., The solar spectral irradiance and its action in the atmospheric photodissociation processes, *Planet. Space Sci.*, **29**, 951–974, 1981a.

NICOLET, M., The photodissociation of water vapor in the mesosphere, *J. Geophys. Res.*, **86**, 5203–5208, 1981b.

NICOLET, M., Photodissociation of molecular oxygen in the terrestrial atmosphere: Simplified numerical relations for the spectral range of the Schumann-Runge bands, *J. Geophys. Res.*, **89**, 2573–2582, 1984.

NICOLET, M. and S. CIESLIK, The photodissociation of nitric oxide in the mesosphere and stratosphere, *Planet. Space Sci.*, **28**, 105–115, 1980.

NICOLET, M. and W. PEETERMANS, Atmospheric absoption in the O_2 Schumann-Runge band spectal range and photodissociation rates, in the stratosphere and mesosphere, *Planet. Space Sci.*, **28**, 85–103, 1980.

NICOLET, M., R. R. MEIER, and D. E. ANDERSON, Jr., Radiation field in the troposphere and stratosphere-II, Numerical analysis, *Planet. Space Sci.*, **30**, 935–983, 1982.

NICOLI, M. P. and G. VISCONTI, Impact of coupled perturbations of atmospheric trace gases on earth's climate and ozone, *Pure Appl. Geophys.*, **120**, 626–641, 1982.

NISHI, K., Observation of the absolute intensity of the sun in the vacuum ultraviolet region, *Solar Phys.*, **42**, 37–42, 1975.

NORMAN, C., Satellite data indicate ozone depletion, *Science*, **213**, 1088–1089, 1981.

NORTON, R. B. and C. A. BARTH, Theory of nitric oxide in the earth's upper atmosphere, *J. Geophys. Res.*, **75**, 3903–3909, 1970.

NOXON, J. F., Stratosphere NO_2, 2. global behavior, *J. Geophys. Res.*, **84**, 5067–5076, 1979.

NOXON, J. F., W. A. TRAUB, N. P. CARLETON, and P. CONNES, Detection of O_2 dayglow emission from Mars and the Martian ozone abundance, *Astrophy. J.*, **207**, 1025–1035, 1976.

NOXON, J. F., R. B. NORTON, and W. R. HENDERSON, Observation of atmospheric NO_3, *Geophys. Res. Lett.*, **5**, 675–678, 1978.

NOXON, J. F., E. C. WHIPPLE, and R. S. HYDE, Stratospheric NO_2 1. Observational method and behavior at mid-latitude, *J. Geophys. Res.*, **84** 5047–5065, 1979.

NOXON, J. F., R. B. NORTON, and E. MAROVICH, NO_3 in the troposphere, *Geophys. Res. Lett.*, **7**, 125–128, 1980.

O'BRIEN, R. S. and W. F. J. EVANS, Rocket measurements of the distribution of water vapor in the stratosphere at high latitudes, *J. Geophys. Res.*, **86**, 12,101–12,107, 1981.

OGAWA, M., Absorption cross sections of O_2 and CO_2 continua in Schumann and far-UV regions, *J. Chem. Phys.*, **54**, 2550–2556, 1971.

OGAWA, T., Chemistry of stratospheric chlorine, *J. Met. Soc. Japan*, **54**, 294–307, 1976.

OGAWA, T. and T. SHIMAZAKI, Diurnal varitations of odd nitrogen and ionic desities in the mesosphere and lowere thermosphere: Simultaneous solution of photochemical-diffusive equations, *J. Geophys. Res.*, **80**, 3945–3960, 1975.

OGAWA, T., K. SHIBASAKI, and K. SUZUKI, Balloon observation of the stratospheric NO_2 profile by visible absorption spectroscopy, *J. Met. Soc. Japan*, **59**, 410–416, 1981.

OHRING, G., The radiation budget of the stratosphere, *J. Met.*, **15**, 440–451, 1958.

OLIVER, R. C., E. BAUER, H. HIDALGO, K. A. GARDNER, and W. WASHLKIWSKYJ, Aircraft emission: potential effects on ozone and climate, a review and progress report, Federal Aviation Administration Report, FA-EQ-73-3, 1977.

O'NEILL, A., R. L. NEWSON, and R. J. MURGATROYD, An analysis of the large-scale features of the upper troposphere and stratosphere in a global three-dimensional general circulation model, *Quart. J. Roy. Met. Soc.*, **108**, 25–53, 1982.

OORT, A. H. and E. M. RASMUSSEN, On the annual variation of the monthly mean meridional circulation, *Mon. Wea. Rev.*, **98**, 432–442, 1970.

OORT, A. H. and E. M. RASMUSSEN, Atmospheric circulation statistics, NOAA professional paper, **5**, U. S. Dept. Commerce, 1971.

ORAN, E. S., P. S. JULIENNE, and D. F. STROBEL, The aeronomy of odd nitrogen in the thermosphere, *J. Geophys. Res.*, **80**, 3068–3076, 1975.

OWEN, T., K. BRIEMANN, D. R. RUSHNECK, J. E. BILLER, D. W. HOWARTH, and A. L. LAFLEUR, The composition of the atmosphere at the surface of Mars, *J. Geophys. Res.*, **82**, 4635–4639, 1977.

PAETZOLD, H. K., New experimental and theoretical investigations on the atmospheric ozone layer *J. Atmos. Terr. Phys.*, **7**, 128–140, 1955.

PAETZOLD, H. K., F. PISCALAR, and H. ZSCHORNER, Secular variation of the stratospheric ozone layer over middle Europe during the solar cycles from 1951 to 1972, *Nature. Phys. Sci.*, **240**, 106–107, 1972.

PARK, J. H. and J. LONDON, Ozone photochemistry and radiative heating of the middle atmosphere, *J. Atmos. Sci.*, **31**, 1898–1916, 1974.

PARKINSON, T. D. and D. M. HUNTEN, Spectroscopy and aeronomy of O_3 on Mars, *J. Atmos. Sci.*, **29**, 1380–1390, 1972.

PARRY, H. D., Ozone depletion by chlorofluoromethanes? Yet another look, *J. App. Meteorol.*, **16**, 1137–1148, 1977.

PATEL, C. K. N., E. G. BURKHARDT, and C. A. LAMBERT, Spectroscopic measurements of stratospheric nitric oxide and water vapor, *Science,* **184**, 1173–1176, 1974.

PAUKERT, T. T. and H. S. JOHNSTON, Spectra and kinetics of the hydroperoxyl force radicals in the gas phase *J. Chem Phys.*, **56**, 2824–2838, 1972.

PAULSEN, D. E., R. E. HUFFMAN, and J. C. LARRABEE, Improved photoionization rates of $O_2(^1\Delta_g)$ in the *D*-region, *Radio Sci.*, **7**, 51–55, 1972.

PELON, J. and G. MEGIE, Ozone monitoring in the troposphere and lower stratosphere: Evaluation and operation of a ground-based lidar station, *J. Geophys. Res.*, **87**, 4947–4955, 1982.

PENFIELD, H., M. M. LITVAK, C. A. GOTTLIEB, and A. E. LILLEY, Mesospheric ozone measured from ground-based millimeter-wave observations, *J. Geophys. Res.*, **81**, 6115–6120, 1976.

PENKETT, S. A., N. J. D. PROSSER, R. A. RASMUSSEN, and M. A. K. KHALL, Atmospheric measurements of CF_4 and other fluorocarbons containing the CF_3 grouping, *J. Geophys. Res.*, **86**, 5172–5178, 1981.

PENNER, J. E. and J. S. CHANG, Possible variations in atmospheric ozone related to the eleven-year solar cycle, *Geophys. Res. Lett.*, **5**, 817–820, 1978.

PENZIAS, A. A. and C. A. BURRUS, Millimeter-wavelength radio astronomy techniques, *Ann. Rev. Astron. Astrophys.*, **11**, 51–71, 1973.

PERNER, D., D. H. EHHALT, H. W. PATZ, U. PLATT, E. P. ROTH, and A. VOLZ, OH-radicals in the lower troposphere, *Geophys. Res. Lett.*, **3**, 466–468, 1976.

PETRONELLI, P., G. FIOCCO, and A. MUGNAI, Annual variation of the effects of diffuse radiation on the photodissociation of ozone, *Pure Appl. Geophys.,* **118**, 20–34, 1980.

PICK, D. R. and J. T. HOUGHTON, Measurements of atmospheric infrared emission with a balloon-borne multi-filter radiometer, *Quart. J. Roy. Met. Soc.*, **95**, 535–543, 1969.

PIEROTTI, D., L. E. RASMUSSEN, and R. A. RASMUSSEN, The Sahara as a possible sink for trace gases, *Geophys. Res. Lett.*, **5**, 1001–1004, 1978.

PLATT, U. and D. PERNER, Direct measurements of atmospheric CH_2O, HNO_2, O_3, NO_2, and SO_2 by differential optical absorption in the near UV, *J. Geophys. Res.*, **85**, 7453–7458, 1980.

POLLOCK, W., L. E. HEIDT, R. LUEB, and D. H. EHHALT, Measurement of stratospheric water vapor by cryogenic collection, *J. Geophys. Res.*, **85**, 5555–5568, 1980.

POPPOFF, I. G., R. C. WHITTEN, R. P. TURCO, and L. A. CAPONE, An assessment of the effect of supersonic aircraft operations on the stratospheric ozone content, NASA Ref. Publ. RP-1026, 1978.

PORTER, G. and F. J. WRIGHT, Studies of free-radical reactivity by the methods of flash photolysis. The photochemical reaction between chlorine and oxygen, *Disc. Faraday Soc.*, **14**, 23–34, 1953.

POTEMA, T. A., Ionizing radiation affecting the lower ionosphere, *ELF-VLF Radio Wave Propagation*, ed. by J. A. Holtet, pp. 21–37, D. Reidel Pub. Co., 1974.

PRABHAKARA, C., B. J. CONRATH, R. A. HANEL, and E. J. WILLIAMSON, Remote sensing of atmospheric ozone using the 9.6 band, *J. Atmos. Sci.*, **27**, 689–697, 1970.

Prabhakara, C., G. Dalu, and V. G. Kunde, A search for global and seasonal variation of methane from Nimbus 4 Iris measurements, *J. Geophys. Res.*, **79**, 1744–1749, 1974.

Prabhakara, C., E. B. Rodgers, B. J. Conrath, R. A. Hanel, and V. G. Kunde, The Nimbus 4 infrared spectroscopy experiment 3. Observations of the lower stratospheric thermal structure and total ozone, *J. Geophys. Res.*, **81**, 6391–6399, 1976.

Pratt, P. F. and eleven authers, Effect of increased nitrogen fixation on stratospheric ozone, *Climate Change*, **1**, 109–135, 1977.

Preston, K. F. and R. F. Barr, Primary processes in the photolysis of nitrous oxide, *J. Chem. Phys.*, **54**, 3347–3348, 1971.

Prinn, R. G., F. N. Alyea, and D. M. Cunnold, Stratospheric distributions of odd nitrogen and odd hydrogen in a two-dimensional model, *J. Geophys. Res.*, **80**, 4997–5004, 1975.

Pyle, J. A. and C. F. Rogers, A modified diabatic circulation model for stratospheric tracer transport, *Nature*, **287**, 711–714, 1980a.

Pyle, J. A. and C. F. Rogers, Stratospheric transport by stationary planetary waves—the importance of chemical processes, *Quart. J. Roy. Met. Soc.*, **106**, 421–446, 1980b.

Ramanathan, V., Greenhouse effect due to chlorofluorocarbons: climate implications, *Science*, **190**, 50–52, 1975.

Ramanathan, V., Radiative transfer within the earth's troposphere and stratosphere: a simplified radiative-convective model, *J. Atmos. Sci.*, **33**, 1330–1346, 1976.

Randhawa, J. S., An investigation of solar eclipse effect on the subpolar stratosphere, *J. Geophys. Res.*, **78**, 7139–7144, 1973.

Rangarajan, S., Effect of solar activity on atmospheric ozone, *Nature*, **206**, 497–498, 1965.

Rao-Vupputuri, R. K., Numerical experiments on the steady state meridional structure and ozone distribution in the stratosphere, *Mon. Wea. Rev.*, **101**, 510–527, 1973.

Rao-Vupputuri, R. K., The steady-state structure of the natural stratosphere and ozone distribution in a 2-D model incorporating radiation and *O-H-N* photochemistry and the effects of stratospheric pollutants, *Atmosphere*, **14**, 214–236, 1976.

Rao-Vupputuri, R. K., The structure of the natural stratosphere and the impact of chlorofluoromethanes on the ozone layer investigated in a 2-D time dependent model, *Pure Appl. Geophys.*, **117**, 448–485, 1978/1979.

Rasmussen, R. A., S. A. Penkett, and N. Prosser, Measurement of carbontetrofluoride in the atmosphere, *Nature*, **277**, 549–551, 1979.

Rasmussen R. A., L. E. Rasmussen, M. A. K. Khalil, and R. W. Dalluge, Concentration distribution of methyl chloride in the atmosphere, *J. Geophys. Res.*, **85**, 7350–7356, 1980a.

Rasmussen, R. A., M. A. K. Khalil, S. A. Penkett, and N. J. O. Prosser, $CHClF_2$ (F-22) in the earth's atmosphere, *Geophys. Res. Lett.*, **7**, 809–812, 1980b.

Rasmussen, R. A. and M. A. K. Khalil, Global atmosphere distribution and trend of methylchloroform (CH_3CCl_3), *Geophys. Res. Lett.*, **8**, 1005–1007, 1981.

Ratner, M. I. and J. C. G. Walker, Atmospheric ozone and the history of life, *J. Atmos. Sci.*, **29**, 803–808, 1972.

Reagan, J. B., R. E. Megerott, R. W. Nightingale, R. C. Gunton, R. G. Johnson, J. E. Evans, W. L. Imhof, D. F. Heath, and A. J. Kreuger, Effects of the August 1972 solar particle events on stratospheric ozone, *J. Geophys. Res.*, **86**, 1473–1494, 1981.

Rebbert, R. E. and P. J. Ausloos, Photodecomposition of $CFCl_3$ and CF_2Cl_2, *J. Photochem.*, **4**, 419–434, 1975.

Rebbert, R. E. and P. J. Ausloos, Gas-phase photodecomposition of carbon tetrachloride, *Ibid.*, **6**, 265–276, 1976/1977.

Reber, C. A. and F. T. Huang, Total ozone-solar activity relationship, *J. Geophys. Res.*, **87**, 1313–1318, 1982.

Reed, E. J., A night measurement of mesospheric ozone by observations of ultraviolet airglow, *J. Geophys. Res.*, **73**, 2951–2957, 1968.

Reed, R. J. and E. F. Danielsen, Fronts in the vicinity of the tropopause, *Arch. Meteorol. Geophys. Bioklimatol., Ser. A*, **11**, 1–17, 1959.

Reed, R. J. and K. E. German, A contribution of the problem of stratospheric diffusion by lamp scale mixing, *Mon. Wea. Rev.*, **93**, 313–321, 1965.

Regener, V. H., On a sensitive method for the recording of atmospheric ozone, *J. Geophys. Res.*, **65**, 3975–3977, 1960.

Regener, V. H., Measurement of atmospheric ozone with the chemiluminescent method, *Ibid.*, **69**, 3795–3800, 1964.

Reid, G. C., Ice clouds at the summer polar mesopause, *J. Atmos. Sci.*, **32**, 523–535, 1975.

Reid, G. C., The production of water-cluster positive ions in the quiet daytime *D*-region, *Planet. Space Sci.,* **25**, 275–290, 1977.

Reid, G. C., J. R. McAfee, and P. J. Crutzen, Effects of intense stratospheric ionization events, *Nature*, **275**, 489–492, 1978.

Reiter, E. R., The detailed structure of the wind field near the jet stream, *J. Met.*, **18(1)**, 9–30, 1961.

Reiter, E. R., *Atmospheric Transport Processes*, part 2: Chemical tracers, 382 pp., U. S. Atomic Energy Com., 1971.

Reiter, E. R., *Atmospheric Transport Processes*, part 3: Hyedrodynamic tracers, 212 pp., U. S. Atomic Energy Com., 1972.

Richtmyer, R. D. and K. W. Morton, *Different Methods for Initial Value Problems*, 405 pp., Interscience, New York, 1967.

Ridley, B. A., H. I. Schiff, A. Shaw, and L. R. Megill, *In situ* measurements of stratospheric nitric oxide using a balloon-borne chemiluminescent instrument, *J. Geophys. Res.*, **80**, 1925–1929, 1975.

Ridley, B. A. and H. I. Schiff, Stratospheric odd nitrogen: nitric oxide measurements at 32°N in autumn, *J. Geophys. Res.*, **86**, 3167–3172, 1981.

Riegler, G. R., S. K. Atreya, T. M. Donahue, S. C. Liu, B. Wasser, and J. F. Drake, UV stellar occultation measurements of nighttime equatorial ozone, *Geophys. Res. Lett.*, **4**, 145–148, 1977.

Rishbeth, H. and O. K. Gariott, *Introduction to Ionospheric Physics*, 331 pp., Academic Press, New York, 1969.

Robbins, D. E., Ultraviolet photoabsorption cross-sections for methyl chloride and methyl bromide, *Geophys. Res. Lett.*, **3**, 213–216, 1976.

Robinson, E., R. Rasmussen, J. Krasner, D. Pierotti, and M. Jakuboric, Detailed halocarbon measurements across the Alaska tropopause, *Geophys. Res. Lett.*, **3**, 323–326, 1976.

Robinson, E., D. R. Cronn, and D. E. Harsch, Profiles and other characteristics of trace gases in the antarctic troposphere, a paper presented at AGU fall meeting, San Francisco, December, 1979.

Roble, R. G. and P. B. Hays, The nighttime distribution of ozone in the low-latitude mesosphere, *Pure Appl. Geophys.*, **106–108**, 1281–1291, 1973.

Roble, R. G. and P. B. Hays, On determining the ozone number density distribution from OAO-2 stellar occultation measurement, *Planet. Space Sci.*, **22**, 1337–1340, 1974.

Rodgers, C. D. and C. D. Walshaw, The computation of infrared cooling rate in planetary atmospheres, *Quart. J. Roy. Met. Soc.*, **92**, 67–92, 1966.

Rogers, J. W., R. E. Murphy, A. T. Stair, J. C. Ulwick, K. D. Baker, and L. L. Jensen, Rocket-borne radiometric measurements of OH in the auroral zone, *J. Geophys. Res.*, **78**, 7023–7031, 1973.

Rogers, J. W., A. T. Stair, T. C. Degges, C. L. Wyatt, and D. J. Baker, Rocket-borne measurement of mesospheric H_2O in the auroral zone, *Geophys. Res. Lett.*, **9**, 366–368, 1977.

ROMAND, J., Absorption ultraviolette dans la region de Schumann etude de ClH, BrH et IH gazeux, *Ann. Phys. (Paris)*, **4**, 527–592, 1949.

ROPER, R. G. and W. G. ELFORD, Periodic wind components at meteor heights, NASA Rep. X605-65-86, 1965.

ROSCOE, H. K., Tentative observation of stratospheric N_2O_5, *Geophys. Res. Lett.*, **9**, 901–902, 1982.

ROSCOE, H. K., J. R. DRUMMOND, and R. F. JARNOT, Infrared measurements of stratospheric composition, III. The daytime changes of NO and NO_2, *Proc. R. Soc. London. A*, **375**, 507–528, 1981.

ROSEN, J. M., D. J. HOFMANN, and J. LADY, Stratospheric aerosol measurements II: The worldwide distribution, *J. Atmos. Sci.*, **32**, 1457–1462, 1975.

ROSENBERG, T. J. and L. J. LANZEROTTI, Direct energy input to the middle atmosphere, in *Middle Atmosphere Electrodynamics*, NASA CP-2090, pp. 43–70, 1979.

ROTTMAN, G. J., Disc values of the solar ultraviolet flux, 1150Å to 1850Å, *Trans. AGU*, **56**, (abstract), 1157, 1974.

ROTTMAN, G. J., C. A. BARTH, R. J. THOMAS, G. H. MOUNT, G. M. LAWRENCE, D. W. RUSCH, R. W. SANDERS, G. E. THOMAS, and J. LONDON, Solar spectral irradiance, 120–190 nm, October 13, 1981-January, 1982, *Geophys. Res. Lett.*, **9**, 587–590, 1982.

ROWLAND, F. S. and M. J. MOLINA, Chlorofluoromethanes in the environment, *Rev. Geophys. Space Sci.*, **13**, 1–35, 1975.

ROWLAND, F. S., J. E. SPENCER, and M. J. MOLINA, Stratospheric formation and photolysis of chlorine nitrate, *J. Phys. Chem.*, **80**, 2711–2713, 1976.

ROWLAND, F. S., S. C. TYLER, D. C. MONTAGUE, and Y. MAKIDE, Dichlorodifluoromethane CCl_2F_2 in the earth's atmosphere, *Geophys. Res. Lett.*, **9**, 481–484, 1982.

ROY, C. R., Atmospheric nitrousoxide in the mid-latitudes at the southern hemisphere, *J. Geophys. Res.*, **84**, 3711–3718, 1979.

RUDERMAN, M. A. and J. W. CHAMBERLAIN, Origin of the solar modulation of ozone: its implications for stratospheric NO injection, *Planet. Space Sci.*, **23**, 247–268, 1975.

RUDOLPH, J., D. H. EHHALT, and A. TÖNNISSEN, Vertical profiles of ethane and propane on the stratosphere, *J. Geophys. Res.*, **86**, 7267–7272, 1981.

RUDOLPH, R. N. and E. C. Y. INN, OCS photolysis and absorption in the 200 to 300 nm region, *J. Geophys. Res.*, **86**, 9891–9894, 1981.

RUNDEL, R. D., Determination of diurnal average photodissociation rate, *J. Atmos. Sci.*, **34**, 639–641, 1977.

RUNDEL, R. D., D. M. BUTLER, and R. S. STOLARSKI, Uncertainty propagation in a stratospheric model 1. Development of a concise stratospheric model, *J. Geophys. Res.*, **83**, 3063–3073, 1978.

RUSCH, D. W., A. I. STEWART, P. B. HAYS, and J. H. HOFFMAN, The NI(5200Å) dayglow, *J. Geophys. Res.*, **80**, 2300–2304, 1975.

RUSCH, D. W., J. C. GERARD, S. SOLOMON, P. J. CRUTZEN, and G. C. REID, The effect of particle precipitation events on the neutral and ion chemistry of the middle atmosphere, 1, Odd nitrogen, *Planet. Space Sci.*, **29**, 767–774, 1981.

RUSCH, D. W., G. H. MOUNT, C. A. BARTH, G. J. ROTTMAN, R. J. THOMAS, G. E. THOMAS, R. W. SANDERS, G. M. LAWRENCE, and R. S. ECKMAN, Ozone densities in the lower mesosphere measured by a limb scanning ultraviolet spectrometer, *Geophys. Res. Lett.*, **10**, 241–244, 1983.

RUSSELL, D. A., The enigma of the extinction of the dinosaurs, *Ann. Rev. Earth Planet. Sci.*, **7**, 163–182, 1979.

SAMAIN, D. and P. C. SIMON, Solar flux determination in the spectral range 150–210 nm, *Solar Phys.*, **49**, 33–41, 1976.

SANDFORD, M. C. W. and A. J. GIBSON, Laser rader measurements of the atmospheric sodium layer, *J. Atmos. Terr. Phys.*, **32**, 1423–1430, 1970.

SANDORFY, C., Review paper: UV absorption of fluorocarbons, *Atmos. Env.*, **10**, 343–351, 1976.

SASAMORI, T., J. LONDON, and D. V. HOYT, Radiation budget of the southern hemisphere, *Meteorological Monographs*, **13**, 9–23, Amer. Meteor. Soc., 1972.

SCHMELTEKOPF, A. L., P. D. GOLDAN, W. R. HENDERSON, W. J. HARROP, T. L. THOMPSON, F. C. FEHSENFELD, H. I. SCHIFF, P. J. CRUTZEN, I. S. A. ISAKSEN, and E. E. FERGUSON, Measurements of stratospheric $CFCl_3$, CF_2Cl_2 and N_2O, *Geophys. Res. Lett.*, **2**, 393–396, 1975.

SCHMELTEKOPF, A. L., D. L. ALBRITTON, P. J. CRUTZEN, P. D. GOLDAN, W. J. HARROP, W. R. HENDERSON, J. R. MCAFEE, M. MCFARLAND, H. I. SCHIFF, T. L. THOMPSON, D. J. HOFMANN, and N. T. KJOME, Stratospheric nitrous oxide altitude profiles at various latitudes, *J. Atmos. Sci.*, **34**, 729–736, 1977.

SCHMIDT, S. S., R. C. AMME, D. G. MURCRAY, A. GOLDMAN, and F. S. BONOMO, Ultraviolet absorption by nitric acide vapor, *Nature (London), Phys. Sci.*, **238**, 109, 1972.

SCHMIDT, U., Molecular hydrogen in the atmosphere, *Tellus*, **26**, 78–90, 1974.

SCHOEBERL, M. R. and D. F. STROBEL, The zonally averaged circulation of the middle atmosphere, *J. Atmos. Sci.*, **35**, 577–591, 1978a.

SCHOEBERL, M. R. and D. F. STROBEL, The response of the zonally averaged circulation to stratospheric ozone reductions, *Ibid.*, **35**, 1751–1757, 1978b.

SCHOLZ, T. G. and D. OTTERMANN, Measurement of neutral atmospheric composition at 85–150 km by mass spectrometer with cryoion source, *J. Geophys. Res.*, **79**, 307–310, 1974.

SCHULTZ, E. D., A. C. HOLLAND, and F. F. MARMO, A congeries of absorption cross sections for wavelengths less than 3000 Å, Geophys. Corp. Amer. Tech. Rep. No. 62-15-N, 97 pp., 1962.

SCHÜRGERS, M. and K. H. WELGE, Absorption koeffizient von H_2O_2 und N_2O_4 zwischen 1200 und 2000 Å, *Z. Naturforsch*, **23a**, 1508–1510, 1968.

SCHWENTEK, H., The sunspot cycle 1958/70 in ionosphere absorption and stratospheric temperature, *J. Atmos. Terr. Phys.*, **33**, 1839–1852, 1971.

SEELEY, J. S. and J. T. HOUGHTON, Spectroscopic observations of the vertical distribution of some minor constituents of the atmosphere, *Infrared Phys.*, **1**, 116–132, 1961.

SEERY, D. J. and D. J. BRITTON, The continuous absorption spectrum of chlorine, bromine, bromine chloride, iodine chloride, and iodine bromide, *J. Phys. Chem.*, **68**, 2263–2266, 1964.

SEILER, W. and P. WARNECK, Decrease of the carbon monoxide mixing ratio at the tropopause, *J. Geophys. Res.*, **77**, 3204–3214, 1972.

SELWYN, G., J. PODOLSKE, and H. S. JOHNSTON, Nitrous oxide ultraviolet absorption spactrum at stratospheric temperatures, *Geophys. Res. Lett.*, **4**, 427–430, 1977.

SHAH, G. M., The effect of aerosol attenuation on determination of atmospheric ozone from measurements with a Dobson spectrophotometer, *J. Atmos. Sci.*, **33**, 2462–2464, 1976.

SHEFOV, N. N., Hydroxyl emission of the upper atmosphere III Diurnal variations, *Planet. Space Sci.*, **19**, 129–136, 1971.

SHEKHTER, F. N., On the calculation of radiant heat fluxes in the atmosphere, *Proc. Main Geophys. Obs*, No. 22, 84, 1950.

SHEMANSKY, D. E., CO_2 extinction coefficient 1700–3000 Å, *J. Chem. Phys.*, **56**, 1582–1587, 1972.

SHETTLE, E. P. and J. A. WEINMAN, The transfer of solar irradiance through inhomogeneous turbid atmospheres evaluated by Eddington's approximation, *J. Atmos. Sci.*, **27**, 1048–1055, 1970.

Shimabukuro, F. I., P. L. Smith, and W. J. Wilson, Estimation of the ozone distribution from millimeter wavelength absorption measurements, *J. Geophys. Res.*, **80**, 2957–2959, 1975.

Shimazaki, T., Dynamic effects on atomic and molecular oxygen density distributions in the upper atmosphere: a numerical solution to equations of motion and continuity, *J. Atmos. Terr. Phys.*, **29**, 723–743, 1967.

Shimazaki, T., Dynamical effects on height distributions of neutral constituents in the Earth's upper atmosphere: a calculation of atmospheric model between 70 km and 500 km, *J. Atmos. Terr. Phys.*, **30**, 1279–1292, 1968.

Shimazaki, T., On the boundary conditions in theoretical model calculations of the distributions of minor neutral constituents in the upper atmosphere, *Radio Sci.*, **7**, 695–702, 1972.

Shimazaki, T., A two-dimensional model for stratospheric ozone denisty distributions in the meriodional plane, II. Effects of chemistry and dynamics, *Proc. 2nd International Conf. on Environ. Impacts of Aerospace Operation in High Atmosphere*, pp. 155–162, San Diego, CA, American Met. Soc., 1974.

Shimazaki, T., *The Stratospheric Ozone* (in Japanese), 184 pp., Univ. of Tokyo Press, 1979.

Shimazaki, T., A model of temporal variations in ozone density in the Martian atmosphere, *Planet. Space Sci.*, **29**, 21–33, 1981.

Shimazaki, T., The photochemical time constants of minor constituents and their families in the middle atmosphere, *J. Atmos. Terr. Phys.*, **46**, 173–191, 1984.

Shimazaki, T. and A. R. Laird, A model calculation of the diurnal variation in minor neutral constituents in the mesosphere and lower thermosphere including transport effects, *J. Geophys. Res.*, **75** 3221–3235, 1970 (see also **77**, 276–277, 1972).

Shimazaki, T. and A. R. Laird, Seasonal effects on distributions of minor neutral constituents in the mesosphere and lower themosphere, *Radio Sci.*, **7**, 23–43, 1972.

Shimazaki, T. and R. D. Cadle, Theoretical model of vertical distributions of CO and CH_4 in the mesosphere and upper stratosphere, *J. Geophys. Res.*, **78**, 5352–5361, 1973.

Shimazaki, T., D. J. Wuebbles, and T. Ogawa, A two-dimensional theoretical model for stratospheric ozone density distributions in the meridional plane, AIAA/AMS International Conf. Environ. Impacts of Aerospace Operations in High Atmosphere, Denver, Colo., June, 1973, AIAA paper 73–541; and also NOAA Tech. Rep. ERL 279-OD 9, 29 pp., 1973.

Shimazaki, T. and D. J. Wuebbles, On the theoretical model for vertical ozone density distributions in the mesosphere and upper stratosphere, *Pure Appl. Geophys.*, **106–108**, 1446–1463, 1973.

Shimazaki, T. and T. Ogawa, A theoretical model of minor constituent distributions in the stratosphere including diurnal variations, *J. Geophys. Res.*, **79**, 3411–3423, 1974a.

Shimazaki, T. and T. Ogawa, On the theoretical model of vertical distributions of minor neutral constituents concentrations in the stratosphere, NOAA Tech. Memo. TM ERL OD-20, 35 pp., 1974b.

Shimazaki, T. and T. Ogawa, Theoretical modeling of minor constituents distribution in the stratosphere and the impacts of the SST exhaust gases, *Proc. Int. Conf. of Structure, Composition and General Circulation of the Upper and Lower Atmosphere and Possible Anthropogenic Perturbations*, Melbourne, Australia, (sponsored by IAMAP), pp. 1062–1092, 1974c.

Shimazaki, T. and R. C. Whitten, A comparison of one-dimensional theoretical models of stratospheric minor constituents, *Rev. Geophys. Space Phys.*, **14**, 1–12, 1976.

Shimazaki, T., T. Ogawa, and B. C. Farrell, Simplified methods for calculating photodissociation rates of various molecules in Schumann-Runge band systems in the upper atmosphere, NASA Tech. Note TN D-8399, 39 pp., 1977.

SHIMAZAKI, T. and M. SHIMIZU, The seasonal variation of ozone density in the Martian atmosphere, *J. Geophys. Res.*, **84**, 1269–1276, 1979.

SHIMAZAKI, T. and L. C. HELMLE, A simplified method for calculating the atmospheric heating rate by absorption of solar radiation in the stratosphere and mesosphere, NASA Tech. Paper, 1398, 31 pp., 1979.

SHVED, G. M. and I. V. TSARITSYNA, Infrared absorption functions for the mesosphere and upper stratosphere, Leningrad Univ. Press, 1963, (NASA Technical Transl., TTF-184, 106–169, 1964).

SIMON, P. C., Irradiation solar flux measurements between 120 and 400 nm; current position and future needs, *Planet. Space Sci.*, **26** 355–365, 1978.

SIMONAITIS, R. and J. HEICKLEN, Temperature dependence of the reactions of HO_2 with NO and NO_2, *Int. J. Chem. Kinetics*, **10**, 67–87, 1978.

SISSENWINE, N., D. D. GRANTHAM, and H. A. SALMELA, Mid-latitude humidity to 32 km, *J. Atmos. Sci.*, **25**, 1129–1140, 1968.

SMITH, D., N. G. ADAMS, and M. J. CHURCH, Mutual neutralization rates of ionospherically important ions, *Planet. Space Sci.*, **24**, 697–703, 1976.

SMITH, D., N. G. ADAMS, and E. ALGE, Ion-ion mutual neutralization and ion-neutral switching reactions of some stratospheric ions, *Ibid.*, **29**, 449–454, 1981.

SMITH, F. L. and C. SMITH, Numerical evaluation of Chapman's grazing incidence integral Ch(X, χ), *J. Geophys. Res.*, **77**, 3592–3597. 1972.

SMITH, W. S., C. C. CHOU, and F. S. ROWLAND, The mechanism for ultraviolet photolysis of gaseous chlorine nitrates at 302.5 nm, *Geophys. Res. Lett.*, **4**, 517–519, 1977.

SOLOMON, S. and P. J. CRUTZEN, Analysis of the August 1972 solar proton event including chlorine chemistry, *J. Geophys. Res.*, **86**, 1140–1146, 1981.

SOLOMON, S., D. W. RUSCH, J. C. GERARD, G. C. REID, and P. J. CRUTZEN, The effect of particle precipitation on the neutral and ion chemistry of the middle atmosphere, II. Odd hydrogen, *Planet. Space Sci.*, **29**, 885–893, 1981.

SOLOMON, S., G. C. REID, D. W. RUSCH, and R. J. THOMAS, Mesospheric ozone depletion during the solar proton event of July 13,1982, Part II. Comparison between theory and measurements, *Geophys. Res. Lett.*, **10**, 257–260, 1983.

SPENCER, J. E. and F. S. ROWLAND, Bromine nitrate and its stratospheric significance, *J. Phys. Chem.*, **82**, 7–9, 1978.

SRIDHARAN, U. C., L. X. QIU, and F. KAUFMAN, Kinetics of the reaction $OH+HO_2 \rightarrow H_2O+O_2$ at 296 K, *J. Phys. Chem.*, **85**, 3361–3363, 1981.

STAIR, A. T., J. C. ULWICK, D. J. BAKER, C. L. WYATT, and K. D. BAKER, Altitude profiles of infrared radiance of O_3(9.6 μm) and CO_2(15 μm), *Geophys. Res. Lett.*, **1**, 117–118, 1974.

STARR, W. L., R. A. CRAIG, M. LOEWENSTEIN, and M. E. MCGHAN, Measurements of NO, O_3 and temperature at 19.8 km during the total eclips of 26 February 1979, *Geophys. Res. Lett.*, **7**, 553–555, 1980.

STEWART, R. W. and M. I. HOFFERT, A chemical model of the troposphere and stratosphere, *J. Atmos. Sci.*, **32**, 195–210, 1975.

STOCKWELL, W. R. and J. G. CALVERT, The near ultraviolet absorption spectrum of gaseous HONO and N_2O_3, *J. Photochem.*, **8**, 193–203, 1978.

STOLARSKI, R. S. and R. J. CICERONE, Stratospheric chlorine: a possible sink for ozone, *Can. J. Chem.*, **52**, 1610–1615, 1974.

STOLARSKI, R. S. and R. D. RUNDEL, Fluorine photochemistry in the stratosphere, *Geophys. Res. Lett.*, **2**, 443–444, 1975.

STROBEL, D. F., Odd nitrogen in the mesosphere, *J. Geophys. Res.*, **76**, 8384–8393, 1971.

STROBEL, D. F., Photochemical-radiative damping and instability in the stratosphere,

Geophys. Res. Lett., **4**, 424–426, 1977.

Strobel, D. F., Parameterizaton of the atmospheric heating rate from 15 to 120 km due to O_2 and O_3 absorption of solar radiation, *J. Geophys. Res.*, **83** 6225–6230, 1978.

Swider, W., The determination of the optical depth at large solar zenith distances, *Planet. Space Sci.*, **12**, 761–782, 1964.

Swider, W. and M. E. Gardner, On the accuracy of Chapman function approximations, *App. Opt.*, **8**, 725, 1969.

Swider, W., and T. J. Keneshea, Decrease of ozone and atomic oxygen in the lower mesosphere during a PCA event, *Planet. Space Sci.*, **21**, 1969–1973, 1973.

Swider, W. and R. S. Narcisi, Auroral *E*-region: ion composition and nitric oxide, *Planet. Space Sci.*, **25**, 103–116, 1977.

Sze, N. D., Variational methods in radiation transfer problems, *J. Quart. Spectrosc. Radiat. Transfer*, **16**, 763–780, 1976.

Sze, N. D. and W. F. Wu, Measurements of fluorocarbons 11 and 12 and model variation: An assessment, *Atmos. Environ.*, **10**, 1117–1125, 1976.

Sze, N. D., Stratospheric fluorine: a comparison between theory and measurements, *Geophys. Res. Lett.*, **5**, 781–783, 1978.

Takahashi, H., B. R. Clemensha, and Y. Sahai, Nightglow OH(8,3) band intensities and rotational temperature at 23°S, *Planet. Space Sci.*, **22**, 1323–1329, 1974.

Takahashi, H., Y. Sahai, B. R. Clemensha, P. P. Batista, and N. R. Teixeira, Diurnal and seasonal variations of the OH(8,3) airglow band and its correlation with OI 5577 Å, *Planet. Space Sci.*, **25**, 541–547, 1977.

Temps, F. and H. G. Wagner, Studies on the $OH+HO_2 \rightarrow H_2O+O_2$ reaction with a laser magnetic resonance spectrometer (in German), *Ber. Bunsenges Phys. Chem.*, **86**, 119–125, 1982.

Thacker, D. L., C. J. Gibbins, P. R. Schwartz, and R. M. Bevilacqua, Ground-based microwave observations of mesosphere H_2O in January, April, July and September, 1980, *Geophys. Res. Lett.*, **8**, 1058–1062, 1981.

Thekaekara, M. P., Solar energy outside the earth's atmosphere, *Solar Energy*, **14**, 109–127, 1973.

Thomas, L., Mesospheric temperatures and the formation of water cluster ions in the *D*-region, *J. Atmos. Terr. Phys.*, **38**, 1345–1350, 1976.

Thomas, R. J. and R. A. Young, Measurement of atomic oxygen and related airglows in the lower thermosphere, *J. Geophys. Res.*, **86**, 7389–7393, 1981.

Thomas, R. J., C. A. Barth, G. J. Rottman, D. W. Rusch, G. H. Mount, G. M. Lawrence, R. W. Sanders, G. E. Thomas, and L. E. Clemens, Ozone density distributions in the mesosphere (50–90 km) measured by the SME limb scanning near infrared spectrometer, *Geophys. Res. Lett.*, **10**, 245–248, 1983a.

Thomas, R. J., C. A. Barth, G. J. Rottman, D. W. Rusch, G. H. Mount, G. M. Lawrence, R. W. Sanders, G. E. Thomas, and L. E. Clemens, Mesospheric ozone depletion during the solar proton event of July 13, 1982, Part I. Measurement, *Geophys. Res. Lett.*, **10**, 253–255, 1983b.

Thompson, B. A., P. Harteck, and R. P. Reeves, Ultraviolet absorption coefficients of CO_2, CO, O_2, H_2O, N_2O, NH_3, NO, SO_2, and CH_4 between 1850 and 4000 Å, *J. Geophys. Res.*, **68**, 6431–6436, 1963.

Thornton D. and N. Niazy, Effects of solution mass transport on the ECC ozonesonde background current, *Geophys. Res. Lett.*, **10**, 148–151, 1983.

Thrush, B. A. and J. P. T. Wilkinson, The rate of reaction of HO_2 radicals with HO and with NO, *Chem. Phys. Lett.*, **81**, 1–3, 1981.

Tinsley, B. A., Hydrogen in the upper atmosphere, *Fund. Cosmic Res.*, **1**, 201–300, 1974.

TISONE, G. C., Measurements of the absorption by O_2 and O_3 in the 2150-A region, *J. Geophys. Res.*, **77**, 2971–2974, 1972.

TIWARI, S. N., Models for infrared atmospheric radiation, *Adv. Geophys.*, **20**, 1–85, 1978.

TOHMATSU, T. and N. IWAGAMI, Measurement of nitric oxide abundance in equatorial upper atmosphere, *J. Geomag. Geod., Japan*, **28**, 343–358, 1976.

TOLSON, R. H., Spatial and temporal variations of monthly mean total columnar ozone derived from 7 years of BUV data, *J. Geophys. Res.*, **86**, 7312–7330, 1981.

TOON, O. B., R. P. TURCO, P. HAMILL, C. S. KIANG, and R. C. WHITTEN, A one-dimensional model describing aerosol formation and evolution in the stratosphere: II, Sensitivity studies and comparison with observations, *J. Atmos. Sci.*, **36**, 718–736, 1979.

TOON, O. B. and J. B. POLLACK, Atmospheric aerosols and climate, *Am. Sci.*, **68**, 268–278, 1980.

TORR, M. R., D. G. TORR, and H. E. HINTEREGGER, Solar flux variability in the Schumann-Runge continuum as a function of solar cycle 21, *J. Geophys. Res.*, **85**, 6063–6068, 1980.

TORRES, A. L. and A. R. BANDY, Performance characteristics of the electrochemical concentration cell ozonesonde, *J. Geophys. Res.*, **83**, 5501–5504, 1978.

TOTH, R. A., C. B. FARMER, R. A. SCHINDLER. O. F. RAPER, and P. W. SCHAPER, Detection of nitric oxide in the lower atmosphere, *Nature Phys. Sci.*, **244**, 7–8, 1973.

TRAUB, W. A. and K. V. CHANCE, Stratospheric HF and HCl observations, *Geophys. Res. Lett.*, **8**, 1075–1077, 1981.

TREOR, P. L., G. BLACK, and J. R. BARKER, Reaction rate constant for $OH+HOONO_2 \rightarrow$products over the temperature range 246 to 324K, *J. Phys. Chem.*, **86**, 1661–1669, 1982.

TRINKS, H., Ozone measurements between 90 and 110 km altitude by mass spectrometer, *Geophys. Res. Lett.*, **2**, 99–102, 1975.

TUCK, A. F., Numerical model studies of the effect of injected nitrogen oxides on stratospheric ozone, *Proc. Roy. Soc., London, A*, **355**, 267–299, 1977.

TUCKER, A. W., M. BIRNBAUM, and C. L. FINCHER, Atmospheric NO_2 determinaion by 442 nm laser induced fluorescence, *App. Opt.*, **14**, 141–1422, 1975.

TURCO, R. P., Photodissociation rates in the atmosphere below 100 km, *Geophys. Survey*, **2**, 153–192, 1975.

TURCO, R. P. and R. C. WHITTEN, Chlorofluoromethans in the stratosphere and some possible consequences for ozone, *Atmos. Environ.*, **9**, 1045–1061, 1975.

TURCO, R. P. and R. C. WHITTEN, The NASA Ames Research Center one- and two-dimensional stratospheric models. Part I: The one-dimensional model, NASA Tech. Paper, TP-1002, 20 pp., 1977.

TURCO, R. P. and R. C. WHITTEN, A note on the diurnal averaging of aeronomical models. *J. Atmos. Terr. Phys.*, **40**, 13–20, 1978.

TURCO, R. P., R. C. WHITTEN, I. G. POPPOFF, and L. A. CAPONE, SSTs, nitrogen fertilizer and stratospheric ozone, *Nature*, **276**, 805–807, 1978.

TURCO, R. P., P. HAMILL, O. B. TOON, R. C. WHITTEN, and C. S. KIANG, A one-dimensional model describing aerosol formation and evolution in the stratosphere: I. Physical processes and mathematical analogs, *J. Atmos. Sci.*, **36**, 699–717, 1979.

TURCO, R. P., R. C. WHITTEN, O. B. TOON, J. B. POLLACK, and P. HAMILL, OCS, stratospheric aerosols and climate, *Nature*, **283**, 283–286, 1980.

TURCO, R. P., R. C. WHITTEN, O. B. TOON, E. C. Y. INN, and P. HAMILL, Stratospheric hydroxyl radical concentrations: New limitations suggested by observations of gaseous and particulate sulfur, *J. Geophys. Res.*, **86**, 1129–1139, 1981.

TWOMEY, S., On the deduction of the vertical distribution of ozone by ultraviolet spectral measurements from a satellite, *J. Geophys. Res.*, **66**, 2153–2162, 1961.

TYSON, B. J., J. F. VEDDER, J. C. ARVESEN, and R. B. BREWER, Stratospheric measurements of CF_2Cl_2 and N_2O, *Geophys. Res. Lett.*, **5**, 369–372, 1978a.

TYSON, B. J., J. C. ARVESEN, and D. O'HARA, Interhemispheric gradients of CF_2Cl_2, $CFCl_3$, CCl_4 and N_2O, *Geophys. Res. Lett.*, **5**, 535–538, 1978b.

UCHINO, O., M. MAEDA, J. KOHNO, T. SHIBATA, C. NAGASAWA, and M. HIRONO, Observation of stratospheric ozone layer by a XeCl laser radar, *App. Phys. Lett.*, **3**, 807–809, 1978.

UREY, H. C., L. H. DAWSEY, and F. O. RICE, The absorption spectrum and decomposition of hydrogen peroxide by light, *J. Amer. Chem. Soc.*, **51**, 1371–1383, 1929.

U. S. STANDARD ATMOSPHERE, 1976, NOAA, NASA and U.S. Air Force, Washington, D.C., 1976.

VANDERHOFF, J. A., Time response characteristics of an aluminum oxide sensor, *J. Geophys. Res.*, **79**, 2207–2214, 1974.

VAN MIEGHEM, J., Scale analysis of large atmospheric motion systems in all latitudes, *Adv. Geophys.*, **20**, 87–130, 1978.

VIDAL-MADJAR, A., Evolution of the solar Lyman-alpha flux during four consecutive years, *Solar Phys.*, **40**, 69–86, 1975.

VIDAL-MADJAR, A., J. E. BLAMONT, and B. PHISSAMEY, Solar Lyman-alpha changes and related hydrogen density distribution at the earth's exobase (1969–1970), *J. Geophys. Res.*, **78**, 1115–1144, 1973.

VIGGIANO, A. A., R. A. PERRY, D. L. ALBRITTON, E. E. FERGUSON, and F. C. FEHSENFELD, The role of H_2SO_4 in the stratospheric negative ion chemistry, *Ibid.*, **85**, 4551–4555, 1980.

VIGGIANO, A. A. and F. ARNOLD, The first height measurements of the negative ion composition of the stratosphere, *Planet. Space Sci.*, **29**, 895–906, 1981a.

VIGGIANO, A. A. and F. ARNOLD, Extended sulfuric acid vapor concentration measurements in the stratosphere, *Geophys. Res. Lett.*, **8**, 583–586, 1981b.

VIGGIANO, A. A., R. A. PERRY, D. L. ALBRITTON, E. E. FERGUSON, and F. C. FEHSENFELD, Stratospheric negative-ion reaction rates with H_2SO_4, *J. Geophys. Res.*, **87**, 7340–7342, 1982.

VIGGIANO, A. A. and F. ARNOLD, Stratospheric sulfuric acid vapor—new and updated measurements, *J. Geophys. Res.*, **88**, 1457–1462, 1983a.

VIGGIANO, A. A., H. SCHLAGER, and F. ARNOLD, Stratospheric negative ions detailed height profiles, *Planet. Space Sci.*, **31**, 813–820, 1983b.

VISCONTI, G., G. PITARI, and D. CUNNOLD, A two-dimensional photochemical model of the stratosphere with Rayleigh scattering, *Pure Appl. Geophys.*, **118**, 1033–1051, 1980.

VOLMAN, D. H., Photochemical gas phase reactions in the hydrogen-oxygen system, *Adv. Photochem.*, **1**, 43–82, 1963.

WALKER, J. C. G., *Evolution of the Atmosphere*, 318 pp., Macmillan Pub. Co., New York, 1977.

WALKER, J. D. and E. I. REED, Behavior of the sodium and hydroxyl nighttime emissions during a stratospheric warming, *J. Atmos. Sci.*, **33**, 118–130, 1976.

WALSHAW, C. D., Integrated absorption by the 9.6 μm band of ozone, *Quart. J. Roy. Met. Soc.*, **83**, 315–321, 1957.

WANG, W. C., Y. L. YUNG, A. A. LACIS, T. MO, and J. E. HANSEN, Greenhouse effect due to man-made perturbations of trace gases, *Science*, **194**, 685–690, 1976.

WANG, P.-H., A. DEEPAK and S.-S. HONG, General formulation of optical paths for large zenith angles in the earth's curved atmosphere, *J. Atmos. Sci.*, **38**, 650–658, 1981.

WANG, P.-H., S.-S. HONG, M.-F. WU, and A. DEEPAK, A model study of the temporal and spatial variations of the zonally-average ozone heating rate, *J. Atmos. Sci.*, **39**, 1398–1409, 1982.

WARK, D. Q., J. H. LIENESCH, and M. P. WEINREB, Satellite observations of atmospheric water vapor, *App. Opt.*, **13**, 507–511, 1974.

Watanabe, K., Ultraviolet absorption processes in the upper atmosphere, *Adv. Geophys.*, **5**, 153–221, 1958.

Watanabe, K. and M. Zelikoff, Absorption coefficients of water vapor in the vacuum ultraviolet, *J. Opt. Soc. Amer.*, **43**, 713–755, 1953.

Watanabe, K., M. Zelikoff, and E. C. Y. Inn, Absorption coefficients of several atmospheric gases, Geophys. Res. Paper, No. 21, AFCRL, Cambridge, Mass., 1953.

Waters, J. W., J. J. Gustincis, R. V. Kakar, A. R. Kerr, H. K. Roscoe, and P. N. Swanson, The microwave limb sounder experiment observations of stratospheric and mesospheric water vapor, NASA Report TMX-73630, NTIS, Springfield, VA, 1977.

Waters, J. W., J. J. Gustincic, R. K. Kakar, H. K. Roscoe, P. N. Swanson, T. G. Phillips, T. DeGraauw, A. R. Kerr, and R. J. Mattauch, Aircraft search for millimeter-wavelength emission by stratospheric ClO, *J. Geophys. Res.*, **84**, 7034–7040, 1979.

Waters, J. W., J. C. Hardy, R. F. Jarnot, and H. M. Pickett, Cholorine monoxide radical, ozone, and hydrogen peroxide: stratospheric measurements by microwave limb sounding, *Science*, **214**, 61–64, 1981.

Watson, R. T., Rate constants for reactions of ClO_x of atmospheric interest, *J. Phys. Chem. Data*, **6**, 871–917, 1977.

Webber, W., The production of free electrons in the ionospheric *D*-layer by solar and galactic cosmic rays and the resultant absorption of radio waves, *J. Geophys. Res.*, **67**, 5091–5106, 1962.

Weeks, L. H., Lyman-alpha emission from the sun near solar minimum, *Astrophy. J.*, **147**, 1203–1205, 1967.

Weeks, L. H., R. S. Cuikay, and J. R. L. Corbin, Ozone measurement in the mesosphere during the solar proton event of 2 November 1969, *J. Atmos. Sci.*, **29**, 1138–1142, 1972.

Wehrbein, W. M., C. W. Hord, and C. A. Barth, Mariner 9 ultraviolet spectrometer experiment: vertical distribution of ozone on Mars, *Icarus*, **38**, 280–299, 1979.

Weinstock, E. M., M. J. Phillips, and J. G. Anderson, *In situ* observations of ClO in the stratosphere: a review of recent results, *J. Geophys. Res.*, **86**, 7273–7278, 1981.

Weiler, K. H., P. Fabian, G. Flentje, and W. A. Matthews, Stratospheric NO measurements: a new balloon-borne chemiluminescent instrument, *J. Geophys. Res.*, **85**, 7445–7452, 1980.

Weiss, R. F., The temporal and spatial distribution of tropospheric nitrous oxide, *J. Geophys. Res.*, **86**, 7185–7195, 1981.

Whitten, R. C. and R. P. Turco, Diurnal variations of HO_x and NO_x in the stratosphere, *Ibid.*, **79**, 1302–1304, 1974a.

Whitten, R. C. and R. P. Turco, Perturbations of the stratosphere and mesosphere by aerospace vehicles, *AIAA J.*, **12**, 1110–1117, 1974b.

Whitten, R. C. and R. P. Turco, The effect of SST emissions on the earth's ozone layer, *Proc. Int. Conf. of Structure, Composition and General Circulation of the Upper and Lower Atmosphere and Possible Anthropogenic Perturbations*, pp. 971–983, Melbourne, Australia, (sponsored by IAMAP), 1974C.

Whitten, R. C., W. J. Borucki, I. G. Poppoff, and R. P. Turco, Preliminary assessment of the potential impact of solid-fueled rocket engines in the stratosphere, *J. Atmos. Sci.*, **32**, 613–619, 1975.

Whitten, R. C., W. J. Borucki, V. R. Watson, T. Shimazaki, H. T. Woodward, C. A. Riegel, L. A. Capone, and T. Becker, The NASA Ames Research Center one- and two-dimensional stratospheric models, II. The two-dimensional model, NASA Tech. Pap., 1003, 1977.

Whitten, R. C., W. J. Borucki, L. A. Capone, C. A. Riegel, and R. P. Turco, Nitrogen fertilizer and stratospheric ozone: latitude effects, *Nature*, **283**, 191–192, 1980.

WHITTEN, R. C., W. J. BORUCKI, H. T. WOODWARD, L. A. CAPONE, C. A. RIEGEL, R. P. TURCO, I. G. POPPOFF, and K. SANTHANAM, Implications of smaller concentrations of stratospheric OH: A two-dimensional model study of ozone perturbations, *Atmos. Environ.*, **15**, 1583–1589, 1981.

WIDHOPF, G. F. and T. D. TAYLOR, Numerical experiments on stratospheric meridional ozone distributions using a parameterized two-dimensional model, *Proc. 3rd Conf. on CIAP, DOT-TSC-OST-74-15*, pp. 376–389, Dept. of Transp., Washinton, D.C., 1974.

WIDGER. W. K. and M. P. WOODALL, Integration of the Planck blackbody radiation function, *Bull, Amer. Meteorol. Soc.*, **57**, 1217–1219, 1976.

WILCOX, R. W., G. D. NASTROM, and A. D. BELMONT, Periodic analysis of total ozone and its vertical distribution, *J. App. Meteor.*, **16**, 290–298, 1977.

WILKES, M. W., A table of Chapman's grazing incidence integral $Ch(x, \chi)$, *Proc. Phys. Soc.*, **B67**, 304–305, 1954.

WILKNISS, F. E., J. W. SWINNERTON, D. J. BRESSAN, R. A. LAMONTAGNE, and R. E. LARSON, CO, CCL_4, Freon-11, CH_4, and Rn-222 concentrations at low altitude over the arctic oceans in January 1974, *J. Atmos. Sci.*, **32**, 158–162, 1975.

WILLETT, H. C., The relationship of total atmospheric ozone to the sunspot cycle, *J. Geophys. Res.*, **67**, 661–670, 1962.

WILLIAMS, W. J., J. N. BROOKS, D. G. MURCRAY, F. H. MURCRAY, P. M. FRIED, and J. A. WEINMAN, Distribution of nitric acide vapor in the stratosphere as determined from infrared atmospheric emission data, *J. Atmos. Sci.*, **29**, 1375–1379, 1972.

WILLIAMS, W. J., J. J. KOSTERS, A. GOLDMAN, and D. G. MURCRAY, Measurement of the stratospheric mixing ratio of HCl using infrared absorption technique, *Geophys. Res. Lett.*, **3**, 383–385, 1976.

WILSON, W. J. and P. R. SCHWARTZ, Diurnal variations of mesospheric ozone using millimeter-wave measurements, *J. Geophys. Res.*, **86**, 7385–7388, 1981.

WINE, P. H., A. R. RAVISHANKARA, N. W. KREUTTER, R. C. SHAH, J. M. NICOVICH, R. L. THOMPSON, and D. J. WUEBBLES, Rate of reaction of OH with HNO_3, *J. Geophys. Res.*, **86**, 1105–1112, 1981.

WMO, *The Stratosphere 1981: Theory and Measurements*,WMO Global ozone research and monitoring project, report No. 11, ed. by R. D. Hudson *et al.*, 1981.

WOFSY, S. C., Temporal and latitudinal variations of stratospheric trace gases: a critical comparison between theory and experiment, *J. Geophys. Res.*, **83**, 364–378, 1978.

WOFSY, S. C., and M. B. MCELROY, HO_x, NO_x, and ClO_x: their role in atmospheric photochemistry, *Can. J. Chem.*, **52**, 1582–1591, 1974.

WOFSY, S. C. and J. A. LOGAN, Recent developments in stratospheric photochemistry, in *Causes and effects of stratospheric ozone reduction*, (Appendix C), pp. 159–205, National Academy Press, Washington D.C., 1982.

WOOD, H. C., W. F. J. EVANS, E. J. LLEWELLYN, and A. VLLANCE JONES, Summer daytime height profiles of $O_2(^1\Delta_g)$ concentration at Fort Churchill, *Can. J. Phys.*, **48**, 852–867, 1970.

WOOD, T. B. V. and M. H. THIEMENS, The fate of the hydroxyl radical in the earth's primitive atmosphere and implications for the production of molecular oxygen, *J. Geophys. Res.*, **85**, 1605–1610, 1980.

WOODWELL, G. M., The carbon dioxide question, *Sci. Amer.*, **238**, No. 1, 34–43, 1978.

WUEBBLES, D. J. and J. S. CHANG, Sensitivity of time-varying parameters in stratospheric modeling, *J. Geophys. Res.*, **80**, 2637–2642, 1975.

WUEBBLES, D. J. and J. S. CHANG, A theoretical study of stratospheric trace spacies variations during a solar eclipse, *Geophys. Res. Lett.*, **6** 179–182, 1979.

WUEBBLES, D. J. and J. S. CHANG, A study of the effectiveness of the ClX catalytic ozone loss mechanisms, *J. Geophys. Res.*, **86**, 9869-9872, 1981.

WUEBBLES, D. J., F. M. LUTHER, and J. E. PENNER, Effect of coupled anthropogenic perturbations on stratospheric ozone, *J. Geophys. Res.*, **88**, 1444–1456, 1983.

YAMAMOTO, G., On a radiaton chart, Sci. Rep. Tohoku Univ., Series 5, *Geophysics*, **4**, 9–23, 1952.

YAMAMOTO, H., H. SEKIGUCHI, T. MAKINO, T. WATANABE, K. S. ZALPURI, and T. OGAWA, A mesospheric ozone profile at sunset, *Adv. Space Res.*, **2**, 197–199, 1983.

YOUNG, D. M., The numerical solution of elliptic and parabolic partial differential equations, in *Survey of Numerical Analysis*, ed. by J. Todd, pp. 380–438, McGraw-Hill Book Co., New York, 1962.

YUNG, Y. L., A numerical method for calculating the mean intensity in an inhomogeneous Rayleigh scattering atmosphere, *J. Quart. Spectrosc. Radiat. Transfer*, **16**, 755–761, 1976.

YUNG, Y. L., W. C. WANG, and A. A. LACIS, Greenhouse effect due to atmospheric nitrous oxide, *Geophys. Res. Lett.*, **3**, 619–621, 1976.

YUNG, Y. L., J. P. PINTO, R. T. WATSON, and S. P. SANDER, Atmospheric bromine and ozone perturbations in the lower stratosphere, *J. Atmos. Sci.*, **37**, 339–353, 1980.

ZANDER, R., Water vapor above 25 km altitude, *Pure Appl. Geophys.*, **106–108**, 1346–1352, 1973.

ZANDER, R., Recent observations of HF and HCl in the upper stratosphere, *Geophys. Res. Lett.*, **8**, 413–416, 1981.

ZANDER, R., H. LECLERCQ, and J. D. KAPLAN, Concentration of carbon monoxide in the upper stratosphere, *Geophys. Res. Lett.*, **8**, 365–368, 1981.

ZANDER, R., G. M. STOKES, and J. W. BRAULT, Simultaneous detection of FC-11, FC-12, and FC-22, through 8 to 13 micrometers IR solar absorptions from the ground, *Geophys. Res. Lett.*, **10**, 521–524, 1983.

ZBINDEN, P. A., M. A. HIDALGO, P. EBERHARDT, and J. GEISS, Mass spectrometer measurements of the positive ion composition in the *D*- and *E*-regions of the ionosphere, *Planet. Space Sci.*, **23**, 1621–1642, 1975.

ZELIKOFF, M. K., K. WATANABE, and E. C. Y. INN, Absorption coefficients of gases in the vacuum ultraviolet, Part II. Nitrous oxide, *J. Chem. Phys.*, **21**, 1643–1647, 1953.

ZIMMERMAN, S. P., The effective vertical turbulent viscosity as measured from radio meteor trails, *J. Geophys. Res.*, **79**, 1095–1098, 1974.

ZIPF, E. C. and R. W. MCLAUGHLIN, On the dissociation of nitrogen by electron impact and E. U. V. photo-absorption, *Planet. Space Sci.*, **26**, 449–462, 1978.

ZMUDA, A. J. and T. A. POTEMA, Bombardment of the polar-cap ionosphere by solar cosmic rays, *Rev. Geophys. Space Phys.*, **10**, 981–991, 1972.

SUBJECT INDEX

AUTHOR INDEX